SECOND EDITION

Materials for Civil and Highway Engineers

Kenneth N. Derucher

George P. Korfiatis

Department of Civil and Ocean Engineering
Stevens Institute of Technology

Prentice Hall
Englewood Cliffs, New Jersey 07632

Library of Congress Cataloging-in-Publication Data

Derucher, Kenneth N.
  Materials for civil and highway engineers.

  Includes bibliographies and index.
  1. Materials. 2. Civil engineering—Equipment
and supplies. I. Korfiatis, George Panayiotis.
II. Title.
TA403.2.D47   1988   620.1'1   87-15282
ISBN 0-13-560509-1

Editorial/production supervision
and interior design: *Carolyn Fellows*
Cover design: *Ben Santora*
Manufacturing buyer: *Rhett Conklin* and *Gordon Osbourne*

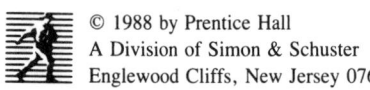 © 1988 by Prentice Hall
A Division of Simon & Schuster
Englewood Cliffs, New Jersey 07632

Printed in the United States of America

10 9 8 7 6 5 4 3

ISBN 0-13-560509-1   025

Prentice-Hall International (UK) Limited, *London*
Prentice-Hall of Australia Pty. Limited, *Sydney*
Prentice-Hall Canada, Inc., *Toronto*
Prentice-Hall Hispanoamericana, S. A., *Mexico*
Prentice-Hall of India Private Limited, *New Delhi*
Prentice-Hall of Japan, Inc., *Tokyo*
Simon & Schuster Asia Pte. Ltd., *Singapore*
Editora Prentice-Hall do Brasil, Ltda., *Rio de Janeiro*

*Dedicated to Dr. C.P. Heins
and our families*

CONTENTS

9 Asphalt Cements   194

10 The Metallic State   214

11 Ferrous Metals   234

Preface

Many books have been written on engineering materials for material scientists, but none have been written for civil engineers. Unfortunately, civil engineers have had to rely on the available books for lack of better or more appropriate texts.

Materials for Civil and Highway Engineers was designed specifically for civil engineers. Unlike other texts, *Materials for Civil and Highway Engineers* covers all materials used in civil engineering. The text discusses the engineering performance of soil, aggregate, cements, concrete, timber, asphalt, metals, and plastics.

Materials for Civil and Highway Engineers is intended for use by second- or third-year students taking introductory courses in civil engineering materials. The text is meant for use at a four-year institute, but it would be equally acceptable at a two-year school. *Materials for Civil and Highway Engineers* complements the laboratories that are a major part of civil engineering materials courses.

The major difference of this revision is that presentations have been added on soils and on proportioning structural concrete mixtures with fly ash and other pozzolans. The chapter on proportioning structural concrete mixtures with fly ash is most beneficial, for it provides a method of utilizing the waste product fly ash in concrete.

The text's wide variety of subjects enables the instructor to cover several topics within one semester. The text provides some theory but is centered around the practical approach. The instructor may use the text to cover the basics and supplement this with personal notes emphasizing relevance where needed. The essay-type homework problems were left as they were in the first edition. Many instructors create their own examples and problems for the classroom based upon specific points they wish to address and emphasize.

Chapter 1 is devoted to soils as used in civil and highway engineering. Chapter 2 concentrates on mineral aggregates as the foundation for portland cement concrete and bituminous concrete. Chapter 3 deals with, in addition to portland cement, many types of cement. Chapters 4, 5, 6, and 7 are concerned with concrete, its general strength, the design of concrete mixes, and its failures. Chapter 8 is devoted to timber classifications, physical characteristics, mechanical properties, decay, durability, and preservation. Chapter 9 covers asphalt and asphaltic cements. Chapters 10 and 11 cover metals, the metallic state, and ferrous metals. Finally, Chapter 12 is concerned with plastics.

To assist the instructor, a number of the more important ASTM standards are presented in the appendix. The student in a first materials course needs to become familiar with such standards, as they will become part of his or her professional career.

The present authors would like to express their appreciation to the family of the late Conrad Heins (who co-authored the first edition) for allowing the text to continue in its second edition.

Kenneth N. Derucher
George P. Korfiatis

1

Soil

Soil encountered close to the earth's surface is one of the most important and widely-used engineering materials. Its frequency of use and the importance of its functions rank it among the top materials in the fields of civil, highway, and architectural engineering.

Soil can be used in engineering works as imported fill or as in-situ material. As imported fill, soil is brought to the site from a different location and is placed in a controlled manner to fill excavations, build earthen structures, or to support various other types of engineering structures. As in-situ material, the soil is used in its original location to perform similar functions.

This chapter is intended to give an overview of soil composition, properties, and use as an engineering material.

ORIGIN AND GEOLOGIC PROPERTIES

Minerals—Rocks—Soils

The outer crust of the earth, which extends 10 to 15 kilometers in depth, is composed of eight predominant chemical elements. The relative abundance of these elements in the earth's crust is shown in Table 1-1.

These elements are found in the soil and rock of the crust, assembled in solid homogeneous chemical compounds of regular architecture called *minerals*. Minerals form rocks which in turn are exposed to *weathering* to produce soil. The five most predominant minerals that are found near the earth's surface are shown in Table 1-2.

TABLE 1-1 RELATIVE ABUNDANCE
OF ELEMENTS IN THE EARTH'S
CRUST (BY WEIGHT)

Element	Percent
Oxygen	46.0
Silicon	28.0
Aluminum	8.0
Iron	6.0
Magnesium	4.0
Calcium	2.4
Potassium	2.3
Sodium	2.1

Geologists classify rocks according to their origin into three groups:

1. *Igneous rocks.* These are rocks that are formed by the solidification of molten magma ejected to the surface of the earth by volcanic eruption. The magma cools as it moves through areas of lower temperature and solidifies when it comes into contact with the atmosphere. The rate of magma cooling determines the type of rock that will be formed. Some of the most commonly encountered igneous rocks are granite, basalt, gabbro, rhyolite, obsidian, pumice, and diorite.

2. *Sedimentary rocks.* These rocks are formed by compaction and cementation of soil deposits formed by weathering, or by various chemical processes. Pressure is exerted by the weight of overlying deposits (*overburden pressure*) and cementation is caused by cementing agents that are found dissolved in groundwater. Some of the most commonly encountered sedimentary rocks are shale, sandstone, mudstone, limestone, gypsum, and dolomite.

TABLE 1-2 PREDOMINANT ROCK MINERALS

Mineral	Percentage	Chemical Composition
1. Feldspars	30	
Plagioclase (calcium or sodium feldspar)		$Ca(Al_2Si_2O_3), Na(AlSi_3O_8)$
Orthoclase (potassium feldspar)		$K(AlSi_3O_8)$
2. Quartz (silicon dioxide)	28	SiO_2
3. Micas	18	
Biotite		$K(Mg,Fe)_3AlSi_3O_{10}(OH)_2$
Muscovite		$KAl_3Si_3O_{10}(OH)_2$
4. Calcium carbonates		
Calcite	9	$CaCO_3$
Dolomite	9	$CaMg(CO_3)_2$
5. Iron oxides	4	

3. *Metamorphic rocks.* These rocks are formed by a process called *metamorphism* (change of form). This entails the transformation of all types of rocks—igneous, sedimentary and metamorphic—by heat and pressure. Some of the most commonly encountered metamorphic rocks are gneiss from metamorphism of granite, marble from limestone, and dolomite slate from shale.

All three types of rocks can be subjected to weathering, a process by which rock breaks down to smaller grains to form soil. Weathering can take place due to changes in mineral composition (*chemical weathering*), changes in temperature (*mechanical weathering*), or a combination of the two.

Soils are also classified with respect to the mode by which they are transported.

1. Residual soils: Formed by weathering and remaining at their place of origin.
2. Transported soils or sediments:
   A. Soils transported by water
      Alluvial: transported by streams and rivers.
      Deltaic: transported by rivers and deposited in delta areas.
      Marine: formed by deposition and transport in the sea.
      Lacustrine: deposited in lake bottoms.
   B. Soils transported by ice (glacial deposits): transported and desposited by glacial movement.
   C. Soils transported by wind (aeolian deposits): such as loess and dune sand.
   D. Gravity-transported soils (colluvial deposits): transported by gravity during slope failures and landslides.

The various geologic processes that take place with time bring certain changes to the character of residual soils. These changes are responsible for the development of a natural stratification of surface soils, which is called the *soil profile,* or weathering profile. In general, there are three distinct zones or horizons in the natural soil profile.

The *A horizon* is the top surficial soil layer which includes humus (topsoil).

The *B horizon* is the subsurface layer composed mostly of fine colloidal soil particles transported from the A horizon.

The *C horizon* contains the less-weathered soils which extend into the bedrock.

Each horizon has a particular importance in highway, airfield, and building design and construction.

SOIL COMPOSITION AND PROPERTIES

Particle Size

Soils are classified into four major categories in relation to their particle (grain) size: *gravel, sand, silt,* and *clay.*

TABLE 1-3 GRAIN SIZE LIMITS FOR SOIL IDENTIFICATION

	Grain Size—mm and (in.)			
Soil	*USC*[a]	*MIT*[b]	*AASHTO*[c]	*USDA*[d]
Gravel[e]	76.2 (3) to 4.75 (0.2)	> 2 (0.08)	76.2 (3) to 2 (0.08)	> 2 (0.08)
Sand	4.75 (0.2) to 0.075 (0.003)	2 (0.08) to 0.06 (0.002)	2 (0.08) to 0.075 (0.003)	2 (0.08) to 0.05 (0.002)
Silt		0.06 (0.002) to 0.002 (8×10^{-5})	0.075 (0.003) to 0.002 (8×10^{-5})	0.05 (0.002) to 0.002 (8×10^{-5})
	Fines			
Clay	<0.075 (0.003)	<0.002 (8×10^{-5})	<0.002 (8×10^{-5})	<0.002 (8×10^{-5})

[a]Unified Soil Classification System (U.S. Army Corp of Engineers).
[b]Massachusetts Institute of Technology.
[c]American Association of State Highway and Transportation Officials.
[d]U.S. Department of Agriculture.
[e]Rock fragments larger than gravel size are called cobbles or boulders.

Several agencies have developed grain size limits to identify these soil categories. Table 1-3 shows the limits developed by four agencies and which are most frequently encountered in literature concerning grain size limits.

The Unified Soil Classification system is the most widely used in the United States and abroad.

Particle Size Distribution

Soil in its natural state consists of a conglomeration of solid grains of various sizes. The distribution of the particle size range is a very important characteristic of a given material. This distribution is generally depicted graphically on a curve called the *gradation* or *grain size distribution curve*. The grain size distribution curve is a semilogarithmic plot of the grain diameter (log scale) versus the weight expressed as a percent of the dry weight. Typical grain size distribution curves are shown in Figure 1-1.

There are two laboratory methods used to determine the grain size distribution of soils: *mechanical* or *sieve analysis* and *hydrometer analysis*.

Mechanical (Sieve) Analysis. This test is performed on soils with particle sizes larger than 0.075 mm in diameter. It consists of shaking the dried soil through a series of sieves of progressively smaller diameter openings. The weight of the soil retained in each sieve is measured and the percent of soil passing each sieve is computed based on the total dry weight of the sample. The grain size distribution curve is then plotted. The U.S. standard sieve numbers and size of the openings are given in Table 1-4.

The grain size distribution curve is very important for soil identification and clas-

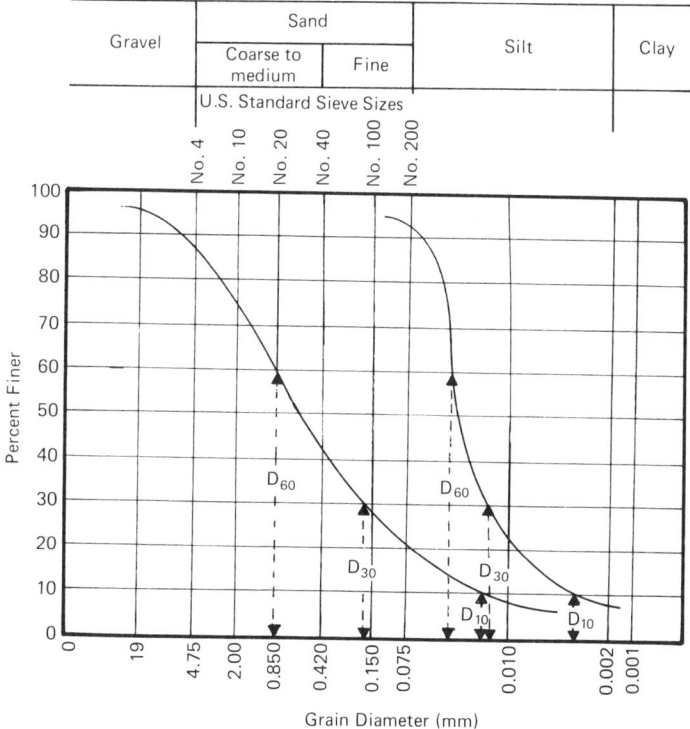

Figure 1–1 Typical grain size distribution curves.

sification of granular soils. Three parameters that are used to describe the grain size distribution curve are:

1. The effective size, D_{10}.
2. The coefficient of gradation,

$$C_c = \frac{(D_{30})^2}{D_{60} D_{10}}$$

3. The uniformity coefficient,

$$C_u = \frac{D_{60}}{D_{10}}$$

where D_{10}, D_{30}, and D_{60} = the diameters corresponding to 10%, 30%, and 60% finer, respectively, in the grain size distribution curve.

These parameters are used for classification of granular soils.

TABLE 1-4 U.S. STANDARD SIEVES
AND SIZE OF OPENINGS

Sieve Size or Number	Opening Size (mm)
Size (inches)	
4	101.60
3	76.10
2	50.80
$1\frac{1}{2}$	38.10
$1\frac{1}{4}$	32.00
1	25.40
$\frac{3}{4}$	19.00
$\frac{1}{2}$	16.00
$\frac{3}{8}$	9.51
Number	
4	4.750
6	3.350
8	2.360
10	2.000
16	1.180
20	0.850
30	0.600
40	0.425
50	0.300
60	0.250
80	0.180
100	0.150
140	0.106
170	0.088
200	0.075
270	0.053

Hydrometer Analysis.   This method is valid for soil with particle diameters of less than 0.075 mm. It is based on Stoke's law, the principle of the rate of settling of suspended solids in a liquid solution. Stoke's law of settling is the basis for the computation of grain size distribution. According to Stoke's law, the settling velocity of suspended particles is proportional to the square of their diameter.

The hydrometer test entails the measurement of the specific gravity of a dispersed soil suspension, contained in a sedimentation cylinder, by a hydrometer bulb immersed in the suspension. As larger soil particles settle, the specific gravity of the suspension decreases, approaching the specific gravity of water. The hydrometer readings are then correlated to the percent finer (equivalent to the percent passing in the sieve analysis) and the grain size distribution curve is plotted similarly to the method used in sieve analysis.

The standard test methods for both sieve and hydrometer analyses are described in detail in ASTM Designation D 422-63.

The grain size distribution of soils is a very important parameter used to classify soil and to estimate several of its properties, such as compaction and drainage characteristics.

Weight-Volume Relationships

Soil as found in nature is a three-phase medium, which makes it a unique engineering material. It is composed of solid mineral grains, water (moisture), and air. The water and air (or vapor) occupy the openings (voids or pores) which surround the solid grains.

In order to visualize and better understand the basic definitions that will follow, consider the elementary unit volume shown in Figure 1-2, which shows the three phases as if segregated.

It is clear that the total volume (V) of this element is

$$V = V_s + V_w + V_a = V_s + V_v \tag{1.1}$$

where V = total element volume
      V_s = volume of solids
      V_w = volume of water
      V_a = volume of air
      V_v = volume of voids

4th volume (?) - ice

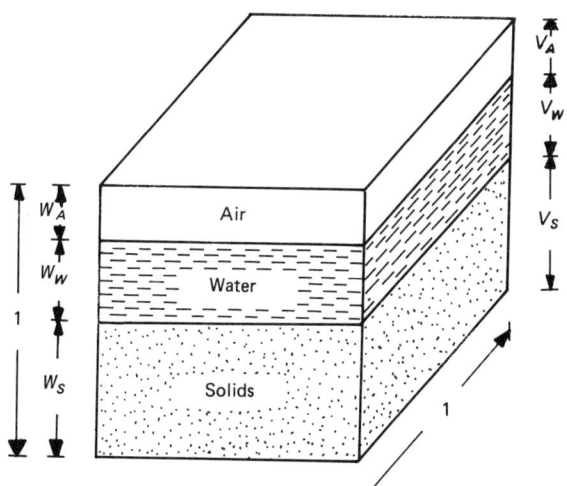

Figure 1–2 Elementary unit volume.

Similarly, by considering the weight of the air to be negligible, the total weight can be expressed as

$$W = W_s + W_w + W_a \text{ or } W = W_s + W_w \qquad (1.2)$$

where W_s = weight of solids
W_w = weight of water

The following definitions can be made based on these elementary relationships:

1. *Void ratio, e,* is defined as the volume of voids divided by the volume of solids.

$$e = \frac{V_v}{V_s} \qquad (1.3)$$

2. *Porosity, n,* is defined as the volume of voids divided by the total volume.

$$n = \frac{V_v}{V} \qquad (1.4)$$

3. *Water* or *moisture content, w,* is defined as

$$\text{percent } w = \frac{W_w}{W_s} \times 100 \qquad (1.5)$$

4. *Degree of saturation, S,* is defined as

$$\text{percent } S = \frac{V_w}{V_v} \times 100 \qquad (1.6)$$

5. *Unit weight, γ,* is defined as

$$\gamma = \frac{W}{V} \qquad (1.7)$$

The unit weight, γ, is sometimes called moist, wet, total or in-situ unit weight since it contains the weight of the moisture present in the voids. In two special cases the unit weight is of particular importance.

A. When the soil is free of any moisture the *dry unit weight, γ_d,* is defined as

$$\gamma_d = \frac{W_s}{V} \qquad (1.8)$$

B. When the soil is fully saturated (all voids are filled with water) equation 1.7 yields the *saturated unit weight, γ_{sat}.*

In analogy with the unit weight the density of the soil ρ is defined as

$$\rho = \frac{M}{V} \qquad (1.9)$$

and

$$\rho_d = \frac{M_s}{V} \qquad (1.10)$$

Copy for Geotech.

TABLE 1-5 EXPRESSIONS FOR WEIGHT–VOLUME RELATIONS FOR SATURATED SOILS (AFTER JUMIKIS, 1962)

Given Parameters	Dry Unit Weight γ_d	Equations For — Saturated Moisture Content, w %	Void Ratio e	Specific Gravity G_s	Porosity n	Saturated Unit Weight γ_{sat}
$\gamma_w,\ \gamma_{sat},\ w$	$\dfrac{\gamma_{sat}}{1+w}$		$\dfrac{w\gamma_{sat}}{\gamma_w - w(\gamma_{sat}-\gamma_w)}$	$\dfrac{\gamma_{sat}}{\gamma_w - w(\gamma_{sat}-\gamma_w)}$	$\dfrac{w\gamma_{sat}}{(1+w)\gamma_w}$	
$\gamma_w,\ \gamma_{sat},\ e$	$\gamma_{sat} - \dfrac{e}{1+e}\gamma_w$	$\dfrac{e\gamma_w}{\gamma_{sat} + e(\gamma_{sat}-\gamma_w)}$		$(1+e)\dfrac{\gamma_{sat}}{\gamma_w} - e$	$\dfrac{e}{1+e}$	
$\gamma_w,\ \gamma_{sat},\ n$	$\gamma_{sat} - n\gamma_w$	$\dfrac{n\gamma_w}{\gamma_{sat} - n\gamma_w}$	$\dfrac{n}{1-n}$	$\dfrac{\gamma_{sat} - n\gamma_w}{(1-n)\gamma_w}$		
$\gamma_w,\ G_s,\ \gamma_d$		$\left(\dfrac{1}{\gamma_d} - \dfrac{1}{G_s\gamma_w}\right)\gamma_w$	$\dfrac{G_s\gamma_w}{\gamma_d} - 1$		$1 - \dfrac{\gamma_d}{G_s\gamma_w}$	$\left(1 - \dfrac{1}{G_s}\right)\gamma_d + \gamma_w$
$\gamma_w,\ G_s,\ w$	$\dfrac{G_s}{1+wG_s}\gamma_w$		wG_s		$\dfrac{wG_s}{1+wG_s}$	$\left(\dfrac{1+w}{1+wG_s}\,G_s\right)\gamma_w$
$\gamma_w,\ G_s,\ e$	$\dfrac{G_s}{1+e}\gamma_w$	$\dfrac{e}{G_s}$			$\dfrac{e}{1+e}$	$\dfrac{G_s + e}{1+e}\gamma_w$
$\gamma_w,\ G_s,\ \gamma_{sat}$	$\dfrac{\gamma_{sat} - \gamma_w}{G_s - 1}\,G_s$	$\dfrac{G_s\gamma_w - \gamma_{sat}}{(\gamma_{sat}-\gamma_w)G_s}$	$\dfrac{G_s\gamma_w - \gamma_{sat}}{\gamma_{sat}-\gamma_w}$		$\dfrac{G_s\gamma_w - \gamma_{sat}}{(G_s-1)\gamma_w}$	
$\gamma_w,\ G_s,\ n$	$G_s(1-n)\gamma_w$	$\dfrac{n}{G_s(1-n)}$	$\dfrac{n}{1-n}$			$[G_s - n(G_s - 1)]\,\gamma_w$
$\gamma_w,\ w,\ n$	$\dfrac{n}{w}\gamma_w$		$\dfrac{n}{1-n}$	$\dfrac{n}{(1-n)w}$		$n\left(\dfrac{1+w}{w}\right)\gamma_w$
$\gamma_w,\ w,\ e$	$\dfrac{e}{(1-e)w}\gamma_w$			$\dfrac{e}{w}$	$\dfrac{e}{1+e}$	$\dfrac{e}{w}\left(\dfrac{1+w}{1+e}\right)\gamma_w$
$\gamma_w,\ \gamma_d,\ w$			$\dfrac{\gamma_d w}{\gamma_w - \gamma_d w}$	$\dfrac{\gamma_d}{\gamma_w - w\gamma_d}$	$w\dfrac{\gamma_d}{\gamma_w}$	$(1+w)\gamma_d$
$\gamma_w,\ \gamma_d,\ e$		$\left(\dfrac{e}{1+e}\right)\left(\dfrac{\gamma_w}{\gamma_d}\right)$		$\dfrac{(1+e)\gamma_d}{\gamma_w}$	$\dfrac{e}{1+e}$	$\dfrac{e\gamma_w}{1+e} + \gamma_d$
$\gamma_w,\ \gamma_d,\ \gamma_{sat}$		$\dfrac{\gamma_{sat}}{\gamma_d} - 1$	$\dfrac{\gamma_{sat} - \gamma_d}{\gamma_w + \gamma_d - \gamma_{sat}}$	$\dfrac{\gamma_d}{\gamma_w + \gamma_d - \gamma_{sat}}$	$\dfrac{\gamma_{sat} - \gamma_d}{\gamma_w}$	
$\gamma_w,\ \gamma_d,\ n$		$\dfrac{n\gamma_w}{\gamma_d}$	$\dfrac{n}{1-n}$	$\dfrac{\gamma_d}{(1-n)\gamma_w}$		$\gamma_d + n\gamma_w$

TABLE 1-6 TYPICAL SPECIFIC
GRAVITY VALUES FOR SELECTED
SOIL MINERALS

Mineral	Specific Gravity
Bentonite	2.15
Calcite	2.90
Chlorite	2.80
Biotite (mica)	3.00
Muscovite (mica)	2.80
Hornblende	3.30
Kaolinite	2.62
Illite	2.60
Quartz	2.60
Gibbsite	2.40
Feldspar	2.50
Montmorillonite	2.40
Anhydrite	3.00
Dolomite	2.90

where M = total mass (water and solids)

M_s = mass of solids

6. *Specific gravity* of the soil solids (G_s) is defined as

$$G_s = \frac{\gamma_s}{\gamma_w} = \frac{W_s}{V_s \gamma_w} \tag{1.11}$$

where γ_s = unit weight of soil solids,

γ_w = unit weight of water (62.4 lb/ft^3).

Combinations of these definitions (Equations 1.1 through 1.11) yield numerous expressions for the computation of the various parameters. Table 1-5 is a compilation of these expressions for saturated soils. Table 1-6 shows the specific gravity of selected minerals. The specific gravity of the most commonly occurring soil types ranges from 2.6 to 2.7.

Cohesive and Cohesionless Soils

Cohesion is an inherent physical property of a soil which causes individual grains to stick to each other when subjected to wetting or drying. As a result a certain force is required to break the soil apart when it is dry. Soils that exhibit this property are called *cohesive* soils. Soils in which the individual grains are not held together by any forces when dry are called *cohesionless*.

The presence of clay minerals gives soil deposits their cohesive properties. The amount of cohesion depends on the grain size distribution of the soil and the amount of clay minerals present. In general, most soils with the majority of grains passing the No. 200 sieve will exhibit cohesion.

The forces responsible for keeping fine soil grains together are ionic bonds, van der Waals bonds, hydrogen bonds, and gravitational attraction.

Clays and Clay Minerals

As shown earlier in Table 1-3, clays are defined as soils having particle diameters of less than 0.002 mm. This is not, however, the only requirement. For a soil to be classified as clay it must also be able to exhibit plasticity, that is, to behave as a plastic medium for a certain range of moisture content.

Clay minerals are mainly aluminum, iron, or magnesium silicates arranged in crystalline structures composed of two basic units, silica tetrahedron and aluminum octahedron. The silica tetrahedron unit [Figure 1-3(a)] consists of a silicon atom surrounded by four oxygen atoms. The octahedron unit is composed of six hydroxyls surrounding either a magnesium atom or an aluminum atom as shown in Figure 1-3(b). A combination of silica tetrahedron units composes a *silica sheet*. Sheets composed of octahedron units are called *gibbsite* sheets when the metallic atom is aluminum or *brucite* sheets when the metallic atom is magnesium.

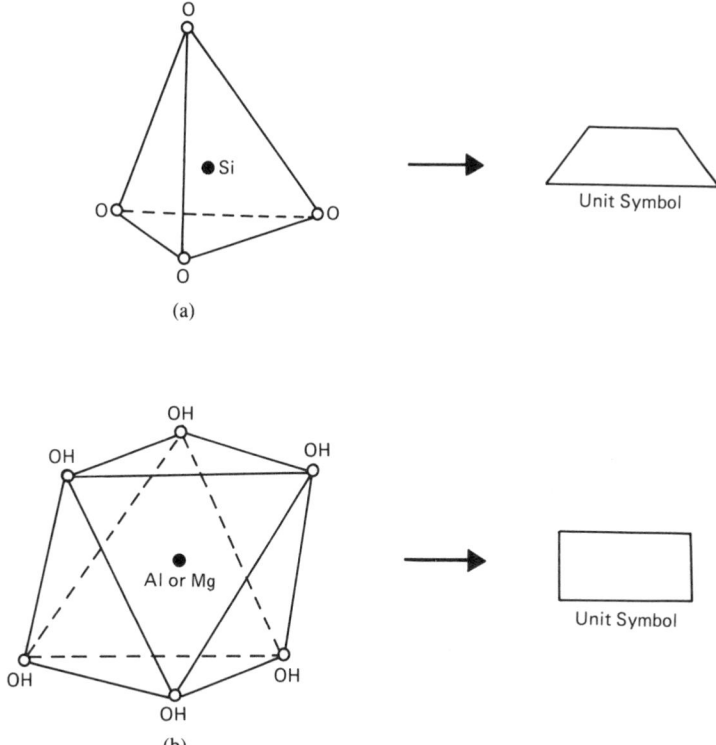

Figure 1-3 Clay mineral basic units: (a) tetrahedron unit (b) octahedron unit.

A physical characteristic of clay minerals is the *specific surface*, which is defined as the surface area of the mineral per unit mass.

There are three predominant clay minerals.

1. *Kaolinite*. Kaolinite consists of silica and gibbsite sheets stacked in alternating layers as shown in Figure 1-4(a). Each layer has a thickness of approximately 7 Angstroms (1 Angstrom (Å) = 1×10^{-10} m). The specific surface of kaolinite is approximately 15 m²/g.

2. *Illite*. Illite consists of a gibbsite sheet embedded between two silica sheets with potassium ions located between each layer as shown in Figure 1-4(b). Illite is derived primarily from micaceous minerals and is sometimes called *clay mica*. The thickness of each illite layer is 10 Å and the specific surface is about 80 m²/g. Another clay mineral in the illite family is *vermiculite*. It contains a brucite sheet in place of the gibbsite and calcium and magnesium ions along with potassium at the layer interfaces.

3. *Montmorillonite*. This mineral is composed of an octahedron sheet sandwiched between two tetrahedra sheets [Figure 1-4(c)]. Several different metallic atoms such as zinc, iron, lithium, or magnesium can be found in place of the aluminum atom in the octahedra sheets and aluminum can be found in place of silica in the tetrahedra sheets. The thickness of the montmorillonite unit is approximately 10 Å and its surface area is 800 m²/g. Bentonite is a member of the montmorillonite mineral group which has found wide applications in soil drilling explorations due to its expansive properties when exposed to water.

Clay particles in their natural state are always surrounded by water. The binding of water in the solid surface of clay minerals is called *adsorption*. This phenomenon greatly influences the physical properties of the clay. The primary cause of adsorption is the

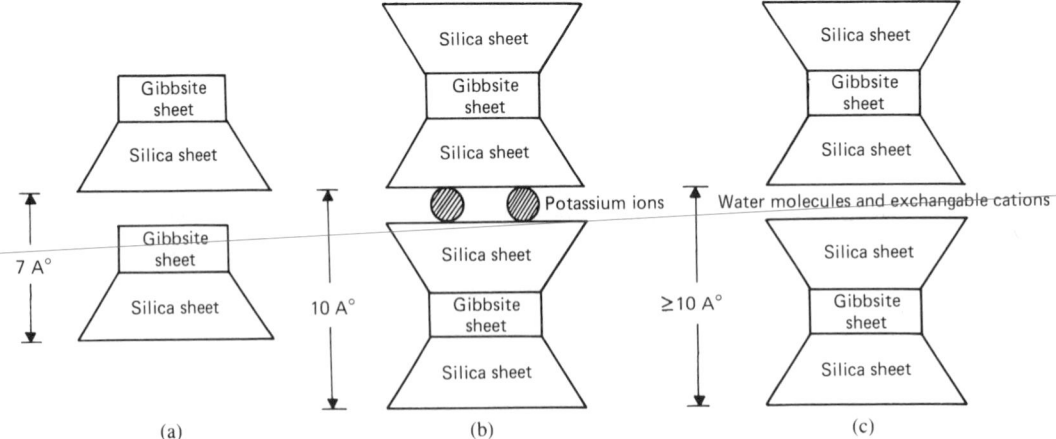

Figure 1–4 Predominant clay minerals: (a) kaolinite (b) illite (c) montmorillonite.

unbalanced electrostatic charge that exists in the mineral surfaces. The clay mineral surfaces generally have a net negative charge which may vary in intensity.

The water molecule is a neutral but also dipole molecule, with one pole having a positive charge and the other a negative charge. The water dipole is aligned when it is within the influential zone of the mineral's electrostatic field. The orientation is such that the positively charged pole of the water molecule is attracted by the negatively charged mineral surface. The attraction is strongest closest to the mineral surface and it diminishes farther away until it becomes so weak that the orientation of the dipoles ceases to exist and water molecules are randomly distributed. The zone which contains oriented water molecules is called the *diffuse double layer*. The larger the specific surface of the mineral, the larger is the negative charge at its surface, and therefore the thicker is the diffuse double layer.

Besides being bonded to the adsorbed water, clay minerals are bonded to each other by hydrogen bonding and van der Waals forces. If an external force squeezes some of the adsorbed water out and brings the particles closer together, the attraction becomes larger and the inter-particle bond becomes stronger. This is the property of plasticity which is a characteristic of clay minerals.

Soil Consistency and Atterberg Limits

As was described in the previous section, the water retained in the voids of fine-grained soils can greatly affect their engineering properties. Depending on the amount of water present in the voids the soil consistency can change from that of a brittle solid to a liquid state. In order to delineate the different states of soil consistency the Swedish soil scientist A. Atterberg proposed five consistency limits called *Atterberg Limits*. The consistency limits are based on the water content of the soil and are as follows:

1. The *liquid limit* (*LL*) is defined as the water content above which the soil behaves as a viscous liquid (its shearing strength is negligible).

   In soils engineering the liquid limit is measured in the laboratory using a liquid limit machine. It is arbitrarily defined as the water content at which a standard trapezoidal groove, cut in moist soil contained in a special cup, closes after 25 blows in the liquid limit machine.

2. The *plastic limit* (*PL*) is defined as the water content below which the soil ceases to behave as a plastic medium, and begins to exhibit the properties of a semisolid medium. The plastic limit is determined in the laboratory as the water content at which the soil cracks and breaks apart when rolled by hand into a thread having a diameter of $\frac{1}{8}$ inch.

3. The *shrinkage limit* (*SL*) is defined as the water content at which the soil reaches its minimum theoretical volume and no more volume reduction takes place with continued drying.

The plastic, liquid, and shrinkage limits have found wide applications in soil mechanics. Two more limits introduced by Atterberg, the *cohesion limit* and *sticky limit,* are utilized

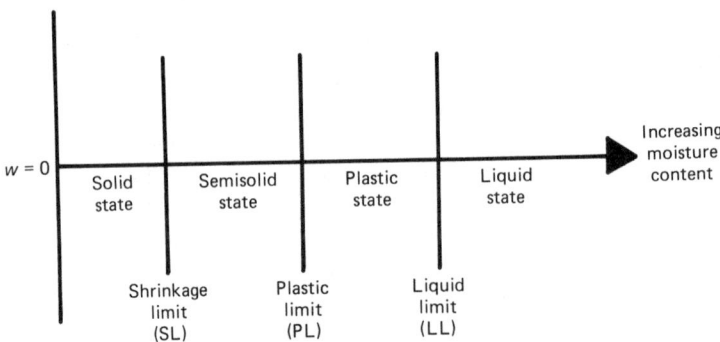

Figure 1–5 Atterberg limits and the corresponding states of soil consistency.

mostly in agricultural applications. Figure 1-5 presents the various states of soil consistency and their corresponding consistency limits.

The *plasticity index* (*PI*) was defined by Atterberg to describe the range of water content over which the soil behaves as a plastic medium. The plasticity index is defined as

$$PI = LL - PL \tag{1.12}$$

Another index used to relate the water content of the soil to the liquid and plastic limit is the *liquidity index* (LI). The liquidity index can be used to describe the engineering behavior of soil when subjected to a shearing force and is defined as

$$LI = \frac{w - PL}{PI} \tag{1.13}$$

where w = the natural water content of the soil.

The following material behavior patterns can be correlated to the liquidity index.

Liquidity index	Engineering behavior
LI < 0	Brittle solid
0 < LI ≤ 1	Plastic
LI > 1	Viscous liquid

The Atterberg limits and indices are very valuable for the classification of soils and they have been empirically correlated to several soil properties with wide applications in soil mechanics. Another parameter often used to express the plasticity of clay minerals is the *activity* (*A*), which is defined as:

$$A = \frac{PI}{\text{Percentage of clay size particles } (\% < 0.002 \text{ mm})} \tag{1.14}$$

will be on the test

all that is accountable in this chapter

The activity is also a measure of the water retention capaci~~ty~~
can be used to identify them.

Some typical values for the predominant clay mineral

Mineral	Activity
Kaolinite	< 1
Illite	0.5–1.~~5~~
Montmorillonite	1–6
Bentonite	>1 –>76 (greater than —

(handwritten annotations: "test also.", "good", "good", "good", "bad")

Standard laboratory test procedures for the liquid and plastic limits are described in the
ASTM Designation D 4318-84.

Structure of Cohesive Soils

The structure of cohesive soils is largely controlled by the presence of clay minerals and
the various forces acting on them. Examination of the soil structure from the microscopic
point of view reveals the microstructure of the soil. The microstructure of clay is dependent
on the geologic history of the material, including type of deposition, stress history, and
deposition environment. There are two general types of clay structure.

> *Dispersed* structures [Figure 1-6(a)] are formed when the repulsive forces between
> grains cause them to be deposited in an oriented fashion with maximum possible
> distances separating each of them.

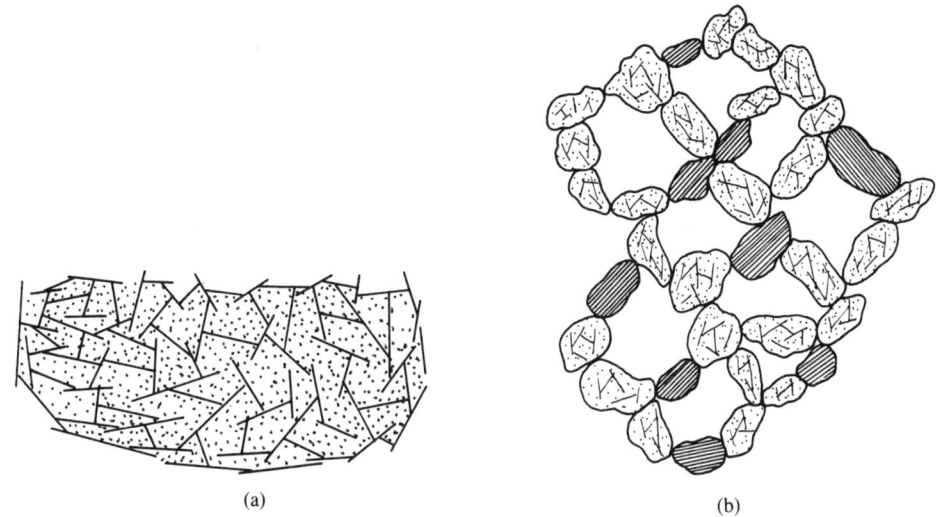

(a)   (b)

Figure 1–6   Structure of Clay Soil: (a) dispersed structure (b) flocculated structure.

Flocculated structures [Figure 1-6(b)] are formed when soil deposition occurs in a chemical environment which aids the flocculation of initially dispersed grains. Flocculation is the process by which individual particles loose their repulsive forces when they are in suspension and attract each other to form clusters called *flocs*. The loss of the repulsive forces is caused by the presence of an electrolyte, such as salt water. As flocs become larger they settle under gravity to form flocculent structures. Clay deposits formed in marine environments are usually highly flocculent.

Examination of the soil from the macroscopic point of view reveals the soil's macrostructure. The macrostructure is used to determine patterns of soil deposition, discontinuities in the soil structure caused by cracking, and the effects of external forces on the soil's structural behavior.

Structure of Granular Soils

Cohesionless soils are generally encountered as single-grain structures. The most important characteristic of a single-grain structure is the arrangement of the individual grains, or *packing*. A soil's state of denseness or looseness depends on three factors: the grain shape, the grain size distribution, and the grain relative positions (packing).

Grain shape can range from angular to very round. This affects the packing of the grains and therefore the void ratio. A wide range of void ratios can be obtained by changing the mode of grain packing. This can be illustrated by packing perfectly spherical particles as shown in Figure 1-7. In Figure 1-7(a), the spheres are packed in a cubical arrangement with a void ratio of $e = 0.91$. In Figure 1-7(b), the same spheres are packed in a pyramidal arrangement with a void ratio of $e = 0.35$.

The grain size distribution is also an important factor affecting the packing of granular soils. A soil is well graded if there is an assortment of small particles to fill the

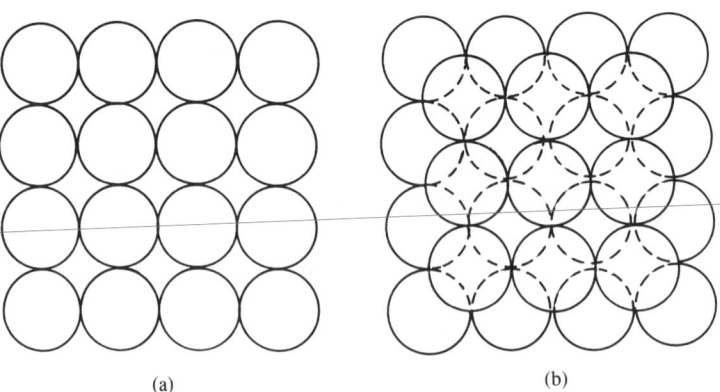

(a)   (b)

Figure 1-7 (a) Cubical grain packing with a void ratio of $e = 0.91$ (b) pyramidal grain packing with a void ratio of $e = 0.35$.

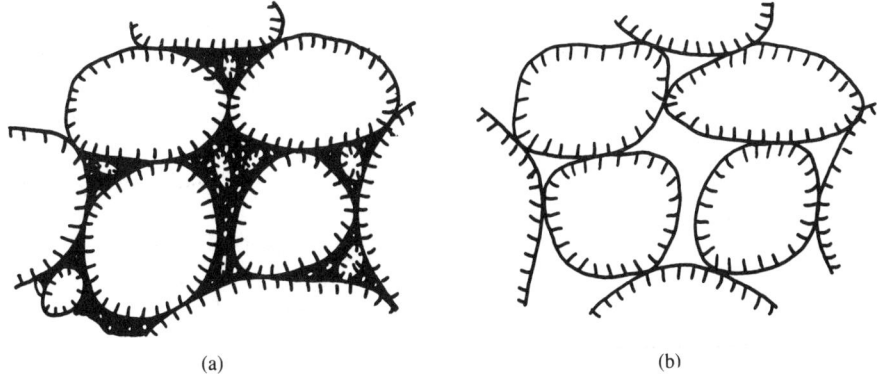

Figure 1-8   Soil gradation: (a) well-graded soil (b) poorly graded soil.

voids formed by larger particles, thus allowing for denser packing [Figure 1-8(a)]. On the other hand, the soil is poorly graded if the voids are not occupied by grains yielding a looser sample with higher void ratio [Figure 1-8(b)]. The *relative density* is used as a measure of the in-situ void ratio of a soil. It is defined as

$$D_r = \frac{e_{max} - e}{e_{max} - e_{min}} \tag{1.15}$$

where e = in-situ void ratio,

$e_{max,\ min}$ = maximum and minimum void ratios as determined in the laboratory.

The relative density can also be expressed in terms of the dry unit weight of the soil as

$$D_r = \frac{\gamma_d - \gamma_{dmin}}{\gamma_{dmax} - \gamma_{dmin}} \frac{\gamma_{dmax}}{\gamma_d} \tag{1.16}$$

where γ_d = in-situ dry unit weight,

$\gamma_{dmax,\ min}$ = maximum and minimum dry unit weights
as determined in the laboratory.

SOIL CLASSIFICATION

Soil classification is the grouping of different types of soils with similar properties. There are several soil classification systems. These systems provide a common language used to briefly and accurately communicate the type of a particular soil as related to its engineering use.

The most widely used classification systems in engineering are the Unified Soil Classification system (USC) and the American Association of State Highway and Transportation Officials (AASHTO) soil classification system. The basis for soil classification in both systems are the index properties of the soil—the grain size distribution and plasticity.

1. *Unified Soils Classification System.* This classification system was first proposed by Casagrande in 1942 in airfield construction works performed by the Army Corps of Engineers. It has since been slightly modified and today is used widely in geotechnical engineering. It is based on grouping and subgrouping of similar soils based on their basic index properties. The USC system is shown in Table 1-7. The soil grouping of the USC system can be described as follows:

TABLE 1-7 UNIFIED SOILS CLASSIFICATION SYSTEM (AFTER ASTM)

Major Divisions			Group Symbols	Typical Names
Coarse-Grained Soils *More than 50% retained on No. 200 sieve*	*Gravels* *50% or more of coarse fraction retained on No. 4 sieve*	*Clean Gravels*	GW	Well-graded gravels and gravel-sand mixtures, little or no fines
			GP	Poorly graded gravels and gravel-sand mixtures, little or no fines
		Gravels with Fines	GM	Silty gravels, gravel-sand-silt mixtures
			GC	Clayey gravels, gravel-sand-clay mixtures
	Sands *More than 50% of coarse fraction passes No. 4 sieve*	*Clean sands*	SW	Well-graded sands and gravelly sands, little or no fines
			SP	Poorly graded sands and gravelly sands, little or no fines
		Sands with Fines	SM	Silty sands, sand-silt mixtures
			SC	Clayey sands, sand-clay mixtures
Fine-Grained Soils *50% or more passes No. 200 sieve*	*Silts and Clays* *Liquid limit 50% or less*		ML	Inorganic silts, very fine sands, rock flour, silty or clayey fine sands
			CL	Inorganic clays of low to medium plasticity, gravelly clays, sandy clays, silty clays, lean clays
			OL	Organic silts and organic silty clays of low plasticity
	Silts and Clays *Liquid limit greater than 50%*		MH	Inorganic silts, micaceous or diatomaceous fine sands or silts, elastic silts
			CH	Inorganic clays of high plasticity, fat clays
			OH	Organic clays of medium to high plasticity
Highly Organic Soils			PT	Peat, muck, and other highly organic soils

TABLE 1-7 *(Continued)*

Classification on Basis of Percentage of Fines				Classification Criteria	
Classification on Basis of Percentage of Fines	*GW, GP, SW, SP* *Less than 5% pass No. 200 sieve*	*GM, GC, SM, SC* *More than 12% pass No. 200 sieve*	*Borderline classification requiring use of dual symbols* *5% to 12% pass No. 200 sieve*	$C_u = D_{60}/D_{10}$ Greater than 4 $C_c = \dfrac{(D_{30})^2}{D_{10} \times D_{60}}$ Between 2 and 3	
				Not meeting both criteria for GW	
				Atterburg limits plot below "A" line and plasticity index less than 4	Atterberg limits plotting in hatched area are borderline classifications requiring use of dual symbols
				Atterberg limits plot above "A" line and plasticity index greater than 7	
				$C_u = D_{60}/D_{10}$ Greater than 6 $C_c = \dfrac{(D_{30})^2}{D_{10} \times D_{60}}$ Between 1 and 3	
				Not meeting both criteria for SW	
				Atterberg limits plot below "A" line or plasticity index less than 4	Atterberg limits plotting in hatched area are borderline classifications requiring use of dual symbols
				Atterberg limits plot above "A" line and plasticity index greater than 7	

PLASTICITY CHART

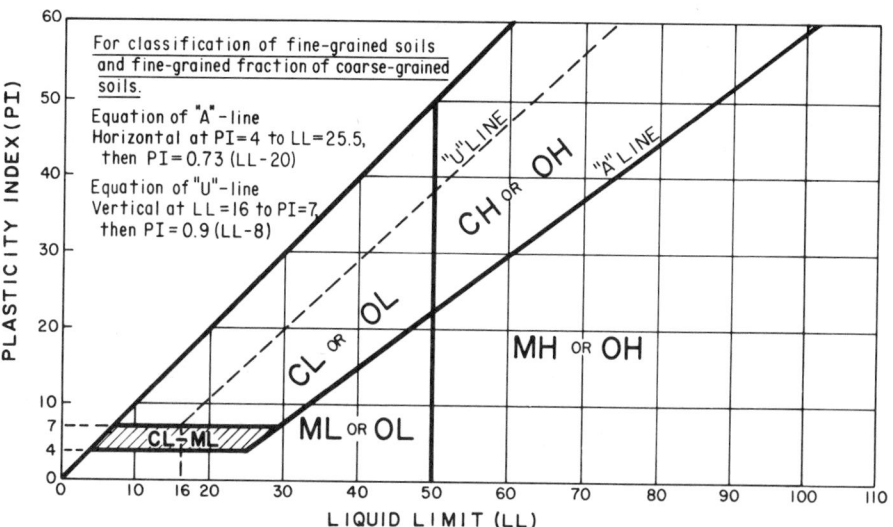

A. *Coarse Grained Soils:* Composed mainly of gravels and sands with less than 50% passing the No. 200 sieve. The letter symbol prefixes used are: G to describe gravels, (50% or more grains are retained on No. 4 sieve) and S to describe sand (more than 50% passes through No. 4 sieve). The suffixes W, P, M, and C are used to describe well graded, poorly graded, silt, and clay content, respectively, of the gravels or sand mixtures.

B. *Fine Grained Soils:* Consisting of silts and clays with 50% or more passing the No. 200 sieve. The prefixes M, C, and O are used to describe inorganic silts, inorganic clays, and organic silts and clays, respectively. The suffix L is used to describe fines of low plasticity (liquid limit of 50% or less) and the suffix H is used to describe fines of high plasticity (liquid limit greater than 50%). The plasticity chart shown in Table 1-7 is used for the classification of fine-grained soils.

C. *Highly Organic Soils:* The symbol PT for peat is used to describe soils of high organic content. For more details on the Unified Soils Classification system, the student should refer to ASTM D 2487-85.

2. *AASHTO Classification System.* The AASHTO Classification system was originally developed by the Bureau of Public Roads. It has been revised several times and was brought to its present form by the Highway Research Board. It is described in ASTM Designation D 3282-83 and AASHTO method M-145 and is shown in Table 1-8. The soils under this system are classified into seven major categories, A-1 through A-7. The soils in groups A-1, A-2, and A-3 are granular soils with 35% or less passing the No. 200 sieve and the soils in groups A-4, A-5, A-6, and A-7 are fine-grained soils with more than 35% passing the No. 200 sieve. The AASHTO system is more frequently used to classify soils used as subgrade materials in highway construction.

SOIL-WATER INTERACTION

As was pointed out in previous sections, soil in its natural state contains water. The degree of saturation in soils can range from a few percent in the upper soil layers, which are subject to evaporation, to 100% within the groundwater table. When the voids in the soil are partially filled with water the soil is said to be partially saturated. When the voids are completely filled, the soil is said to be fully saturated.

The water or moisture present in the soil can be either static (*stationary*) or dynamic (*flowing*). Flow can take place both in partially saturated and fully saturated soils. The interaction of both stationary and flowing water with soil will be examined in this section.

The Concept of Effective Stress

In order to analyze several soil engineering problems we need to know the stress distribution along a soil profile. Consider the soil profile shown in Figure 1-9.

Let us first find what is the vertical stress at point *A*. This point is subjected to the

TABLE 1-8 AASHTO SOIL CLASSIFICATION SYSTEM

General Classification	Granular Materials (35 % or less passing No. 200)							Silt-Clay Materials (More than 35 % passing No. 200)			
	A-1		A-3	A-2				A-4	A-5	A-6	A-7
Group classification	A-1-a	A-1-b		A-2-4	A-2-5	A-2-6	A-2-7				A-7-5, A-7-6
Sieve analysis, percent passing:											
No. 10 (2.00 mm)	50 max										
No. 40 (425 μm)	30 max	50 max	51 min								
No. 200 (75 μm)	15 max	25 max	10 max	35 max	35 max	35 max	35 max	36 min	36 min	36 min	36 min
Characteristics of fraction passing No. 40 (425 μm):											
Liquid limit			...	40 max	41 min	40 max	41 min	40 max	41 min	40 max	41 min
Plasticity index	6 max		N.P.	10 max	10 max	11 min	11 min	10 max	10 max	11 min	11 min[A]
Usual types of significant constituent materials	Stone fragments, gravel and sand		Fine sand	Silty or Clayey Gravel and Sand				Silty Soils		Clayey Soils	
General rating as subgrade	Excellent to good							Fair to poor			

[A]Plasticity index of A-7-5 subgroup is equal to or less than LL minus 30. Plasticity index of A-7-6 subgroup is greater than LL minus 30.

Reprinted with permission of American Association of State Highway and Transportation Officials.

Ground surface

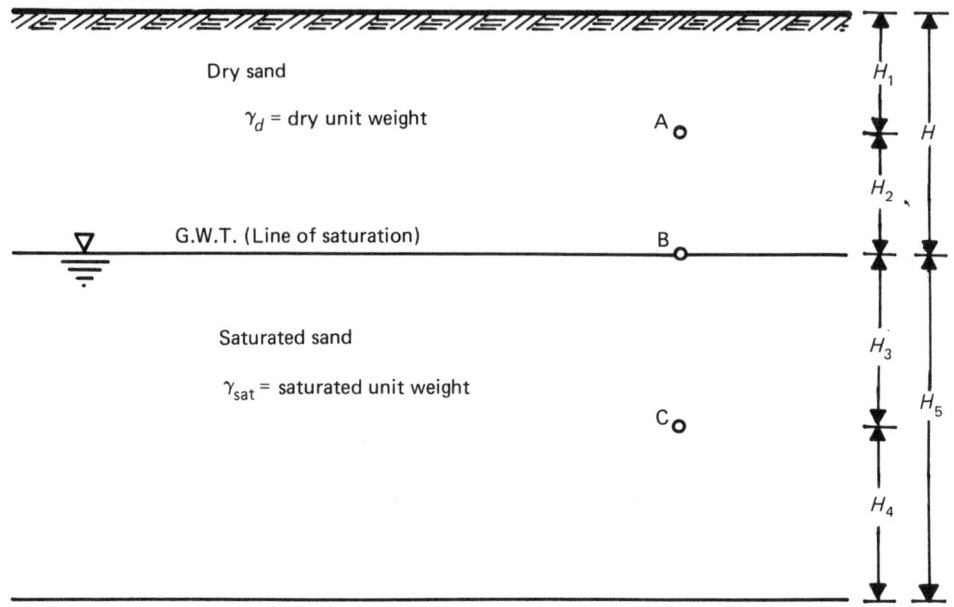

Figure 1–9 Soil profile.

total weight of the soil column having height H_1. The *vertical stress* at this point is therefore

$$\sigma_A = \gamma_d H_1 \tag{1.17}$$

similarly, at point B,

$$\sigma_B = \gamma_d(H_1 + H_2) = \gamma_d H \tag{1.18}$$

The vertical stress due to the weight of the overlying soil is often called *geostatic pressure* or *overburden pressure*.

Let's now consider point C. The total stress at point C is

$$\sigma_c = \gamma_d H + \gamma_{sat} H_3 \tag{1.19}$$

The vertical stress at point C consists of two parts. One part is carried by the water, which at this location fills all the interconnected voids and is acting as a continuum. This portion is called the *neutral stress* or *pore water pressure* and is equal to the *hydrostatic stress*, or

$$u = \gamma_w H_3 \tag{1.20}$$

where γ_w is the unit weight of water.

The other part is carried by the solid structure and is transmitted from grain to grain. This is the *intergranular* or *effective stress* ($\overline{\sigma}$).

The effective stress at point C therefore is

$$\bar{\sigma} = \sigma_c - u = (H\gamma_d + H_3\gamma_{sat}) - H_3\gamma_w$$

or

$$\bar{\sigma}_c = H\gamma_d + H_3(\gamma_{sat} - \gamma_w)$$

or

$$\bar{\sigma}_c = H\gamma_d + H_3\gamma_{sub} \tag{1.21}$$

where γ_{sub} has been defined as the submerged unit weight $= \gamma_{sat} - \gamma_w$.

The microscopic view of the soil arrangement at the vicinity of point C shown in Figure 1-10 will aid in the visualization of the concept of the effective stress. Here we see that the hydrostatic forces are acting in all directions with the same intensity and therefore only the forces transmitted from grain to grain at the points of contact need to be considered. In the case of dry soil the neutral stress is zero, so the effective stress is equal to the total stress.

In this case we considered a situation where the groundwater was stationary. If the water is moving through the pores, seepage forces must also be considered. In almost all engineering analyses involving overburden stresses the effective stresses are used.

Flow of Water in Soil: Darcy's Law

In the context of soils engineering, the property of soil that controls the passage of water through its pores is called *permeability*. It follows from this definition that soils with large interconnected pores will have higher permeability than soils with smaller pores. If we therefore assume that all other parameters in a soil are kept constant, we expect that permeability will increase with increasing void ratio.

The factors that influence the flow of fluids through soils are:

1. The size, shape and arrangement of the pore spaces.
2. The relative density of the soil.

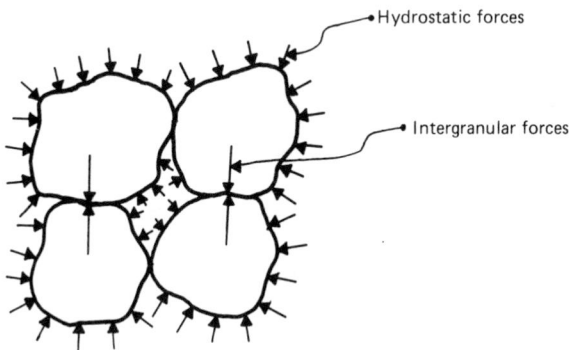

Figure 1–10 Microscopic view of the soil arrangement and forces at the area surrounding point C in Figure 1–9.

3. The density and viscosity of the fluid.

4. The mineralogical and electrochemical properties of the soil particles.

Flow in soils takes place due to the existence of a pressure (*hydraulic head*) difference between two points in the flow field. The direction of flow is from regions of high pressure to low pressure, and the velocity of flow depends on the relative magnitude of the hydraulic head difference and the permeability of the soil.

In the mid-eighteenth century, the French scientist H. Darcy performed experiments to investigate the flow of water through sand filters. The arrangement of Darcy's experiment is shown in Figure 1-11.

Flow is passed through a cylindrical column of inside diameter, D, containing a soil sample having length, L, at a rate, q. Darcy observed that the flow rate per unit time, q, is proportional to the change in hydraulic head drop, Δh, and the cross sectional area of the soil sample and inversely proportional to the length, L, over which the head drop, Δh, takes place. Referring to Figure 1-10:

$$q = \frac{V}{t} = k\frac{\Delta h}{L}A \tag{1.22}$$

where q = volumetric flow rate per unit time
$\quad\ V$ = volume
$\quad\ t$ = time
$\quad\ k$ = coefficient of permeability

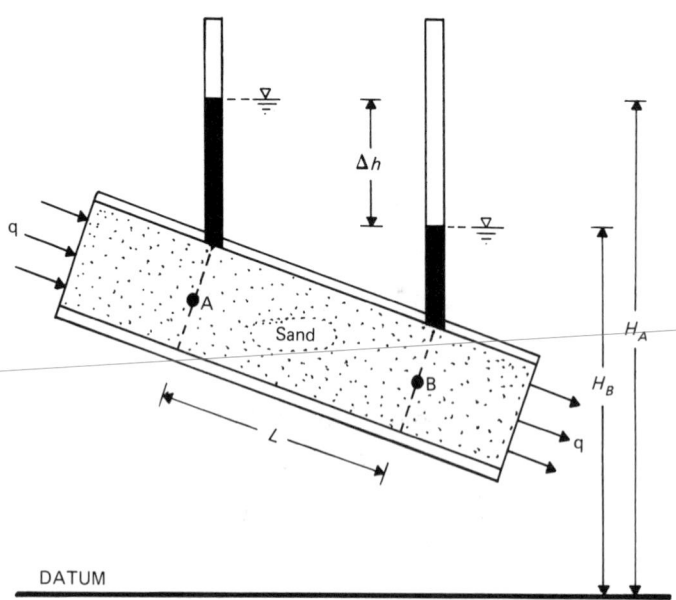

Figure 1–11 Set-up of Darcy's experiment.

Δh = hydraulic head difference between points A and B
A = cross sectional area of soil sample
L = length of soil sample

The ratio $\Delta h/L$ is called the hydraulic gradient, i, and Eq. 1.22 can be written as

$$q = kiA \qquad (1.23)$$

The coefficient of permeability, k, has units of velocity (length per time) and is often called *hydraulic conductivity*. Following the principle of continuity the velocity of flow through the soil is

$$v = \frac{q}{A} = ki \qquad (1.24)$$

The velocity is often called the *Darcy* or *discharge velocity*. It is an average velocity since it is obtained by dividing the flow rate by the gross cross-sectional area of the sample, which includes both soil grains and pores. Since flow takes place through the pores only, the actual or interstitial velocity is

$$v_{act} = \frac{v}{n}$$

where n is the soil porosity.

Permeability is one of the most important soil properties and is used to estimate the flow through soils in several civil and highway engineering applications, such as seepage through earthen dams, site dewatering, and groundwater pollution.

It must be noted here that permeability is a tensor parameter and its value often depends on the direction along which it is measured. Soils can be classified as follows with respect to permeability:

homogeneous: if the coefficient of permeability is independent of position within a soil formation.

heterogeneous: if the coefficient of permeability is dependent on position within a soil formation.

isotropic: if the coefficient of permeability is independent of the direction of measurement at a point in a soil formation.

anisotropic: if the coefficient of permeability varies with the direction of measurement at a point in a soil formation.

Some typical values of the coefficient of permeability for various types of soils are given in Table 1-9.

Permeability Testing.   Permeability is one of the most widely varying properties of engineering materials. As shown in Table 1-9, the coefficient of permeability can take values varying over several orders of magnitude. It is therefore necessary to be able to

TABLE 1-9 PERMEABILITY RANGES FOR VARIOUS TYPES OF SOILS
(AFTER TERZAGHI AND PECK, 1967)

Soil Type	Coefficient of Permeability (cm/sec)	Relative Permeability
Gravel	Greater than 10^{-1}	High
Sandy Gravel	10^{-1}–10^{-3}	Medium
Sand	10^{-3}–10^{-5}	Low
Silty Clay	10^{-5}–10^{-7}	Very Low
Clay	Less than 10^{-7}	Practically Impermeable

measure the coefficient of permeability with relative accuracy in order to reliably predict
the rate of flow through soils. Permeability testing can either be performed in the laboratory
or in the field.

Laboratory testing can be performed quickly and is relatively inexpensive, especially
for granular soils. Laboratory tests, however, often do not produce reliable results, mainly
because of the dependence of the permeability on the in-situ structure of the soil. Test

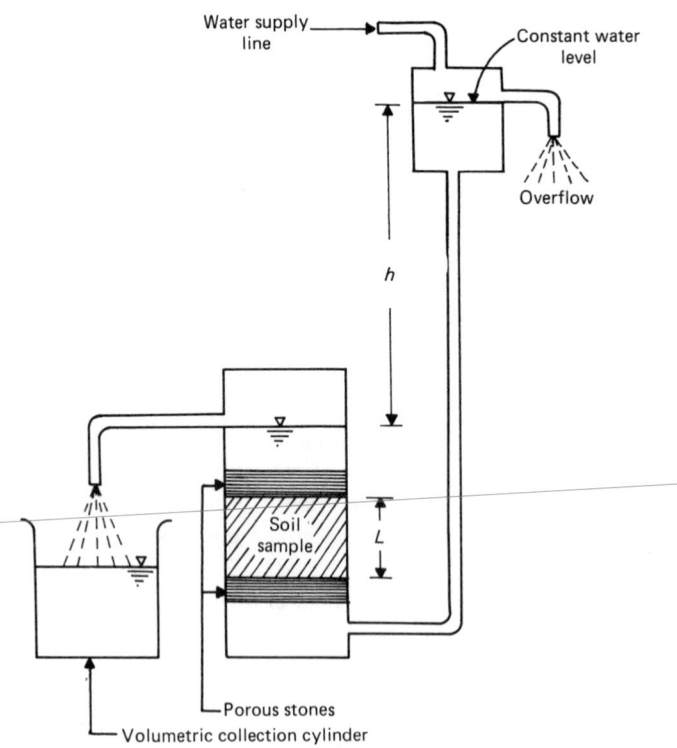

Figure 1–12 Constant head permeameter.

specimens recovered from the field are often disturbed, resulting in an alteration of the in-situ soil structure stratification, grain orientation, and relative density. Such disturbances and remolding of the soil often result in laboratory permeability values different from the in-situ values.

In such situations field tests produce more reliable results but are generally more time consuming and more expensive.

Laboratory Tests. In the laboratory, permeability is measured with devices called *permeameters* by maintaining a flow through a relatively small soil sample. The flow rate and head loss are measured to determine the coefficient of permeability.

Two types of tests are usually employed, the constant head and falling head tests. The constant head permeameter shown in Figure 1-12 is used to measure the permeability of granular soils under low head conditions. Water flows upward through the soil sample under constant head and is collected in a volumetric cylinder at the effluent port. The coefficient of permeability can be computed by

$$k = \frac{VL}{tAh} \qquad\qquad (1.25)$$

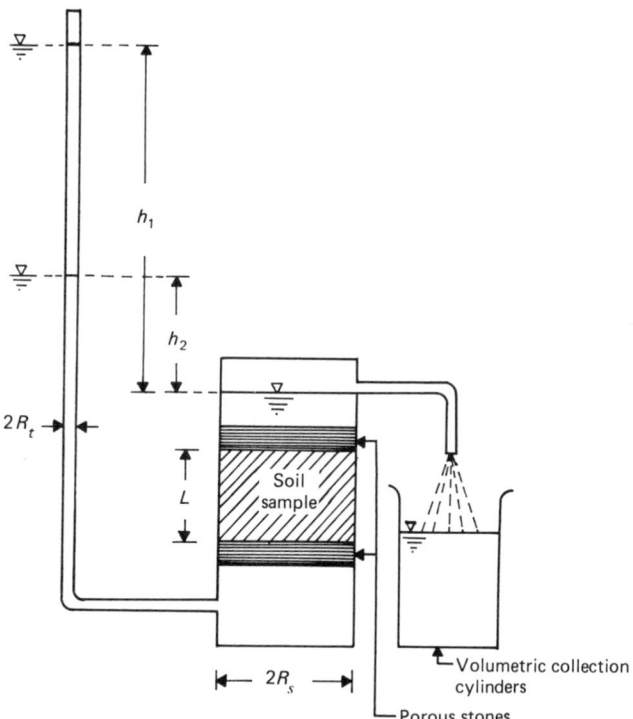

Figure 1–13   Falling head permeameter.

where V = volume of fluid passed through the sample during time t

       L = length of soil specimen

       A = cross sectional area of soil specimen

       h = hydraulic head difference across soil sample

The constant head test is described in ASTM designation D 2434-68.

     The falling head permeameter is shown in Figure 1-13. In this apparatus, water supplied by a long tube flows upward through the soil and is collected at the effluent port in a volumetric cylinder. The head drop per time in the tube is recorded and the coefficient of permeability is computed as

$$ k = \frac{R_t^2 \, L}{R_s^2 \, t} \, \ell n \, \frac{h_1}{h_2} \tag{1.26}$$

where R_t = radius of the tube

       R_s = radius of the soil sample

       L = length of the soil sample

       h_1 = initial head in the tube

       h_2 = final head in the tube

       t = time required for the water level to drop from h_1 to h_2

Falling head permeability tests are performed on soils having relatively low permeability, such as silts.

     Field Tests.   Various field methods are available for in-situ measurement of soil permeability. These include pumping tests, tracer tests, and borehole tests. The advantage of field tests is that the soil is tested under undisturbed in-situ conditions. Various parameters that could influence the results of the test, such as the location of the groundwater table, the overburden stresses, and the soil stratification are not altered as in the case of laboratory tests.

     Some cased borehole testing methods are included in this book. These tests are performed in a cased boring drilled in the ground. The water level variations with time, as water is pumped into or out of the boring, are used to compute the soil permeability. The schematic of four types of borehole tests and the respective equations used for determination of the coefficient of permeability are shown in Table 1-10.

Capillary Rise in Soils

Capillarity is a phenomenon that is attributed to a fluid property which is called *surface tension*. In soils, surface tension occurs between the interfaces of water, soil grains, and air and results from the differences in the attractive forces between molecules at the surfaces of these materials. As a result of surface tension, the surface of the water bends, forming a curved surface called the *meniscus* as shown in Figure 1-14. Due to the pore water surface tension, water will rise to a certain height above the surface of saturation. This is illustrated in Figure 1-15, which depicts a typical soil profile with a groundwater table. The height of capillary rise is often called the *capillary fringe*. In analogy with

TABLE 1-10 METHODS FOR PERFORMING CASED BOREHOLE PERMEABILITY TESTS
(Reprinted from McCarthy by permission)

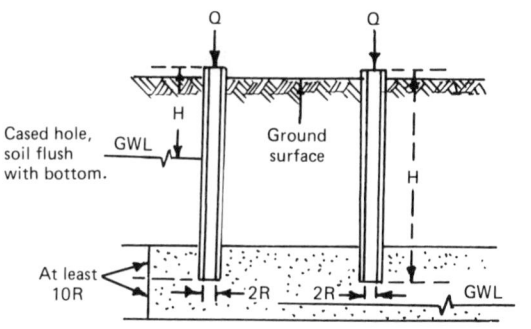

Cased hole, soil flush with bottom.

$$k = \frac{0.024Q}{RH}$$

where Q = gal/min
R = feet
H = feet
k = ft/min

Used for permeability determinations when water is above or below bottom of casing. Q is quantity of water to keep casing filled.

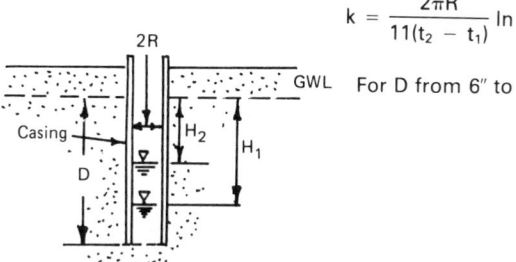

Cased hole, soil flush with bottom.

$$k = \frac{2\pi R}{11(t_2 - t_1)} \ln \frac{H_1}{H_2}$$

For D from 6″ to 60″

Used for permeability determination at shallow depths below the water table. May yield unreliable results in falling head test with silting of bottom of hole.

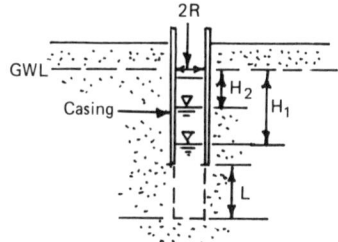

Cased hole, uncased or perforated extension of length L.″

$$k = \frac{R_2}{2L(t_2 - t_1)}$$

$$\ln \left(\frac{L}{R}\right) \ln \left(\frac{H_1}{H_2}\right)$$

For $\frac{L}{R} > 8$

Used for permeability determinations at greater depths below water table.

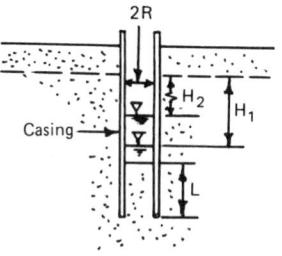

Cased hole, column of soil inside casing to height L.″

$$k = \frac{2\pi R + 11L}{11(t_2 - t_1)} \ln \left(\frac{H_1}{H_2}\right)$$

Principal use is for permeability in vertical direction in anisotropic soils.

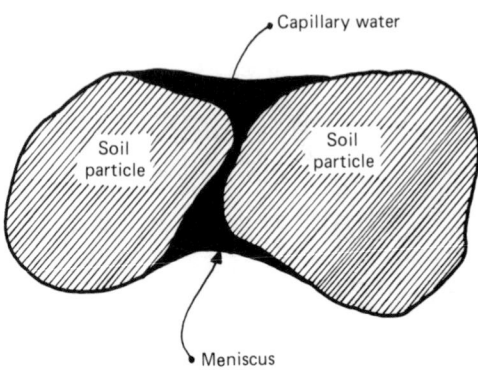

Figure 1–14 Capillary water held between two
soil particles due to surface tension.

capillary rise in thin tubes the following expression can be used to estimate the height
of capillary rise in soils.

$$h_c = \frac{0.31}{0.2D_{10}} \text{ (cm)} \tag{1.27}$$

where h_c = height of capillary rise in cm
$\quad D_{10}$ = 10% particle size of the soil from the grain size
$\qquad$ distribution curve in cm

It is observed from this expression that the height of capillary rise is inversely proportional
to the effective grain size, D_{10}, which is in turn directly related to the pore size. It follows
that smaller effective grain sizes will correspond to smaller pore openings and larger
capillary rises. Typical heights of capillary rise for various types of soils are shown in
Table 1-11.

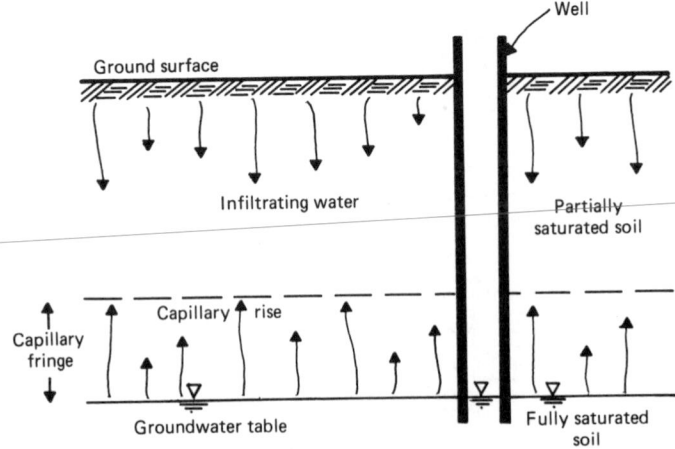

Figure 1–15 Typical soil profile illustrating capillary rise.

TABLE 1-11 TYPICAL VALUES OF
CAPILLARY RISE FOR VARIOUS SOILS

Soil	Capillary Rise (cm)
Coarse gravel	0.5
Fine gravel	7
Coarse sand	15
Medium sand	25
Fine sand	45
Silt	110
Clay	>200

Frost Action in Soils

When the ambient air temperature falls below the freezing point soil will begin to freeze. The temperature profile in the soil will be similar to that depicted in Figure 1-16. The temperature will be lowest at the soil surface and will increase with depth until it reaches the residual subsurface above-freezing temperatures.

Freezing temperatures in the subsurface environment cause the formation of *ice lenses* in the pore spaces. Water in the soil pores expands as it freezes, creating a volume increase in the soil mass called *frost heave*. One would expect that the total volume change in a soil mass would be equal to the volume increase due to transformation of water from the liquid to the solid state, which is approximately 10%. But field experiments in the 1920s showed that frost heave is significantly larger than that attributed to the volumetric expansion of freezing.

Research in the past 50 years has concentrated on the mechanisms of ice lense formation to explain and quantify frost heave. Observations and tests have indicated that

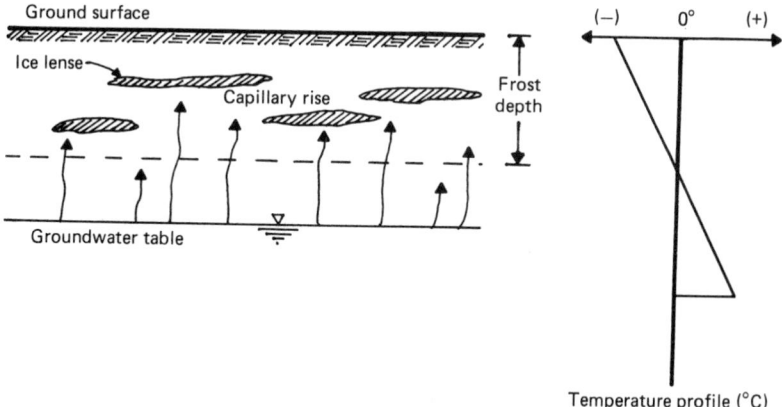

Figure 1–16 Typical soil profile illustrating capillary rise under freezing conditions and variation of soil temperature with depth.

TABLE 1-12 FROST SUSCEPTIBILITY OF VARIOUS TYPES OF SOILS

Soil	USC Classification	Frost Susceptibility
Gravels and Sands	GW, GP, SW, SP	None to low
Silty and Clayey Gravels	GM, GC	Low to medium
Silty and Clayey Sands	SM, SC	Low to high
Inorganic and Organic Clays of High Plasticity	CH, OH	Medium
Inorganic and Organic Silts of High Plasticity and Inorganic Clays	OL, MH, CL	Medium to high
Inorganic Silts of Low Plasticity	ML	Medium to very high

as ice lenses are formed they attract capillary water from the surrounding soil and from the underlying water table and they increase in size. The mechanisms of lense formation and growth are rather complicated, but in general the effects are detrimental to the overall soil structure. Fine grained soils are more susceptible to frost action than coarse grained soils.

Two conditions that can be detrimental to structures supported by soils are associated to frost action. First, the frost heave can induce large differential movements that can lift foundations supporting various structures. Second, the moisture content increases dramatically after thawing, thus reducing soil strength.

Frost penetration can vary from zero inches in the southern portions of the United States to up to 110 inches in the northern states. Structural foundations must always be placed below the maximum frost penetration depth to prevent frost heave. Building codes give the minimum depth of soil cover for various regions of the country.

The effect of frost heave and soil strength reduction is evident in roadway pavements during spring thaw, when potholes appear. Table 1-12 gives the degree of frost susceptibility of various types of soils.

SOIL COMPRESSIBILITY

As with any other engineering material, soil will undergo a certain deformation (strain) when subjected to a load. *Elastic* materials deform immediately upon load application. In certain soils the stress-strain relationship is time dependent. Materials that exhibit such behavior are called *visco-elastic*. There are two major mechanisms responsible for soil deformation.

1. Change in the grain shape (*distortion* or elastic deformation).
2. Change in the soil volume (*compressibility*).

Distortion (elastic deformation) takes place immediately upon the load application and is the main deformation mechanism in coarse grained soils. Compressibility is the volume change due to the expulsion of water from the pores of the soil when subjected to a

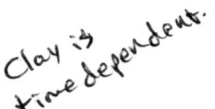
Clay is time dependent.

stress. This time-dependent behavior, which is characteristic of fine grained soils, is called *consolidation.*

When a saturated soil mass is subjected to an increase in stress the pressure in the pore water will suddenly increase. This increase in the pore water pressure will cause flow to occur from regions of high pore water pressures to regions of lower pressures until a pressure equilibrium is reached. In sandy and gravely soils, which have high permeability, flow will take place quickly and equilibrium will be reached almost immediately after the stress application. The elastic deformation and consolidation therefore take place simultaneously. In silty and clayey soils, which have low permeability, the excess pore water pressure created upon application of the load will dissipate over a long period of time.

The expulsion of water from the soil pores during the consolidation process causes a decrease in the soil void ratio. This results in settlement of the soil mass called *consolidation settlement.* The consolidation settlement and the elastic (immediate) settlement are very important parameters in the design of foundations.

In order to compute the consolidation settlement the compressibility characteristics of the soil must be known. The consolidation process is simulated in the laboratory in devices called *consolidometers* or oedometers. A typical consolidometer is shown in Figure 1-17. A relatively undisturbed compressible soil specimen is placed on the brass confining ring and is trimmed carefully. Two porous stones having diameter slightly smaller than the ring diameter are placed on the top and bottom of the sample. The entire assembly is then placed in the holding container. Increments of load are then applied to the specimen via a loading plate and the vertical deformation is measured with a deflection dial.

The data collected from a consolidation test are presented in two types of plots. The first is a plot of dial reading versus the logarithm of time (*time-deformation curve*). One such curve is constructed for each loading increment. A typical time-deformation curve is shown in Figure 1-18. The time-deformation curves are used to compute the coefficient of consolidation, a parameter which is used to determine the time rate of consolidation.

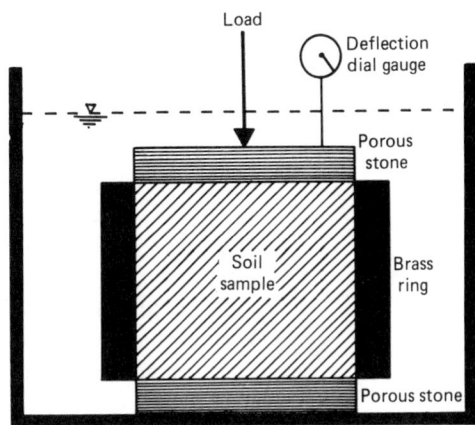

Figure 1–17   Typical consolidometer apparatus.

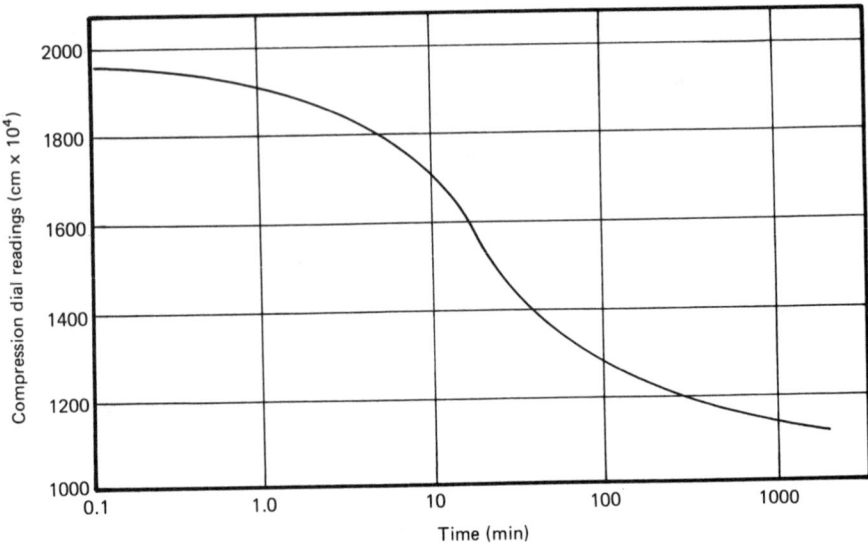

Figure 1-18 Typical time-deformation curve.

The second curve is a plot of void ratio or strain versus the logarithm of the stress applied to the specimen (*stress-deformation curve*). A typical stress-deformation curve is shown in Figure 1-19. The stress-deformation curve is used to determine the compression index C_c which in turn is used to compute the ultimate consolidation settlement.

The compression index is defined as

$$C_c = \frac{\Delta e}{\Delta(\log \overline{\sigma})} \tag{1.28}$$

where Δe = the change of void ratio
$\Delta(\log \overline{\sigma})$ = the corresponding change in consolidation stress

It is observed from this equation that the compression index is the slope of the virgin compression curve as shown in Figure 1-19.

It must be noted that consolidation is not a reversible process. Upon release of the load the sample will rebound slightly but it will never return to its original void ratio, as is shown by the unloading curve in Figure 1-19. This indicates that the past history of the soil sample will have an effect on its compressibility characteristics at present. Based on this observation compressible soils can be classified in two categories:

1. *Normally consolidated.* Soils that have never experienced larger stresses than the existing overburden stresses. Theoretically, such soils should not undergo any consolidation until the consolidation stress exceeds the existing overburden stress. When the consolidation stress becomes larger than the overburden stress the soil mass will be consolidated for the first time, known as virgin compression (Figure 1-19). The difference between the theoretical expected stress-deformation curve

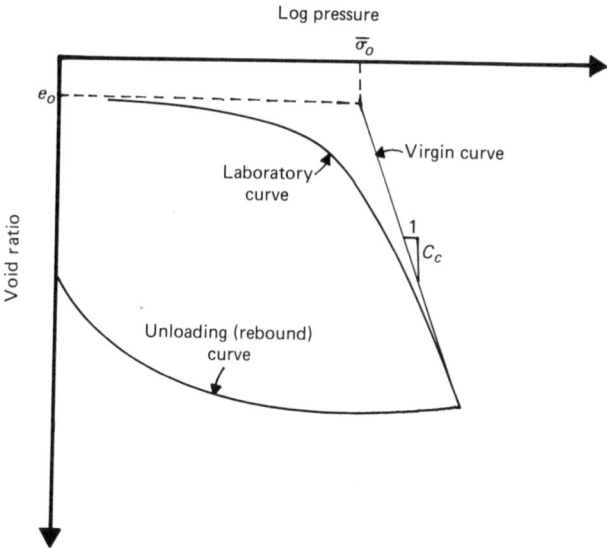

Figure 1–19 Typical stress-deformation curve.

and the actual laboratory behavior depicted in Figure 1.19 is due to the soil dis-
turbance during sampling and handling.

2. *Overconsolidated.* Soils that have been subjected in the past to larger stresses than
the existing overburden stresses. This excess stress is called *preconsolidation stress.*
The most common cause of overconsolidation of natural clay deposits is the extremely high pressure applied at the earth's surface by glaciers.

The past stress history of a compressible soil sample is reflected in the results of a
consolidation test and the preconsolidation stress can be estimated from the stress-
deformation curve.

The ultimate consolidation settlement of a normally consolidated compressible soil
layer that is subjected to a stress increase of magnitude $\Delta\sigma$ is

$$S = \frac{H}{1 + e_o} C_c \log \left(\frac{\overline{\sigma}_o + \Delta\sigma}{\overline{\sigma}_o} \right) \qquad (1.29)$$

where S = ultimate consolidation settlement

    e_o = initial void ratio

    H = thickness of the compressible layer

    C_c = compression index

    $\overline{\sigma}_o$ = effective overburden stress

    $\Delta\sigma$ = consolidation stress due to application of a load

To compute the ultimate consolidation settlement for overconsolidated soils the preconsolidation stress must be considered.

The standard procedure for the consolidation test is described in the ASTM Designation D 2435-80.

SOIL STRENGTH

The ability of soil to resist imposed loads is derived from its shear strength. Loads can be imposed on the soil by foundations of various structures or by the soil's own weight. The soil shearing strength is a very important parameter that is used in the design of shallow and deep foundations, earth retaining structures, slope stability, and numerous other aspects of civil and highway engineering.

Several tests have been developed for the determination of soil shearing strength both in the laboratory and in the field. Laboratory tests are conducted on soil samples recovered from the construction site, in either an undisturbed or a reconstructed state. Field tests are performed in-situ on soils in their natural undisturbed state. In general, shear strength depends on the type of soil, degree of saturation, void ratio, and stress history.

Shearing Strength of Cohesionless Soils

It is known from basic mechanics that a rigid body of weight, N, resting on a horizontal surface and subjected to a horizontal force, H, will start moving when the resisting frictional force, F, developed at the interface, is exceeded (Figure 1-20).

At the point of impending movement the resultant resisting force, R, forms an angle, α, with the vertical direction. This is called the *friction angle*. The tangent of angle α is the coefficient of friction. In granular soils the failure mechanism is quite similar to the principle of sliding friction. When a dry cohesionless soil mass is subjected to an increasing load, it will undergo shear deformation and eventually a failure plane will be created. The shearing strength is developed along the failure surface due to the friction and interlocking of grains to resist the movement.

The shear stress-shear strain behavior of the soil will depend on the initial density of the sample. Figure 1-21 shows typical shear stress-shear strain curves for two samples

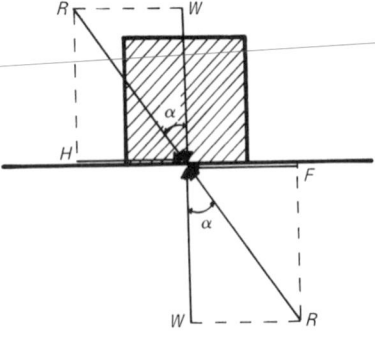

Figure 1–20 Development of friction and impending motion of a rigid body on a horizontal surface.

Soil Strength

$Void \atop ratio = \dfrac{V_{void}}{V_{solids}} = e$

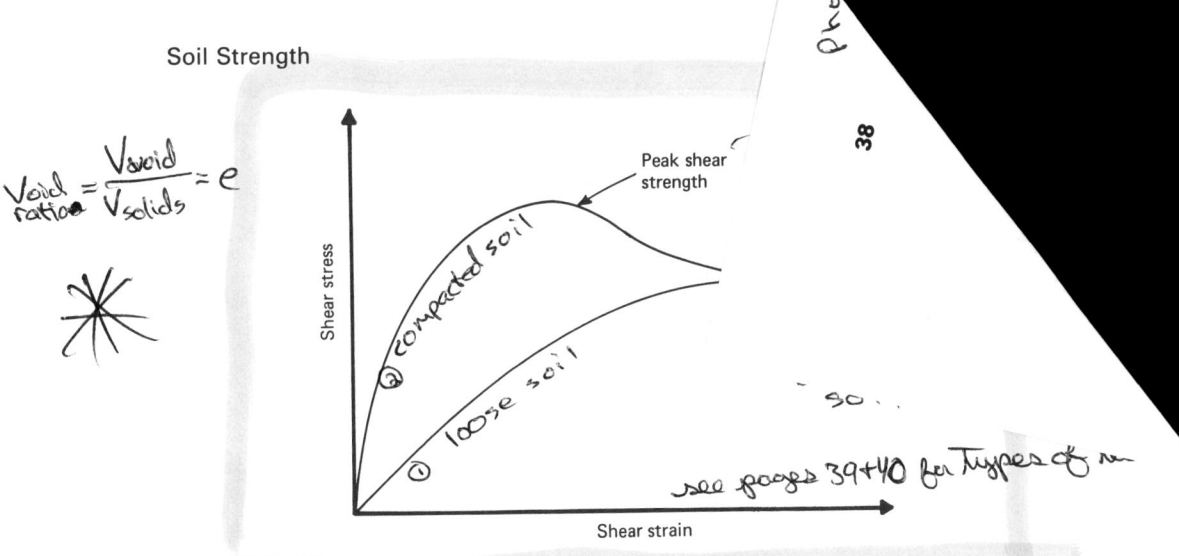

Figure 1–21 Typical shear stress–shear strain curves for soils with different void ratios.

of the same soil having different initial densities (void ratios). It is observed that the two samples have the same ultimate strength but the denser sample exhibits a peak strength which is higher than the ultimate. This difference is due to the higher degree of particle interlocking in the denser sample.

The relationship of the normal stress to the shear stress is

$$\tau = \sigma \tan \phi \qquad (1.30)$$

where τ = shear stress
σ = normal stress
ϕ = angle of internal friction

A plot of Equation 1.30 is shown in Figure 1-22. The equation plots a line which defines a failure envelope called the *Mohr-Coulomb failure envelope*. In terms of Mohr's

Figure 1–22 Mohr-Coulomb envelope for dry cohesionless soils.

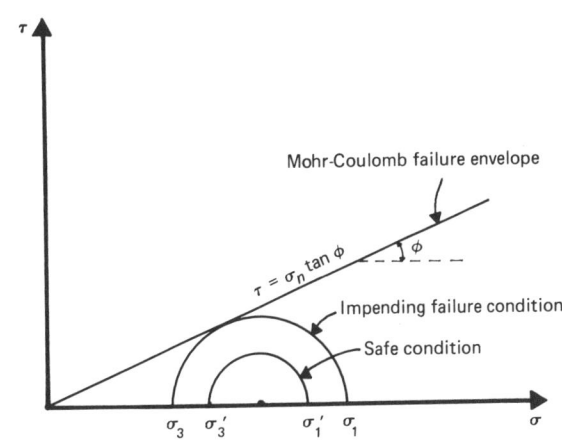

TABLE 1-13 REPRESENTATIVE
VALUES OF THE ANGLE OF INTERNAL
FRICTION FOR COHESIONLESS SOILS

Soil Type	ϕ (degrees)
Gravel and Sand-gravel Mixtures	34–38
Sands (well graded)	32–35
Sands (poorly graded)	29–33
Sand-silt mixtures	26–32
Nonplastic Silt	25–30

circle analysis, a stress condition for which the Mohr's circle touches the failure envelope indicates an impending failure condition, whereas a Mohr's circle that is below the failure envelope line indicates that the full shearing strength has not been mobilized (safe condition). The angle of internal friction, ϕ, depends on the type of soil. Some representative values of ϕ are given in Table 1-13.

Moist cohesionless soils exhibit a strength relationship of the form

$$\tau = \sigma \tan \phi + c_{ap} \tag{1.31}$$

where c_{ap} = apparent cohesion

The apparent cohesion results from the capillary stresses in the soil and it disappears when the soil dries out.

Shearing Strength of Cohesive Soils

The shearing strength of cohesive soil is mainly derived from its cohesion. The two major factors that influence the strength-deformation characteristics of cohesive soils are the stress history of the soil and the drainage characteristics during loading. Since most cohesive (clay) soils are saturated in their natural state, the effective stress is considered

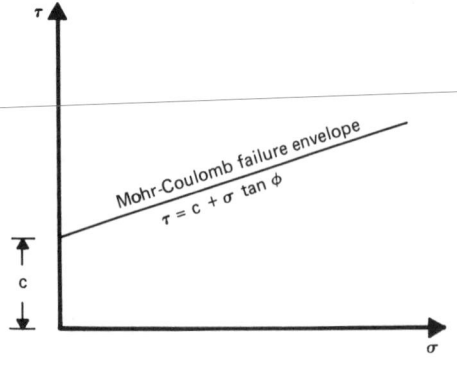

Figure 1–23 Morh-Coulomb envelope for saturated soils

TABLE 1-14 REPRESENTATIVE
VALUES OF COHESION *c*
FOR CLAYS

Clay Condition	c (psf)
Soft	250–500
Medium	500–1000
Stiff	1000–2000
Very Stiff	2000–4000
Hard	>4000

in the computation of the shearing strength. The Mohr-Coulomb envelope for saturated soils is defined as

$$\tau = \bar{\sigma} \tan \phi + c \qquad (1.32)$$

where $\bar{\sigma}$ = effective vertical stress (total
 stress minus pore water pressure)
 c = cohesion

A plot of Equation 1.32 is shown in Figure 1-23. The cohesion of inorganic silts and sands is zero. The cohesion of normally consolidated clays is approximately zero and for overconsolidated clays is it greater than zero. Some representative values of cohesion for various types of clays are given in Table 1-14.

Laboratory Testing The three most frequently used laboratory tests for shear strength are: direct shear, unconfined compression, and triaxial.

Direct Shear Test. (ASTM D 3080-72) In this test a soil sample is placed on a shear box arranged as shown in Figure 1-24. The shear box is split in two halves. The sample is compressed by two porous stones and a constant normal force is applied to the top of the sample via a loading plate. The bottom half of the box is kept stationary while the top half is subjected to a shear force until the sample fails. A shear stress–shear

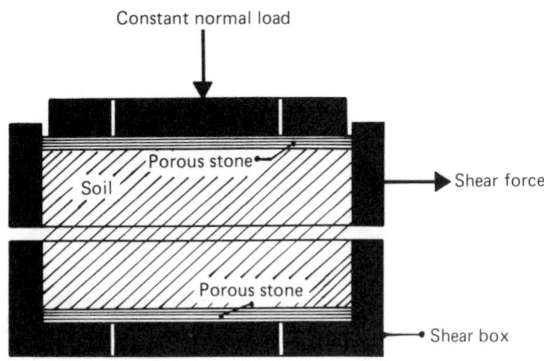

Figure 1–24 Direct shear test arrangement.

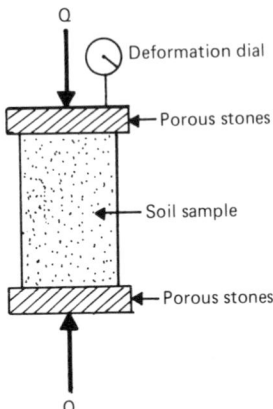

Figure 1–25   Unconfined compression test arrangement.

displacement graph similar to the one shown in Figure 1-21 is plotted. The test is repeated for different normal forces and the normal stresses are plotted against the corresponding shear stresses at failure to obtain the Mohr-Coulomb strength envelope.

One disadvantage of the direct shear test is that the soil is forced to fail along a predetermined plane, which may not necessarily be the weakest plane.

The direct shear test can be used to test both granular and cohesive soils.

Unconfined Compression Test.   (ASTM D 2166-85) This test is used to determine the unconfined compressive strength of cohesive soils. Soils can be treated in their original undisturbed state, remolded, or compacted. A cylindrical soil specimen is placed on a compression machine and is subjected to a strain-controlled axial load, Q, as shown in Figure 1-25. The load is applied at a relatively fast rate so that the pore water does not have adequate time to drain and therefore consolidate the sample. This condition is called *undrained unconsolidated*. The deformation of the sample is recorded and the loading is continued until the sample fails in shear or experiences excessive deformation. A stress-deformation curve is plotted and the ultimate strength (q_u) is recorded. Since the sample is not confined laterally, the total minor principal stress at

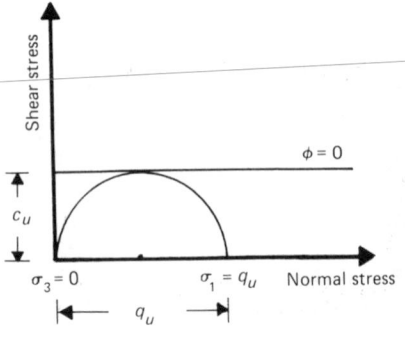

Figure 1–26   Mohr's circle for unconfined compression test.

failure is zero and the total major principal stress is σ_1. The Mohr circle for this condition is plotted as shown in Figure 1-26. It follows that the undrained shearing strength is half of the ultimate strength:

$$c_u = \frac{q_u}{2} \tag{1.33}$$

The unconfined compression test has the advantage of providing quickly and inexpensively an approximate compressive strength for soils that possess enough cohesion. A disadvantage of this test is that the way the soil specimen is tested is not representative of the in-situ conditions, due to the absence of a confining pressure that represents the overburden stress.

Triaxial Shear Test. The triaxial shear test is the most reliable method to determine the shear strength of soils. In this test a cylindrical soil sample is encased in a thin rubber membrane and is placed in a chamber containing water. The water in the chamber is pressurized to apply a confining pressure to the sample. Shear failure is caused by an axial stress applied at the top of the sample. The axial stress is applied by incremental application of axial loads or by application of a constant rate deformation (strain controlled test). The triaxial test apparatus is equipped with drainage lines and devices to measure the pore pressure variations in the sample and sample deformation. The stress condition that a sample is subjected to during triaxial testing is shown in Figure 1-27. The axial stress applied to the sample is called the *deviator stress*. The following three standard triaxial tests are usually performed.

1. Unconsolidated—Undrained Triaxial Compression Test. (UU test) (ASTM D 2850-82) During this test, drainage of the pore water from the sample is not permitted. The axial load is applied quickly and the undrained shearing strength is obtained. The

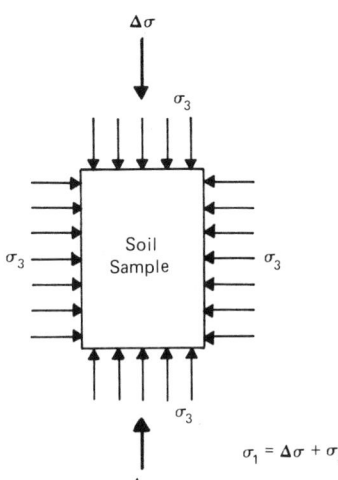

Figure 1–27 State of stress in a soil sample subjected to triaxial testing.

undrained shearing strength is used for design in cases where soil is subjected to large loads very suddenly.

    2. Consolidated—Undrained Triaxial Compression Test. (CU test) During this test, the sample is first consolidated by applying a confining stress and allowing drainage. After the sample has been consolidated the drainage lines are closed and the axial stress is applied until the sample fails.

    3. Consolidated—Drained Triaxial Compression Test. (CD test) In this test the sample is first consolidated by applying a confining stress. After consolidation is complete, and with the drainage lines remaining open, the deviator stress is applied at a very slow rate so that complete dissipation of the pore water pressure is obtained. This test results in determination of the drained shearing strength of the soil and is representative of field situations where in-situ soils are subjected to slow rates of loading.

SOIL IMPROVEMENT

In many areas soil deposits are not suitable to provide support for various engineering structures. Such deposits may be composed of soft clays, organic soils and loose sands, or silts. In many cases the option of relocating the structure may not be feasible. The choices in such situations are limited to two. The unsuitable soils can be excavated and replaced by good quality soils or they can be improved in-situ.

    In the first case soil is used as construction material to fill the excavation. When placed properly and compacted, a soil mass can provide adequate support for engineering structures. As a construction material soil is not limited to filling excavations, but also can be used to build various earthen structures such as highway embankments, earth dams and foundations for buildings, roadways, and parking areas. When soil is used as a construction material it is called *compacted fill* or *structural fill*.

    In the second case the soil is stabilized in-situ. Stabilization is generally defined as the improvement of the engineering properties of soils in-situ by mechanical or chemical means or both. There are numerous soil stabilization methods and techniques currently in practice. The type of technique to be chosen for a particular site depends on a variety of factors, including the type of soil to be stabilized, the extent of required stabilization, the type of structure to be built, the availability of materials, and the environmental effects.

    In this section the most commonly used methods of soil improvement will be summarized.

Soil Compaction

Compaction is the densification of loose soils by mechanical means. Densification results in a decrease of void ratio and therefore in an increase of shearing strength and volume stability. The soil improvements realized are reduction or elimination of settlements, increase in the soil bearing capacity, and reduction or elimination of volume changes that can occur due to frost heave or swelling.

Soil compaction is divided in two categories on the basis of the extent of treatment: shallow compaction and deep compaction. Shallow compaction is performed on soft in-situ soils or on earth fills. Deep compaction is performed on in-situ soils at greater depths.

R.R. Proctor developed the fundamentals of soil compaction in the early 1930s. He established that the degree of compaction of a soil depends on the soil dry unit weight, moisture content, and grain size distribution and on the magnitude of the mechanical energy imparted on the soil. When water is added to a soil during compaction it acts as a lubricating agent, causing soil grains to slip on each other and attain more efficient packing (higher dry unit weight). As the water content is increased, a point is reached beyond which any addition of water to the soil mass will result in a decrease of the dry unit weight (dry density). This is because water will occupy void spaces otherwise occupied by solid grains. The moisture content at which the maximum dry unit weight is obtained for a given compactive energy is called the *optimum moisture content*. A representative dry unit weight-moisture content relationship is shown in Figure 1-28.

In the laboratory the dry unit weight–moisture content relationship can be established by a test called the *Proctor Compaction Test*. In this test the soil is compacted on a compaction mold having a volume of $1/30$ ft^3 by a hammer falling from a certain height.

Two types of tests are commonly used–the Standard Proctor Test and the Modified Proctor Test. In the Standard Proctor Test (ASTM D 698-78) the soil is compacted in three layers of equal height. Each layer is subjected to 25 blows by a hammer that weighs 5.5 lbs (2.49 kg) and free falls 12 inches (305 mm). The test is repeated with the soil at different moisture contents until enough points are obtained to plot the dry unit weight–moisture content curve. With known moisture content and the weight of the compacted soil in the mold, the dry unit weight is computed by

$$\gamma_d = \frac{W}{\left(1 + \dfrac{w}{100}\right) V_m} \tag{1.34}$$

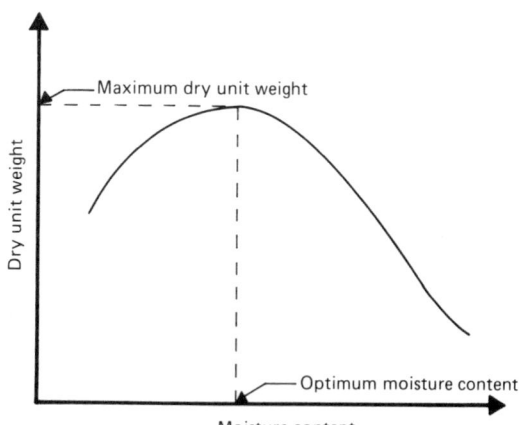

Figure 1–28 Variation of dry unit weight of soil with moisture content in a compaction test.

where W = weight of compacted soil in the mold

w = moisture content in percent

V_m = volume of compaction mold

The Modified Proctor Test (ASTM D 1557-78) is similar in execution to the standard test. In the modified test the soil is subjected to a higher impact energy. The soil is compacted in five equal layers by a hammer having a weight of 10 lb (4.54 kg) falling 18 in (457 mm). For a given moisture content the maximum dry unit weight is obtained when there is no air contained in the pores of the soil. The maximum dry unit weight for the zero air-void condition is computed by

$$\gamma_{zav} = \frac{\gamma_w}{\dfrac{w}{100} + \dfrac{1}{G_s}}$$

where γ_w = unit weight of water

G_s = specific gravity of soil solids

w = moisture content in percent

The zero air-void curve is usually plotted on the same plot with the compaction curve and it serves as a guideline. The compaction curve should never cross the zero air-void curve. A compaction curve and zero air-void curve for a silty soil is shown in Figure 1-29.

In addition to the moisture content, compaction is affected by the soil type and the

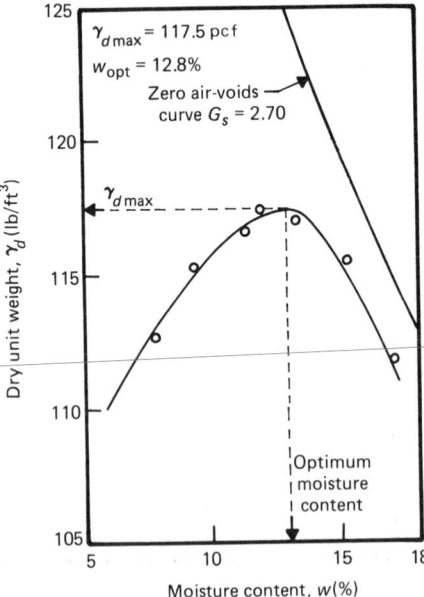

Figure 1–29 Standard Proctor compaction test results for a silty soil.

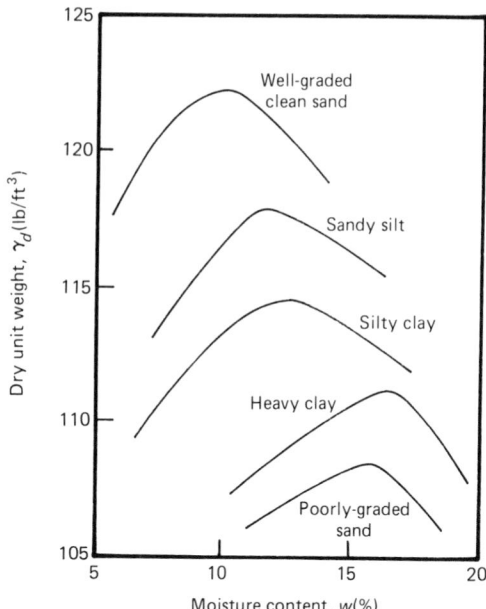

Figure 1–30 Typical compaction curves of various types of soil.

compactive energy. The soil characteristics that most influence compaction are the grain size distribution, content of clay minerals, and specific gravity. Well-graded soils compact more efficiently than poorly-graded soils. Figure 1-30 shows typical compaction curves of various soil types.

The effect of the compactive energy is shown in Figure 1-31, where the compaction curves obtained using the standard and modified Proctor tests are plotted for a silty sand. Since in the standard test less energy is imparted on the sample, the resulting maximum

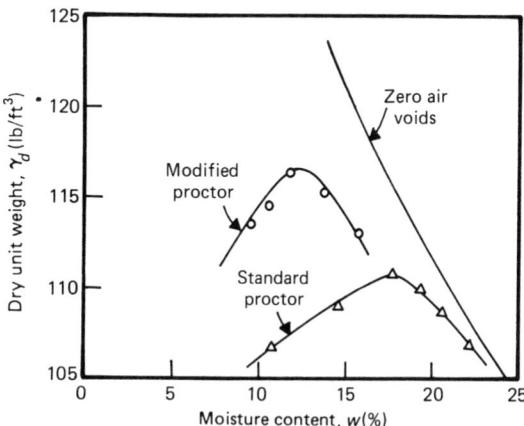

Figure 1–31 Standard and modified Proctor test results for a silty sand.

dry unit weight is less than the one obtained by the modified Proctor test. Besides impact compaction, other methods of compaction in the laboratory include kneading, vibration, and static compaction. For more information on these types of compaction the reader should refer to a soil mechanics book.

In the field, shallow compaction is performed by heavy compacting equipment called *rollers*. The type of roller used for a particular job depends on the type of soils to be compacted. The most commonly used rollers are:

1. *Pneumatic rubber tire rollers*. These rollers are effective in compacting both cohesive and cohesionless soils. They are heavy equipment supported on rows of closely spaced rubber tires. They compact primarily by kneading, and the contact pressure under the tires can range from 80 to 100 lb/m^2 (585 to 690 kg/m^2). Pneumatic rollers are usually fitted with ballast containers to vary the weight of the machine and thus the contact pressure.

2. *Sheep's-foot rollers*. These compactors consist of a steel drum manufactured with small projections. They compact soil by a combination of tamping and kneading. In some types of rollers the drum can be filled with water to provide greater weight. As rolling takes place, the surfaces of the projections knead the soil. Constant pressures under the drum can range from 100 psi (700 kN/m^2) to 600 psi (4200 kN/m^2). Sheep's-foot rollers are suitable for compaction of silty and clayey materials.

3. *Vibratory rollers*. These rollers consist of vibratory drums and are manufactured as self propelled units or tow-back units. Vibration in the drum is generated by a motor that drives a set of eccentric weights located inside the drum. Both smooth surface and sheep's foot drums are available. Vibratory rollers are most efficient in compacting granular soils.

Fill placement and compaction generally requires procedures that include laboratory testing and field inspection and testing. The dry unit weight–moisture content relationship of the fill soil is initially determined in the laboratory. Either the standard or modified Proctor test is used according to the job specifications. The fill is usually placed in layers (lifts) having a loose thickness of twelve inches. The job specification in most cases requires that the soil be compacted to a dry unit weight higher than 90 to 95 percent of the maximum dry unit weight as determined by the laboratory compaction curves. Each lift is compacted by several passes of a compaction roller. Field tests are then performed to determine the in-situ (compacted) soil density. If the in-situ dry unit weight is not within the specified percentage of the maximum dry unit weight, the soil is compacted more. The number of roller passes required to result in acceptable compaction is determined. As long as the fill moisture content and the lift thickness remain the same the compaction is controlled by making sure that the roller makes a sufficient number of passes over each lift. Periodically, in-situ dry unit weight tests are performed to verify that the in-situ fill dry weight is within specifications.

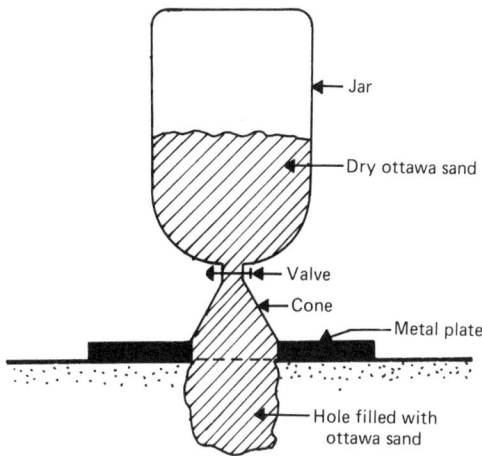

Figure 1–32 Sand cone test arrangement.

There are three methods commonly used to determine the in-situ soil density.

1. The *Sand Cone Method* (ASTM D 1556). This test consists of a glass or plastic jar with a metal cone attached to its top. The jar is filled with dry Ottawa sand of very uniform grain size with known dry unit weight. The weight of the sand in the jar and the sand that is required to fill the cone is measured. In the field a rectangular metal plate with a circular hole having diameter equal to the diameter of the cone is used as a guide to excavate a small hole in the soil to be tested. All of the loose soil is recovered from the excavated hole and its weight and moisture content are determined. The sand cone assembly is then placed on the metal plate as shown in Figure 1-32. The sand is allowed to "flow" into the hole. The weight of sand required to fill the hole is determined and is divided by the dry unit weight of the sand to give the volume of the hole. The in-situ dry unit weight of the soil is then determined by

$$\gamma_d = \frac{\text{dry weight of soil excavated from hole}}{\text{volume of hole}} \qquad (1.36)$$

The test procedure is described in detail in the ASTM Standard D 1556.

2. The *Rubber Balloon Method* (ASTM D 2167): This method is based on the same principle as the sand cone method. In this test the volume of the hole is determined by the volume of water, contained in a rubber balloon, that is required to fill the hole.

3. *Nuclear Density Method:* Nuclear density meters have been developed for the direct determination of the compacted dry unit weight of soils. By detecting emitted gamma-rays passing through the soil mass they measure the density and moisture content of the soil.

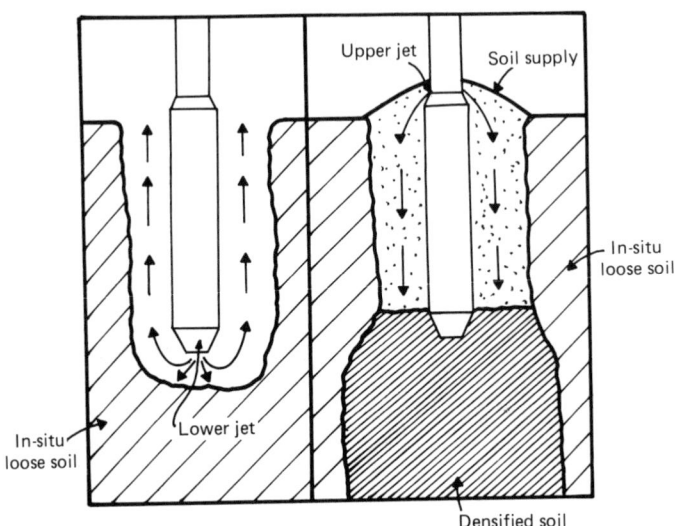

Figure 1–33 Principle of vibroflotation (a) advancement of the vibrofloat (b) retraction of the vibrofloat.

Deep Compaction

The most popular method of massive in-situ soil densification is *vibroflotation*. This technique involves the use of a device called vibrofloat. The vibrofloat is a metal cylinder, usually 6 feet (183 cm) long and 16 inches (40.6 cm) in diameter, which is suspended from a crane. An eccentric set of weights inside the cylinder creates the vibration, which is used to propel the vibrofloat into the soil. Jet ports are located at the top and bottom of the vibrofloat from which water under pressure of 60 psi (444 kPa) can be ejected. The vibrofloat is propelled into the loose soil by the action of the vibration and by the water jet action at its lower end. Once the desired depth is reached, the lower water jets are shut off, the upper jets are turned on, and the vibrofloat is slowly retracted. A cavity is formed around the vibrofloat due to vibration as the surrounding soil is densifying. The cavity is filled with sand shoveled from the ground surface. The radial extend of the densification can range from 3 feet (1 m) to 10 feet (3 m). Figure 1-33 shows the sequence of vibroflotation. This method of deep compaction is effective in soils having less than 20% fines and no more than 3% active clays. Other methods of deep compaction are also available.

Dynamic consolidation is a method by which soils are densified by heavy tamping. This is obtained by dropping heavy weights on the soil. Weights varying from 5 to 40 tons are dropped by a crane from heights ranging from 20 to 100 feet (6 to 30 m) at a predetermined pattern to densify the soil.

Piles are also used to densify soils. Piles are driven into the soil and densification is obtained by displacement and vibration.

Soil Stabilization

Stabilization is the process by which a stabilizing agent (*admixture*) is added to the natural soil deposit to improve its engineering behavior. One or more soil properties can be improved by admixture stabilization. There are several types of admixtures used for soil stabilization purposes. One of the most common stabilization methods used in highway construction is the blending of coarse grained materials with fine grained soils (called binders) to produce a soil mixture that possesses both cohesion and friction. The appropriate blending of soils of different grain size distribution can produce a soil having much better engineering properties.

Portland cement is also commonly used as a stabilizing agent. An increase in soil strength occurs from cementation of the soil particles upon hydration of the cement. The soil is mixed with the cement in-situ and the blend is compacted by rollers. The resulting mixture is called soil cement and usually contains 7% to 15% cement by weight. Similar stabilization is obtained by using asphalt cement as the admixture agent. Both portland and bituminous cement stabilization have found wide applications in roadway and parking lot construction.

Lime and calcium chloride are used as stabilizing agents for improvement of fine grained soils. During lime stabilization the soil is mixed with lime and, after curing for a few days, is remixed and compacted. This results in improvement of the soil strength and reduction of the swelling characteristics of the clay.

Several types of chemicals have been used as cementing agents for soils. The most frequently used agents are silicates. Grouting is the most common method of chemical stabilization. Grouting refers to injection of a liquid stabilizing agent into soil. Pressurized grout is forced to flow through a natural soil deposit and fill the pores. The grout solidifies, bonding the soil grains. An advantage of grouting is that the soil is stabilized in its natural state without being disturbed. A limitation of grouting is that it is effective only in soils with relatively high permeability. Often the purpose of chemical grouting is to minimize the soil permeability and close fractures in rock formations.

A special method of soil stabilization is earth reinforcement. Fabrics composed of synthetic fibers (geotextiles) or metal strips are embedded into soil to improve its strength. Several types of fabrics and reinforcing strips are currently available. They have been used extensively to stabilize slopes and to build embankments, dams, and earth retaining walls.

PROBLEMS

1.1. The moist unit weight for a sandy soil is 18.9 kN/m^3. If the moisture content is 13.5% and the specific gravity of soil solids is 2.69, find:

    a. dry unit weight

    b. void ratio

    c. porosity

    d. degree of saturation.

1.2. A laboratory study showed that a soil has the following physical properties:

Seive No.	Soil, Percent Passing
4	100
10	100
20	98.1
40	96.5
60	95
100	90.4
200	85.5

moisture content = 31%
moist unit weight = 20 kN/m^3
specific gravity of the solids = 2.75
plastic limit = 48.65%
liquid limit = 24.21%

a. Determine the plasticity and liquidity indices and properties.
b. Classify the soil using the Unified Soils Classification system.

1.3. The results of grain size analysis for a soil sample are as follows:

Seive No.	Soil, Percent Finer
4	100
10	99.44
20	86.07
40	36.07
60	20.30
100	16.70
200	12.80

a. Plot the gradation curve.
b. Calculate C_u, and C_c.
c. Classify the soil, using the USC and AASHTO tables.

1.4. Which phenomenon is responsible for capillary rise in soils?

1.5. Daytona Beach in Florida is often used as a car race track. What phenomenon makes the sandy beach so compact as to support moving vehicles?

1.6. What property is responsible for clay soils forming chunks after they dry out?

1.7. The following data are collected from a constant head permeability test:
Sample diameter = 6.5 cm
Sample length = 15.4 cm
Total saturated weight = 1000 g
Test duration = 5 min
Average volume of water collected = 775 cm^3
Constant head = 46 cm
Compute the coefficient of permeability.

1.8. Explain the difference between elastic and consolidation settlement in soils.

1.9. Explain why well-graded soils compact more efficiently than poorly-graded soils.

1.10. Laboratory results from direct shear tests on 6 samples of silty soil are:

No.	Normal Stress σ (psi)	Max Shear Stress τ_{max} (psi)
1	3.27	5.54
2	7.25	7.68
3	9.1	8.81
4	11.51	10.09
5	14.64	11.8
6	16.2	12.8

   a. Plot the Mohr-Coulomb envelope.

   b. Determine the angle of internal friction ϕ and the cohesion c.

1.11. Explain the soil properties responsible for shearing strength.

REFERENCES

American Society of Civil Engineers, *"Soil Improvement, History Capabilities and Outlook,"* *Report by the Committee on Placement and Improvement of Soils,* Geotechnical Engineering Division, ASCE, 1978.

American Society for Testing and Materials, *Annual Book of Standards,* Volume 04.08, Soil and Rock; Building Stones, 1986.

BOWLES, J. E., *Physical and Geotechnical Properties of Soils,* 2nd ed., McGraw-Hill Book Company, New York, 1984.

DAS, B. M., *Advanced Soil Mechanics,* McGraw-Hill Book Company, New York, 1983.

DAS, B. M., *Principles of Geotechnical Engineering,* Prindle, Webber, and Schmidt, A Division of Waldsworth, Boston, 1985.

FREEZE, R. A. and CHERRY, F. A., *Groundwater,* Prentice-Hall Inc., Englewood Cliffs, N.J., 1979.

HOLTZ, R. D. and KOVACS, W. D., *An Introduction to Geotechnical Engineering,* Prentice-Hall, Englewood Cliffs, N.J., 1981.

JUMIKIS, A. R., *Soil Mechanics,* D. Van Nostrand Co., Inc., Princeton, N.J., 1962.

McCARTHY, D. F., *Essentials of Soil Mechanics and Foundations,* 2nd ed., Reston Publishing Co., 1982.

SOWERS, G. F., *Soil Mechanics and Foundations: Geotechnical Engineering,* 4th ed., Macmillan Publishing Co., Inc., New York, 1979.

TERZAGHI, K. and PECK, R. B., *Soil Mechanics in Engineering Practice,* 2nd ed., John Wiley & Sons, Inc., New York, 1967.

TODD, D. K., *Groundwater Hydrology,* 2nd ed., John Wiley & Sons, Inc., New York, 1980.

WINTERKORN, H. F., and FANG, H. Y., *Foundation Engineering Handbook,* Van Nostrand Reinhold Co., New York, 1978.

2

Mineral Aggregates

Aggregates play a very important role in the design and construction of highway and airport pavements, as the underlying material upon which the pavement rests. They are also important as an ingredient in rigid (concrete) and flexible (asphalt) pavements or structures.

Aggregates are the most important factor in the cost of pavement construction, accounting for more than 30 percent of the total cost. Aggregates make up approximately 65 to 85 percent of a concrete structure and 92 to 96 percent of an asphalt structure.

AGGREGATE TYPES AND PROCESSING

Aggregate is a combination of sand, gravel, crushed stone, slag, or other material of mineral composition, used in combination with a binding medium to form such materials as bituminous and portland cement concrete, macadam, mastic, mortar, and plaster, or alone, as in railroad ballast, filter beds, and various manufacturing processes such as fluxing.

Aggregates may be further classified as natural or manufactured. *Natural aggregates* are taken from natural deposits without change in their nature during production, with the exception of crushing, sizing, grading, or washing. In this group, crushed stone, gravel, and sand are the most common, although pumice, shells, iron ore, and limerock may also be included. *Manufactured aggregates* include blast furnace slag, clay, shale, and lightweight aggregates.

A further classification would be to divide the aggregate into two types: fine and coarse. According to ASTM C125 (Concrete and Concrete Aggregates), *fine aggregate* is defined as aggregate passing a 3/8-in. (9.5-mm) sieve and almost entirely passing a No. 4 (4.75-mm) sieve and predominantly retained on the No. 200 (75-μm) sieve or that portion of an aggregate passing the No. 4 (4.75-mm) sieve and retained on the No. 200 (75-μm) sieve. *Coarse aggregate* is defined as aggregate predominantly retained on the No. 4 (4.75-mm) sieve or that portion of an aggregate retained on the No. 4 (4.75-mm) sieve. These definitions are for concrete aggregates; for bituminous concrete mixtures the dividing line between fine and coarse aggregate is the No. 8 (9.5-mm) or the No. 10 (11.8-mm) sieve.

Processing

The main fundamental rule of good aggregate processing is to obtain aggregates of the highest quality at the least cost. Each process is completed with these objectives in mind. The processes include, but are not limited to, excavation, transportation, washing, crushing, and sizing. Processing begins with excavation and quarrying of the material and ends upon being stockpiled or delivered to the site.

In the excavation process the overburden is removed (if applicable), as its presence in aggregate in the form of silt or clay cannot be tolerated. The removal of the overburden is carried out through the use of power shovels, draglines, or scrapers. Overburden removal is usually considered only if there is a depth of 50 ft (15.24 m) or more. If the overburden is light, it will wash out in the processing of the aggregate.

After the aggregate is excavated, it is transported by rail, truck, or conveyor belt to the processing plant.

At the processing plant, unacceptable (deleterious) materials are removed. A deleterious material is a material that may prove harmful to the final product for which the aggregate is to be used. One method of removing deleterious materials (clay, mud, leaves, etc.) is to wash the raw material. Sometimes conveyor belts are used to haul the aggregate through flumes which are flushed with water.

The next process is to reduce the size of the stone or gravel. In this process many types of crushers are used. The oldest is the jaw crusher, which consists of a fixed jaw and a reciprocating jaw, which are suited for hard rocks of all types. Newer crushers have a higher capacity than the jaw crusher, but this is the only disadvantage of the jaw crusher. The usual practice is to reduce the size of the rock at a ratio of 1 : 6 or less.

For sizing, vibratory sieves are used for coarse material and hydraulic classification devices for fine material. The screens vary in design, capacity, and efficiency. In the screening process about 70 percent of the material will pass through the screen, so that the goals of high efficiency and capacity are met. In most cases some removal of oversize particles, called scrapping, will take place.

Particles

The screened aggregate particles may be rounded or angular. Gravel consists of naturally rounded particles resulting from disintegration and abrasion of rock or processing of weakly bound conglomerate. Sand consists of rock particles that have been disintegrated naturally; the grains are generally angular but have been subjected to weathering. Sand is fine aggregate resulting from natural disintegration and abrasion of rock or processing of completely friable sandstone. Crushed stone is a product of the artificial crushing of rocks, boulders, or large cobblestones, substantially all faces of which result from the crushing operation. Stone sand is a finely crushed rock corresponding to sand in size. Gravel and crushed stone are considered to be coarse aggregate.

Aggregate particles vary considerably in texture. Gravel particles have a very smooth texture, whereas crushed stone has a rough surface texture. Aggregate particles have considerable variations in porosity. Usually, crushed stone and gravel have low porosity.

Gradation and Aggregate Blending

The gradual gradation in size from coarse to fine is a key property of aggregates. Aggregate gradation affects the workability of portland cement concrete mixes and the stability and durability of bituminous concrete mixes, as well as the stability, drainage, and frost resistance of base courses. Therefore, aggregates should be tailored to their proposed usage: in concrete or bituminous or as base courses. For specific examples, refer to succeeding chapters.

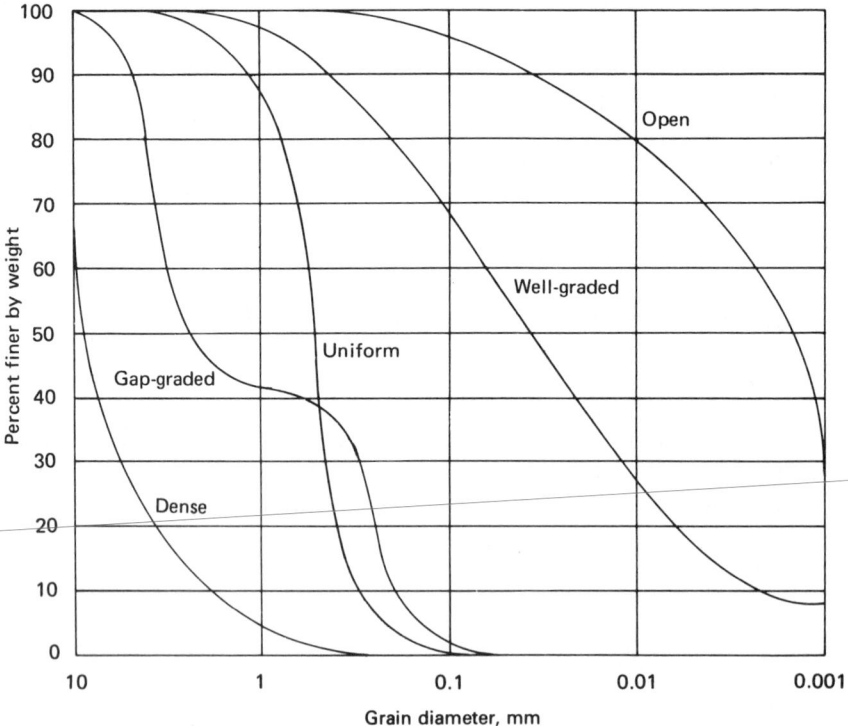

Figure 2–1 Five types of gradation.

Aggregates may be dense, gap-graded, uniform, well-graded, or open-graded. The terms "dense" and "well-graded" are essentially the same, as are "gap," "uniform," and "open-graded." Figure 2-1 illustrates the five types of gradations. Other methods of expressing size distribution have been developed. One such method is the Fuller-Thompson method. An empirical formula that can be used to determine the gradation:

$$P = 100 \left(\frac{d}{D}\right)^N$$

where P = percent finer than the sieve
$\quad d$ = sieve size in question
$\quad D$ = maximum size aggreagate to be used (top size)
$\quad N$ = coefficient of adjustment, which adjusts the curve for fineness or coarseness

Figure 2-2 shows a typical plot utilizing this formula at values of N equal to 0.3, 0.5, and 0.7. The maximum size of the aggregate is 1 in. (2.54 cm). Notice that a fine gradation is represented by $N = 0.3$ and a coarse-graded material is represented by $N = 0.7$. Therefore, a dense material would be $N = 0.5$, as the Fuller-Thompson experiments indicate. Originally, the work of Fuller and Thompson included the cement in a cement-aggregate mixture, but the relationship (equation) has become known as *Fuller's maximum density curve*, regardless of the constituents. The maximum density curve is only an approximation, since the actual gradation may depend on the nature of the material.

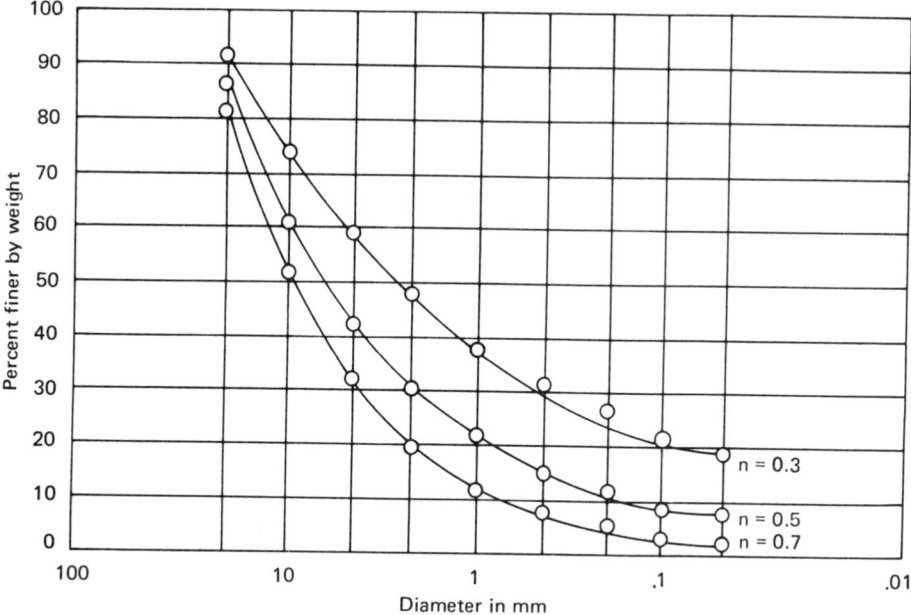

Figure 2-2 Fuller-Thompson curves.

However, if employed properly, it can be a valuable tool as a point of beginning in designing aggregate blends for maximum density.

Other research to determine maximum density of aggregates has been conducted. One theory states that if aggregates were screened into three sizes, coarse, medium, and fine, the combination with the greatest density would be a mixture of approximately 2 parts coarse aggregate and 1 part fine aggregate, with no medium fraction. Further, research showed that high densities could be obtained by using equal portions of each size.

Manufactured Aggregates

Manufactured aggregates are man-made aggregates. One such aggregate, air-cooled blast-furnace slag, has become one of the most common manufactured aggregates to be used in highway construction. ASTM C125 (Concrete and Concrete Aggregates) defines *air-cooled blast-furnace slag* as the material resulting from solidification of molten blast-furnace slag under atmospheric conditions; subsequent cooling may be accelerated by application of water to the solidified surface. *Blast-furnace slag* is defined as the non-metallic product, consisting essentially of silicates and aluminosilicates of calcium and other bases, that is developed in a molten condition simultaneously with iron in a blast furnace. It is not uncommon to find this manufactured aggregate used in industrial areas, especially if steel mills are present, as in the states of Alabama, Indiana, Maryland, New York, Ohio, and Pennsylvania. Slag is lighter in weight than are natural aggregates. Slags have a specific gravity between 2.0 and 2.5, whereas natural aggregates have a specific gravity between 2.3 and 3.2.

Although the definition of manufactured aggregate given here is sufficient for most uses, others further classify manufactured aggregates as those man-made aggregates that have resulted as a direct product rather than as a by-product. In this sense, slag would not be a manufactured aggregate but an artificial aggregate. Thus, an artificial aggregate is a man-made aggregate that results as a by-product from the manufacturing of some other product. Lightweight aggregates are considered artificial aggregates. Lightweight aggregates may be cinders, clay, shale, shells, or slag. These type of aggregates are used to produce lightweight concrete in a structure where dead weight is important.

AGGREGATE AS A BASE-COURSE MATERIAL

The importance of aggregates in a base course lying between the compacted subgrade and the portland cement concrete slab or the bituminous concrete slab cannot be overstated. An improper design of a base course can lead to structural failure of the slab. Base courses may also include several types of undercourses, such as subbases, filter beds, and leveling courses. Base courses serve a variety of purposes depending upon construction practices and the environment. They serve to provide structural capacity to bituminous concrete slabs, drainage for portland cement concrete slabs, and low susceptibility to frost. As previously indicated, gradation is the key factor in the success of aggregates as a base course.

Gradation

The gradation of the aggregate can affect structural capacity, drainage, and frost susceptibility. Thus, the control over gradation is a principal concern for most engineers. Related to this control is the hardness of the aggregate as soft, weak, or friable articles undergo aggregate degradation (a process whereby fines are generated by aggregate breakdown during the placement or the use of the material).

According to Krebs and Walker, three general types of aggregate mixtures can be recognized with respect to fines:

1. Aggregates only, no fines.
2. Fines just filling the voids of aggregate fraction.
3. Fines overfilling the voids of aggregate fraction.

In the first case the aggregate derives its strength from grain-to-grain contact of the aggregate particles. In this situation the base-course material would be unstable unless it was confined, but it does provide excellent drainage and is completely non-frost-susceptible.

In the second case, also, the aggregate derives its strength from grain-to-grain contact of the aggregate particles. However, in this situation the base-course material would be stable even if unconfined, because of the inherent cohesive properties of the fine content that fills the voids between the aggregate particles. Further, the drainage is adequate and can be non-frost-susceptible.

In the final case, the strength is derived from the grain-to-grain contact of the fines rather than of the aggregate particles; thus, a strength reduction occurs. The drainage characteristics of this base course would be poor and it would be very frost-susceptible.

In most highway construction techniques the base course falls between types 1 and 2 for best practice. Thus, a base-course material should have sufficient fines to just fill the voids within the aggregate particles, with a gradation curve approaching that of Fuller's maximum density curve.

Particle Strength, Shape, and Texture

To resist the stress of repeated loads and to avoid aggregate degradation, base-course aggregates must exhibit strength and toughness to function for their intended purpose. Open-graded base courses are more susceptible to degradation than is dense-graded material. In addition, the hardness of the aggregate adds strength to the base course. Sandstones and shale generally degrade easily, although the individual particles may be strong.

Table 2-1 gives the average value for the physical properties of the principal types of rocks: igneous, sedimentary, and metamorphic. In each group the specific gravity,

TABLE 2-1 AVERAGE VALUES FOR PHYSICAL PROPERTIES OF THE PRINCIPAL TYPES OF ROCKS[a]

Type of Rock	Bulk Specific Gravity	Absorption (%)	Loss by Abrasion% Deval[c]	Loss by Abrasion% Los Angeles[d]	Hardness[c]	Toughness[f]
Igneous						
Granite	2.65	0.3	4.3	38	18	9
Syenite	2.74	0.4	4.1	24	18	14
Diorite	2.92	0.3	3.1	—	18	15
Gabbro	2.96	0.3	3.0	18	18	14
Peridotite	3.31	0.3	4.1	—	15	9
Felsite	2.66	0.8	3.8	18	18	17
Basalt	2.86	0.5	3.1	14	17	19
Diabase	2.96	0.3	2.6	18	18	20
Sedimentary						
Limestone	2.66	0.9	5.7	26	14	8
Dolomite	2.70	1.1	5.5	25	14	9
Shale	1.85–2.5	—	—	—	—	—
Sandstone	2.54	1.8	7.0	38	15	11
Chert	2.50	1.6	8.4	26	19	12
Conglomerate	2.68	1.2	10.0	—	16	8
Breccia	2.59	1.8	6.4	—	17	11
Metamorphic						
Gneiss	2.74	0.3	5.9	45	18	9
Schist	2.85	0.4	5.5	38	17	12
Amphibolite	3.02	0.4	3.9	35	16	14
Slate	2.74	0.5	4.7	20	15	18
Quartzite	2.69	0.3	3.3	28	19	16
Marble	2.63	0.2	6.3	47	13	6
Serpentinite	2.62	0.9	6.3	19	15	14

[a]From Bureau of Public Roads manual, *The Identification of Rock Types* (1950).
[b]After immersion in water at atmospheric temperature and pressure.
[c]AASHTO T3 or ASTM D289.
[d]AASHTO T96 or ASTM C131.
[e]Dorry hardness test, *U.S. Dept, Arg. Bull. 949*.
[f]AASHTO T5 or ASTM D3a.

hardness, and toughness are given together with other properties. The purpose of the table is to give the reader an indication of which materials would serve best as a base-course material.

Of equal importance to base-course material other than strength is particle shape and toughness. Table 2-1 also showed the toughness of this rock. These two properties are very important to base-course materials. Angular, nearly equidimensional particles rough in texture are extremely preferred for base-course material. The angularity contributes to aggregate interlocking, and a rough surface texture prevents movement of one particle upon another. Rounded particles tend to role over one another as they do not

TABLE 2-2 SUMMARY OF ENGINEERING PROPERTIES OF ROCKS[a]

Type of Rock	Mechanical Strength	Durability	Chemical Stability	Surface Characteristics	Presence of Undesirable Material	Crushed Shape
Igneous						
Granite, syenite, diorite	Good	Good	Good	Good	Possible	Good
Felsite	Good	Good	Questionable	Fair	Possible	Fair
Basalt, diabase, gabbro	Good	Good	Good	Good	Seldom	Fair
Peridotite	Good	Fair	Questionable	Good	Possible	Good
Sedimentary						
Limestone, dolomite	Fair	Fair	Good	Good	Possible	Good
Sandstone	Fair	Fair	Good	Good	Seldom	Good
Chert	Good	Poor	Poor	Fair	Likely	Poor
Conglomerate, breccia	Fair	Fair	Good	Good	Seldom	Fair
Shale	Poor	Poor		Good	Possible	Fair
Metamorphic						
Gneiss, schist	Good	Good	Good	Good	Seldom	Good
Quartzite	Good	Good	Good	Good	Seldom	Fair
Marble	Fair	Good	Good	Good	Possible	Good
Serpentinite	Fair	Fair	Good	Fair	Possible	Fair
Amphibolite	Good	Good	Good	Good	Seldom	Fair
Slate	Good	Good	Good	Fair	Seldom	Poor

[a]From Bureau of Public Roads manual, *The Identification of Rock Types* (1950).

interlock with one another. Smooth-textured aggregate particles allow slippage when they are in contact with each other. Thus, rounded aggregate with a smooth surface texture is the least preferred for base-course usage. Table 2-2 summarizes the engineering properties of various aggregate types.

AGGREGATES FOR PORTLAND CEMENT CONCRETE

Aggregate properties for portland cement concrete are in many cases different from aggregates used for base courses or for use in bituminous concrete. Aggregate gradation becomes a key factor as it controls the workability of the plastic concrete. Further, the aggregates used in portland cement concrete are a blend of fine and coarse aggregate to achieve an economical mix. For coarse aggregate the same definitions apply as previously given. For fine aggregate the same definitions as previously given apply, but we introduce a new term, fineness modulus. The *fineness modulus*, which denotes the relative fineness of the sand, is defined as one-hundredth of the sum of the cumulative percentages held on the standard sieves in a sieve test of sand. Six sieves are used in the determination,

Nos. 4, 8, 16, 30, 50, and 100. The smaller the value of the fineness modulus, the finer the sand. The fineness modulus for a good sand should range between 2.25 and 3.25.

In portland cement concrete the strength of the aggregate is not as important as it would be if it was used as a base-course material. The aggregate in portland cement concrete acts as a filler so that not as much sand, cement, and water are needed.

The bond between the aggregate and the cementing materials of portland cement concrete is influenced by surface texture. In most cases the bond strength is improved by an aggregate with a rough surface texture. The degree of texture need not be great, but some texture is desired.

In most portland cement concrete mixtures, the effect of surface texture is minimal. If one was to compare two types of aggregates, a smooth-surfaced gravel and a rough-surfaced crushed stone, to be utilized in two separate concrete mixes of equal cement factor and consistency, the smooth-surfaced gravel mix would require less water. The strength achieved from using the smooth-surface gravel would be the same as the rough-surfaced crushed stone. Thus, in the smooth-surfaced gravel mix, less water resulted n a lower water-cement ratio increasing its strength. The mix containing rough-surfaced crushed stone required more water, increasing the water-cement ratio, resulting in a weaker concrete. Thus, the two effects balanced out and the resulting strengths were equal.

Particle shape influences the workability of the concrete mix, but the interlocking characteristic needed for base-course material is not important here. Angular aggregates require more mortar to fill voids and separate aggregate particles for workability, which results in a higher water-cement ratio for a given cement factor and consistency.

Freezing and Thawing

Probably the single most important reason why concrete fails is due to the effects of freezing and thawing. Other reasons are deleterious materials and chemical reactions. The latter two reasons will be discussed in more detail in Chapter 7. The freezing and thawing phenomenon is also discussed in Chapter 7, but some detail will be given here.

The freezing and thawing of concrete is much more important at present than it was several years ago with respect to aggregates. Sources of aggregates of known satisfactory performance are being depleted through one means or another. This fact can only lead to three major possibilities:

1. Production of synthetic aggregate.
2. Beneficiation of unsuitable material.
3. Use of manufactured and waste materials as supplements and replacements for conventional aggregates in construction.

Therefore, the need for an understanding of the influence of aggregates on the resistance of concrete to freezing and thawing is becoming increasingly important. If the aggregate is unsound, three major classifications of distress result:

1. Pitting and popouts.
2. D-line cracking deterioration.
3. Map cracking.

Pitting, the disintegration of weak, friable pieces of aggregates due to frost action, usually occurs by the gradual deterioration of deleterious particles near the surface. *Popouts* are caused by the rapid disruption of harder but saturated pieces of rocks or aggregates. Pitting and popouts themselves do not affect the structure in terms of strength, but they allow water to enter the structure through accesses provided by the pitting and popout and ultimately result in deterioration.

D-line cracking is caused by the change in volume of coarse-aggregate particles breaking the bond with the mortar matrix. It is aided by the entrance of water. Ordinarily, the cracks are filled with a calcareous deposit. D-line cracking appears first along transverse joints, then later along longitudinal joints.

Map cracking is a form of disintegration in which random cracks develop in a well-distributed pattern over an entire surface instead of concentrated along joints or free edges as in D-line cracking. D-line cracking and map cracking can aid structural failure of pavements by causing blowups.

As one views the three major classifications of distress, the question arises as to what materials cause these harmful effects. Several aggregates cause deterioration of concrete, including chert, limestone, and shale. But what is the specific nature of these materials that cause deterioration? Chert is a fine-grained hard rock that will usually pass conventional specifications. However, chert is sufficiently weathered as to be deleterious to concrete exposed to freezing and thawing. Shale and argillaceous limestone cause considerable damage to concrete exposed to severe weathering.

The different responses of aggregates to freezing when saturated depend upon the pore characteristics of the aggregate and the cement paste. Saturated aggregates of low porosity may accommodate pore-water freezing by simple elastic expansion. Saturated aggregates of moderate to high porosity may fail because the particle dimension exceeds a certain critical size or may cause failure in the paste immediately adjacent to the aggregate particles because of aggregate pore-water displacement. As for the shale and argillaceous limestones, these materials have small interconnected voids and are capable of attaining a high degree of saturation.

The disruption of concrete by aggregates is a result of hydraulic pressures. The hydraulic pressure is a result of the degree of saturation (proportional to total void space filled with water) and the permeability and size of the aggregate particles. Upon freezing, water expands 9 percent, and if the degree of saturation of the aggregate is critical, 91.7 percent water will be expelled into the paste surrounding the aggregate particles, and potentially destructive hydraulic pressure may develop there also. So the properties of paste, its permeability, air content, and porosity are also involved in the problem. Three additional factors, composition, texture, and structure, also play important roles in the freezing and thawing of concrete. As shown by Powers, maximum hydraulic pressure is greatest when the aggregate has a low coefficient of permeability. At a given porosity, the smaller the mean size of the capillaries, the lower the coefficient of permeability and the higher the resistance to flow at a given ratio of efflux. Thus,smallness of capillaries is one characteristic of an unfavorable capillary system. In short, low water capacity and high permeability (i.e., low porosity and coarse texture) is a favorable combination, resulting in a relatively large critical size. Critical size varies from rock to rock (because it is dependent on capillary size), so no value is given here. The limit of water capacity for elastic accommodations is 0.3 percent. It should be noted that a wide range of critical sizes exist at a given strength.

The critical saturation point is 91.7 percent, as shown by Powers. One would immediately try to keep the water below the 91.7 percent level, but it must be remembered that water is not uniformly distributed in an aggregate particle at the time of freezing. For an aggregate that is not uniformly distributed, the critical saturation may be something less. Critical diameter is usually considered 4 μm (0.004 mm). The surface texture, which is directly related to the porosity of the material, is considered important in aggregates as well as the permeability.A rough, porous texture will produce high loading strengths and a favorable critical size. Irregular particles need more mortar to fill the void space, and they also have a greater demand for water—because more mortar, more sand; more fines, more water. Another aspect to remember is that some aggregates expand when they come in contact with water, which may result in deterioration.

If the situation is such that the aggregate might become critically saturated, the paste is sure to become saturated, and when it does it can quickly become destroyed by freezing, regardless of the kind of cement of which it is made or the extent of curing.

When a specimen of saturated cement paste becomes frozen, it, too, has a critical thickness of about three thousandths of an inch. Pastes are highly porous and have an extremely fine texture. The structure of the paste is such that 72 percent is solid and 28 percent accessible to evaporated water. The cavities in the cement paste are seldom completely saturated at the time of freezing, and therefore hydraulic pressure does not necessarily develop at the time of freezing. Nevertheless, it is possible for the paste to become dilated.

Concrete freezes from the outside to the inside. When freezing starts, the void spaces contain water. If the water is above the critical saturation point of 91.7 percent, the capillaries will freeze and expel water to the surrounding area—probably into the cement paste because of its small capillaries. Since the paste and aggregate cannot accommodate this expansion of 9 percent, the hydraulic pressure is so great that cracking or popouts result. Once an opening is formed in the concrete, water is allowed to enter and deterioration is underway.

The most acceptable explanation of why concrete may fail during freezing and thawing was given by Powers. Concrete contains many air-filled cavities, consisting of entrained air bubbles, accessible pores in the aggregate particles, and thin fissures under the aggregate particles. All empty cavities of the type mentioned, especially air bubbles, are difficult to fill with water. They cannot be filled by capillary action because a liquid cannot flow from a small capillary to fill a larger one. However, pressures caused by freezing water are more than sufficient to free unfrozen water into such spaces. The resistance to movement of water must be the primary source of pressures, for practically all concrete contains enough air-filled space to accommodate the water and its volume of expansion when frozen.

It is therefore apparent that test procedures must be used or developed to determine the soundness of the aggregates used in the making of concrete.

AGGREGATES FOR BITUMINOUS MIXTURES

The influence of aggregates on the properties and performance of bituminous mixes is great. The ideal aggregate for a bituminous mix would have proper gradation and size, be strong and tough, and be angular in particle shape. Other properties would consist of low porosity, surfaces that are free of dirt, rough texture, and hydrophobic nature. Tables 2-1 and 2-2, which illustrate aggregate porosity abrasion, hardness, toughness, strength, durability, surface characteristics, shape, and so on, also apply to bituminous mixes. The aggregate gradation and size, strength, toughness, and shape are important considerations for stability of the structure. The porosity and the surface characteristics are important to the aggregate-bitumen interaction. The asphalt cement or its product must adhere to the aggregate and at the same time coat all the aggregate particles. If the aggregate particle has a low porosity and is smooth, the asphalt cement will not adhere to the aggregate. Adhesion becomes an extremely important property during periods when the mix is exposed to water. If the aggregate wets easily, the water will compete with the bitumen for adsorption onto the aggregate surface, and the aggregate will separate from the bitumen, which is known as slippage.

Aggregate Gradation and Size

Depending upon the specific use of the bituminous mix, the size and gradation of the aggregate varies tremendously. A high-quality bituminous mix that is to be used as a pavement for heavy traffic will generally utilize a dense-graded aggregate (a well-graded aggregate from coarse to fine). In this particular case one would not utilize Fuller's maximum density curve because it does not leave sufficient room for the asphalt cement. Therefore, the best procedure would be to open the grading somewhat more than the maximum. This opening of the gradation is achieved by the addition of fines (material less than the No. 200 sieve).

Strength, Toughness, Shape, and Porosity

The aggregate in a bituminous mix, unlike that in portland cement concrete, supplies most of the stability and thus should have a certain amount of strength and toughness; otherwise loss of stability will result. Open-graded mixes are subject to greater mechanical breakdown than a dense-graded mix. Thus, if the material selected for a bituminous mix is of minimal strength, a denser mix of the same material would be utilized. ASTM C131 (Resistance to Degradation of Small Size Coarse Aggregate by Abrasion and Impact in the Los Angeles Machine) attempts to measure the effective strength and toughness of an aggregate. Table 2-1 shows the abrasive results of various aggregates subjected to the Los Angeles machine. The test shows very little correlation with the field performance of the aggregates. It does, however, have some use in distinguishing aggregates that are unsuitable in surface treatment.

Particle shape is a property of the aggregate that is more important than gradation and size, strength, and toughness when it comes to bituminous mixes. When rounded aggregates are used in an open-mix gradation, very little stability is achieved. Thus, when an open-mix gradation is used, angular aggregate should be used. If rounded aggregate has to be used in a bituminous mix, it should be crushed, thus resulting in a fractured plane.

Porosity of the aggregate strongly affects the economics of a mix. In each mix the aggregate should have a certain amount of porosity. In general, the higher the porosity, the more asphalt will be absorbed into the aggregate, thus requiring a higher percent of asphalt in the mix design. Thus, an aggregate that is porous gives rise to the possibility of selective absorption. In *selective absorption* the oily portions of the asphalt are selectively absorbed, leaving a hard residue on the surface of the aggregate particle. This process could lead to stripping of the aggregate from the asphalt cement.

AGGREGATE BENEFICIATION

In various parts of the country, sound aggregates are scarce and transportation costs for this good aggregate are expensive. In those areas, the aggregate that is available may have certain deleterious materials which prevent these aggregates from passing specifications, or these aggregates have an adverse field performance record. In a few of these cases the aggregate can be beneficiated so that it becomes a useful economical product.

In the beneficiating of aggregates, several processes have been developed:

1. Washing.
2. Heavy-media separation.
3. Elastic fractionation.
4. Jigging.

Washing

Undesirable aggregates are washed to remove the particle coatings or to change the gradation. Fine content can be removed by exposing the aggregate to streams of water while it moves over screens or in special wash tanks. Stokes' law can also be utilized to remove certain fractions of fine aggregate by differential settlement. For water at 77°F (25°C) and using 2.65 as the specific gravity of sand, Stokes' law is

$$V = 9000D^2 \tag{2.2}$$

where V = velocity of settlement
$\quad$ D = particle diameter

Thus, if sand is introduced near the top of an elongated water tank, various sized fractions can be removed from pockets spaced away from the sand-water stream, which is entering horizontally.

Heavy-Media Separation

Heavy-media separation utilizes the principle that the specific gravity of much deleterious material is lighter than the specific gravity of sound aggregate. In this method, referred to as the *sink float method*, the suspension is composed of water and magnetite or ferrosilicon, or a combination of the two. The suspension is maintained at a specific gravity that will allow the deleterious material to float to the top and the sound material to sink.

Elastic Fractionation

Elastic fractionation is a procedure whereby heavy but soft particles can be removed. Aggregates fall on an inclined plate, and their quality is measured by the distance they bounce from the surface. The bouncing stones are collected in three different compartments, placed so as to collect particles in categories of their bouncing characteristics. Poor, soft, or friable particles bounce only a short distance, whereas the harder, sound aggregate particles bounce much farther. The elastic fractionation process removes only those particles having elastic properties that cause them to bounce poorly. It does not remove deleterious particles that have a high modulus of elasticity. Thus, heavy-media separation should be used in conjunction with elastic fractionation to remove all the deleterious material.

Jigging

Jigging is a specific-gravity method of removing light particles such as coal, lignite, or sticks. Upward pulsations created by air tend to hinder settlement of lighter particles, which are removed by shimming devices. The advantage of the process is that either fine or coarse aggregate may be used.

ASTM TEST SPECIFICATIONS FOR AGGREGATES

In the following section we will look at various ASTM test specifications, with special emphasis on the following categories:

1. Tests concerning the general quality of aggregates.
2. Tests concerning deleterious materials in aggregates.
3. Tests used in the design of portland cement concrete and bituminous mix design.

In each case we will look at the purpose of the test, followed by a brief description of the test and the authors' opinions and conclusions concerning the procedure, results, and validity of the test.

Tests Concerning the General Quality of Aggregates

ASTM C131 (Resistance to Degradation of Small Size Coarse Aggregate by Abrasion and Impact in the Los Angeles Machine)   The purpose of this specification is to test coarse aggregate smaller than 1.5 in. (3.81 cm.) for resistance to abrasion using the Los Angeles testing machine and to evaluate base-course aggregates for possible degradation.

In this procedure, the test sample is placed in the Los Angeles testing machine after it has been prepared for testing in accordance with this specification. The machine is rotated at a speed of 30 to 33 rpm for 500 revolutions. The material is discharged from the machine and a preliminary separation of the sample is made on a sieve coarser than the No. 12. The finer portion is sieved using the No. 12 sieve in a manner conforming to the specification. The material coarser than the No. 12 sieve is washed and oven-dried at 221 to 230°F (105 to 110°C) to constant weight and weighed to the nearest gram. The difference between the original weight and the final weight of the test sample is expressed as a percent of the original weight. The value is reported as a percent of wear.

According to the specifications, backlash or slip in the driving mechanism is very likely to furnish results that are not duplicated by other laboratories—an apparent disadvantage. In 1937, Woolf compared the Los Angeles abrasion results with the service records of coarse aggregates and concluded that the Los Angeles test gives accurate indications of the quality of the material under test and that its use in specifications controlling the acceptance of coarse aggregates is warranted.

However, lack of sufficient data from any one test procedure makes it impossible to suggest limits or specifications for the abrasion resistance on any type of concrete surface. Also, different surfaces may require different abrasion values. For example, a sidewalk would not have to have a high-wearing resistance as compared to one on which heavy roller cars are constantly passing over. Some materials can pass this test but ultimately are dangerous to the concrete.

Unless someone places limits on wearing-resistance values of material such as limestone, blast-furnace slag, hard quartz, and other materials, the test is only a fair representation of what may happen. It is a good test for base-course aggregates, in that it can give an idea of degradation characteristics.

ASTM C88 (Soundness of Aggregates by Use of Sodium Sulfate or Magnesium Sulfate)   The purpose of this specification is to determine the potential resistance of an aggregate to weathering.

In this procedure a specific weight of an aggregate having a known sieve analysis is immersed in a solution of sodium or magnesium sulfate for 16 to 18 hours. Next, it is placed in an oven at 230°F (110°C) and dried to constant weight. The procedure is repeated for the desired period (usually 5 or 10 cycles); the sample is cooled, washed, and dried to constant weight; then sieved, weighed, and recorded as the percent of weight lost.

This method furnishes information helpful in judging the soundness of aggregates subjected to weathering action, particularly when adequate information is not available from service records of the material exposed to actual weathering conditions. Attention is called to the fact that test results by the use of the two salts differ considerably, and care must be exercised in fixing proper limits in any specifications that may include requirements for these tests.

ASTM C666 (Resistance of Concrete to Rapid Freezing and Thawing) The purpose of this specification is to determine how concrete will react under continuous cycles of freezing and thawing and to rank aggregates.

In this procedure two methods are used. Method A involves rapid freezing and thawing in water, and method B involves rapid freezing in air and thawing in water. Immediately after curing, the specimen is brought to a temperature within $-2°F$ and $+4°F$ ($-1.1°C$ and $2.2°C$) of the target thaw temperature that will be used in the freeze-thaw cycle and tested for fundamental transverse frequency, weighed, and measured in accordance with ASTM C215 (Fundamental Transverse, Longitudinal, and Torsional Frequencies of Concrete Specimens). The specimen is protected against loss of moisture between the time of removal from curing and the start of the freeze-thaw test.

The freezing and thawing test is started by placing the specimens in the thawed water at the beginning of the thawing phase of the cycle. After each interval the specimen is tested for the fundamental transverse frequency, weighed, and returned to the apparatus. For procedure A, the container is rinsed out, clean water is added, and the specimen is returned to the freezer. This procedure is continued for 300 cycles or until the relative dynamic modulus of elasticity reaches 60 percent of the initial modulus. If the test is interrupted, the specimen is stored in the frozen condition or in a wet condition in a refrigerator. For procedure B it is undesirable to store the specimens in the thawed conditions for more than 2 cycles.

For procedure A, the specimen must have at least $\frac{1}{8}$ in. (0.32 cm) of water all around it. In order to do this, it must be kept in a container. If the container is made of metal, the water will freeze from the top down, and as it expands there is no place for the water to go except into the concrete pores. This situation results in scaling and may cause misleading interpretations. If the container is made of rubber or other material that will expand as the water expands, the problem of scaling in the initial stages will not occur. With scaling there is an eventual weight loss, and in the early stages, especially in a metal container, the results may be misleading. Another problem may be the way the specimen is cured: whether it is cured in water or cured in air and saturated. Another problem arises because of the rate of freezing. The Corps of Engineers freeze concrete at a rate of 12 cycles per day; other agencies freeze at a rate of 4 to 6 cycles per day; and still others freeze at a rate of 1 cycle per day. Thus, the thermal properties of the aggregates begin to play an important role in the process. The test should be used to rank various aggregates rather than to determine their performance characteristics when used in concrete.

ASTM C215 (Fundamental Transverse, Longitudinal, and Torsional Frequencies of Concrete Specimens)   The purpose of this test is to determine the relationship between strength loss and cycles of freezing and thawing.

Test specimens are made in accordance with ASTMC192 (Making and Curing Concrete Test Specimens in the Laboratory). The weight and the average length will be determined. The most important test for portland cement concrete is the transverse frequency. The specimen is placed on supports such that it may vibrate without restrictions in a free transverse mode. The specimen is forced to vibrate at various frequencies. Record the frequency of the test specimen that results in maximum indication having a well-defined peak on the indicator and at which observation of nodal points indicates fundamental transverse vibration as the fundamental transverse frequency. *Young's modulus* is then calculated as follows:

$$\text{durability factor} = \text{DF}_{300} = \frac{PN}{M} = \frac{(\text{relative } E)\,(N \text{ cycles})}{\text{duration of test}} \qquad (2.3)$$

In this test it is not necessary to perform the test for 300 cycles of freezing and thawing. It is only necessary to perform the test for 150 cycles and then calculate the durability factor at 50% and the DF_{300} can be calculated. The only problem with this test is to make sure that the specimen is vibrating at its fundamental transverse mode. The test is a good one in that it is nondestructive and gives a relationship between strength loss and the number of cycles of freezing and thawing.

ASTM C597 (Pulse Velocity through Concrete)   The major purpose of this specification is to check the uniformity in mass concrete, to indicate characteristic changes in concrete, and in the survey of field structures estimate the severity of deterioration, cracking, or both.

In this procedure a sound wave is transmitted through the concrete mass and the length of time it takes to travel from one end to the other is recorded. Knowing the time and the path length, the velocity can be computed.

According to the ASTM, results obtained from this test should not be considered as a means of measuring strength or as adequate for establishing the compliance of the modulus of elasticity of field concrete. The primary use of the test is to check dams for weak spots. There have been complaints about its lack of precision, and so far it has not proven to be completely satisfactory.

ASTM C671 (Critical Dilation of Concrete Specimens Subjected to Freezing)   The purpose of this specification is to determine the test period of frost immunity of concrete specimens measured by the water immersion time required to produce critical dilation when subjected to a prescribed slow-freezing procedure.

In this procedure, the test specimen is molded and cured as prescribed by ASTM C192 (Making and Curing Concrete Test Specimens in the Laboratory). Once the test specimen is prepared and conditioned, the test starts. The test cycle consists of cooling the specimen in water-saturated kerosene from 35 to 15°F (1.67 to -9.44°C) at a rate of 5 ± 1°F (-2.8 ± 0.5°C) per hour; followed immediately by returning the specimen to the 35°F (1.67°C) water bath, where the specimen will remain until the next cycle.

Normally, one test cycle would be carried out every 2 weeks. The length changes are measured during the cooling process. The test is continued until critical dilation is exceeded or until the period of interest is over.

Basically, this is one of the most poorly written procedures in the ASTM specifications. If a good aggregate is utilized, the aggregate may never exceed its critical dilation.

ASTM C682 (Evaluation of Frost Resistance of Coarse Aggregates in Air-entrained Concrete by Critical Dilation Procedure) The purpose of this procedure is to evaluate the frost resistant of coarse aggregates in air-entrained concrete.

This procedure is basically the same as that for the preceding specification. The only difference is that the sample is prepared in accordance with ASTM C295 (Petrographic Examination of Aggregates for Concrete). The aggregate is graded in accordance with field use; otherwise, equal portions of the No. 4, $3/8$-in., $1/2$-in., and 1-in. sieves are used. Further, the aggregate should be used in this test as it is used in the field. Portland cement should meet the specifications of ASTM C150 (Portland Cement), and the fine aggregate should meet the specification of ASTM C33 (Concrete Aggregates). The mix proportion should be in accordance with the ACI method of mix design with an air content of 6 percent and a slump of 2.5 ± 0.5 in.

This test is supposed to simulate the field performance. The significance of the results in terms of potential field performance will depend upon the degree to which field conditions can be expected to correlate with those employed in the laboratory. Thus,the field conditions must be assessed. Obviously, one of the main problems is to simulate field conditions. Some problems are: degree of saturation, age of concrete when the first freeze comes, length of freezing season, amount of water available during freezing, curing procedure, and condition of the aggregate when it enters the mixer. All of these conditions must be duplicated and thus there are definite chances for error. Also, the project should be planned well in advance and the conditions known before the test is run.

ASTM C672 (Scaling Resistance of Concrete Surfaces Exposed to Deicing Chemicals) The purpose of this specification is to evaluate the effect of mix design, surface treatment, curing, or other variables of concrete subjected to scaling due to freezing and thawing, and to determine the resistance to scaling of a horizontal concrete surface subjected to freezing and thawing in the presence of deicing chemicals.

In this specification the concrete at the age of 28 days, after proper curing, is covered with approximately 0.25 in. (0.64 cm) of calcium chloride and water solution having a concentration such that each 100 ml of solution contains 4 g of anhydrous calcium chloride. The specimen is then placed in a freezing chamber for 16 to 18 hours. The specimen is then removed and placed in air at $75 \pm 3°F$ ($23 \pm 1.7°C$) with a relative humidity of 45 to 55 percent for 6 to 8 hours. Water is added to the chamber between each cycle to maintain the depth of the solution. The procedure is repeated daily, and at the end of five cycles the surface of the concrete is flushed thoroughly. A visual inspection of the concrete is made with the ratings given in Table 2-3. These ratings are recorded and the test continues.

Generally, this test is performed up to 50 cycles before final evaluation is made.

TABLE 2-3   CONCRETE SCALING
RATINGS

Rating	Condition of Surface
0	No scaling
1	Very slight scaling
2	Slight to moderate scaling
3	Moderate scaling
4	Moderate to severe scaling
5	Severe scaling

The only problem with this test is that of the rating system. What represents moderate scaling to one person may not to another.

ASTM C295 (Petrographic Examination of Aggregates for Concrete)   The purpose of this specification is to screen the good from the bad aggregates. It has eight specific purposes:

1. Preliminary determination of quality.
2. Establishing properties and probable performance.
3. Correlating samples with aggregates previously tested and used.
4. Selecting and interpreting other tests.
5. Detecting contamination.
6. Determining effects of processing.
7. Determining physical and chemical properties.
8. Describing and classifying constituents.

In this specification the procedure should be carried out by a geologist utilizing x-ray diffraction, differential thermal analysis, electron microscopy, electron diffraction, electron probe, infrared spectroscopy microscope, and the naked eye.

The main purpose of petrographic examinations is to determine physical and chemical properties of aggregates. The relative abundance of specific types of rocks and minerals is established as well as particle shape, surface texture, pore characteristics, hardness, and potential alkali reactivity. Coatings are identified and described; and the presence of contaminating substances is determined.

If the petrographic examination predicts potential alkali reactivity of the aggregate, it is very helpful, in that the time required for this test is substantially shorter than that of the freeze-thaw test or most other available tests.

This specification also sets down a fundamental principle regarding aggregates: if an unfamiliar source is found, it can be compared to known data.

Petrographic examination is the best method by which deleterious and extraneous substances can be detected and determined quantitatively.

ASTM D1075 (Effect of Water on Cohesion of Compacted Bituminous Mixtures)   The purpose of this specification is to measure the loss of cohesion resulting from the action of water on compacted bituminous mixtures containing penetration-grade asphalts. In other words, it evaluates the stripping properties of aggregates.

In this specification a 4-in. (10.16-cm) cylindrical specimen 4 in. (10.16 cm.) high is tested in accordance with ASTM D1074 (Compressive Strength of Bituminous Mixtures). Then the bulk specific gravity of each specimen is determined. Each set of six test specimens is sorted into two groups of three specimens each so that the average bulk specific gravity is the same in each group. Group 1 is tested in accordance with procedure A and group 2 in accordance with procedure B. In test procedure A, the test specimens are brought to the test temperature of 77 $\pm$ 1.8°F (25 $\pm$ 1°C) by storing them in an air bath maintained at the test temperature for not less than 4 hours, their compressive strength determined in accordance with ASTM D1074. In test procedure B, the test specimen is immersed in water for 4 days at 120 $\pm$ 1.8°F (49 $\pm$ 1°C). The specimen is transferred to a second water bath at 77 $\pm$ 1.8°F (25 $\pm$ 1°C) and stored for 2 hours. At that time the compressive strength is determined and the numerical index of resistance of bituminous mixtures to the detrimental effect of water as the percentage of the original strength that is retained after the immersion period is calculated as follows:

$$\text{index of retained strength (\%)} = \frac{S_2}{S_1} \times 100 \qquad (2.4)$$

where S_1 = compressive strength of dry specimens (group 1)
 S_2 = compressive strength of immersed specimens (group 2)

This test gives an excellent indication of the retained strength but sets no definite values as to what is a good or a bad limit.

Tests Concerning Deleterious Materials in Aggregates

ASTM C33 (Concrete Aggregates)   The purpose of this specification is to ensure that satisfactory materials are used in concrete. The specification covers both fine and coarse aggregates but does not cover lightweight aggregates.

The specification establishes definitions for fine and coarse aggregate and places restrictions on grading, deleterious substances, and soundness. The specification also establishes methods for testing and sampling.

This specification is good in that it lays down the fundamental rules for fine and coarse aggregates used in concrete as well as the sampling and testing methods to be followed.

ASTM C142 (Clay Lumps and Friable Particles in Aggregates)   The purpose of this specification is to measure only particles that might cause unsightly blemishes in concrete surfaces. It is an approximate method for the determination of clay lumps and friable particles in natural aggregates.

Aggregates for this test consist of the material remaining after the completion of ASTM C117 [Materials Finer Than No. 200 (75-μm) Sieve in Mineral Aggregates by Washing]. The aggregate is dried to a constant weight at a temperature of 230°F (105 ± 5°C). Weigh the test sample and spread it into a thin layer on the bottom of the container, cover it with water and examine it for clay lumps or friable particles. Particles that can be broken down with the fingers into finely divided particles are classified as friable particles, provided that they can be removed by wet sieving. The residue is removed and weighed. The amount of clay lumps and friable particles in fine aggregate or individual sizes of coarse aggregate is computed as follows:

$$P = \frac{W - R}{W} \times 100 \tag{2.5}$$

where P = percent of clay lumps or friable particles
W = weight of test sample passing the layer of sieves but coarser than the No. 16 sieve
R = weight of particles retained on designated sieve

This test is not widely used because of the physical limitations of sorting through all the particles, and also because these particles are merely a symptom of inadequate processing, which can be remedied by improved washing techniques. Also, the techniques of breaking the particles with the fingers vary from one person to another (i.e., the pressure applied by one person is different from that applied by another). Some clay lumps are hard and cannot be broken apart with the fingers. Thus, inaccurate results will be obtained.

ASTM C117 [Materials Finer than No. 200 (75-μm) Sieve in Mineral Aggregates by Washing]   The purpose of this test is to determine the amount of material finer than a No. 200 (75-μm) sieve in aggregate by washing. Clay particles and other aggregate particles that are dispersed by the wash water as well as water-soluble materials will also be removed from the aggregate during the test.

A sample of the aggregate is washed in a prescribed manner and the decanted wash water containing suspended and dissolved materials is passed through a No. 200 (75-μm) sieve. The loss in weight resulting from the wash treatment is calculated as weight percent of the original sample and is reported as the percentage of material finer than a No. 200 (75-μm) sieve by washing.

This test provides a measure of fines, including clay and silt, in concrete aggregates. The test has been criticized as not furnishing an indication of harmful clays, which may increase mixing-water requirements and volume-change tendencies of concrete. Experiments show that individual operators should have a 95 percent probability of checking their results within 0.3 percent. The amount of material passing a No. 200 (75-μm) sieve, by washing to the nearest 0.1 percent, is calculated as follows:

$$A = \frac{B - C}{B} \times 100 \tag{2.6}$$

where A = percentage of material finer than a No. 200 (75-μm) sieve, by washing

B = original dry weight of sample, grams

C = dry weight of sample, after washing, grams

Excessive quantities of fines in portland cement concrete detract from the quality of the mix by increasing the mixing requirement. In bituminous mixtures, asphalt demand may increase, although the problem is less serious than that in portland cement concrete. In base-course aggregates, this test can be especially important in determining potential susceptibility to frost action. For this reason, this is an important test.

ASTM C123 (Light Weight Pieces in Aggregates) The purpose of this speci-fication is to determine the approximate percentage of lightweight pieces in aggregates by means of sink-float separation in a heavy liquid of suitable specific gravity.

In this procedure the fine aggregate is allowed to dry and cooled to room temperature after following the procedure prescribed in ASTM D75. The material is sieved using a No. 50 sieve and then brought to saturated-surface dry conditions. It is put into a heavy liquid such as kerosene with 1,1,2,2,-tetrabromethane. The particles will be separated by the float-sink method provided that the specific gravities are different enough to permit separation. The liquid is poured off into a second container and passed through a skimmer. Care is taken that only the floating pieces are poured off with the liquid and that none of the sand is decanted onto the skimmer. The liquid is returned to the first container, agitated again, and the decanting process is repeated until the liquid is free of friable particles. The pieces are dried and the weight determined. For a coarse aggregate the particles are sieved using a No. 4 sieve and the foregoing process is repeated.

The materials or chemicals used in combination to form the heavy liquid for the separation process are very toxic and must be handled with great care. The test eliminates particles that might produce concrete of low durability when exposed to freezing and thawing. Experiments have shown that test procedures using gravity levels up to about 2.50 result in adequate aggregate. Coal and lignite are separated at a specific gravity of 2.0. Potentially harmful chert is separated at a specific gravity of 2.35.

ASTM C40 (Organic Impurities in Sands for Concrete) This specification covers an approximate determination of the presence of injurious organic compounds in natural sands that are to be used in cement mortar or concrete. The principal value of the test is to furnish a warning that further tests of the sands are necessary before they are approved for use.

The procedure for this specification involves a color test. The sand and a 3 percent solution of sodium hydroxide are mixed vigorously in a graduate and allowed to stand for 24 hours. The color of the liquid is then compared to the color of a solution of potassium dichromate in sulfuric acid. If the solution of the sand and sodium hydroxide is darker than the potassium dichromate, organic impurities are present in the sand.

This is a good quick test, and if impurities are indicated, the mortar strength test should be performed to determine if the impurities are deleterious. Certain types of organic matter, principally tannic acid and its compounds derived from the decay of vegetable matter, interfere with the hardening and strength development of cement. This test detects this type of material but unfortunately also reacts to other organics, such as bits of wood, which might not be harmful to strength. A negative test is conclusive evidence of freedom from harmful organic matter, but a positive test may or may not fortell difficulty. This is an excellent test in that it is a warning of possible dangers.

ASTM C227 [Potential Alkali Reactivity of Cement-Aggregate Combinations (Mortar-Bar Method)]

The purpose of this specification is to determine if an aggregate will react with the alkalies in the cement. This test is basically to predict the alkali-silica reaction.

The test method consists of molding bars of mortar 1 in. $\times$ 1 in. $\times$ 12 in. (2.54 cm $\times$ 2.54 cm $\times$ 30.5 cm) in which the aggregate in question is combined with a cement that is to be used in the field. The proportions should be 1 part cement to 225 parts of graded aggregate by weight. Use enough water to develop a flow of 105 to 120 in accordance with ASTM C109 (Compressive Strength of Hydraulic Cement Mortar), with the exception that the flow table drops $\frac{1}{2}$ in. (1.27 cm) for 10 trips in 6 seconds. After 24 hours in the molds, the lengths of the bars are measured, and they are stored at a constant temperature of 100°F (37.8°C) in sealed containers containing a small amount of water in the bottom but not in contact with the specimens. Length changes are to be measured after 1, 2, 3, 6, 9, and 12 months and, if necessary, every 6 months after. If expansion is less than 0.04 percent in 6 months, it is considered nonreactive. If the expansion is between 0.04 and 0.07 percent, the aggregate is suspicious. If the expansion is between 0.07 and 0.1 percent, the aggregate is reactive.

The obvious disadvantage to this test is the length of time involved to perform it. The test would seem to be reliable in rejecting poor aggregate. The limits given are judgments based on reports through research and petrographic examination. The limits are somewhat low for the purpose of conservatism.

ASTM C289 [Potential Reactivity of Aggregates (Chemical Method)]

The purpose of this test procedure is to determine the potential reactivity of an aggregate with alkalies in portland cement concrete in a very short time. This is a test for the alkali-silica reaction and is not intended for the alkali-carbonate reaction.

In this procedure the material is ground to the point when it is finer than the No. 50 sieve but coarser than the No. 100 sieve. Twenty-five grams of the material are mixed with 25 ml of a 1 N solution of NaOH in a steel vessel about 2 in. (5.08 cm) in diameter and $2\frac{1}{2}$ in. (6.35 cm) high. The vessel is sealed at a temperature of 176°F (80°C) for 24 hours and then the liquid is filtered and tested for alkalinity and dissolved silica.

The results are presented on semilog paper as dissolved silica vs. reduction in alkalinity. If a considerable amount of silica is dissolved and there is a considerable amount of reduction in alkalinity, we have a potentially reactive material. If there is little dissolved silica and little reduction in alkalinity, the aggregate is of good quality.

This test has the advantage of being quick and accurate, as the results can be run within 24 hours. This test is based upon field experience and should only be used as a screening device. The test will not give satisfactory results with carbonate rocks such as ferrous and magnesium rocks. In these two aggregate types, a reduction in ions results and thus the results are no longer reliable.

Tests Used in the Design of Concrete Mixes (Portland Cement or Bituminous)

ASTM D75 (Sampling Aggregates) The purpose of this test is to sample fine and coarse aggregate for the following purposes:

1. Preliminary investigation of the potential source of supply.
2. Control of the product at the source of supply.
3. Control of the operations at the site of use.
4. Acceptance or rejection of the materials.

In this procedure, sampling plans and acceptance and control tests vary with the type of construction in which the material is used. Samples for preliminary investigation tests are obtained by the party responsible for development of the potential source. The sampler must use every precaution to obtain samples that will show the true nature and condition of the materials they represent. Samples must be inspected and sampling taken from conveyor belts, flowing aggregate stream, or stockpiles. The number of samples taken depends on the variations and the properties measured.

Sampling is as important as testing; the test results depend on the sampling. If the sampler is inexperienced with the techniques involved, the entire test becomes questionable. Thus, in this specification, a person familiar with sampling should perform all sampling procedures.

ASTM C136 (Sieve or Screen Analysis of Fine and Coarse Aggregates) The purpose of this specification is to determine the particle size of fine and coarse aggregates to be used in various tests.

In this procedure, a weighed sample of dry aggregate is separated through a series of sieves or screens of progressively smaller openings for determination of particle-size distribution.

In this specification, the results are dependent upon individual technique. The test is placed in two categories: mechanical sieving and hand sieving. This excellent test determines the gradation of the aggregates, which is so important in mix design procedures.

ASTM C127 (Specific Gravity and Absorption of Coarse Aggregate) The purpose of this specification is to ultimately determine the solid volume of coarse aggregate and the unit volume of the dry rodded aggregate such that a weight-volume characteristic can be determined so that a concrete design mix can be determined. The bulk specific gravity is used to determine the volume occupied by the aggregate.

In this procedure, approximately 5 kg of the aggregate is selected after quartering. After the aggregate is thoroughly washed to remove dust or other coatings from the surface of the particles, the sample is dried to constant weight at a temperature of 212 to 230°F (100 to 110°C) and cooled in air at room temperature for 1 to 3 hours. After cooling, the sample is immersed in water at room temperature for a period of 24 ± 4 hours. Next, the specimen is removed from the water and rolled in a large absorbent cloth towel until all visible films of water are removed. The large particles are wiped by hand. Care is taken not to allow evaporation of water from aggregate pores during the operation of surface drying. The sample is weighed in the saturated-surface-dry condition. Then the sample is weighed in water, making sure that the entrapped air is removed. The sample is dried at 212 to 230°F (100 to 110°C) cooled at room temperature for 1 to 3 hours, and weighed. The bulk and apparent specific gravity and the percent absorption are determined as follows:

$$\text{bulk specific gravity} = \frac{A}{B - C} \tag{2.7}$$

where A = weight of oven-dry specimen in air, grams
B = weight of saturated-surface-dry specimen in air, grams
C = weight of saturated specimen in water, grams

$$\text{bulk specific gravity (saturated-surface-dry)} = \frac{B}{B - C} \tag{2.8}$$

$$\text{apparent specific gravity} = \frac{A}{A - C} \tag{2.9}$$

$$\text{absorption} = \frac{B - A}{A} \times 100 \tag{2.10}$$

The specific gravity of aggregates is important, as it is used to determine the aggregate for use in a concrete mix.

ASTM C128 (Specific Gravity and Absorption of Fine Aggregate)　　The purpose of this specification is to determine the bulk and apparent specific gravity of fine aggregate as well as the absorption.

In this procedure, 500 g of fine aggregate is immersed in a pycnometer which is filled with water to 90 percent of capacity. The pycnometer is rolled, inverted, and agitated to eliminate air bubbles. The temperature is adjusted to 73.4 ± 3°F (23 ± 1.7°C). The total weight of the pycnometer, sample, and water is determined. The fine aggregate is removed, dried to a constant weight at 212 to 230°F (100 to 110°C), cooled at room temperature for 0.5 to 1.5 hours, and weighed. The weight of the pycnometer is determined and the bulk specific gravity, bulk saturated-surface-dry specific gravity, apparent specific gravity, and the absorption are calculated as follows:

$$\text{bulk specific gravity} = \frac{A}{B + 500 - C} \tag{2.11}$$

where A = weight of oven-dry specimen in air, grams

$\quad\quad B$ = weight of pycnometer filled with water, grams

$\quad\quad C$ = weight of pycnometer with specimen and water to calibration mark, grams

$$\text{bulk saturated-surface-dry specific gravity} = \frac{500}{B + 500 - C} \quad\quad (2.12)$$

$$\text{apparent specific gravity} = \frac{A}{B + A - C} \quad\quad (2.13)$$

$$\text{absorption} = \frac{500 - A}{A} \times 100 \quad\quad (2.14)$$

This is an important test because the volume of aggregate is determined for the concrete mix. The results of this test are used in all concrete-mix design procedures, whether portland cement or bituminous.

ASTM C29 (Unit Weight of Aggregate)   This method covers the determination of the unit weight of fine, coarse, or mixed aggregate.

In this procedure, the sample is dried to constant weight in an oven at 220 to 230°F (105 to 110°C) and thoroughly mixed. A cylindrical metal bucket is calibrated using water (knowing that water weighs 62.4 lb/ft^3). The measure is filled one-third full and the surface is leveled with the fingers. The layer of aggregate is rodded 25 times with a tamping rod. The strokes are applied evenly over the sample. This procedure is repeated at two-thirds full and at full. The measure is leveled, weighed, and multiplied by the volume of the bucket. This method applies to aggregates of 1.5 in. (3.81 cm) or less. For aggregate over 1.5 in. (3.81 cm), use the jigging method.

The results of this test should check within 1 percent when duplicated. The results of this procedure are important, as they are used in the mix design procedure for both portland cement and bituminous concrete.

PROBLEMS

2.1. What is the difference between a natural aggregate and a manufactured aggregate?

2.2. Aggregates may be classified as fine aggregate or coarse aggregate; explain the difference.

2.3. Explain how aggregates are processed for use as a portland cement concrete ingredient or as a bituminous concrete ingredient.

2.4. How does particle shape affect the use of aggregate in base-course materials? In portland cement concrete? In bituminous concrete?

2.5. Explain the use of Fuller's maximum density curve.

2.6. Why is gradation important in portland cement concrete?

2.7. Which type of aggregates (igneous, sedimentary, or metamorphic) would you expect to be most suitable as a base-course material? Why?

2.8. Review various references on the subject of freezing and thawing and write a short report on how they eventually lead to concrete failure.

2.9. Review several references and explain why aggregate beneficiation is necessary. Include in your report the methods used for aggregate beneficiating.

2.10. Review the ASTM specifications for tests concerning the general quality of aggregates, deleterious materials in aggregates, and the specifications used in the design of portland cement and bituminous concrete mixes, and write a short report on the purposes, procedures, and reasons for the tests.

REFERENCES

ABDUN-NUR, E. A., "Concrete and Concrete Making Materials," *ASTM Spec. Tech. Publ. 169-A,* 1966, pp. 7–17.

ALLEN, C. W., "Influence of Mineral Aggregates on the Strength and Durability of Concrete," Symposium on Mineral Aggregate, *ASTM Spec. Tech. Publ. 83,* 1948.

BATEMAN, J. H., *Materials of Construction,* Pitman Publishing Corp., New York, 1950, pp. 23–74.

FULLER, W. B., and THOMPSON, S. E., "The Laws of Proportioning Concrete," *Trans. Am. Soc. Civil Engrs., 59,* 1907.

Highway Research Board, *Bibliography on Mineral Aggregates,* Washington, D.C., 1949 (Bibliography No. 6).

JACKSON, F. H., "The Durability of Concrete in Service," *Proc. Am. Concr. Inst., 43* (1942), p. 165.

KREBS, R. D., and WALKER, R. D., *Highway Materials,* McGraw-Hill Book Company, New York, 1971.

MINOR, C. E., "Degradation of the Mineral Aggregates," *ASTM Spec. Tech. Publ. 277,* 1960, pp. 109–121.

NORDBERG, B., "Canada's Most Modern Gravel Plant," *Rock Prod., 55* (Aug. 1952).

PARSONS, W. H., and INSLEY, H., "Observations on Alkali-Aggregate Reaction," *Proc. Am. Concr. Inst., 44* (1948), p. 625.

Pit and Quarry Handbook, 41st ed., Complete Service Publishing Co., Chicago, 1948.

POWERS, T. C., "Basic Considerations Pertaining to Freezing and Thawing Tests," *ASTM Proc., 55* (1955) p. 1132.

STANTON, T. E., "California Experience with the Expansion of Concrete through Reaction between Cement and Aggregate," *J. Am. Concr. Inst., 39* (Jan. 1942).

SWEET, H. S., "Physical and Chemical Tests of Mineral Aggregates and Their Significance," Symposium on Mineral Aggregates, *ASTM Spec. Tech. Publ. 83,* 1948.

VERBECK, V., "Osmotic Studies and Hypothesis Concerning Alkali-Aggregate Reactions," *ASTM Proc., 55* (1955), pp. 1110–1127.

WOOLF, D. O., "Methods for the Determination of Soft Pieces in Aggregates," *Public Roads, 26,* (Apr. 1951), p. 148.

WOOLF, D. O., "Methods for the Determination of Soft Pieces in Aggregate," *Public Roads, 26,* (Apr. 1937).

3

Cements

Cements are materials that exhibit characteristic properties of setting and hardening when mixed to a paste with water. They are a class of products that can be very complex and of somewhat variable composition and constitution.

Cements are divided into two classifications: hydraulic and non-hydraulic. This division is based upon the way in which the cement sets and hardens. The *hydraulic cements* have the ability to set and harden under water. Hydraulic cements include, but are not limited to, the following: hydraulic limes, pozzolan cements, slag cements, natural cements, portland cements, portland-pozzolan cements, portland blast-furnace-slag cements, alumina cements, expansive cements, and a variety of others (white portland cements, colored cements, oil-well cements, regulated cements, waterproofed cements, hydrophobic cements, antibacterial cements, barium and strontium cements).

Nonhydraulic cements do not have the ability to set and harden under water but require air to harden. The main example of a nonhydraulic cement would be a lime.

LIME

Lime, one of the oldest known cementing materials, is readily available and rather inexpensive. Lime is produced by burning limestone (calcium carbonate) with impurities such as magnesia, silica, iron, alkalies, alumina, and sulfur. This burning process takes place in either a vertical or a rotary kiln at a temperature of 1800°F (980°C). Calcium carbonate is decomposed into calcium oxide and carbon dioxide according to the following reaction:

$$CaCO_3 \rightarrow CaO + CO_2 \tag{3.1}$$

The calcium oxide that is formed is called *quicklime,* which, when in the presence of water, reacts to form calcium hydroxide together with a great evolution of heat:

$$CaO + H_2O \rightarrow Ca(OH)_2 + heat \qquad (3.2)$$

This process is called *slaking* and the product calcium hydroxide is called *slaked lime* or *hydrated lime.* The rate of reaction depends mainly on the purity of the lime. The higher the purity of lime, the greater its reactivity with water. Commercial quicklime is classified into three groups: quick, medium, and slow slaking.

Depending upon the amount of water added during the slaking process, lime putty or hydrated lime may be formed. Hydrated lime is produced by adding just enough water (one-third of its weight) to quicklime. Lime putty is formed when an overextended amount of water is added to the quicklime.

Both lime putty and hydrated lime are always mixed with mortar sand in proportions of 1 part lime to 3 parts sand by volume, to prevent excessive shrinkage.

The setting of lime mortar is the result of the loss of water either by absorption of water by the block, brick, or whatever, or by evaporation. The hardening is caused by the reaction of carbon dioxide in the air with the hydrated lime as follows:

$$Ca(OH)_2 + CO_2 \rightarrow CaCO_3 + H_2O \qquad (3.3)$$

This results in the formation of calcium carbonate crystals, which bind the heterogeneous mixture into a coherent mass. The hardening process is slow and may take several years to develop its full strength. However, it needs the free circulation of air to provide the necessary carbon dioxide to penetrate the innermost portion of the mortar for hardening to take effect.

HYDRAULIC LIMES

Hydraulic limes are made by burning siliceous or argillaceous limestone whose clinker after calcination (in a continuous kiln) contains a sufficient percentage of lime silicate to give hydraulic properties to the product, but which normally contain so much free lime that the mass of clinker will slake on the addition of water.

As the content of alumina and silica in the lime increase, the rate of slaking and heat evolution decrease to a point where no reaction occurs between water and lime. At high temperature, the alumina and silica combine with calcium oxide, calcium silicates, and aluminates, which do not easily combine with water when it is in lump form. Therefore, quicklime is added in the slaking process and the large lumps are broken up into the fine powder due to the expansion of the quicklime. The final product consists of lime silicate and about one-fourth hydrated lime. The material in the fine form can now readily combine with water. Hydraulic limes do exhibit hydraulic properties but are not suited to subaqueous construction because they require free access of air during hardening. The air is necessary to ensure carbonation of free calcium hydroxide present in large amounts in calcium carbonate; otherwise, the full strength of the lime cannot be developed.

Hydraulic limes are used for browning plaster coats or for stucco and similar uses.

POZZOLAN CEMENTS

A *pozzolan,* according to ASTM C595 (Blended Hydraulic Cements), is a siliceous or siliceous and aluminous material which, in itself, possesses little or no cementitious value but will, in finely divided form and in the presence of moisture, react chemically with calcium hydroxide at ordinary temperatures to form compounds possessing cementitious properties. Pozzolans are further classified as natural pozzolans or artificial pozzolans. Natural pozzolans are further classified into two groups. The first group is made up of pumicite, obsidian, scoria, tuff, santorin, and trass, which are derived from volcanic rocks. The second group of natural pozzolans contains large quantities of finely dispersed, amorphous silica which react with lime in the presence of water to form hydrated silicates, which accounts for their hydraulic properties.

Artificial pozzolans include fly ash, boiler slag, and by-products from the treatment of bauxite ore.

Pozzolan cements are manufactured by direct grinding of the volcanic rocks or by calcining and grinding clays, shales, and diatomaceous earth. Pozzolan cements include all cementing materials that are made by the incorporation of pozzolans with hydrated lime in which no subsequent calcination is needed.

The requirements for pozzolan use in blended cements are given in ASTM C595 (Blended Hydraulic Cements) and are repeated in part here in Table 3-1. Whenever a pozzolan is used in cement, it must conform to the requirements of ASTM C311 (Sampling and Testing Fly Ash or Natural Pozzolans for Use as a Mineral Admixture in Portland Cement Concrete) and/or ASTM C618 (Specifications for Fly Ash and Raw or Calcinated Natural Pozzolan for Use as a Mineral Admixture in Portland Cement Concrete).

Thus far, very little use has been found for pozzolan cements in the area of structural concrete. Their main use has been where mass is required rather than strength. It has also found limited use when mass concrete is needed with little heat of hydration. However, Chapter 6 will provide a discussion on proportioning such mixes.

SLAG CEMENTS

Slag cements are hydraulic cements consisting mostly of an intimate and uniform blend of granulated blast-furnace slag and hydrated lime in which the slag constituent is at least 60 percent of the weight of the slag cement. The mixture is often not calcined.

TABLE 3-1 PHYSICAL REQUIREMENTS FOR POZZOLANS

Fineness	
Amount retained when wet-seived on No. 325 (45-μm) sieve, max. (%)	20.0
Pozzolan activity index with portland cement, at 28 days, min. (%)	75.0

Basically, two types of slag cement exist. The first is designated Type S slag cement by ASTM C595 (Blended Hydraulic Cements). Type S slag cement may be used in combination with portland cement in making concrete and in combination with hydrated lime in making masonry mortar. The second type of slag cement is designated Type SA slag cement by ASTM C595 (Blended Hydraulic Cements). Type SA slag cement, which is air-entrained slag cement, has the same general uses as Type S.

Blast-furnace slags are a nonmetallic product, consisting essentially of silicates and aluminosilicates of calcium and of other bases, developed in a molten condition simultaneously with iron in a blast furnace. The blast-furnace slags, which are suitable for use in slag cements, are fusible lime silicates derived as waste products from the operation of blast furnaces in smelting iron from its ore.

When slag cements are manufactured, they go through a variety of operations, such as granulation (which not only renders the slag more hydraulic, but at the same time reduces the harmful sulfides), drying of the slag, preparation of the hydrated lime, proportioning the mix, mixing, and final grinding.

Slag cements are of limited importance in structural concrete, but may find success in projects requiring large masses of concrete masonry where weight and bulk are more important than strength. It may also find use as a masonry cement in that it does not have a staining effect because of its low alkali content.

NATURAL CEMENTS

A natural cement was at one time defined by ASTM C10 (Natural Cements) as a hydraulic cement produced by calcining a naturally occurring argillaceous limestone at a temperature below the sintering point and then grinding to a fine powder. The amounts of silica, alumina, and iron oxide present are sufficient to combine with all the calcium oxide to form the corresponding calcium silicates and aluminates, which account for the hydraulic properties of natural cements.

Two types of natural cement existed according to ASTM C10 (Natural Cements), Type N and Type NA. Type N natural cement is for use with portland cement in general concrete construction. Type NA natural cement is air-entrained cement and has the same uses as Type N.

Natural cements are made by the calcination of a natural clay limestone, which is made up of clay material (13 to 35 percent), silica (10 to 20 percent), and a balance of alumina and iron oxide. The clay material gives the cement its hydraulic properties.

After calcination and possible slaking to remove the free lime, the clinker is ground into a fine powder known as a natural cement, with the following average composition:

$$\begin{array}{ccccc} SiO_2 & CaO & MgO & Fe_2O_3 & Al_2O_3 \\ 22\text{--}29\% & 31\text{--}57\% & 1.5\text{--}22\% & 1.5\text{--}3.2\% & 5.2\text{--}8.8\% \end{array}$$

Thus, a quick view of the composition would indicate the possibilities of wide variation among mechanical properties. The physical requirements of natural cements are shown in Table 3-2.

Natural cements should not be used in exposed areas but may be used as a substitute

TABLE 3-2   PHYSICAL REQUIREMENTS OF NATURAL CEMENTS

Properties	Type N	Type NA
Fineness, specific surface (cm^2/g)		
Air permeability apparatus		
Average value, min.	6000	6000
Minimum value, any one sample	5500	5500
Soundness		
Autoclave expansion, of blend of 75 percent natural cement and 25 percent portland cement by weight, max. (%)	0.8	0.8
Time of setting, Vicat test		
Time (min)	30	30
Air entrainment		
No agent used, max. (vol.%)	12	
Using air-entrainment agent (vol. %)	—	19 ± 3
Compressive strength, min. [psi (MPz)]		
Compressive strength of mortar cubes, composed of 1 part natural cement and 1 part standard sand by weight; must be equal to or higher than the values specified for the following ages:		
1 day in moist air, 6 days in water	500 (3.4)	500 (3.4)
1 day in moist air, 27 days in water	1000 (6.9)	1000 (6.9)

for portland cements in mortars and concrete when the stresses encountered will never be high or in situations in which weight and mass are more essential than strength.

PORTLAND CEMENTS

Portland cement is one of the most widely used construction materials and is the most important hydraulic cement. It is used in concrete, mortar, plaster, stucco, and grout. It is used in all types of structural concrete (walls, floors, bridges, tunnels, subways, etc.), whether reinforced or not. It is further used in all types of masonry (foundations, footings, dams, retaining walls, and pavements). When portland cement is mixed with sand and lime, it serves as mortar for laying brick or stone, or as plaster or stucco for interior or exterior walls. When portland cement is mixed with coarse aggregate (aggregate larger than the No. 4 sieve) and fine aggregate (sand) together with enough water to ensure a good consistency, concrete results.

Portland cement is defined, according to ASTM C150, as a hydraulic cement produced by pulverizing clinker consisting essentially of hydraulic calcium silicates, usually containing one or more of the forms of calcium sulfate as an interground addition. The approximate proportions for portland cement are as follows:

Lime (CaO) 60–65%
Silica (SiO$_2$) 20–25%
Iron oxide and alumina (Fe$_2$O$_3$ and Al$_2$O$_3$) 7–12%

History of Portland Cement

The name *portland cement* was proposed by Joseph Aspdin in 1824. The name came about because the powdery material, which he patented, set up with water and sand and resembled a natural limestone quarried on the Isle of Portland in England.

The first portland cement manufactured in the United States was produced by David Saylor at Coplay, Pennsylvania, in 1875 by calcining in vertical kilns at a high temperature a mixture of argillaceous limestone rock with a pure limestone. With the increase in demand for quantity and quality, rotary kilns came into production in 1899. The production of portland cement in the United States has increased from 10 million barrels per year in 1900 to 400 million barrels per year in 1970. World production is approximately six times that of the United States.

Raw Materials

The raw materials of portland cement may be classified into three groups: calcareous, argillocalcareous, and argillaceous. Table 3-3 illustrates the materials that make up each group. From the table and a little knowledge of geology, it is evident that the essential constituents of portland cement are lime, silica, and alumina. Lime does not occur in nature but is found in a suitable form in a carbonate. Silica and alumina are found free in nature in the form of clay, shale, or slate.

Limestone (calcium carbonate) contains impurities of magnesia, silica, iron, alkalies, and sulfur. Magnesia in the form of carbonate of magnesia often occurs in limestone and, if present in the amount of 5 percent or more, will make the limestone unsuitable. Silica by itself does not combine with lime in the kiln; thus, small quantities of free silica in the limestone makes the limestone unacceptable. However, if silica is combined with alumina in the limestone, it will combine with lime in the kiln and is acceptable.

Iron in limestone can occur either as an oxide (Fe_2O_3) or as a sulfide (FeS_2). If the iron is in the form of an oxide, it acts as a flux in combining the lime and silica in the kiln. As a sulfide it reacts very strangely and can prove to be quite injurious to the production of portland cement. If the iron in the form of iron sulfide is present in limestone by 4 percent or greater, the limestone should be rejected.

The makeup of limestone is such that it contains alkalies in the form of soda and potash. These alkalies are not harmful and are usually driven off in the kiln.

One of the major roles of alumina in limestone is to combine with silica such that the limestone will combine with the lime in the kiln and thus make the limestone an acceptable product.

TABLE 3-3 RAW MATERIALS IN PORTLAND CEMENTS

Calcareous ($CaCO_3 > 75\%$)	Argillocalcareous ($CaCO_3 = 40\ to\ 75\%$)	Argillaceous ($CaCO_3 < 40\%$)
Limestone	Clayey limestone	Slate
Chalk	Clayey chalk	Shale
Shells	Clayey marl	Clay

Sulfur, the final impurity in limestone, exists in two forms: lime sulfate and iron pyrite. If each is present in amounts of 3 percent or more, the limestone should be rejected.

Chalk is a variety of limestone formed from pelagic, or floating, organisms that are very fine-grained, porous, and friable. It is white or very light-colored and consists almost entirely of calcite. The rock is made up of the calcite shells of microorganisms partially cemented by structureless calcite. The best known chalks are those of the Cretaceous, exposed in cliffs on both sides of the English Channel. The Selma (Cretaceous) chalk of Alabama, Mississippi, and Tennessee and the Niobrara chalk of the same age in Nebraska are well-known deposits in the United States.

Marl is an argillaceous, nonindurated calcium carbonate deposit that is commonly gray or blue-gray. It is somewhat friable, and in some respects resembles chalk, with which it is interbedded in some localities. It is formed in some freshwater lakes, partially by the action of aquatic plants. The clay content of marls varies, and all gradations between small amounts of clay (marly limestone) and large amounts (marly clay) are found.

Slates are clays that have been solidified in a laminated structure and have the property of splitting into thin sheets. They have limited application in the production of portland cement.

Shales are clays that have become hardened by pressure. They have been formed from deposits of sedimentary clay. Shales are preferred over soft clays in the production of portland cement because segregation of the shale and limestone is less likely to occur.

Clays are formed from the debris resulting from the decay of rocks. Clays take the form of three groups with respect to methods of transportation: residual, sedimentary, and glacial. Clays left where the rocks decayed are *residual*. Clays that have been transported and deposited by stream action are *sedimentary*. *Glacial* clays are those deposited by glacial movement. In any clay the silica content should not be less than 55 to 65 percent, and the amount of alumina and iron oxide combined should be between one-third and one-half the amount of silica.

Manufacture of Portland Cement

Portland cement is the most important hydraulic cement used in construction for mortars, plasters, grouts, and concrete. The manufacture of portland cement occurs through a series of steps (quarrying, crushing, grinding, mixing, calcining, addition of retarder, and packing). Portland cement is made by burning an intimate mixture composed mainly of calcareous, argillocalcareous, or argillaceous materials, at a clinker temperature of 2800°F (1550°C). This partially sintered clinker is then ground to a very fine powder with a very small amount of gypsum (2 to 4 percent) as a retarder. The cement is then packaged into 94-lb (43-kg) bags or into a hypothetical barrel which contains four sacks [376 lb (170 kg)] and is approximately 4 ft^3 (0.1 m^3) loose volume or 1.912 ft^3 (0.05 m^3) solid volume with a specific gravity of 3.15; or it may be bulk-stored. The cement-making process shown in Figure 3-1 is the process as it appears today, that is, as most industries are set up.

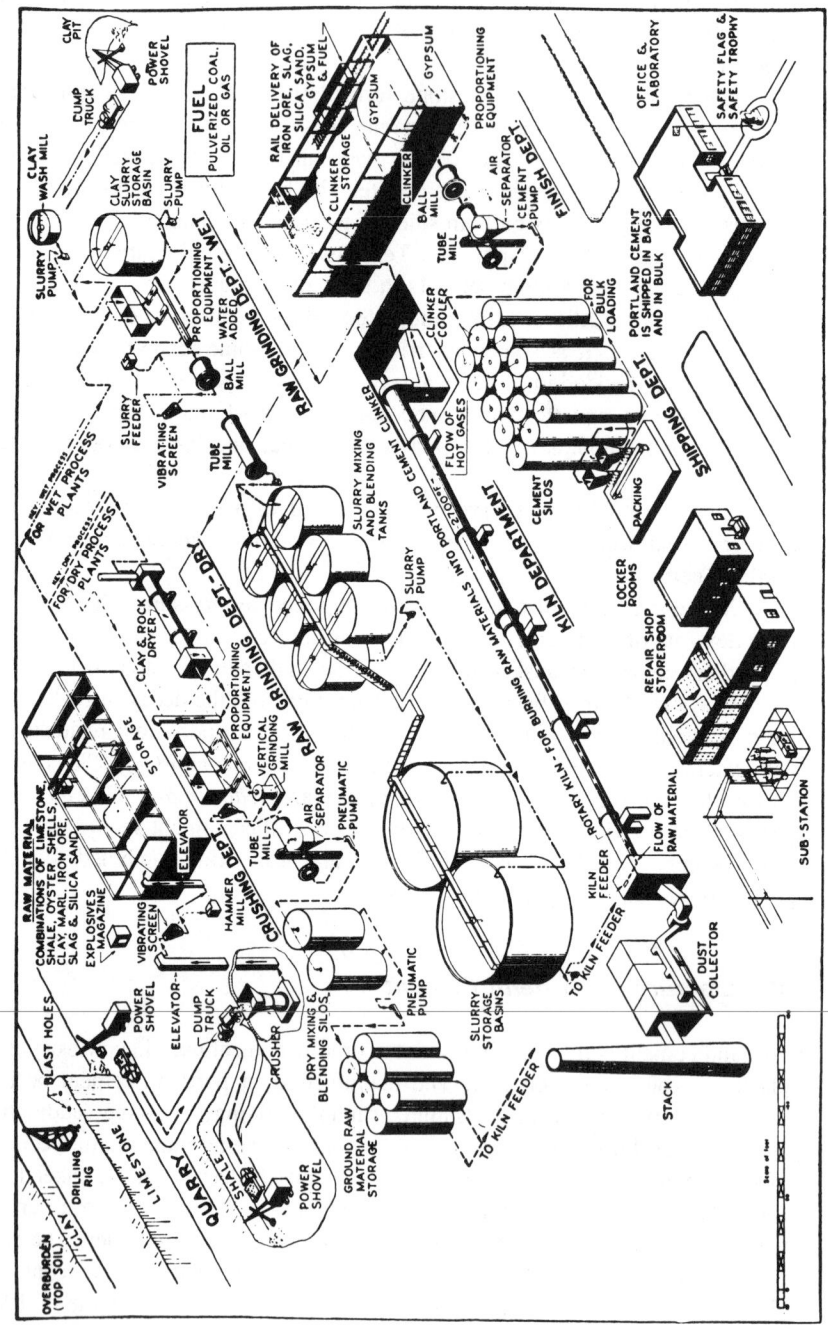

Figure 3–1 Flowchart of manufacturing process of portland cement. (Courtesy of PCA)

Composition of Portland Cement

Eight types of portland cement are recognized by the ASTM under specification ASTM C150:

Type I.	For use when the special properties specified for any other type are not required.
Type IA.	Air-entraining cement for the same uses as Type I, when air entrainment is desired.
Type II.	For general use, more especially when moderate sulfate resistance or moderate heat of hydration is desired.
Type IIA.	Air-entraining cement of the same uses as Type II, where air entrainment is desired.
Type III.	For use when high early strength is desired.
Type IIIA.	Air-entraining cement for the same use as Type III, where air entrainment is desired.
Type IV.	For use when a low heat of hydration is desired.
Type V.	For use when high sulfate resistance is desired.

If a chemical analysis were to be performed on any one of the eight portland cements, the composition would be calcium oxide, silica, alumina, iron oxide, magnesium oxide, sulfur trioxide, and others. The chemical analysis would further reveal that these oxides exist in portland cement as calcium silicates and aluminates, such as tricalcium silicate ($3CaO \cdot SiO_2$), dicalcium silicate ($2CaO \cdot SiO_2$), tricalcium aluminate ($3CaO \cdot Al_2O_3$), and tetracalcium aluminoferrite ($4CaO \cdot Al_2O_3 \cdot Fe_2O_3$). In the nomenclature of the cement in-

TABLE 3-4   STANDARD CHEMICAL REQUIREMENTS

	Cement Type				
	I and IA	II and IIA	III and IIIA	IV	V
Silicon dioxide (SiO_2), min. (%)	—	20	—	—	—
Aluminum oxide (Al_2O_3), max. (%)	—	6	—	—	—
Ferric oxide (Fe_2O_3), max. (%)	—	6	—	6.5	—
Magnesium oxide (MgO), max. (%)	6	6	6	6	6
Sulfur trioxide (SO_3), max. (%)	—	—	—	—	—
When ($3CaO \cdot Al_2O_3$) is 8 percent or less	3	3	3.5	2.3	2.3
When ($3CaO \cdot Al_2O_3$) is more than 8 percent	3.5	—	4.5	—	—
Loss on ignition, max. (%)	3.0	3	3	2.5	3
Insoluble residue, max. (%)	0.75	0.75	0.75	0.75	0.75
Tricalcium silicate ($3CaO \cdot SiO_2$), max. (%)	—	—	—	35	—
Dicalcium silicate ($2CaO \cdot SiO_2$), max. (%)	—	—	—	40	—
Tricalcium aluminate ($3CaO \cdot Al_2O_3$), max. (%)	—	8	15	7	5
Tetracalcium aluminoferrite plus twice the tricalcium aluminate	—	—	—	—	20

TABLE 3-5   OPTIONAL CHEMICAL REQUIREMENTS

	Cement Type					
	I and *IA*	*II and* *IIA*	*III and* *IIIA*	*IV*	*V*	*Remarks*
Tricalcium aluminate ($3CaO \cdot Al_2O_3$), max. (%)	—	—	8	—	—	Moderate sulfate resistance
Tricalcium aluminate ($3CaO \cdot Al_2O_3$), max. (%)	—	—	5	—	—	High sulfate resistance
Sum of tricalcium silicate and tricalcium aluminate, max. (%)	—	58	—	—	—	Moderate heat of hydration
Alkalies ($Na_2O + 0.658K_2O$), max.(%)	0.6	0.6	0.6	0.6	0.6	Low-alkali cement

dustry, these four compounds are written as follows: C_3S, C_2S, C_3A, and C_4AF, respectively. Tables 3-4 and 3-5 show the standard chemical requirements and the optional chemical requirements for portland cement.

Properties of the Main Compounds

As indicated, portland cement is made up of four main compounds: tricalcium silicate, dicalcium silicate, tricalcium aluminate, and tetracalcium aluminoferrite. These com-

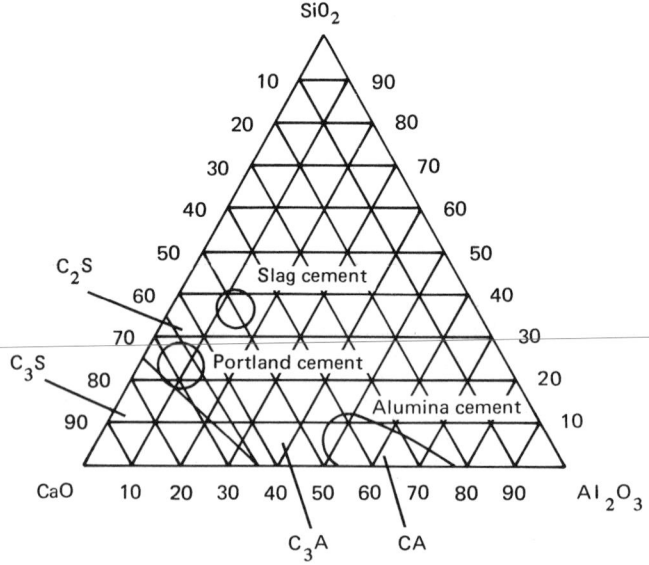

Figure 3–2   Phase diagram.

TABLE 3-6   CHARACTERISTICS OF THE MAJOR COMPOUNDS IN PORTLAND CEMENT

Values	Tricalcium Silicate, $3CaO \cdot SiO_2$ (C_3S)	Dicalcium Silicate, $2CaO \cdot SiO_2$ (C_2S)	Tricalcium Aluminate, $3CaO \cdot Al_2O_3$ (C_3A)	Tetracalcium Aluminoferrite, $4CaO \cdot Al_2O_3 \cdot Fe_2O_3$ (C_4AF)
Cementing value	Good	Good	Poor	Poor
Rate of reaction	Medium	Slow	Fast	Slow
Amount of heat liberated	Medium	Small	Large	Small

pounds are present in the clinker in the form of interlocking crystals. Figure 3-2 shows a phase diagram CaO-SiO_2-Al_2O_3 (C_3S-C_2S-C_3A) in which the field of portland cement lies between. Also shown in the field are aluminous and blast-furnace-slag cements. Table 3-6 lists the characteristics of the major compounds in portland cement.

The most desirable constituent is that of tricalcium silicate (C_3S), because it hardens rapidly and accounts for the high early strength of the cement. When water is added to tricalcium silicate, a rapid reaction occurs as follows:

$$3CaO \cdot SiO_2 + H_2O = 2CaO \cdot SiO_2 \cdot xH_2O + Ca(OH)_2 \qquad (3.4)$$

The results indicate a less basic amorphous hydrated calcium silicate and a crystalline calcium hydroxide. The calcium silicate that is formed is the product to which early strength is attributed.

Dicalcium silicate hardens slowly and contributes largely to strength increase at ages beyond 1 week. In the presence of water, dicalcium silicate ($2CaO \cdot SiO_2$) hydrates slowly and forms a hydrated calcium silicate ($2CaO \cdot SiO_2 \cdot xH_2O$).

Tricalcium aluminate liberates a large amount of heat during the first few days of hardening. It also contributes to early-strength development. Tricalcium aluminate ($3CaO \cdot Al_2O_3$) hydrates with water to form a hydrated tricalcium aluminate ($3CaO \cdot Al_2O_3 \cdot 6H_2O$). If gypsum is added it acts as a retarder, and the heat of evolution is less and the setting occurs more slowly. This is due to the fact that gypsum, when

TABLE 3-7   TYPICAL COMPOUND COMPOSITION FOR VARIOUS PORTLAND CEMENTS

Types of Portland Cement	Compound Composition (%)			
	C_3S	C_2S	C_3A	C_4AF
I. Normal	50	24	11	8
II. Moderate	42	33	5	13
III. High early strength	60	13	9	8
IV. Low heat	26	50	5	12
V. Sulfate resisting	40	40	4	9

present, results in the formation of calcium sulfoaluminate ($3CaO \cdot Al_2O_3 \cdot 3CaSO_4$) rather than hydrated tricalcium aluminate.

Tetracalcium aluminoferrite formation reduces the clinkering temperature, thereby assisting in the manufacture of portland cement. It hydrates rather rapidly but contributes very little to strength. Table 3-7 shows typical compound composition for the various portland cements.

Types of Portland Cement

As previously mentioned, eight types of portland cement are recognized by ASTM under Specification ASTM C150. The standard five types of portland cement (this excludes the three that are air-entrained) are also recognized by the Canadian Standards Association (CSA) and given specific names. We next investigate each type separately.

ASTM Type I or CSA normal portland cement is a general-purpose cement. It is used when the special properties specified for any other type are not required. It is used where there would be no severe climate changes or severe exposure to sulfate attack from water or soil. Its uses include reinforced-concrete buildings, pavements, sidewalks, bridges, railings, tanks, reservoirs, floors, curbs, culverts, and retaining walls. In general, it is used in nearly all situations calling for portland cement.

ASTM Type II or CSA moderate portland cement is a general-purpose cement to be used when moderate sulfate resistance or moderate heat of hydration is desired. It is used in structures of considerable mass, such as abutments and piers and retaining walls. Its use also minimizes temperature rise when concrete is placed in warm weather.

ASTM Type III or CSA high-early-strength portland cement is used when high early strength is desired, usually less than 1 week. It is usually used when a structure must be put into service as quickly as possible. This cement is made by changing the proportions of raw materials, by finer grinding, and by better burning, such that the dicalcium silicate is less and the tricalcium silicate is greater.

ASTM Type IV or CSA low-heat-of-hydration portland cement is used when a low heat of hydration is required. This type of cement develops strength at a slower rate than

TABLE 3-8 APPROXIMATE RELATIVE STRENGTHS OF CONCRETE AS AFFECTED BY TYPE OF CEMENT

Types of Portland Cement	Compressive Strength (% of Normal Portland Cement)				
	1 Day	3 Days	7 Days	28 Days	3 Months
I. Normal	100	100	100	100	100
II. Modified	75	80	85	90	100
III. High early strength	190	190	120	110	100
IV. Low heat	55	55	55	75	100
V. Sulfate resisting	65	65	75	85	100

does the ASTM Type I. However, it is intended for mass structures such as large gravity dams, where the temperature rise on a continuous pour is great. If the temperature were not minimized, large cracks or flaws would appear and the structure might prove to be unsound.

ASTM Type V or CSA sulfate-resisting portland cement is used when high sulfate resistance is desired. It is used when concrete is to be exposed to severe sulfate action by soil or water.

The three types of air-entraining cements, Types IA, IIA, and IIIA, as given by ASTM C150, are used in concrete for improved resistance to freezing and thawing action and to action of salt scaling by chemical attack. Typical air-entraining agents include Vinsol resin, Airolon, and Darex AEA.

Table 3-8 compares the strengths of the various types of cements with Type I at five different moist-curing intervals.

Properties of Portland Cement

Most specifications for portland cement, such as ASTM specifications, place specific chemical composition and physical property requirements on the cement, as shown in Tables 3-9 and 3-10. Next, we will discuss the fineness, soundness, time of setting, compressive strength, heat of hydration, loss of ignition, and specific gravity of portland cement. Most of these tests are covered by ASTM specifications.

The *fineness* of the cement affects the rate of hydration. The greater the cement fineness, the greater the rate of hydration and hence the greater the strength development during the first 7 days. To measure the fineness of the cement, the Wagner turbidimeter or the Blaine air-permeability apparatus is used.

Soundness of hardened cement paste is a measure of the potential expansion of the several constituent parts or the ability to retain its volume after setting. Lack of soundness (unsoundness) is attributed to excessive amounts of hard-burned free lime or magnesia. This free lime takes on water and at some later date develops expansive forces. Most specifications call for the use of an autoclave (high-pressure steam boiler) to indicate the soundness or unsoundness of the cement.

Time of setting is measured by the Gilmore and/or Vicat apparatus, which is used to determine the rate at which portland cement hardens: in other words, to determine if a cement paste remains plastic long enough to permit normal placing of the concrete. The length of time that concrete remains plastic is dependent upon the chemical composition, fineness, water content, and temperature.

Compressive strength of portland cement is determined by mixing the cement specimen with a uniform silica sand and water in prescribed proportions and molding the mixture into 2 in. $\times$ 2 in. $\times$ 2 in. (5.08 cm $\times$ 5.08 cm $\times$ 5.08 cm) cubes. The cubes are cured and then tested in compression to give an indication of the strength-developing characteristics of the portland cement.

Heat of hydration is the heat generated when cement and water react. The amount of heat generated is dependent chiefly on the chemical composition, fineness of the cement, and the temperature of curing time.

TABLE 3-9 OPTIONAL CHEMICAL REQUIREMENTS

	Cement Type[a]					
	I and IA	*II and IIA*	*III and IIIA*	*IV*	*V*	*Remarks*
Tricalcium aluminate $(3CaO \cdot Al_2O_3)^b$, max. (%)	—	—	8	—	—	For moderate sulfate resistance
Tricalcium aluminate $(3CaO \cdot Al_2O_3)^b$, max. (%)	—	—	5	—	—	For high sulfate resistance
Sum of tricalcium silicate and tricalcium aluminate[b], max. (%)	—	58[c]	—	—	—	For moderate heat of hydration
Alkalies (Na_2O + $0.658K_2O$), max. (%)	0.6[d]	0.6[d]	0.6[d]	0.6[d]	0.6[d]	Low-alkali cement

[a] Attention is called to the fact that cements conforming to the requirements for all of these types may be carried in stock in some areas. In advance of specifying the use of other than Type I cement, it should be determined whether the proposed type of cement is or can be made available.

[b] The expressing of chemical limitations by means of calculated assumed compounds does not necessarily mean that the oxides are actually or entirely present as such compounds.

When the ratio of percentages of aluminum oxide to ferric oxide is 0.64 or more, the percentages of tricalcium silicate, dicalcium silicate, tricalcium aluminate, and tetracalcium aluminoferrite should be calculated from the chemical analysis as follows:

$$\text{tricalcium silicate} = (4.071 \times \text{percent CaO}) - C7.600 \times \text{percent SiO}_2) - (6.718 \times \text{percent Al}_2O_3) - (1.430 \times \text{percent Fe}_2O_3) - (2.852 \times \text{percent SO}_3)$$

$$\text{dicalcium silicate} = (2.867 \times \text{percent SiO}_2) - (0.7544 \times \text{percent C}_3S)$$

$$\text{tricalcium aluminate} = (2.650 \times \text{percent Al}_2O_3) - (1.692 \times \text{percent Fe}_2O_3)$$

$$\text{tetracalcium aluminoferrite} = 3.043 \times \text{percent Fe}_2O_3$$

When the alumina–ferrite oxide ratio is less than 0.64, a calcium aluminoferrite solid solution [expressed as $ss(C_4AF + C_2F)$] is formed. Contents of this solid solution and of tricalcium silicate should be calculated by the following formulas:

$$ss(C_4AF + C_2F) = (2.100 \times \text{percent Al}_2O_3) + (1.702 \times \text{percent Fe}_2O_3)$$

$$\text{tricalcium silicate} = (4.071 \times \text{percent CaO}) - (7.600 \times \text{percent SiO}_2) - (4.479 \times \text{percent Al}_2O_3) - (2.859 \times \text{percent Fe}_2O_3) - (2.852 \times \text{percent SO}_3)$$

No tricalcium aluminate will be present in cements of this composition. Dicalcium silicate shall be calculated as previously shown.

In the calculation of C_34, the values of Al_2O_3 and Fe_2O_3 determined to the nearest 0.01 percent should be used. In the calculation of other compounds, the oxides determined to the nearest 0.1 percent shall be used.

All values calculated as described in this note should be reported to the nearest 1 percent.

[c] This limit applies when moderate heat of hydration is required and tests for heat of hydration are not requested.

[d] This limit may be specified when the cement is to be used in concrete with aggregates that may be deleteriously reactive.

Loss of ignition of portland cement is determined by heating a cement sample of known weight to a full red heat of 1652°F (900°C) until a constant weight is obtained. The weight loss of the sample is then determined. Ignition is the indication of prehydration and carbonation, which may be caused by improper or prolonged storage.

The *specific gravity* of portland cement is generally about 3.15.

TABLE 3-10 STANDARD PHYSICAL REQUIREMENTS

	Cement Type[a]							
	I	*IA*	*II*	*IIA*	*III*	*IIIA*	*IV*	*V*
Air content of mortar[b], (vol. %)								
Maximum	12	22	12	22	12	22	12	12
Minimum	—	16	—	16	—	16	—	—
Fineness, specific surface (cm^2/g) (alternative methods)[c]								
Turbidimeter test, min.	160	160	160	160	—	—	160	160
Air permeability test, min.	280	280	280	280	—	—	280	280
Autoclave expansion, max. (%)	0.8	0.8	0.8	0.8	0.8	0.8	0.8	0.8
Compressive strength [psi (MPa)], not less than the values shown for the following ages:[d]								
1 Day	—	—	—	—	1800 (12.4)	1450 (10.0)	—	—
3 Days	1800 (12.4)	1450 (10.0)	1500 (10.3) 1000[e] (6.9)[e]	1200 (8.3) 800[e] (5.5)[e]	3500 (24.1)	2800 (19.3)	— —	2200 (15.2)
7 Days	2800 (19.3)	2250 (15.5)	2500 (17.2) 1700[f] (11.7)[f]	2000 (13.8) 1350[f] (9.3)[f]	—	—	1000 (6.9)	2200 (15.2)
28 Days	—	—	—	—	—	—	2500 (17.2)	3000 (20.7)
Time of setting (alternative methods)[f]								
Gilmore test								
Initial set (min), not less than	60	60	60	60	60	60	60	60
Final set (hr), not more than	10	10	10	10	10	10	10	10

TABLE 3-10 STANDARD PHYSICAL REQUIREMENTS (Continued)

	Cement Type[a]							
	I	*IA*	*II*	*IIA*	*III*	*IIIA*	*IV*	*V*
Vicat test						45	45	45
Initial set (min), not less than	45	45	45	45	45			
Final set (min) not more than	375	375	375	375	375	375	375	375

[a]Attention is called to the fact that cements conforming to the requirements for all of these types may not be carried in stock in some areas. In advance of specifying the use of other than Type I cement, it should be determined whether the proposed type of cement is or can be made available.

[b]Compliance with the requirements of this specification does not necessarily ensure that the desired air content will be obtained in concrete.

[c]Either of the two alternative fineness methods may be used at the option of the testing laboratory. However, in case of dispute, or when the sample fails to meet the requirements of the air-permeability test, the turbidimeter test shall be used, and the requirements in this table for the turbidimeter method shall govern.

[d]The strength at any age shall be higher than the strength at any preceding age.

[e]When the optional heat of hydration on the chemical limit or the sum of the tricalcium silicate and tricalcium aluminate is specified.

[f]The purchaser should specify the type of setting-time test required. In case it is not specified or in case of dispute, the requirements of the Vicat test only shall govern.

PORTLAND-POZZOLAN CEMENTS

Portland-pozzolan cements are hydraulic cements consisting of an intimate and uniform blend of portland cement or portland blast-furnace-slag cement and fine pozzolan produced either by intergrinding portland-cement clinker and pozzolan, by blending portland cement or portland blast-furnace-slag cement and finely divided pozzolan, or a combination of intergrinding and blending, in which the pozzolan constituent is between 15 and 40 percent by weight of the portland-pozzolan cement.

Portland-pozzolan cements consists of two types: Type IP, and Type P, each with three optional provisions: Type IP portland-pozzolan cement and option Type IP-A air-entrained portland-pozzolan cement are used in general concrete construction. Type P portland-pozzolan cement and option Type P-A air-entrained portland-pozzolan cement are used in concrete construction where high strengths at early stages are not required. For Types IP and IP-A cement, moderate sulfate resistance or moderate heat of hydration or both may be specified by adding the suffixes (MS) or (MH) or both to the selected type designation. For Types P and PA, moderate sulfate resistance or low heat of hydration or both may be specified by adding the suffixes (MS) or (LH) or both to the selected type designation.

Portland-pozzolan cements require more water for a given consistency but exhibit greater shrinkage upon drying. Further, they exhibit less strength prior to 28 days of curing, but greater strength after 28 days of curing when compared to normal portland cement concrete.

Portland-pozzolan cements may be used for mass concrete where mass and weight are more important than strength. It also exhibits excellent sulfate resistance and hence is good for seawalls. Portland-pozzolan cements are also used in dam construction because of their low heat of hydration.

PORTLAND BLAST-FURNACE-SLAG CEMENT

Portland blast-furnace slag cement, according to ASTM C595 (Blended Hydraulic Cements), is a hydraulic cement consisting of an intimate and uniform blend of portland cement and fine, granulated blast-furnace slag produced either by intergrinding portland cement clinker and granulated blast-furnace slag or by blending portland cement and finely ground granulated blast-furnace slag, in which the slag constituent is between 25 and 70 percent of the weight of portland blast-furnace-slag cement.

Portland blast-furnace-slag cement exists as one type, with three optional provisions. Type IS portland blast-furnace-slag cement is for use in general concrete construction. Option Type IS-A, air-entrained portland blast-furnace-slag cement, is also used in general concrete construction. Moderate sulfate resistance or moderate heat of hydration or both may be specified by adding the suffixes (MS) or (MH) or both to the selected type designation.

ALUMINA CEMENTS

Alumina cement has a high alumina content because it consists primarily of calcium aluminates. Aluminate cement is sometimes referred to as "high-alumina cement." Alumina cements have a different chemical composition and constitution from portland cement. However, they have several outstanding properties, such as high early strength (usually setting and hardening to full strength at 48 hours, compared to 28 days for portland cement), excellent refractoriness, and good resistance against chemical attacks (hence, resistance to disintegrating action of seawater).

They are somewhat limited because elevated temperatures can produce permanent strength reductions when moisture is present.

The raw materials used for the manufacture of alumina cement are limestone and bauxite. The two are ground together and then placed in a kiln until the mixture melts at 2900°F (1600°C). The clinker is cooled, ground, and gypsum is added.

Alumina cements are used where high early strength is required and moderate temperatures are to be maintained.

EXPANSIVE CEMENTS

An *expansive cement* is a hydraulic cement containing a constituent that, during the process of hydration, setting, or hardening, undergoes an expansion (increase in volume) but remains sound and eventually develops into satisfactory strength. An expansive cement is normally used in a situation where shrinkage of the concrete cannot be tolerated; hence, it compensates for the shrinkage that will take place.

There are generally three types of expansive cements: Types M (Soviet), K (Klein), and S (Portland Cement Association). Expansive cements can achieve high strength because of the high-alumina or portland cement components.

SPECIAL PORTLAND CEMENTS

A variety of *special cements* exists (white portland, colored cements, oil-well cements, regulated cements, waterproofed cements, hydrophobic cements, antibacterial cements, barium and strontium cements) that are limited for specific uses and purposes. Each type will be mentioned briefly and the reader should keep in mind that this list is by no means complete.

White portland cement is used for decorative displays. It makes an excellent base when colored aggregates are used. White portland cement is low in iron and manganese, which gives the cement its white look, compared to normal portland cement, which is gray. Hence, by reducing iron and manganese in normal portland cement, a white cement is produced.

Colored cements are made by intergrinding a chemically inert pigment such as metallic oxide in the amount of 3 to 10 percent to portland cement. Colored cements, like white portland cement, are used for decorative purposes. However, they have one disadvantage in that they have a tendency to fade over the years.

Oil-well cements are slow-setting cements which are used to seal deep wells. The cement is made in a slurry and pumped to depths within the well under high temperature and pressure before it is allowed to set. These types of cements are governed by the American Petroleum Institute for each of eight classes.

Regulated cements are rapid-setting and -hardening cements. They are used in the manufacture of blocks, pipes, prestressed and precast concrete, and, of course, for patch work. In strength, they are comparable to portland cement Types I, II, and III.

A *waterproofed cement* is a portland cement interground with a water-repellent material, such as calcium stearate. The purpose is to reduce the water permeability of the concrete.

Hydrophobic cements are similar to waterproofed cements, in that portland cement is interground with a hydrophobic (water-repellent) material. However, the purpose is to prolong the life of the cement during storage or while it is being transported long distances.

An *antibacterial cement* is a portland cement interground with an antibacterial agent with the intention of reducing harmful microorganisms. It is used in food-processing plants to minimize deterioration caused by fermentation.

Barium and strontium cements are portland cements in which the calcium oxide is replaced completely or in part by barium oxide or strontium oxide. Their purpose is to act as a concrete shield in which the barium and strontium absorb x-rays and gamma rays.

PROBLEMS

3.1. Explain the difference between hydraulic and nonhydraulic cements.

3.2. Why is lime important in the manufacturing process of portland cement?

3.3. Discuss the uses of the following: pozzolan cements, slag cements, natural cements, portland cements.

3.4. List eight types of portland cement and explain their uses.

3.5. Explain how the following compounds affect the character of portland cement: (a) tricalcium silicate, (b) dicalcium silicate, (c) tricalcium aluminates, (d) tetracalcium aluminoferrite, and (e) alumina.

3.6. Why are portland-pozzolan cements important?

3.7. What special property of alumina cement makes its use attractive?

3.8. List several special cements and discuss their uses.

3.9. Overall, which types of cements do you view as exceptional? Explain.

3.10. After reviewing outside reference materials, discuss the manufacturing process of portland cements.

REFERENCES

American Society for Testing and Materials, *Book of Standards,* Part 14, 1979.

BAUER, E. E., *Plain Concrete,* 3rd ed., McGraw-Hill Book Company, New York, 1949.

BROWN, L. S., "Tricalcium Aluminate and the Microstructure of Portland Cement Clinker," *Proc. ASTM, 37,* Part II (1937).

DAVIS, R. E., KELLEY, J. W., TROXELL, G. E., and DAVIS, H. E., "Properties of Mortars and Concretes Containing Portland-Pozzolan Cement," *Am. Concr. Inst, 32* (Sept.–Oct. 1935), p. 80.

KREBS, R. D., and WALKER, R. D., *Highway Materials,* McGraw-Hill Book Company, New York, 1971.

LARSON, T. D., *Portland Cement and Asphalt Concretes,* McGraw-Hill Book Company, New York, 1963.

MILLS, A. P., HAYWARD, H. W., and RADER, L. F., *Materials of Construction,* 6th ed., McGraw-Hill Book Company, New York, 1955.

POPOVICS, S., *Concrete-making Materials,* McGraw-Hill Book Company, New York, 1979.

4

Strength
of Concrete

Whether used in buildings, bridges, pavements, or any other of its numerous areas of service, concrete must have strength, the ability to resist force. The forces to be resisted may result from applied loads, from the weight of the concrete itself, or, more commonly, from a combination of these. Therefore, the strength of concrete is taken as an important index of its general quality. Hence, tests to determine strength are undoubtedly the most common type made to evaluate the properties of hardened concrete. There are three reasons for this: (1) the strength of concrete, in compression, tension, shear, or a combination of these, has, in most cases, a direct influence on the load-carrying capacity of both plain and reinforced structures; (2) of all the properties of hardened concrete, those concerning strength can usually be determined most easily; and (3) by means of correlations with other more complicated tests, the results of strength tests can be used as a qualitative indication of other important properties of hardened concrete.

The results of tests on hardened concrete are usually not known until it would be very difficult to replace any concrete that is found to be faulty. These tests, however, have a policing effect on those responsible for construction and provide essential information in cases when the concrete forms a vital structural element of any building. The results of tests on hardened concrete, even if they are known late, help to disclose any trends in concrete quality and enable adjustments to be made in the production of future concrete.

COMPRESSIVE STRENGTH

Significance of Compressive Strength

Concrete is used in many ways and is subject to a variety of different loading conditions, and so different types of stress develop. Often the dominant stress is compressive in nature, since concrete has long been known to exhibit its best strength characteristics when subjected to compressive loading. The compressive strength of concrete is one of its most important and useful properties and one of the most easily determined. The compressive strength of concrete is indicated by the unit stress required to cause failure of a test specimen. Concrete also exhibits tensile and shear strength, for which compressive strength is frequently used as a measure. The tensile strength of concrete is roughly 10 to 12 percent of the compressive strength, and the flexural strength of plain concrete, as measured by the modulus of rupture, is about 15 to 20 percent of the compressive strength.

In addition to being a significant indicator of load-carrying ability, strength is also indicative of other elements of quality concrete in a direct or indirect manner. In general, strong concrete will be more impermeable, better able to withstand severe exposure, and more resistant to wear. On the other hand, strong concrete may have greater shrinkage and susceptibility to cracking than a weaker material.

Finally, the concrete-making properties of the various ingredients of the mix are usually measured in terms of the compressive strength.

Specimens

Specimens to determine the compressive strength of concrete are generally obtained from four different sources: (1) cylinders made in the laboratory, (2) cylinders made in the field, (3) cores of hardened concrete cut from structures, and (4) portions of beams broken in flexure. Each type of specimen has a specific purpose or purposes.

ASTM C192 (Method of Making and Curing Concrete Compression and Flexure Test Specimens in the Laboratory) describes in detail methods for preparation and examination of the constituent material; proportioning and mixing of concrete; determining the consistency of the mix; and molding, curing, and capping of the specimens. Cylinders made in the laboratory constitute a large portion of the compression specimens. There are three reasons for this: (1) in research, to determine the effect of variations in materials or conditions of manufacture, storage, or testing on the strength and other properties of concrete; (2) as control tests in conjunction with (a) tests on plain or reinforced concrete members or structures or (b) tests to determine other properties of hardened concrete; and (3) to evaluate mix designs for laboratory field use. In the making of such specimens, large variations can be introduced into the results of the compression test if great care is not taken in the manufacture of the test specimen. These variations may be attributed to the character of the cement, conditions of mixing, character and grading of the aggregate, size of the aggregate, size and shape of the specimen, curing and aging, temperature, and moisture content at time of testing.

ASTM C31 (Method of Making and Curing Concrete Compression and Flexure

Test Specimens in the Field) describes the detailed method of making standard cylinders in the field. When cylinders are prepared in the field, they should be made from the same concrete used on the job. In addition, the same curing process, or a process as similar as possible, should be used. The purpose of cylinders made in the field may be to check the adequacy of the laboratory mix design, to determine when a structure may be put in service, or to measure and control the quality of the concrete.

Compressive test results of cored hardened concrete usually result in lower compressive strength than anticipated. ASTM C42 (Methods of Obtaining and Testing Drilled Cores and Sawed Beams of Concrete) covers the procedure for securing and testing the cylindrical cores that are most commonly used for determining compressive strength. Cores are only drilled when results of the standard cylinder test are questionable or when investigations are made of old structures.

Finally, ASTM C116 (Test for Compressive Strength of Concrete Using Portions of Beams Broken in Flexure) describes the procedure and apparatus necessary to determine the compressive strength of concrete from broken portions of beams tested in flexure. This test is extremely useful where beam specimens are made to determine the modulus of rupture, as in highway construction, of which one would like an appropriate value of the compressive strength. The method is not meant to be used as a comparison with laboratory cylinder tests. When the method is used and a correlation attempt is made, it becomes necessary to apply a correlation factor.

Making Specimens

The compressive strength of concrete depends primarily on the water-cement ratio. However, other factors, such as character of the cement, conditions of mixing, character and grading of the aggregate, size of the aggregate, size and shape of the specimen, curing and aging, temperature, and moisture content at the time of testing also have a bearing on the compressive strength.

The characterization of the cement for a given water-cement ratio plays an important role in the early compressive strength development of concrete. All portland cements behave more or less similarly, although the gain in strength with age is not always the same. Some cements gain their strength more rapidly at first, whereas others show greater increase at later periods. This applies not only to the eight types covered by the ASTM specifications, but to some extent to different cements within a single group. Tests have shown that the strength for a given water-cement ratio shows the greatest difference among cements at the early ages. For 90 days and later the differences are much less.

The importance of thorough mixing for the development of strength and for uniformity throughout the batch has long been recognized. The earliest studies in concrete showed increases in strengths with continued mixing, but the increase became slight after an initial rapid rise. The size of the batch, the type and consistency of the concrete, and the type of mixer are all involved in fixing the period in which gain in strength with time of mixing is significant. The time of mixing is governed by ASTM method C94.

Surface conditions, the size and shape of the particles, and the gradation are the characteristics of the aggregate that are of principal concern to the strength of the concrete.

The surface conditions of the aggregate affect the adhesion of the cement paste to the aggregate particles. The presence or absence of adherent dirt or clay, the roughness, and texture affect the adhesion. These characteristics have greater effect on flexural strength than on compressive strength.

The shape of the particles influence the strength of the concrete by affecting the quality and the amount of paste that is required for workability with a given mixture. Also, the bond with the cement paste may be weakened where relatively large surface areas of the flat pieces of aggregate occur, especially when they happen to be combined in planes of shear and tension.

When the water-cement ratio is the same and the mixtures are plastic and workable, considerable changes in grading will affect the strength of the concrete only to a small degree. The principal effect of changing the aggregate grading is to change the amount of cement and water needed to make the mixture workable with the desired water-cement ratio.

In general, as the maximum size of the aggregate is increased, lower water-cement ratios can be used for suitable workability and, therefore, greater strengths are obtained for a given cement content. In the high strength range, over 4500 psi (31 MPa), higher compressive strengths are usually obtained at a given water-cement ratio with smaller maximum sizes of aggregates. Data from compression tests of concrete containing very large aggregates, 4 in. (10.16 cm), are conflicting because of limitations in the size of test specimens.

Compression tests of concrete are ordinarily conducted on cylindrical specimens with height equal to twice the diameter, so that surface rupture, produced upon fracture, will not intersect the end bearings. The ends of the cylinders should be carefully formed to give parallel, smooth surfaces so as to obtain uniform distribution of stress. A uniformly stressed cylinder that has been properly molded will break in the shape of a double cone with vertex in the center of the cylinder.

It is generally accepted that the diameter of the specimen should be at least three times the nominal size of the coarse aggregate. A 6 in. × 12 in. (15.24 cm × 30.48 cm) cylinder is the standard for aggregate smaller than 2 in. (5.08 cm). If the aggregate is too large for the size of mold available, the oversize aggregate may be removed by set screening. If a mold having a diameter less than three times the maximum size of the aggregate is used, the indicated compressive strength will be lowered. However, a large mold may be used; in some cases, molds as large as 36 in. (91.44 cm) in diameter have been used for concrete containing very large aggregates, such as those used in dam construction. Attention must be called to the fact that the size of the cylinder itself affects the observed compressive strength; for example, the strength of a cylinder 36 in. (91.44 cm) in diameter by 72 in. (182.88 cm) high may be only about 82 percent of that of a standard 6 × 12 in. (15.24 cm × 30.48 cm) cylinder. A reduction in the size of the specimen below that of the standard 6 in. × 12 in. (15.24 × 30.48 cm) cylinder will yield a somewhat greater indicated compressive strength.

Unless the specimens are carefully molded, errant and irregular results will be obtained. Generally, a cylinder of poorly compacted concrete will have a lower strength than one that is properly compacted. Thus, it is necessary for the standard methods for making specimens, Methods C31 and C192, to specify procedures for compacting, rodding, or vibrating the concrete in the mold. If the specifications under which the work is being performed do not state the method of consolidation, the choice is determined by the slump. Concrete with a slump greater than 3 in. (7.62 cm) should be rodded. If the slump is between 1 and 3 in. (2.54 and 7.62 cm), the concrete may be either vibrated or rodded. When the slump is less than 1 in. (2.54 cm), the specimens must be consolidated by vibration. When the concrete is to be rodded, it should be placed in the cylinder in three layers and rodded 25 strokes per layer if the cylinder is 3 to 6 in. (7.62 to 15.24 cm) in diameter, 50 if 8 in. (20.32 cm), 75 if 10 in. (25.40 cm). When vibrating, the mold is filled and vibrated in two layers. Care must be exercised to vibrate only long enough to obtain proper consolidation. Overvibration tends to cause segregation. These methods are specified in order to permit reproducibility of results by different technicians.

Cylinder molds should be of nonabsorbent material and are generally steel; however, cardboard molds are often used in the field. ASTM C470 [Specification for Single-Use Molds for Forming 6 × 12 in. (15.24 × 30.48 cm) Concrete Compression Test Cylinders] defines adequate paper molds as well as lightweight sheet-steel molds. Although the cardboard is heavily paraffined, in most cases it absorbs part of the water in the concrete mixture. The use of cardboard molds may lower the observed compressive strength about 3 percent on the average; reductions as great as 9 percent have been noted.

Capping procedures are standardized under ASTM Method C617 (Capping Cylindrical Concrete Specimens). Any material that is sufficiently strong and can be molded can be used for capping. The most common capping materials are neat portland cement paste, high-strength gypsum plaster, and sulfur compounds. The cap should be as thin as possible and the plane surface of the cap at either end of the specimen should be truly at right angles to the axis of the cylindrical specimen. All surfaces that depart from a plane by more than 0.002 in. (0.005 cm) should be capped and all caps should be checked for this by means of a feeler gauge and steel straightedge. Slight irregularities can be fixed by scraping if the capping material has not set too hard. Further, according to the ASTM specification, the surface of the capping must not depart by more than 0.5 degree from perpendicularity with the axis of the cylinder, or a cant of 1 in 96.

If neat portland cement is used for capping, it should be mixed about 3 hours before use to minimize any harmful effect due to shrinkage. The capping procedure should be carried out at least 3 days before testing.

Gypsum plaster can, according to the ASTM specification, be used for capping, provided that it has a strength of over 5000 psi (31 MPa) when tested as a 2-in. (5.08-cm) cube. Suitable mixtures of sulfur (melted) and granular materials applied about 2 hours prior to testing are also recommended, but care is necessary to avoid overheating the melted mixture to prevent loss of rigidity.

Concrete can gain in strength only as long as moisture is available and used for hydration. The term "curing" is used in reference to the maintenance of a favorable environment for the continuation of the chemical reactions that take place. It is through the early curing process that the internal structure of the concrete is built up to provide strength and water tightness. While simply retaining moisture within the concrete may be sufficient for low to moderate cement contents, mixes that are rich in cement generate considerable heat of hydration, which may expel moisture from the concrete in the period immediately after setting. The standard curing conditions require that the specimen be held at a temperature of 73.4 ± 3°F (23 ± 1.7°C) and in the "moist conditions" until the time of the test (ASTM C192). Any variation from this procedure may produce a specimen having a different strength from that which would be produced under standard conditions.

Cylinders to be used for quality control should be cured according to the standard conditions; however, cylinders made in the field and tested to measure the strength of the concrete in the structure should be cured in the same manner as the structure. Two methods of field curing are presently used: (1) those which interpose a source of water, in the form of ponding, or a wet material to prevent or counteract evaporation; and (2) those which minimize loss of water by interposing an impermeable medium or by other means. A third method of curing that is used in the manufacture of concrete products is the artificial application of heat while the concrete is maintained in a moist condition.

Curing and aging cannot be separated; an increase in age provides for further chemical combinations if the conditions are favorable for continued reaction. Provided that favorable conditions are met, concrete gains strength with age.

Temperature also plays an important role in the curing process of concrete. The chemical reactions proceed more rapidly at higher temperatures. Tests of specimens sealed against loss of moisture show higher early strength, but lower strengths at later ages as the temperature is increased in the range of 40 to 115°F (4.4 to 26.1°C). The U.S. Bureau of Reclamation has found that for job control, specimens cured at 70°F (21.2°C) or lower temperatures at the time of casting and for a few hours thereafter give higher strengths at 1 to 3 months. The rapid stiffening in the first few hours under the higher temperature is apparently detrimental to the later development in strength.

Test Procedure

Once the specimen is made, the method by which it is tested affects the strength obtained. Two of the more important influences are the rate of loading and the eccentricity of loading.

The rate of loading has a definite effect on compressive strength, although the effect is usually fairly small over the ranges of speed used in ordinary testing. The results of tests on concrete indicate that the relationship between strength and rate of loading is approximately logarithmic; the more rapid the rate, the higher the indicated strength. A rapid rate of loading may indicate as much as a 20 percent increase in the apparent compressive strength. For this reason, ASTM method C39, which applies also to testing of cores, specifies that the rate of loading for screw-powered machines shall be 0.05 in./1 min (0.11 cm/min) and for hydraulic machines 20 to 50 psi/sec (1.41 to 3.52 kg/cm^2/sec).

The effect of eccentric loading is obvious and the alignment of all machines should be checked. Any eccentricity will tend to decrease the strength of the test specimen, the amount of decrease being greater for low-strength than for high-strength concrete. Therefore, to ensure that a concentric and uniformly distributed load is applied to the specimen, a spherically seated bearing block is required on one end, and the specimen should be carefully centered on this bearing block. The object of using the block is to overcome the effect of a small lack of parallelism between the head of the machine and the end face of the specimen, giving the specimen as even a distribution of initial load as possible. It is desirable that the spherically seated bearing block be at the upper end of the test specimen. In order that the resultant of the forces applied to the end of the specimen is not eccentric with the axis of the specimen, it is important that the center of the spherical surface of this block be in the flat face that bears on the specimen and that the specimen itself be carefully centered with respect to the center of this spherical surface. Due to increased frictional resistance as the load builds up, the spherically seated bearing cannot be relied upon to adjust itself to bending action that may occur during the test.

Significance of Results

The results of a compression test are essentially only comparative, as the value of the compressive strength obtained cannot be regarded as equal to the strength of the concrete deposited in the work. The value will only give an indication of the quality of concrete, because of the various factors that affect the mix, which have already been discussed. This lack of knowledge regarding the relationship between the strengths of concrete in a cylinder and in a structure requires the use of a larger factor of safety than would otherwise be necessary.

Compressive strength may be used as a qualitative measure of other properties of hardened concrete. No exact relationship exists between compressive strength and flexural strength, tensile strength, modulus of elasticity, wear resistance, fire resistance, or permeability. Only an approximation can be made of these properties. Nevertheless, this approximation is very useful to the engineer.

Compressive tests further aid in the selection of ingredients that may be used in making concrete. Compressive strength is a measure of the indirect effect of admixtures, which may be beneficial for one purpose but detrimental for another.

TENSILE AND FLEXURAL STRENGTH

Significance of Tensile and Flexural Strength

Flexural tension is most commonly developed in beams and slabs as the result of loads, temperature changes, shrinkage, and, in some cases, moisture changes. The case of simple uniaxial tension is rarely encountered in structures or members and can be obtained in laboratory tests only with care. However, significant principal tension stresses may be associated with multiaxial states of stress in walls, shells, or deep beams.

Because concrete that has to withstand tensile stresses is normally reinforced, its tensile strength has not received much attention, although it is of great importance in determining the ability of concrete to resist cracking due to shrinkage on drying and thermal movements. The tensile strength develops more quickly than the compressive strength and is usually about one-tenth the compressive strength at ages up to about 14 days, falling to about 5 percent at later ages. Cracking of concrete is usually a tensile failure, and this alone makes the tensile strength of concrete quite important.

Specimens

It is not easy to perform an axial tension test on concrete. It is difficult to apply the load truly axially and to grip the ends of the specimen without imposing high local stresses; a relatively large specimen must be used if a measurable load is to be applied, and a large number of specimens is required to ensure a reliable average. The beam test for flexural tension and the split-cylinder test are the simplest procedures for obtaining an indication of the tensile strength and are the tests usually performed.

Flexural tension tests may be made in several ways, the most common being ASTM C78 [Test for Flexural Strength of Concrete (Using Simple Beam with Third-Point Loading)]. ASTM C293 [Test for Flexural Strength of Concrete (Using Simple Beam with Center-Point Load)] is also used to a limited extent. The second method is intended for use with small specimens and not as an alternative to the test with third-point loading. The results of the flexural tests are expressed by the formula

$$R = \frac{Mc}{I} \tag{4.1}$$

where R = modulus of ruptures
M = maximum bending moment
c = one-half the depth of the beam
I = moment of inertia of the cross section

Another procedure for obtaining an indication of tensile strength is given in ASTM C496 (Test for Splitting Tensile Strength of Cylindrical Concrete Specimens). In this test a standard cylinder is loaded in compression on its side. Fracture occurs along the plane that includes both lines through which the load is applied. While high compressive stresses occur at the lines where load is applied, the plane on which fracture occurs is subjected largely to a uniform tensile stress. The splitting strength is calculated as follows:

$$T = \frac{2P}{\pi l d} \tag{4.2}$$

where T = splitting tensile strength
P = maximum applied load
l = length
d = diameter

Making Specimens

ASTM methods C192 and C31 describe the procedures for making flexural test specimens and cylinders for the splitting tensile strength test in the laboratory and in the field. ASTM method C31 stipulates that the length of the beam should be at least 2 in. (5.08 cm) longer than three times its depth and that its width should be not more than one and one-half times its depth. The minimum depth or width should be at least three times the maximum size of aggregate. A typical specimen used would be 6 in. × 6 in. × 21 in. (15.24 cm × 15.24 cm × 53.34 cm), and is tested under third-point loading on a span of 18 in. (45.72 cm). As the depth of the beam is increased, there is a decrease in the modulus of rupture.

Many of the parameters that affect the compressive strength of concrete also apply to the flexural strength.

Test Procedure

Flexural strength measurements are extremely sensitive to all aspects of specimen preparation and testing procedure.

The principal requirements of the supporting and loading blocks of the apparatus for the flexural strength of concrete are as follows: (1) they should be of such shape that they permit use of a definite and known length of span; (2) the areas of contact with the material under test should be such that unduly high stress concentrations (which may cause localized crushing around bearing areas) do not occur; (3) there should be provision for longitudinal adjustment of the position of the supports so that longitudinal restraint will not be developed as loading progresses; (4) there should be provision for some lateral rotational adjustment to accommodate beams that have a slight twist from end to end, so that torsional stresses will not be induced; and (5) the arrangement of parts should be stable under load.

Apparatus for measuring deflection should be so designed that crushing at the supports, settlement of the supports, and deformation of the supporting and loading blocks or parts of the machine do not introduce serious errors into the results. One method of avoiding these sources of errors is to measure deflections with reference to points on the neutral axis above the supports.

Routine flexure tests are usually simple to conduct. Ordinarily, only the modulus of rupture is required; this is determined from the load at rupture and the dimensions of the specimen. When the modulus of elasticity is required, a series of load-deflection observations are made.

The dimensions of the concrete specimen should be measured to the nearest 0.01 in. (0.025 cm). The supporting and loading blocks should be located with a reasonable degree of accuracy (0.2 percent of the span length). The supports and specimens should be placed centrally in the testing machine and checked to see that they are in proper alignment and can function as intended. Deflectometers and strainometers should be located carefully and checked to see that they operate satisfactorily and are set to operate over the range required.

The rate of load application, unless standardized, may cause considerable variation in the results of flexure tests, the variation being as much as 15 percent for the range of rates that may be obtained in the average laboratory. Concrete beams may be loaded rapidly at any desired rate up to 50 percent of the breaking load, after which loads should be applied at a rate such that the extreme fibers are stressed at 150 psi (0.93 MPa) or less per minute.

Beams may be tested under either center-point or third-point loading. Third-point loading invariably gives lower strengths than center-point loading. Tests indicate the following order of decreasing magnitude of the strength obtained: (1) center loading, with moment computed at center; (2) center loading, with moment computed at point of fracture; and (3) third-point loading. Third-point loading probably gives lower strengths because the maximum moment is distributed over a greater length of the beam; since the concrete is not homogeneous, this loading method seeks the weakest section.

Curing affects the tensile strength in much the same manner as it affects the compressive strength. A beam that has been allowed to dry before testing will yield lower flexural strength than one tested in a saturated condition. Consequently, in tests used to determine or control the quality of concrete, uniformity of results will be assured only if the beams are cured in a standard manner and tested when wet.

The temperature of a beam at the time of testing will also affect the results. As the temperature increases, the strength decreases.

ASTM method C78 is also prescribed for tests of beams sawed from hardened concrete. When such beams are used, primarily as a control of concrete quality, they should be turned on their sides before testing and will usually require capping because of the irregularity of the sawed surface (the sides are not plane and parallel).

The test for splitting tensile strength described in method C496 is simple to make. The effectiveness with which the material in the bearing strips is able to conform to the irregularities of the surface of the specimen and distribute the load affects the results. For uniformity, method C496 specifies that $1/8$-in. (0.32-cm)-thick plywood should be used for bearing strips. Care must be taken to apply the load through a diametrical plane. The load should be applied such that the stress increases between 10 and 200 psi/min (7.03 and 14.1 kg/cm^2/min).

Significance of Results

In the case of pavement construction, it appears that the flexural strength of concrete is at least as important as the compressive strength. It has been suggested that pavements be designed on the basis of flexural strength. The design procedure would be exactly the same except that the tables would have to be prepared on the basis of flexural strength instead of compressive strength. Many agencies that are involved in pavements make only flexural tests. The results of these tests give an indication of when the concrete has gained sufficient strength so that load may be applied or the forms removed.

Splitting tension tests have been used to evaluate the bond-splitting resistance of concrete.

With increased emphasis on the control of cracking in reinforced concrete, appreciation of the importance of tensile strength has increased. However, the three general methods of estimating tensile strength give slightly different results. The splitting-tensile-strength test is the easiest to perform and gives the most uniform results. The tensile strength from the splitting test is about $1^1/_2$ times greater than that obtained in a direct tension test and about two-thirds of the modulus of rupture.

SHEARING AND TORSIONAL STRENGTH

Significance of Shearing Strength

The shearing strength of concrete is a very important property of the material, as it is the primary determining factor in the compressive strength of short columns. However, under certain conditions the strength of concrete beams also depends upon the shearing strength of the material.

The average strength of concrete in direct shear varies from about 0.5 of the compressive strength for rich mixtures to about 0.8 of the compressive strength for lean mixtures.

Torsion

The application to a concrete specimen of torsion alone produces pure shearing stresses on certain planes. However, the case of pure shear acting on a plane is seldom, if ever, encountered in actual structures. Further, failure under these conditions will occur in tension rather than shear. The strength of concrete subjected to torsion is therefore related to its tensile strength rather than to its shearing strength.

COMBINED STRESSES

Significance of Combined Stresses

Concrete in structures is almost never subjected to a single type of stress. Just as nearly all structural members are acted upon by various combinations of moments, shear, and axial load, the concrete in them is usually subjected to some combination of compressive, tensile, and shearing stresses.

ASTM C801 (Recommended Practice for Determining Mechanical Properties of Hardened Concrete under Triaxial Loads) is useful in determining the strength and deformation characteristics of concrete, such as shear strength at various lateral pressures, angle of shearing resistance, strength in pure shear, deformation modulus, and creep behavior. Prior to the application of this relatively new method, extensive research was conducted on cylinders with combinations of axial tension and lateral compression, and torsion and axial compression.

Results of Tests

There are no universally accepted criteria of failure of concrete; hence, there is no single correct way of discussing the behavior of concrete under combined stresses and there are many ways of presenting the results. ASTM method C801 allows four ways in which the data may be presented:

1. Graphical plots of the following equations:

$$f_1 = f'_c + K(f_3)^a \tag{4.3}$$

or, for the strength increase beyond the uniaxial strength:

$$f_1 - f'_c = K(f_3)^a \tag{4.4}$$

where f_1 = largest principal stress
$\quad\ f_3$ = smallest principal stress
$\quad\ f'_c$ = unconfined compressive strength
$\quad\ K,a$ = empirical coefficients

2. A graphical plot of the stress difference versus axial strain. Stress difference is defined as the maximum principal axial stress minus the minimum principal stress. The value of the minimum principal stress should be indicated on the curve.
3. A graphical plot of axial stress versus axial strain for different confining pressures.
4. Mohr stress circles constructed on an arithmetic plot with shear stresses as ordinates and normal stresses as abscissas. At least three triaxial compression tests, each at a different confining pressure, should be made on the same material to define the envelope to the Mohr stress circle.

FATIGUE STRENGTH

Concrete will, when subjected to repeated load, fail at a load smaller than its static ultimate strength. Early work on fatigue was conducted on low-strength concrete. Thus, in the early classic work of Probst, the strength of the concrete tested was only just over 2000 psi (12.4 MPa). His results nevertheless agree with more recent work conducted on concretes of much higher strength. Probst discovered that there was a critical stress below which concrete, subjected to repeated loading, increased in strength and elasticity. His minimum stress was 5 percent of the ultimate and the maximum or critical stress ranged from 47 to 60 percent of the ultimate. Further tests have shown that the fatigue strength at 10 million cycles is about 55 percent of the static flexural strength. The fatigue strength in axial compression is about 55 percent of the static compressive strength. No data are available to establish the fatigue strength in axial tension, in shear, or under combined stresses. Concrete does not have a fatigue or endurance limit, at least at less than 10 million cycles of load, as most body-centered-cubic metals do. Failure under repeated loads is especially important in pavement design.

Frequent rest periods during a fatigue test may raise the fatigue strength as much as 9 percent higher than if there had been no rest periods. The fatigue strength increases as the rest periods are increased in duration to 5 minutes. No additional gain is obtained for longer rest periods.

Tests to determine the fatigue strength of concrete should be made on specimens as large as possible in order to decrease the influence of lack of homogeneity. The cross section of the specimens should be at least three times the maximum nominal size of the aggregate, and even larger dimensions might be desirable in some cases. Because of the large size of specimens used, the large testing machines required are usually capable of applying load at a rate ranging from only a few cycles a day to 500 cycles per minute. Thus, it takes a minimum of 2 weeks to several months to apply as many as 10 million cycles of load. Because of the time involved, the specimens are generally aged and air-dried before being tested in order to prevent gain of strength during the test.

PROBLEMS

4.1. Why are tests to determine the strength of concrete the most frequently used to evaluate the properties of hardened concrete?

4.2. What is the significance of the compressive strength of concrete?

4.3. How are the specimens obtained for the compression test of concrete?

4.4. The compressive strength of concrete depends upon many factors. List them and discuss each in detail.

4.5. In the testing of concrete in compression, what factors can affect the results of the test? Explain.

4.6. What is the significance of flexural strength?

4.7. What is splitting tensile strength?

4.8. What factors during testing affect the final results of the flexure test?

4.9. Explain the significance of the shearing strength; the combined stresses.

4.10. Why is fatigue strength important?

REFERENCES

BERGSTROM, S. G., "Curing Temperature, Age and Strength of Concrete," *Constr. Rev.*, 27, No. 3 (July 1954).

"Curing Concrete Specimens," U.S. Bureau of Reclamation, Spec. Rep. 16, 1954.

DAVIS, H. E., TROXELL, G. E., and WISKOCIL, C. T., *The Testing and Inspection of Engineering Materials*, McGraw-Hill Book Company, New York, 1964.

"Effects of Wet-Screening to Remove Large Size Aggregate Particles on the Strength of the Concrete," Corps Eng., Ohio River Div. Lab., Mariemont, Ohio, Jan. 1953.

FORSSBLAD, L., "Investigations of Internal Vibration of Concrete," *Acta Polytech. Scand., Civil Eng. Build. Constr. Ser. No. 29,* Stockholm, (1965).

GOLDBECK, A. T., "Apparatus for Flexural Tests of Concrete Beams," Report of Committee C-9 on Concrete and Concrete Aggregates, Appendix VIII, *Proc. ASTM, 30,* Part 1 (1930), p. 591.

GONNERMAN, H. F., "Effect of End Condition of Cylinder in Compression Tests of Concrete," *Proc. ASTM, 24,* Part II (1924), p. 1036.

KEIL, F., "Cements," *3rd Int. Symp. Chem. Cement,* London, Sept. 1952.

KELLERMAN, W. F., "Effect of Size of Specimen, Size of Aggregate, and Method of Loading upon the Uniformity of Flexural Strength Tests," *Public Roads, 13,* No. 11 (Jan. 1933), p. 177.

KENNEDY, T. B., "A Limited Investigation of Capping Materials for Concrete Test Specimens," *Proc. Am. Concr. Inst., 41* (1944), p. 117.

KESLER, C. E., "Effect of Length to Diameter Ratio on Compressive Strength-An ASTM Cooperative Investigation," *Proc. ASTM, 59* (1959), p. 1216.

KESLER, C. E., " Statistical Relation between Cylinder, Modified Cube, and Beam Strength of Plain Concrete," *Proc. ASTM, 54* (1954), p. 1178.

KESSLER, C. E., "Strength of Hardened Concrete," "Significance of Tests and Properties of," in "Concrete and Concrete Making Materials," *ASTM Spec. Tech. Publ. 169A,* 1966.

KLIEGER, P., "Effect of Mixing and Curing Temperature on Concrete Strength," *Proc. Am. Concr. Inst., 54* (1958), p. 1063.

KREBS, R. D., and WALKER, R. D., *Highway Materials,* McGraw-Hill Book Company, New York, 1966.

LEA, F. M., "Modern Developments in Cement in Relation to Concrete Practice," *J. Inst. Civil Eng.* (Feb. 1943).

LECAMUS, B., "Research of Fatigue Strength," *J. Am. Concr. Inst., 63,* No. 1 (Jan. 1966).

MATHER, B., "Effect of Type of Test Specimen on Apparent Compressive Strength of Concrete," *Proc. ASTM, 45* (1945), p. 802.

McMILLAN, F. R., "Suggested Procedure for Testing Concrete in Which the Aggregate Is More than One-Fourth the Diameter of the Cylinders," *Proc. ASTM, 30,* Part I (1930), p. 521.

"Methods of End Conditions before Capping upon the Compressive Strength of Concrete Test Cylinders," *Proc. ASTM, 41* (1941), p. 1038.

MILLS, A. P., HAYWARD, H. W., and RADER, L. F., *Materials of Construction,* John Wiley & Sons, Inc., New York, 1955.

MURDOCK, J. W., "A Critical Review of Research on Fatigue of Plain Concrete," *Bull. 975,* Eng. Exp. Sta., Univ. Ill., Urbana, Ill., 1965.

NASH, J. P., "Tests of Concrete Road Aggregates," *Proc. ASTM, 17,* Part II (1917), p. 394.

NEVILLE, A. M., *Properties of Concrete,* John Wiley & Sons, Inc., New York, 1963.

NIELSON, K. E. C., "Effect of Various Factors on the Flexural Strength of Concrete Test Beams," *Mag. Concr. Res.,* No. 15 (Mar. 1954), p. 105.

NORDLY, G. M., "Fatigue of Concrete—A Review of Research," *Proc. Am. Concr. Inst., 55* (1959), p. 191.

OORCHARD, D. F., *Concrete Technology,* John Wiley & Sons, Inc., New York, 1973.

PRICE, W. H., "Factors Influencing Concrete Strength," *Proc. Am. Concr. Inst., 47* (1951), p. 417.

PROBST, E., "Further Investigations of Alternating Loads on Concrete," *Cement Concr. Res. 31* (Mar. 1942).

PROBST, E., "The Influence of Rapidly Alternating Loading on Concrete and Reinforced Concrete," *Struct. Eng., 9,* No. 10 (Oct. 1931), and No. 12 (Dec. 1931).

RICHART, F. E., BRANDIZAEG, A., and BROWN, R. L., "A Study of the Failure of Concrete under Combined Compressive Stresses," *Bull. 185,* Eng. Exp. Sta., Univ. Ill., Urbana, Ill., 1928.

SHANK, J. R., "Plastic Flow of Concrete at High Overload," *Proc. Am. Concr. Inst., 45* (1949), p. 493.

SHIDELER, J. J., and MCHENRY, D., "Effect of Speed in Mechanical Testing of Concrete," in "Speed of Testing of Nonmetallic Materials," *ASTM Spec. Tech. Publ. 185,* 1955.

SMITH, F. C., and BROWN, R. Q., "The Shearing Strength of Cement Mortar," *Bull. 106,* Eng. Exp. Sta., Univ. Wash., Seattle, Wash., 1941.

TALBOT, A. N., "Tests of Concrete: I. Shear; II. Bond," *Bull. 8,* Eng. Exp. Sta., Univ. Ill., Urbana, Ill., 1906.

TALBOT, A. A., and RICHART, F. E., "The Strength of Concrete," *Bull. 137,* Eng. Exp. Sta., Univ. Ill., Urbana, Ill., 1923.

TUTHILL, L. H., and DAVIS, H. E., "Over-vibration and Re-vibration of Concrete," *Indian Concr. J., 20* (1946).

WALKER, S., and BLOEM, D. L., "Studies of Flexural Strength of Concrete, Part 2: Effects of Curing and Moisture Distribution," *Proc. Highway Res. Board, 36* (1957), p. 334; "Part 3: Effects of Variations in Testing Procedures," *Proc. ASTM, 57* (1957), p. 1122.

WATERS, T., "The Effect of Allowing Concrete to Dry before It Has Fully Cured," *Mag. Concr. Res., 7,* No. 20 (July 1955).

WATSTEIN, D., "Effect of Straining Rate on the Compressive Strength and Elastic Properties of Concrete," *Proc. Am. Concr. Inst., 49* (1953), p. 729.

WERNER, G., "The Effect of Type of Capping Material on the Compressive Strength of Concrete Cylinders," *Proc. ASTM, 58* (1958), p. 1166.

WERNER, G., "The Effect of Type of Capping Material on the Compressive Strength of Concrete Cylinders," Report of the Bureau of Public Roads presented to the Sixty-first Annual Meeting of the ASTM, June, 1958.

WRIGHT, P. J. F., "Comments on an Indirect Tensile Test on Concrete Cylinders," *Mag. Concr. Res.,* No. 20 (1955), p. 87.

WRIGHT, P. J. F., and GARWOOD, F., "The Effect of the Method of Test on the Flexural Strength of Concrete," *Mag. Concr. Res.,* No. 15 (Mar. 1954), p. 105.

WRIGHT, P. V. F., "The Design of Concrete Mixes on the Basis of Flexural Strength," *Proc. Symp. Mix Design Quality Control Concr.,* Cement and Concrete Association, London, May 1954.

5

Design Procedure in Making Concrete

Concrete is a composite material made up of inert materials of varying sizes which are bound together by a binding medium. Mortar is made up of a mixture of cement, water, air, and fine aggregate. Concrete contains coarse aggregate in addition to cement, water, air, and fine aggregate. The cement, water, and air combine to form a paste that binds the aggregates together. Thus, the strength of the concrete is dependent on the strength of the aggregate-matrix bond. The entire mass of the concrete is deposited or placed in a plastic state and almost immediately begins to develop strength (harden), a process which, under proper curing conditions, may continue for years. Because concrete is initially in a plastic state, it lends itself to all kinds of construction, regardless of size or shape. One drawback is that the concrete in a plastic condition must be placed within forms, and these forms cannot be removed until the concrete has hardened somewhat.

In types of work where concrete is used to counteract compressive stresses, it is an excellent building material. However, if the concrete must counteract tensile stresses, it must be reinforced with steel, as concrete is weak in tension.

CONCRETE MATERIALS

Cement

Usually, portland cement is specified for general concrete construction work and should conform to the standard specifications of the ASTM. From time to time special cements may be required if the project involves unusual requirements. Chapter 3 discussed the various types of portland cement as well as special cement requirements. The reader is directed to Chapter 3 for a description of the specifications established by the ASTM and also to review the importance and uses of special cements.

Water

There is usually very little trouble in obtaining water for use in concrete. Almost any water that is drinkable may be used to make concrete. Drinking water with a noticeable taste or odor should not be used until it is tested for organic impurities. Impurities in mixing water may cause any one or all of the following:

1. Abnormal setting time.
2. Decreased strength.
3. Volume changes.
4. Efflorescence.
5. Corrosion of reinforcement.

Some of the impurities in mixing water that cause these undesirable effects in the final concrete are:

1. Dissolved chemicals.
2. Seawater.
3. Sugar.
4. Algae.

Dissolved chemicals may either accelerate or retard the set and can substantially reduce the concrete strength. Further, such dissolved chemicals can actively attack the cement-sand bond, leading to early disintegration of the concrete.

Seawater containing less than 3 percent salt is generally acceptable for plain concrete but not for reinforced or prestressed concrete. The presence of salt can lead to corrosion of the reinforcing bars and prestressing of tendons.

If sugar is present in even small amounts, it can cause rapid setting and reduced concrete strength.

Algae can cause a reduction in the strength of concrete by increasing the amount of air captured in the paste and reducing the bond strength between the paste and the aggregate.

Aggregates

The requirements for fine and coarse aggregate for concrete construction were described in Chapter 2.

PRINCIPAL REQUIREMENTS FOR CONCRETE

In the design of concrete mixes, three principal requirements for concrete are of importance:

1. Quality.
2. Workability.
3. Economy.

Quality

The *quality* of concrete is measured by its strength and durability. Hardened concrete must have sufficient strength to resist the stresses from loads as well as the stresses created by its own weight. The compressive and the flexural strength of concrete are both important in the design of concrete structures. The principal factors affecting the strength of concrete, assuming sound aggregates, are the water-cement ratio and the extent to which hydration has progressed. *Hydration* is the chemical reaction that takes place between the water and cement while the concrete is hardening. The strength of concrete at the end of 28 days is the generally accepted standard for evaluating the strength properties of concrete. Laboratory compressive strengths are usually obtained by testing cylinders 6 in. (15.24 cm) in diameter and 12 in. (30.48 cm) in height under gradually increased compressive loading until failure occurs. On the other hand, flexural strengths are generally obtained by loading a suitably configured test specimen transverse to the longitudinal axis at the third points, until failure occurs. That is, the force is applied simultaneously at one-half and two-thirds of the distance between the end supports.

Durability of concrete is the ability of the concrete to resist the forces of disintegration due to freezing and thawing and chemical attack. These two factors will be discussed in more detail in Chapter 7.

Workability

Workability of concrete may be defined as a composite characteristic indicative of the ease with which the mass of plastic material may be deposited in its final place without segregation during placement, and its ability to conform to fine forming detail. Workability is a term for which there is no perfectly satisfactory definition. The size and gradation of the aggregate, the amount of mixing water, the time of mixing, and the size and shape of the forms are all factors that affect workability.

No satisfactory measures of workability have been defined. However, the term *consistency* is generally considered a descriptive term for workability. Consistency measures the fluidity or the lack of it. The most generally used test for the measurement of consistency is ASTM C143 (Slump of Portland Cement Concrete).

Economy

Economy takes into account effective use of materials, effective operation, and ease of handling. The cost of producing good-quality concrete is an important consideration in the overall cost of the construction project.

INFLUENCES OF INGREDIENTS ON PROPERTIES OF CONCRETE

The amount of each ingredient used in concrete must be proportioned very carefully to produce the desired effects so that the concrete may be used for its intended purposes. If the proportions are not as designed, adverse effects may result. Table 5-1 shows the influence of each principal ingredient on the properties of concrete.

Aggregates

Almost any type of aggregate can be used for the making of concrete. The most commonly used aggregates are sand, gravel, crushed stone, and air-cooled blast-furnace slag. These aggregates produce normal-weight concrete ranging from 135 to 160 lb/ft^3. Shales, clay, slag, and slate can also be used in the making of concrete. These aggregates are used to make lightweight concrete weighing between 85 and 115 lb/ft^3.

Aggregates exhibit a variety of physical and chemical characteristics; hence, their influence on concrete mixtures is varied. Physical characteristics include size and shape of the aggregate, surface texture, gradation, and top size of aggregate. Chemical characteristics of aggregates are those which may result in aggregate reactivity with the hardened concrete. In any concrete structure the maximum amount of aggregate should be used.

Water

In many concrete mix design procedures, water is considered an influential part of the total mix, as the water allows the concrete to be handled easily.

In other words, it allows the concrete mixture to be workable. Water is also important

TABLE 5-1 INFLUENCE OF INGREDIENTS
ON THE PROPERTIES OF CONCRETE[a]

Ingredient	Quality	Workability	Economy
Aggregate	Increases	Decreases	Increases
Portland cement	Increases	Increases	Decreases
Water	Decreases	Increases	Increases

[a]From W. A. Cordon, *Properties, Evaluation, and Control of Engineering Materials*, McGraw-Hill Book Company, New York, 1979.

from the standpoint of hydration. Too much water added to the mixture results in poor-quality concrete. Water reacts with the cement particles, resulting in a chemical change that binds the paste to the fine and coarse aggregate particles. The addition of water to the cement forms a paste that acts as a glue. If too much water is added, the glue becomes diluted, and this leads to a weak bond between the paste (matrix) and the aggregate. In addition to this, excess water produces segregation of the aggregate particles from the paste, and this results in a nonuniform mix.

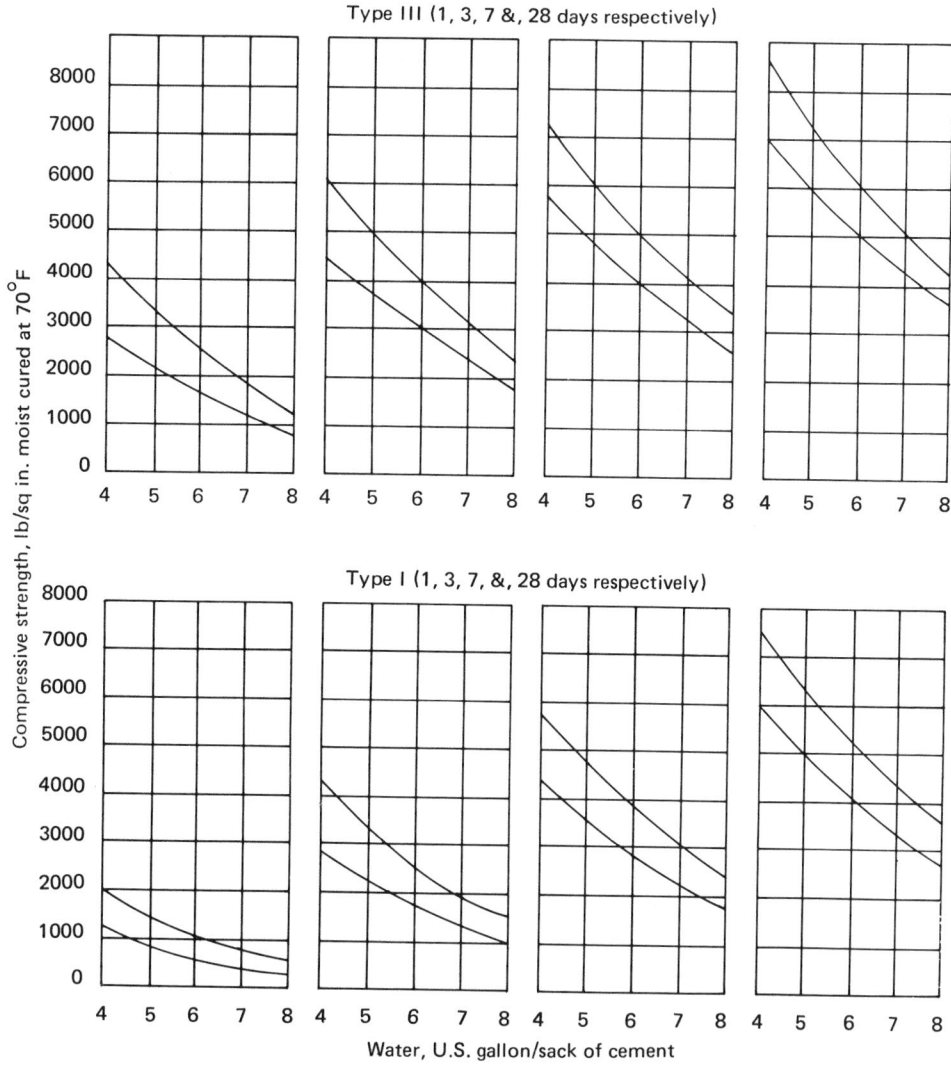

Figure 5-1 Age/Compressive Strength Relationships for Types I and III Portland Cement

TABLE 5-2 WATER CONTENTS SUITABLE FOR VARIOUS CONDITIONS OF EXPOSURE (GALLONS PER BAG OF CEMENT)

Types or Location of Structure	Severe or Moderate Climate: Wide Range of Temperatures, Rain and Long Freezing Spells or Frequent Freezing and Thawing					Mild Climate: Rain, or Rarely Snow or Frost				
	Thin Sections		Moderate Sections		Heavy and Mass Sections	Thin Sections		Moderate Sections		Heavy and Mass Sections
	Reinforced	Plain	Reinforced	Plain	Plain	Reinforced	Plain	Reinforced	Plain	Plain
At the water line in hydraulic or waterfront structures or portions of such structures where complete saturation or intermittent saturation is possible, but not where the structure is continuously submerged										
In seawater	5	5½	6	6		5	5½			
In fresh water	5½	6	6	6½		5½	6	6	6½	
Portions of hydraulic or waterfront structures some distance from the										

118

Condition						
water line, but subject to frequent wetting						
By seawater	5½	6	6	5½	6½	7
By fresh water	6	6½	6½	6	7	7½
Ordinary exposed structures, buildings, and portions of bridges not coming under groups above	6	6½	7	6	7	7½
Complete continuous submergence						
In seawater	6	6½	7	6	6¼	7
In fresh water	6¼	7	7½	6¼	7	7¼
Concrete deposited through water	a	a	5½	a	a	5½
Pavement slabs directly on ground						
Wearing slabs	5½	6	a	6	6½	a
Base slabs	6½	7	a	7	7½	a

Special cases:
(a) For concrete exposed to strong sulfate groundwaters or other corrosive liquids or salts, the maximum water content should not exceed 5 gal per bag.
(b) For concrete not exposed to the weather, such as the interior of buildings and portions of structures entirely below ground, no exposure hazard is involved and the water content should be selected on the basis of the strength and workability requirements.

[a]These sections not practicable for the purpose indicated.

Two little water produces a dry mix which easily crumbles under its design load, resulting in failure of the structure.

A normal bag of cement requires $2\frac{1}{2}$ to 3 gallons of water to produce a mix that hydrates at the appropriate rate, resulting in a concrete mix of good quality.

Portland Cement

When water is added to portland cement a chemical reaction (hydration) takes place and a calcium silicate hydrate is produced. The amount of water needed to complete this reaction is approximately 30 percent of the cement by weight. However, if this is the only amount of water added to the mix, the mix will be stiff and unworkable. Therefore, additional water is added to the mix to make it more plastic and workable. The ratio of water to cement (w/c) determines the quality of the paste and controls the strength of the concrete. The w/c ratio is the most significant item affecting the strength of the concrete mix. A discussion of this fact was published by Duff A. Abrams in 1918 and is frequently called the *Abrams' water-cement ratio law*. Extensive research has proved this law to be valid, and graphs of such experimental work (Figure 5-1) are available for various types and combinations of materials. This type of graph is used as a beginning point in mix proportioning when an individual job design is not justifiable. Table 5-2 gives the recommended water-cement ratios for various types of structures and degrees of exposure.

PROPORTIONING CONCRETE MIXES

In proportioning concrete mixes we will briefly discuss four methods: the trial-batch method, job-curve method, the mortar-voids method, and the method of Goldbeck and Gray.

Trial-Batch Method

The purpose of the *trial-batch method* is to produce a given water-cement ratio such that the factors of quality, workability, and economy are balanced for the most desirable combination of aggregates. The size and gradation of the aggregate, surface texture of the fine and coarse aggregate, and the proportions of fine and coarse aggregate are the most important factors in determining the combination that will give the best quality and most desirable workability at the lowest cost.

A rough procedure for a typical trial-batch method of proportioning follows. Select the desired water-cement ratio. Weigh out a definite amount of cement and place it in the mixing apparatus, with the proper amount of water, to obtain the required water-cement ratio. Weigh out definite quantities of fine and coarse aggregate and place them in a container. Add fine and coarse aggregate from the weighed quantities to the cement-water paste until the desired consistency is obtained for the plastic state.

The remaining fine and coarse aggregate not used to make the plastic mixture is weighed and subtracted from the original quantities. The proportions of cement to sand

TABLE 5-3 RECOMMENDED MIX CONSISTENCIES FOR CEMENT

	Slump (in.)	
Type of Structure	Minimum	Maximum
Massive sections, pavement and floors laid on ground	1	4
Heavy slabs, beams, or walls	3	6
Thin walls and columns, ordinary slabs or beams	4	8

to coarse aggregate (by weight) may be changed to volumetric proportions by dividing the weight of each material by its unit weight.

To measure the consistency of the mix, the slump-cone test is utilized. Table 5-3 shows the recommended consistency for various classes of concrete structures.

After the desired proportion has been established and the desired consistency is satisfactory, further experimenting may be carried out by finding a desirable ratio of fine to coarse aggregate. Concrete mixtures with a low cement-sand mortar (one that does not fill the spaces between pebbles) are hard to work with and result in a honeycombed surface. A concrete mixture with a high cement-sand mortar produces a very porous concrete.

The trial-batch method may be used for making a batch of any size. The accuracy obtained is dependent on the care exercised in producing the batch.

Job-Curve Method

The *job-curve method* utilizes Abrams' water-cement-ratio law but allows variations due to the differences caused by cements and aggregates. Trial batches are prepared utilizing different water-cement ratios, using the cement and fine and coarse aggregate to be used on the job. Various test specimens are molded, cured, and tested. A job curve is plotted showing compressive strength versus gallons of water per sack of cement. If a strength of 3500 psi were required for a specific job after 28 days of curing, one would simply find 3500 psi on the job curve, read across until the job curve was intersected, and read down to the required number of gallons per sack of cement. A further trial batch would be run with the new data to refine the properties of the concrete and the necessary quantities of materials for the design mixture. Figure 5-2 shows a typical job curve in which the desired strength would be 3500 psi and the necessary water in gallons per sack of cement would be 6.

Mortar-Voids Methods

The *mortar-voids method* was developed by A. N. Talbot in 1923. His method involves the proportioning of concrete by the determination of voids in the mortar. It was his belief that the strength of concrete depends upon the composition of the cement paste

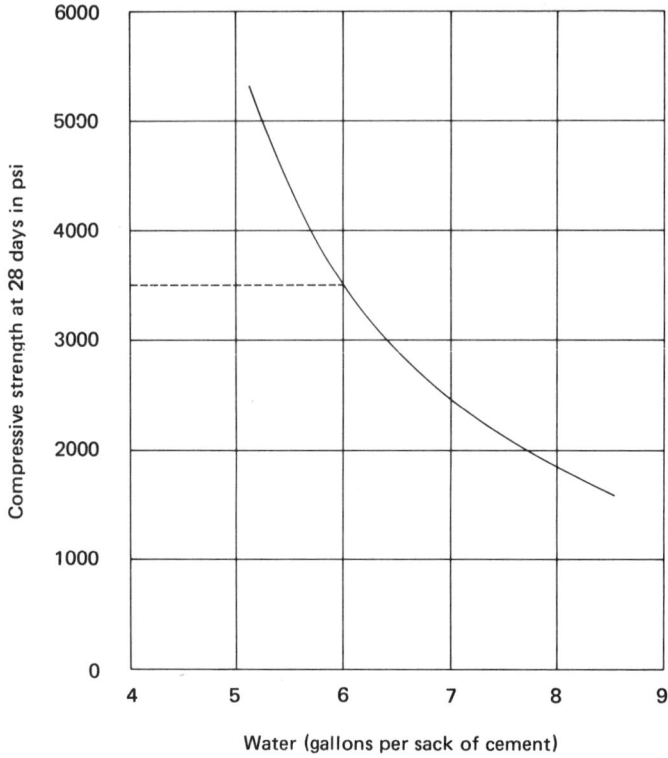

Figure 5-2   Typical Job Curve

(matrix). The matrix occupies all that space not occupied by the aggregate. Therefore, concrete is a composite of a matrix and an aggregate. The method is not used that much but deserves mentioning.

Method of Goldbeck and Gray

The *method of Goldbeck and Gray* has also been called the b/b_0 *method* and is sometimes referred to as the *ACI method*. This method is based on the absolute volumes of the materials in a unit volume of concrete. The method takes into account a b/b_0 ratio. The term b equals the solid volume of coarse aggregate in a unit volume concrete. The b_0 term equals the solid volume of coarse aggregate in a unit volume of dry-rodded coarse aggregate. The term b/b_0 equals the dry-rodded volume (bulk volume) of coarse aggregate in a unit of concrete. The procedure for designing concrete mixtures involves the use of data from Tables 5-4 to 5-6 which have been prepared by Goldbeck and Gray. These tables are utilized for typical materials. Table 5-7 is for pavement concrete and should be used for such.

The steps in performing a concrete mix design will be listed and then an example will be given. The steps are as follows:

1. Perform the following ASTM specifications on the fine and coarse aggregates.
   a. Determine the bulk specific gravities of the fine and coarse aggregate.
   b. Determine the dry-rodded unit weight of the coarse aggregate.
   c. Determine the gradation and fineness modulus of the fine aggregate and the gradation of the coarse aggregate.
2. Compute the solid weights per cubic foot of the cement and the fine and coarse aggregates. This is computed by multiplying the bulk specific gravity by 62.4 lb.
3. Pick the size of coarse aggregate desired along with the determined fineness modulus of sand and enter Table 5-4 to determine b/b_0.
4. Calculate the solid volume of coarse aggregate per cubic foot of dry-rodded materials.

$$b_0 = \frac{\text{dry-rodded weight per cubic foot}}{\text{solid weight per cubic foot}}$$

5. Calculate b, which is equal to $b/b_0 \times b_0$.
6. Knowing the kind and size of the coarse aggregate desired as well as the 28-day compressive strength desired and the chosen slump, determine from Table 5-5 (if non-air-entrained concrete is to be used) or Table 5-6 (if air-entrained concrete is to be used) the cement factor and the water content. Also select from the appropriate tables the percentage of entrapped air or optimum entrapped air and calculate its solid volume per cubic yard of concrete.
7. Determine and sum up the solid volumes of cement, coarse aggregate, water, and air.

TABLE 5-4 DRY-RODDED VOLUME b/b_0 OF COARSE AGGREGATE (ANY TYPE) PER UNIT VOLUME OF CONCRETE[a]

	Fine Sand		Medium Sand			Coarse Sand		
				Fineness Modulus of Sand				
Size of Coarse Aggregate (Square-opening Laboratory Sieves)	2.4	2.5	2.6	2.7	2.8	2.9	3.0	3.1
					Values[b] for b/b_0			
No. 4 to ½ in.	0.59	0.58	0.57	0.56	0.55	0.54	0.53	0.52
No. 4 to ¾ in.	0.66	0.65	0.64	0.63	0.62	0.61	0.60	0.59
No. 4 to 1 in.	0.71	0.70	0.69	0.68	0.67	0.66	0.65	0.64
No. 4 to 1½ in.	0.75	0.74	0.73	0.72	0.71	0.70	0.69	0.68
No. 4 to 2 in.	0.78	0.77	0.76	0.75	0.74	0.73	0.72	0.71
No. 4 to 2½ in.	0.80	0.79	0.78	0.77	0.76	0.75	0.74	0.73

[a]From A. T. Goldbeck and J. E. Gray, "A Method of Proportioning Concrete for Strength, Workability, and Durability," *Bull. 11*, National Crushed Stone Association, Nov. 1953.

[b]For concrete that is to be assisted in place by internal vibration under very rigid inspection, increase tabulated values of b/b_0 approximately 10%.

TABLE 5-5 NON-AIR-ENTRAINING STRUCTURAL CONCRETE CEMENT FACTORS REQUIRED FOR 28-DAY COMPRESSIVE STRENGTHS LISTED[a,b]

Size of coarse Aggregate Square-opening Laboratory Sieves):		*No. 4 to ½ in.*		*No. 4 to ¾ in.*		*No. 4 to 1 in.*		*No. 4 to 1½ in.*		*No. 4 to 2 in.*		*No. 4 to 2½ in.*	
Slump (in.):		*3*	*6*	*3*	*6*	*3*	*6*	*3*	*6*	*3*	*6*	*3*	*6*
Water[c] (gal yd³ of concrete)													
Angular coarse aggregate		42	44	40	42	38	40	36	38	35	37	34	36
Rounded coarse aggregate		38	40	36	38	34	36	32	34	31	33	30	32

28-Day Compressive Strength (psi)	*Cement (sacks/yd³ of concrete)*												
2000	4.6	4.8	4.4	4.6	4.2	4.4	4.0	4.2	3.9	4.0	3.8	3.9
2500	5.0	5.2	4.8	5.0	4.5	4.8	4.2	4.5	4.1	4.3	4.0	4.2
3000	5.4	5.7	5.2	5.4	4.9	5.2	4.6	4.9	4.4	4.7	4.3	4.6
3500	5.9	6.3	5.6	5.9	5.3	5.6	5.0	5.3	4.9	5.2	4.8	5.0
4000	6.5	6.9	6.2	6.5	5.8	6.2	5.5	5.8	5.4	5.7	5.2	5.5
4500	7.2	7.5	6.8	7.1	6.4	6.8	6.1	6.4	5.9	6.3	5.7	6.1
5000	8.1	8.5	7.7	8.1	7.3	7.7	6.9	7.3	6.7	7.1	6.5	6.9

Entrapped air (approx. %)	2.5		2		1.5		1		1		1	

[a]From A. T. Goldbeck and J. E. Gray, "A Method of Proportioning Concrete for Strength, Workability, and Durability," *Bull. 11*, National Crushed Stone Association, Nov. 1953.
[b]For concrete to be assisted in place by internal vibration, use 3-in. slump and decrease tabulated water contents by approximately 4 gal. No reduction in cement factor is suggested.
[c]This is water actually effective as mixing water.

8. Determine the solid volume of sand in a cubic yard of concrete by subtracting from 27 ft³ from item 7.

9. Convert solid volumes to pounds per cubic yard of concrete.

10. Calculate the additional water requirement.

The following example should illustrate the procedure that could be used to design a concrete mix and to calculate the batching weight.

Example 5.1.

Determine the batch quantities for a concrete mix design to have a 28-day compressive strength of 3000 psi using angular aggregate from a No. 4 sieve opening to ½ in. in

TABLE 5-6 AIR-ENTRAINING STRUCTURAL CONCRETE CEMENT FACTORS REQUIRED FOR 28-DAY COMPRESSIVE STRENGTHS LISTED[a,b]

Size of Coarse Aggregate Square-opening Laboratory Sieves):	No. 4 to ½ in.		No. 4 to ¾ in.		No. 4 to 1 in.		No. 4 to 1½ in.		No. 4 to 2 in.		No. 4 to 2½ in.	
Slump (in.):	3	6	3	6	3	6	3	6	3	6	3	6
Water[c] (gal yd³ of concrete) Angular coarse aggregate	38	40	36	38	34	36	32	34	31	33	30	32
Rounded coarse aggregate	35	37	33	35	31	33	29	31	28	30	27	29
28-Day Compressive Strength (psi)	Cement (sacks/yd³ of concrete)											
2000	4.4	4.7	4.2	4.4	3.9	4.2	3.7	3.9	3.6	3.8	3.5	3.7
2500	4.9	5.2	4.6	4.9	4.4	4.7	4.2	4.4	4.0	4.3	3.9	4.2
3000	5.6	5.9	5.3	5.6	5.0	5.3	4.7	5.0	4.5	4.8	4.3	4.7
3500	6.3	6.7	6.0	6.3	5.6	6.0	5.3	5.6	5.1	5.4	4.9	5.3
4000	7.2	7.5	6.8	7.2	6.4	6.8	6.0	6.4	5.8	6.2	5.6	6.0
4500	8.1	8.5	7.6	8.1	7.2	7.6	6.8	7.2	6.6	7.0	6.4	6.8
5000	9.2	9.7	8.7	9.2	8.2	8.7	7.7	8.2	7.4	8.0	7.2	7.7
Optimum entrained-air content (%)	6.0		6.0		5.5		5.0		5.0		4.5	

[a]From A. T. Goldbeck and J. E. Gray, "A Method of Proportioning Concrete for Strength, Workability, and Durability," *Bull. 11,* National Crushed Stone Association, Nov. 1953.
[b]This table should always be used to proportion concrete subject to freezing.
[c]This is water actually effective as mixing water.

size and a medium fine aggregate (sand) with a fineness modulus of 2.70. A slump of 3 in. is desired.

The material descriptions are as follows:

Coarse Aggregate (CA)

Size, No. 4 to ½ in., angular aggregate
Absorption (%) = 0.1
SG (bulk dry) = 2.7
SG (bulk saturated surface dry) = 2.73

TABLE 5-7 PAVEMENT CONCRETE USE: DRY-RODDED VOLUME OF COARSE AGGREGATE AND MIXING WATER REQUIRED[a]

Size of Coarse Aggregate (Square-opening Laboratory Sieve)	Fine Sand	Medium Sand				Coarse Sand			Water[b] (gal/yd^3 of concrete for 2-in. slump)			
									Non-air-entrained[c]		Air-entrained	
	2.4	2.5	2.6	2.7	2.8	2.9	3.0	3.1	Angular	Rounded	Angular	Rounded
	Fineness Modulus of Sand								Coarse Aggregate			
	Values for b/b_o											
No. 4 to 1 in.	0.75	0.74	0.73	0.72	0.71	0.70	0.69	0.68	35	31	31	28
No. 4 to 1½ in.	0.79	0.78	0.77	0.76	0.75	0.74	0.73	0.72	33	29	29	26
No. 4 to 2 in.	0.82	0.81	0.80	0.79	0.78	0.77	0.76	0.75	32	28	28	25
No. 4 to 2½ in.	0.84	0.83	0.82	0.81	0.80	0.79	0.78	0.77	31	27	27	24

[a]From A. T. Goldbeck and J. E. Gray, "A Method of Proportioning Concrete for Strength, Workability, and Durability," *Bull. 11*, National Crushed Stone Association, Nov. 1953.

[b]This is to water actually effective as mixing water.

[c]Since the maximum size of coarse aggregate used in highways is 1½ in. or larger, constant value of 1% may be taken for entrapped air, or 27 ft^3/yd^3 of concrete.

Solid weight (lb/ft³) = SG (bulk dry) × unit weight of water
$$= 2.7 \times 62.4 = 168.48$$
Dry-rodded unit weight (lb/ft³) = 101.5
$$b_0 = \text{dry-rodded unit weight} \div \text{solid weight} = 101.5 \div 168.48$$
$$= 0.60$$

Fine Aggregate (FA)

Fineness modulus (FM) = 2.7
Absorption (%) = 0.8
SG (bulk dry) = 2.5
SG (bulk saturated surface dry) = 2.53
Solid weight (lb/ft³) = SG (bulk dry) × unit weight of water
$$= 2.5 \times 62.4 = 156.00$$

Cement

SG = 3.14
Solid weight (lb/ft³) = SG × unit weight of water
$$= 3.14 \times 62.4 = 195.9$$
Weight per sack (lb) = 94
Solid volume per sack (ft³) = weight per sack ÷ solid weight
$$= 94 \div 195.9 = 0.48$$

Calculations of Proportions

b/b_0 (from Table 5-4) = 0.56
$b = b/b_0 \times b_0 = 0.56 \times 0.60 = 0.34$
Designing for a 3-in. slump and 3000-psi compressive strength for non-air-entrained concrete (2.5%) using Table 5-5.
Cements (sacks/yd³) = 5.4
Water (gal/yd³) = 42

	Solid Volume (ft³/yd³ of concrete)	Quantities (lb/yd³ of concrete)
Cement: 5.4 × 0.48 =	2.59 × 195.9 =	507.38
CA: 0.34 × 27 =	9.18 × 168.5 =	1546.83
Water: 42 ÷ 7.5 =	5.60 × 62.4 =	349.44
Air: 0.025 × 27 =	0.68 × 0 =	0
	Σ 18.05	
FA: 27 − 18.05 =	8.96 × 156.00 =	1397.76

Notice that the weights for fine and coarse aggregate are dry, and thus the water content or amount of water must be increased. This increase in water is equivalent to the amount absorbed by the aggregates. In this example the increase would be as follows:

$$\text{water absorbed by CA} = 0.001 \times 1546.83 = 1.547 \text{ lb}$$

$$\text{water absorbed by FA} = 0.008 \times 1397.76 = 11.180 \text{ lb}$$

$$\text{total} = 12.73 \text{ lb}$$

Thus, the actual amount of mixing water would be increased by 12.73 lb, for a total of 362.17 lb. In most cases the amount of water added is insignificant, but in large jobs it may be significant. In situations in which the aggregate is wet, the mixing water is decreased by the amount of free water.

Keep in mind that pavement concrete is a less workable mix and therefore requires different b/b_0 values and slumps from those given here. Use Table 5-7.

PROPORTIONING MATERIALS

Cement

Cement is usually purchased in bags or sacks. A sack contains 94 lb and is assumed to be 1 ft^3 in volume. Cement may also be purchased in the form of a barrel, which is equal to four sacks weighing 376 lb. Cement may also be purchased by bulk weight. The reason for measuring cement by weight rather than volume is that if cement were purchased by the cubic foot, the weight might vary due to different amounts of compaction.

Water

The accurate weighing of water is important due to the fact that the cement and water form a paste which binds the aggregate and thus influences its strength. In stationary mixing plants, the water is usually weighed out. Volumetric measurements are usually made with modern concrete mixers.

Aggregates

In modern proportioning plants aggregates are weighed, thus eliminating the errors of volumetric measurement. Volumetric measurements are common but are likely to be inaccurate, due to the differences in compaction.

MIXING AND DEPOSITING

Mixing

When the ingredients have been measured in the proper proportion, the next step in the manufacturing of concrete is to mix the ingredients until the aggregate particles are coated with cement and a homogeneous mixture is obtained. Mixing may be done by hand or by a power-driven machine.

Most concrete, regardless of the size of the job, is machine-mixed in batch mixers of 3 to 4 yd^3 capacity. Most batch mixers operate with a revolving drain fitted with blades projecting inward. The drum revolves and the ingredients are carried part way around and turned over as they drop to the bottom, thus producing a thorough mix. The time of mixing varies depending on the consistency of the mix.

Transporting Concrete

After being mixed, concrete is transported to the place of deposit in such a way as to prevent segregation of the ingredients. The means of transportation is a question of economy. The only important requirement is that the concrete arrives at the place of deposit and is of good quality and uniformity. On small projects, wheelbarrels or two-wheeled carts are used for transportation. On large projects, industrial cars, cable cars, conveyor belts, or towers may be used. Sometimes, concrete is pumped through pipelines from the mixer to the forms. Concrete may also be transported and placed by trucks.

Depositing Concrete

The proper placement of concrete in the forms is an important factor in obtaining durable concrete structures. Before the placement of the concrete, the forms should be free of debris and accumulated water. To obtain uniform concrete it should be placed in the forms evenly and without segregation of materials. If the concrete is of a dry consistency, it should be placed in layers 6 to 8 in. thick. If the concrete is of a wet consistency, it may be placed in layers 12 to 15 in. thick. After placement the concrete is spaded to remove any entrapped air, thus producing a smooth surface. The same results may be obtained by vibration, but care should be taken not to segregate the ingredients.

Depositing Concrete under Water

When one places concrete under water, care should be taken to prevent the cement from being washed away or cements from segregating. The common method of depositing concrete under water is by use of a *tremie,* a long pipe about 1 ft (30.5 cm) in diameter with a hopper at the top and a slightly flaring bottom. It is plugged at the bottom, filled, then lowered to position and kept filled at all times. The only problem is that one never knows if the concrete is being placed uniformly.

Vibrated Concrete

Vibration is a mechanical method (high-frequency electric or pneumatic) of puddling concrete which is used extensively for large masses. Vibration is usually recommended for mixes that are stiff in consistency and may need help to effectively consolidate. Vibration may be internal, or may be achieved by vibration of forms, or by vibrating floates on top of the concrete.

Bonding New Concrete to Old Concrete

In most construction projects it is best to try to avoid the bonding of new concrete to old concrete, and thus a continuous pour is recommended. However, it is sometimes impossible to avoid joints, and so precautions should be taken to make the bond between the new concrete and the old concrete as strong as possible.

In massive work with horizontal joints it will probably be sufficient to wet the old concrete and continue with the new concrete. When the walls are thin, the old concrete is roughened and cleaned of foreign matter and laitance and slushed with grout of neat cement or mortar.

Finishing Concrete

After the concrete has been deposited, it may be finished off by several methods: steel trowel, spading, canvas belt, wooden float, burlap, or a grinding tool.

Curing

Concrete sets and hardens as a result of a chemical reaction that takes place between the cement and the water. This process continues as long as water is present. Thus, a strong concrete results if water is present to carry on the reaction. One other factor that is important to curing is temperature. If the temperature is high, undue cracking will occur. If the temperature is too low, the curing process will not continue. Chapter 4 discusses some of the curing techniques.

COMPRESSIVE, TENSILE, AND FLEXURAL STRENGTH

The major factors that affect concrete strength have been discussed in Chapter 4. Only a comparison table is shown here (Table 5-8).

TABLE 5-8 COMPARISON OF COMPRESSIVE, FLEXURAL, AND TENSILE STRENGTH

Strength of Plain Concrete (psi)			Ratio (%)		
Compressive	Modulus of Rupture	Tensile	Modulus of Rupture to Compressive Strength	Tensile Strength to Compressive Strength	Tensile Strength to Modulus of Rupture
1000	230	110	23	11	48
2000	375	200	19	10	53
3000	485	275	16	9	57
4000	580	340	14	8	59
5000	675	400	14	8	60
6000	765	460	13	8	61
7000	855	520	12	7	61
8000	930	580	12	7	62
9000	1010	630	11	7	63

PROBLEMS

5.1. Define the term "concrete."

5.2. Impurities in mixing water may cause undesirable effects in the final concrete. List five undesirable effects.

5.3. List four undesirable impurities in mixing water and discuss each in detail.

5.4. In the design of concrete mixes, three principal requirements for concrete are of importance. Name them and discuss each in detail.

5.5. For the three requirements requested in Problem 5.4, list the ways in which the ingredients of a concrete mix affect each requirement. Explain.

5.6. List four methods of proportioning concrete mixes and discuss each in detail.

5.7. What is meant by the term b/b_0?

5.8. List the steps in performing a concrete mix design according to the method of Goldbeck and Gray.

5.9. In the method of Goldbeck and Gray, how is the b/b_0 ratio affected if the concrete is to be assisted by internal vibration?

5.10. Determine the batch quantities for a concrete mix design that is to have a 28-day compressive strength of 2500 psi using rounded aggregate from a No. 4 sieve opening to 1 in. in size and a medium-fine aggregate with a fineness modulus of 2.65. A slump of 3 in. is desirable. The following descriptions apply to the coarse and fine aggregates.

Coarse Aggregate

Size, No. 4 to 1 in., rounded
Absorption (%) = 0.1
SG (bulk dry) = 2.7
SG (bulk saturated surface dry) = 2.74
Dry-rodded unit weight (lb/ft^3) = 101.5

Fine Aggregate

Fineness modulus = 2.65
Absorption (%) = 0.8
SG (bulk dry) = 2.4
SG (bulk saturated surface dry) = 2.45

REFERENCES

ABRAMS, D. A., Design of Concrete Mixtures, *Lewis Inst. Struct. Mater. Res. Lab Bull. 1*, Chicago, 1918.

American Society for Testing and Materials, "Significance of Tests and Properties of Concrete and Concrete-making Material," *ASTM Spec. Tech. Publ. 169B*, 1979.

BANER, E. E., *Plain Concrete*, 3rd ed., McGraw-Hill Book Company, New York, 1949.

CORDON, W. A., *Properties, Evaluation, and Control of Engineering Materials*, McGraw-Hill Book Company, New York, 1979.

CORDON, W. A., and MERRILL, D., "Requirements for Freezing and Thawing Durability for Concrete," *Proc. ASTM, 63,* (1963), pp. 1026–1035.

FULLER, W. B., and THOMPSON, S. E., "The Laws of Proportioning Concrete," *Trans. Am. Soc. Civil Eng.* (1907), p. 67.

GOLDBECK, A. T., and GRAY, J. E., "A Method of Proportioning Concrete for Strength, Workability and Durability," *Bull. 11,* National Crushed Stone Association, Nov. 1953.

KREBS, R. D., and WALKER, R. D., *Highway Materials,* McGraw-Hill Book Company, New York, 1971.

FULLER, W. B., and THOMPSON, S. E., "The Laws of Proportioning Concrete," *Trans. Am. Soc. Civil Eng.* (1907), p. 67.

GOLDBECK, A. T., and GRAY, J. E., "A Method of Proportioning Concrete for Strength, Workability and Durability," *Bull. 11,* National Crushed Stone Association, Nov. 1953.

KREBS, R. D., and WALKER, R. D., *Highway Materials,* McGraw-Hill Book Company, New York, 1971.

MILLS, A. P., HAYWARD, H. W., and RADER, L. F., *Materials of Construction,* 6th ed., John Wiley & Sons, Inc., New York, 1955.

POWERS, T. C., "The Air Requirements in Frost-resistant Concrete," *Proc. Highway Res. Board* (1949), p. 184.

POWERS, T. C., "The Mechanism of Frost Action in Concrete," Stanton Walker Lecture Series on the Material Sciences, Lecture No. 3, National Ready-Mixed Concrete Association, Silver Spring, Md., 1965.

TALBOT, A. N., "Proposed Method of Estimating Density and Strength of Concrete and Proportioning Materials by Experimental and Analytical Considerations of Voids in Mortar and Concrete," *Proc. ASTM* (1921).

WALKER, S., BLOEM, D. L., and GAYNOR, R. D., "Relationships of Concrete Strength to Maximum Size of Aggregate," *Proc. Highway Res. Board* (1959).

6

Proportioning Structural Concrete Mixtures with Fly Ash and Other Pozzolans*

No history of concrete as it is made and used today would be complete without reference to the expanding use of pozzolans as an ingredient in concrete. It has been stated that pozzolans, along with other finely divided mineral admixtures, are used in mass concrete, structural concrete, pavements, dam locks, canal linings, tunnels, sewage works, water-works, high-rise residential and commercial structures, and residential concrete, including sidewalks, driveways, and parking areas. In short, concrete containing pozzolans as an ingredient is being used in virtually every application for which concrete is used. Literature provides much information on the desirable effects of pozzolans in concrete. The history of the use of pozzolans parallels the history of the recorded works of man. About 2000 years ago Greeks, Romans, and other Mediterranean peoples discovered the value of using pozzolans (fine volcanic ash) with burned lime to build historic structures—some are still in use today. The earliest methods of proportioning with pozzolan in the United States consisted primarily of substituting pozzolans on a pound-for-pound or volume-for-volume basis with portland cement. Modern building practices have required, however, that a new method of proportioning with pozzolans be developed whereby rapid construction can be coupled with long-range owner satisfaction. These procedures are in accord with ACI 211.1-70.

BACKGROUND

The American Society for Testing and Materials (ASTM) defines pozzolan as " . . . a siliceous or siliceous and aluminous material which in itself possesses little or no ce-

*This chapter was reproduced with permission from the American Concrete Institute, where it appeared in their SP 46-8 publication. The authors of this article are C. E. Lovewell and Edward J. Hyland.

mentitious value but will, in finely divided form and in the presence of moisture, chemically react with calcium hydroxide at ordinary temperatures to form compounds possessing cementitious properties." In brief, pozzolans are powders which will harden when combined with lime and water at ordinary temperatures. In concrete the lime is provided as a by-product of the hydration of the portland cement in the mixture. A commonly accepted amount of lime produced by hydration is 15%, by weight, of the portland cement.

There are in general two types of pozzolans: natural and synthetic. Natural pozzolans are found in the earth's crust and include volcanic ashes, pumicites, tuffs, diatomaceous earths, opaline cherts and shales and certain processed clays, shales, and diatomites. Some natural pozzolans are capable of being used in concrete with minimal preparation such as screening and grinding whereas others require calcination and finish grinding to provide satisfactory properties.

Synthetic pozzolan refers almost exclusively to fly ash. Fly ash is defined as " . . . the finely divided residue that results from the combustion of ground or powdered coal and is transported from the combustion chamber by exhaust gasses." Some factors affecting the chemical and physical characteristics of fly ash are the chemical composition or type of coal, duration of time in open-air storage, fineness of grind, and other factors relating to combustion.

Most recognized specifications for pozzolans in the United States are written to cover both natural pozzolans and fly ash, with the properties of each arranged in convenient tables. Some of these specifications are ASTM C-618 "Fly Ash and Raw or Calcined Natural Pozzolans For Use in Portland Cement Concrete," CRD C262-63 "Corps of Engineers Specifications For Pozzolan For Use in Portland Cement Concrete" and SS-P-570b "Federal Specifications for Pozzolans for Use in Portland Cement Concrete." The characteristics of the pozzolan intended for use must be known before any attempt can be made to proportion concrete with it. For example, a favorable particle shape and fineness of the pozzolan are necessary if low-water demand is to be achieved. Coarse pozzolan of unfavorable shape (such as volcanic glass) may cause an increase in mixing water requirements with consequent excessive bleeding and segregation. Other characteristics, such as the tendency of high-carbon fly ash to reduce entrained air should also be known prior to proportioning. Highway Research Board Special Report No. 119 "Admixtures in Concrete" contains much of this type of information helpful to the proportioner. Selected data from that report are shown in Table 6-1.

Early proportioning of concrete with pozzolans in this country was concerned almost exclusively with mass concrete. Since the proportioner of mass concrete is concerned with producing a mixture having a low heat of hydration and since the structure itself will probably not sustain design loads for many months or years and most of the concrete used in the mass will not be exposed to freezing and thawing, certain procedures were developed whereby the quantity of portland cement was held to an absolute minimum. The pozzolan used therefore comprised a very high percentage of the total cementitious ingredients. However, in proportioning the concrete for use in structures, an entirely different set of criteria has to be applied. For example, most structures have a required design strength at 28 days. Further, floors in many high-rise buildings receive higher loadings during the construction period than they ever do when they are turned over to

TABLE 6-1 EFFECT OF TYPE, FINENESS, AND AMOUNT OF POZZOLAN ON WATER CONTENT AND AIR-ENTRAINING ADMIXTURE REQUIREMENT

Cement (lb/cu yd)	Pozzolan (lb/cu yd)	Percent	Designation	Surface (sq cm/g)	Water (lb/cu yd)	Neutralized Vinsol Resin (ml/cu yd)
532	0	0	None	3,550	265	485
346	116	30	Fly Ash I	3,565	247	379
416	103	25	Pumicite I	4,410	277	529
404	156	35	Tuff	10,460	310	800
346	139	35	Obsidian	3,415	266	378
458	91	20	Calcined Shale I	13,685	286	670
423	141	30	Calcined diatomite	10,450	302	1,018
462	64	16	Uncalcined diatomite	12,125	274	540

Source: "Admixtures in Concrete," Highway Research Board, *Special Report 119,* 1971.

the owner. A cube of brick or concrete block resting on concrete which is 7 days old causes more damaging stress than a stenographer seated at a typewriter in a two year old building. Concrete in the structural market is also invariably designed for higher strength than mass concrete. Further, since many of these structures are designed in areas that experience freezing and thawing, the exposed concrete must be air-entrained. Even in those parts of the building not exposed to the elements air-entrainment may be required by the designer because of the volume of reinforcing steel in the thin members and the need to improve the workability of concrete.

Under these conditions it becomes imperative that a new method of proportioning concrete for the structural market be developed.

Lovewell and Washa reported the development of a method of proportioning concrete mixtures using fly ash. Basically, the method showed that with relatively few tests a designer could develop data which provided guidelines for proportioning concrete and fly ash with predictable results. This was further elaborated by Cannon.

The method that follows is based on the references previously cited and includes additional data developed for mixtures containing fly ash and two types of chemical admixtures. The authors have found that the procedures outlined have proved valid with fly ashes conforming to ASTM C-618 and with a great variety of aggregates and brands of portland cement. The procedure has been proven to be satisfactory by the excellent performance of many millions of cubic yards of structural concrete made with fly ash, with and without chemical admixtures. With a minimum number of check tests the procedure can also be applied to natural pozzolans.

PROCEDURE

As is pointed out in ACI 211.1-70, compressive strength is not the only important characteristic of concrete. Workability, durability, wear resistance and other characteristics may be equally or even more important. The ACI "Guide for Use of Admixtures in Concrete" discusses the effects of admixtures in concrete. Assume that based upon this guide and other data the designer requires that a pozzolan and perhaps other admixtures be used. Other requirements are also specified such as strength, slump, air content, maximum size of aggregate, type of portland cement, and type of chemical admixtures. The following steps should be followed:

1. *Selection of slump.* If the designer has not specified a minimum and maximum slump choose values according to ACI.
2. *Selection of aggregate.* The designer may have specified a certain aggregate size and perhaps even shape. However, if this has been left open, the proportioner should in general select the maximum size of coarse aggregate that is economically available and one which can be placed readily in the structure. Other considerations should be made, such as whether the concrete is intended to produce a very high strength, or whether it is intended for pumping through small lines, in which case a smaller aggregate size may be selected and a specific particle shape may be desired. In no case should the maximum size of the aggregate exceed one-fifth of

TABLE 6-2 APPROXIMATE MIXING WATER AND AIR CONTENT REQUIREMENTS FOR DIFFERENT SLUMPS AND MAXIMUM SIZES OF AGGREGATES[a]

	Water, lb/yd³ of Concrete for Indicated Maximum Sizes of Aggregate						
Slump (in)[b]	$\frac{3}{8}$ in.	$\frac{1}{2}$ in.	$\frac{3}{4}$ in.	*1* in.	$1\frac{1}{2}$ in.	*2* in.	*3* in.
Non-air-entrained concrete							
1 to 2	350	335	315	300	275	260	240
3 to 4	385	365	340	325	300	285	265
6 to 7	410	385	360	340	315	300	285
Approximate Amount of Entrapped Air in Non-air-entrained Concrete, %	3	2.5	2	1.5	1	0.5	0.3
Air-entrained Concrete							
1 to 2	305	295	280	270	250	240	225
3 to 4	340	325	305	295	275	265	250
6 to 7	365	345	325	310	290	280	270
Recommended Average Total Air Content, %	8	7	6	5	4.5	4	3.5

[a]These quantities of mixing water are for use in computing cement factors for trial batches. They are maxima for reasonably well-shaped angular coarse aggregates graded within limits of accepted specifications.

[b]The slump values for concrete containing aggregates larger than $1\frac{1}{2}$ in are based on slump tests made after removal of particles larger than $1\frac{1}{2}$ in by wet-screening.

Note: This is a reproduction of Table 5.2.3 of ACI 211.1-70.

the narrowest dimension of the size of the form or one-third of the depth of slabs or three-fourths of the minimum clear spacing between individual reinforcing bars, bundles of bars, or pretensioning strands.

3. *Estimate the mixing water and air content.* Table 6-2 gives an estimate of the mixing water required with a given size and shape of aggregate to produce a given slump. The table also gives an estimate of the amount of entrapped and entrained air normally present in non-air-entrained concrete. This table is for concrete that does not contain pozzolan or a chemical admixture.

If no other data is available the proportioner can use Table 6.2 and apply a correction factor based on other information, such as HRB Report 119, Wallace and Ore, or other available sources. However, the producer or vendor of the pozzolan proposed for use may have useful information of this type. As an example of what may be available to the proportioner, Figure 6-1 and Figure 6-2 give the approximate mixing water requirements of concrete containing fly ash at different slumps made with rounded or angular coarse aggregate of the maximum sizes commonly used in structural concrete. These curves, while not absolute, are sufficiently accurate to provide the proportioner with a figure for developing a first trial mixture. The curves have been developed over many years and much field work.

The use of chemical admixtures, specifically water-reducers or water-reducing-retarders, has become widespread. These admixtures are used with great frequency in all concrete, including mixtures containing pozzolans. The chemical composition of the water-reducing admixture affects the amount of mixing water required, and its characteristics must be understood. Figure 6-3 and Figure 6-4 give estimates of mixing water required for mixtures of cement plus fly ash and a chemical admixture with the salt of a lignosulfonic acid (lignin) as its base. Figure 6-5 and Figure 6-6

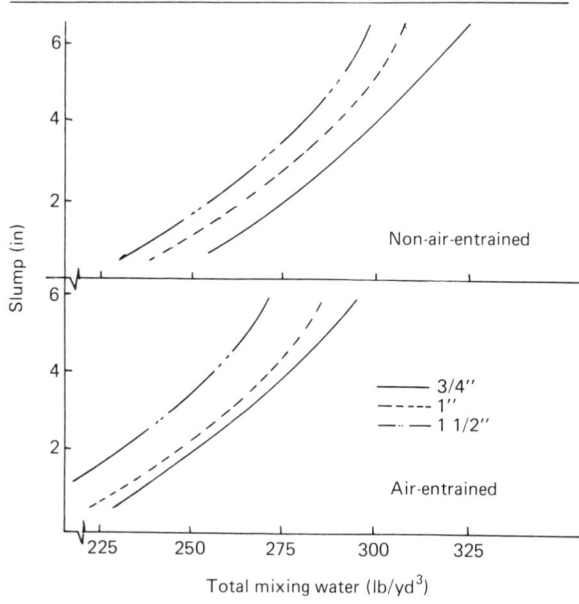

Figure 6-1 Approximate Mixing Water Requirements for Different Slumps and Maximum Sizes of Typical Structural Concrete Aggregates

Materials: Portland Cement–Type I
            Pozzolan–Fly Ash
            Chemical Admixture–None
            Course Aggregate–Rounded

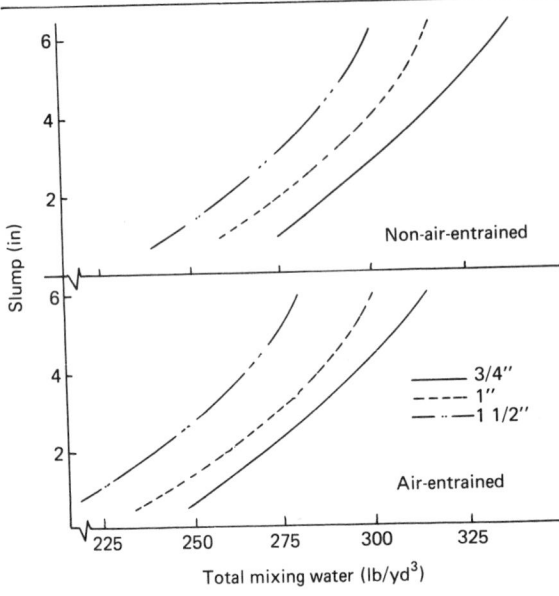

Figure 6-2 Approximate Mixing Water Requirements for Different Slumps and Maximum Sizes of Typical Structural Concrete Aggregates

Materials: Cement–Type I
                  Pozzolan–Fly Ash
                  Chemical Admixture–None
                  Coarse Aggregate–Angular

give similar information, but in this case the chemical admixture has a carbohydrate base.

4. *Selection of compressive strength and air-entrainment.* The designer will ordinarily specify the required average compressive strength and whether or not the concrete is to be air-entrained, along with the required limits of air-entrainment. If the

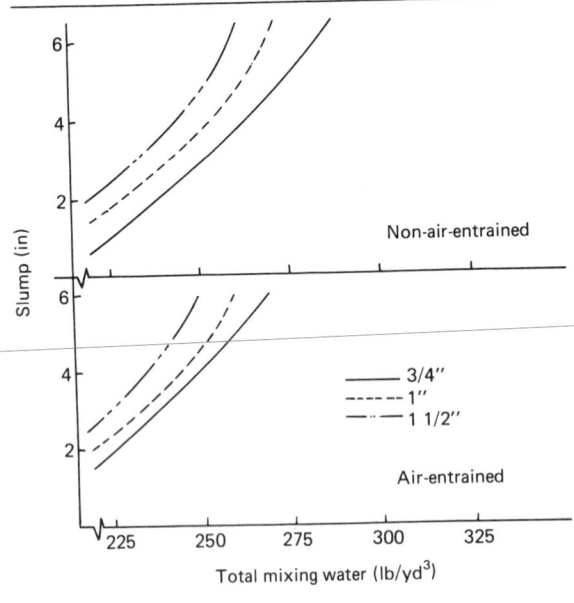

Figure 6-3 Approximate Mixing Water Requirements for Different Slumps and Maximum Sizes of Typical Structural Concrete Aggregates

Materials: Cement–Type I
                  Pozzolan–Fly Ash
                  Chemical   Admixture–Lignin Base
                  Coarse Aggregate–Rounded

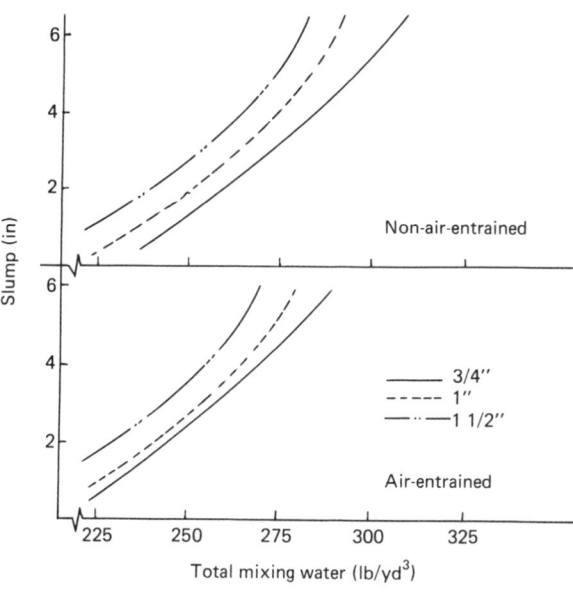

Figure 6-4 Approximate Mixing Water Requirements for Different Slumps and Maximum Sizes of Typical Structural Concrete Aggregates

Materials: Cement–Type I
            Pozzolan–Fly Ash
            Chemical   Admixture–Lignin Base
            Coarse Aggregate–Angular

designer fails to specify average strength and entrained-air content the proportioner may turn to Table 6-3 and Table 6-4.

The durability of concrete containing fly ash or other pozzolans is equal to that of concrete mixtures not containing these materials, provided that the mixtures have equal compressive strength and equal air-entrainment. Thus, although Table 6-3

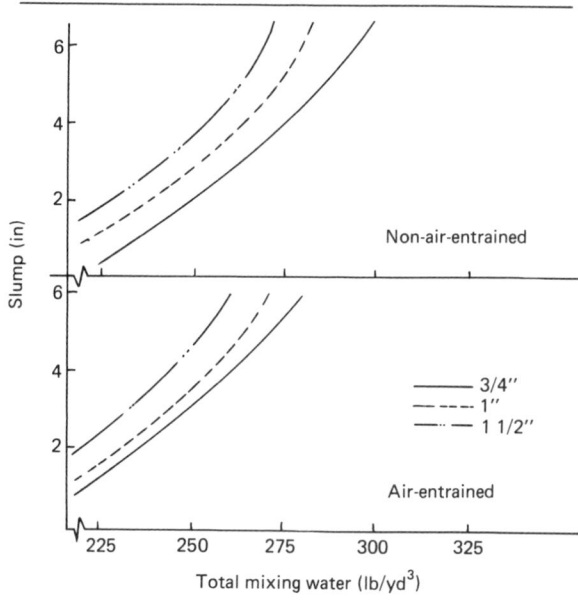

Figure 6-5 Approximate Mixing Water Requirements for Different Slumps and Maximum Sizes of Typical Structural Concrete Aggregates

Materials: Cement–Type I
            Pozzolan–Fly Ash
            Chemical Admixture–Carbohydrate Base
            Coarse Aggregate–Rounded

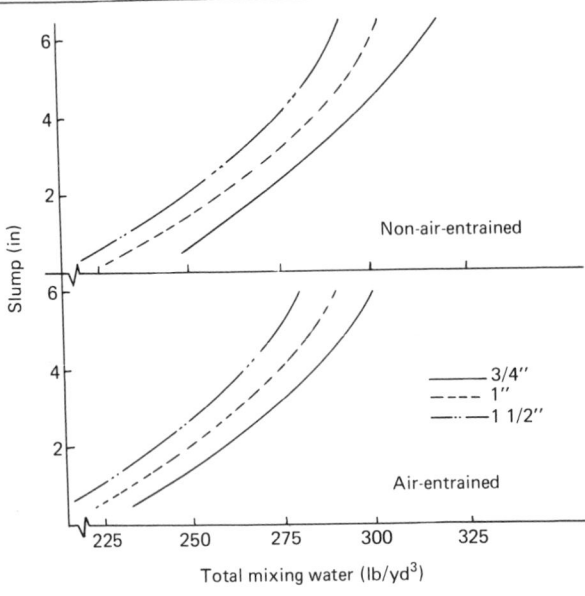

Figure 6-6 Approximate Mixing Water Requirements for Different Slumps and Maximum Sizes of Typical Structural Concrete Aggregates

Materials: Cement–Type I
Pozzolan–Fly Ash
Chemical Admixture–Carbohydrate Base
Coarse Aggregate–Angular

and Table 6-4 were developed for concrete mixtures not containing fly ash or other pozzolans, they may be used safely to determine the required strength and air-entrainment of mixtures containing fly ash and other pozzolans as well.

5. *Calculation of cement content.* Knowing the required average compressive strength, the proportioner now turns to curves which have been previously established or which can be established with relatively few mixtures. An example of such a family of curves is shown in Figures 6-7, 6-8, and 6-9. These figures are averages of hundreds of tests of structural concrete made either in commercial testing laboratories or in laboratories owned and operated by suppliers of ready-mixed concrete. These 3- 7- and 28-day curves are averages of results gained with many different brands of Type I portland cement and various sources of three-quarter inch to one and one-half inch coarse aggregate, both angular and rounded. The water-reducing admixtures conform to ASTM C 494-71 (Type A) and had either lignin or carbohydrate bases. The fly ash came from eight different sources, the loss on ignition ranging from 5% to less than 1%. Concrete cylinders used to obtain data were cast in either waxed cardboard molds of single-use metal molds. Testing procedures were standard. The great many variables in materials, molds and laboratories naturally cause some scatter in the points used to plot the curves, yet the authors feel that the curves are reasonably conservative and represent a good usable average strength of laboratory-fabricated mixtures of structural concrete.

Structural concrete is ordinarily specified on the 28 day compressive strength basis. Therefore selecting the required average 28 day compressive strength of cylinders made, cured, and tested in the laboratory on the ordinate scale and pro-

TABLE 6-3 RELATIONSHIPS BETWEEN WATER-CEMENT RATIO AND COMPRESSIVE STRENGTH OF CONCRETE

Compressive Strength at 28 Days (psi)[a]	Non-air-entrained Concrete	Air-entrained Concrete
6000	0.41	
5000	0.48	0.40
4000	0.57	0.48
3000	0.68	0.59
2000	0.82	0.74

[a]Values are estimated average strengths for concrete containing not more than the percentage of air shown in Table 5.2.3. For a constant water-cement ratio, the strength of concrete is reduced as the air content is increased.

Strength is based on 6 × 12 in. cylinders moist-cured 28 days at 73.4 ± 3F (23 ± 1.7C) in accordance with Section 9(b) of ASTM C 31 for Making and Curing Concrete Compression and Flexural Test Specimens in the Field.

Relationship assumes maximum size of aggregate about $\frac{3}{4}$ to 1 in; for a given source, strength produced for a given water-cement ratio will increase as maximum size of aggregate decreases; see Sections 3.4 and 5.2.2

Note: This is a reproduction of Table 5.2.4 (a) of ACI 211.1-70.

TABLE 6-4 MAXIMUM PERMISSABLE WATER-CEMENT RATIOS FOR CONCRETE IN SEVERE EXPOSURES[a]

Type of Structure	Structure Wet Continuously or Frequently and Exposed to Freezing and Thawing[b]	Structure Exposed to Sea Water or Sulfates
Thin sections (railings, curbs, sills, ledges, ornamental work) and sections with less than 1 in. cover over steel	0.45	0.40[c]
All other structures	0.50	0.45[c]

[a]Based on report of ACI Committee 201, "Durability of Concrete in Service," previously cited.

[b]Concrete should also be air-entrained.

[c]If sulfate resisting cement (Type II or Type V of ASTM C150) is used, permissable water-cement ratio may be increased by 0.05.

Note: This is a reproduction of Table 5.2.4 (b) of ACI 211.1-70.

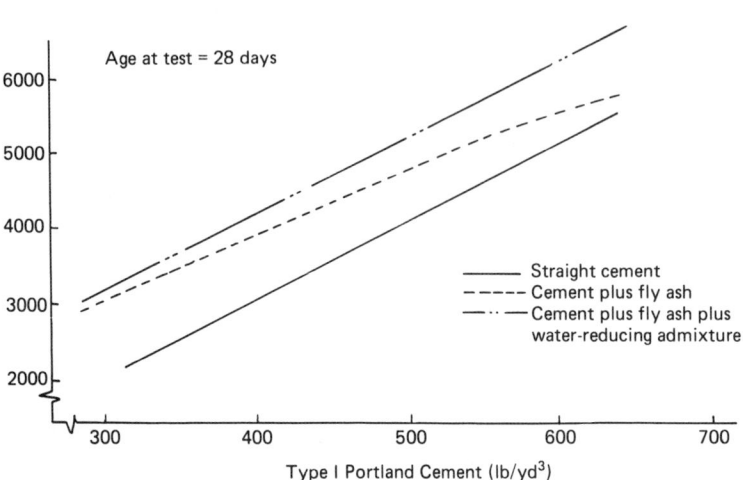

Figure 6-7   Average Compressive Strength of Laboratory-Made, Air-Entrained Concrete Cylinders Containing Maximum Sizes of $^3/_4$ in. to $1^1/_2$ in. Aggregates at 3 in. to 5 in. Slump

ceeding horizontally until intersecting the appropriate curve, and proceeding downward to the abscissa scale, the estimated amount of Type I portland cement required to produce air-entrained concrete can be read.

Figures 6-8 and 6-9 are similar curves developed to show the relationship of the various mixtures at 7 days and at 3 days. Such figures can be invaluable in helping

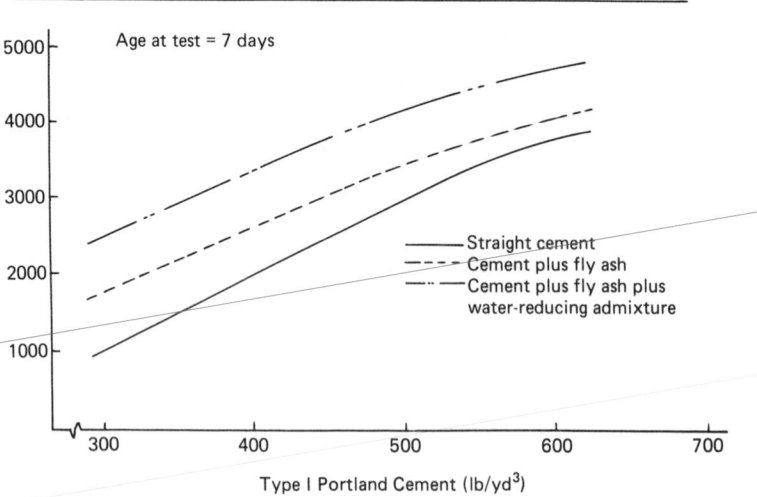

Figure 6-8   Average Compressive Strength of Laboratory-Made, Air-Entrained Concrete Cylinders Containing Maximum Sizes of $^3/_4$ in. to $1^1/_2$ in. Aggregates at 3 in. to 5 in. Slump

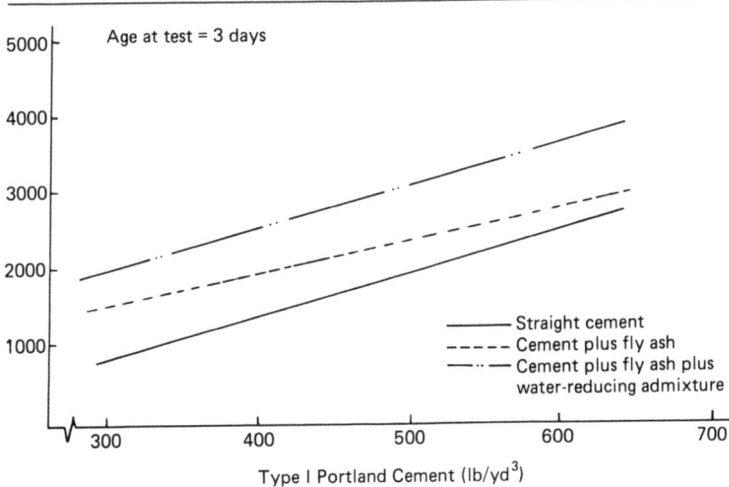

Figure 6-9 Average Compressive Strength of Laboratory-Made, Air-Entrained Concrete Cylinders Containing Maximum Sizes of $^3/_4$ in. to $1^1/_2$ in. Aggregates at 3 in. to 5 in. Slump

the field engineer determine at an early age whether concrete specimens will actually approach the desired strengths. They are also helpful in determining when loads may be applied to "green" concrete.

Figures such as Figures 6-7, 6-8, and 6-9 can also be established for non-air-entrained concrete.

6. *Determination of pozzolan content.* The optimum amount of the pozzolan can be determined by relatively few mixtures and a curve plotted similar to that shown in Figure 6-10, or the producer or vendor of the pozzolan may have such data available. Lovewell and Washa established the curve shown in Figure 6-10 for fly ash, and experience over the years with this material has shown that the curve is valid with only minor modifications. Experience in the field has shown that it is beneficial to increase the value shown in Figure 6-10 by about 20% when the sand available for use is extremely coarse or where the concrete is to be pumped. When the sand selected for use is extremely high in 50 and 100 mesh particles it is also beneficial to decrease the amount shown in Figure 6-10 by about 20% to avoid stickiness. In all cases it should be remembered that the use of figures and tables in any suggested method of proportioning is for the establishment of a trial mix only and the numbers obtained from such references should be considered as estimates and modified when appropriate with good judgment by the proportioner.

7. *Estimation of coarse and fine aggregate contents.* Having determined the cement and pozzolan contents, an estimate of the approximate mixing water requirements, and the entrained-air content (if any), the proportioner is now ready to complete the design by selecting a coarse aggregate by means of Table 6-5. The proportioner may have sufficient data available on the unit weight of fresh concrete containing

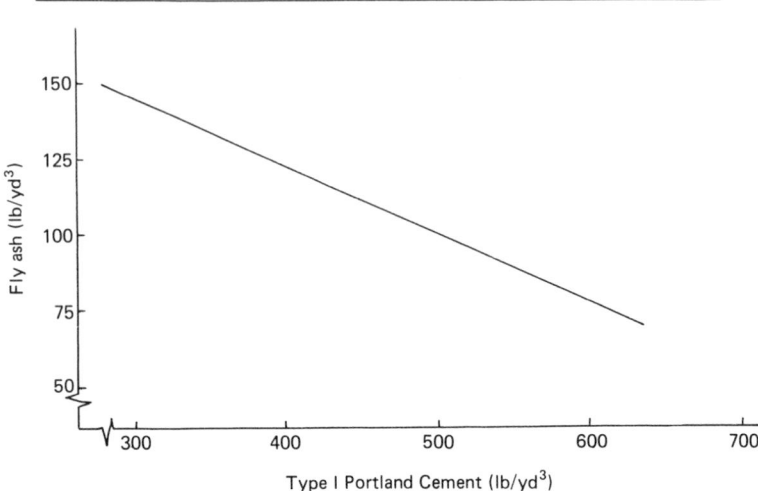

Figure 6-10   Optimum Amount of Fly Ash at Varying Cement Contents of Air-Entrained
Concrete with $^3/_4$ to $1^1/_2$ in. Maximum Size Aggregate

the pozzolan with or without a chemical admixture to allow solving for the fine
aggregate content by the "weight" method, but the probability is that sufficient
background information will not be available and so it is suggested that sand content
be calculated by the absolute volume method. To do this the specific gravity of the
pozzolan in question must be known. Since the specific gravity of one pozzolan
can be widely different from another, no assumption as to specific gravity should

TABLE 6-5   VOLUME OF COARSE AGGREGATE PER UNIT OF VOLUME OF CONCRETE

Maximum size of aggregate (in.)	Volume of Dry-rodded Coarse Aggregate[a] per Unit Volume of Concrete for Different Fineness Moduli of Sand			
	2.40	*2.60*	*2.80*	*3.00*
$\frac{3}{8}$	0.50	0.48	0.46	0.44
$\frac{1}{2}$	0.59	0.57	0.55	0.53
$\frac{3}{4}$	0.66	0.64	0.62	0.60
1	0.71	0.69	0.67	0.65
$1\frac{1}{2}$	0.75	0.73	0.71	0.69
2	0.78	0.76	0.74	0.72
3	0.82	0.80	0.78	0.76
6	0.87	0.85	0.83	0.81

[a]Volumes are based on aggregates in dry-rodding condition as described in ASTM C 29 for Unit Weight of
Aggregate. These volumes are selected from empirical relationships to produce concrete with a degree of
workability suitable for usual reinforced construction. For less workable concrete such as required for concrete
pavement construction they may be increased about 10 percent. For more workable concrete, such as may
sometimes be required when placement is to be by pumping, they may be reduced up to 10 percent.

Note: This is a reproduction of Table 5.2.6 of ACI 211.1-70.

be made. Exact information should be furnished by the producer of the pozzolan or it should be determined by laboratory tests.

8. *Adjustment.* ACI 211.1-70 shows the methods employed to adjust the first trial mix. The methods of the standard are equally valid when applied to mixtures containing pozzolans with or without chemical admixtures.

APPLICABILITY TO LIGHTWEIGHT CONCRETE

In general, the method of selecting the proportions of cement and pozzolan with or without the chemical admixture previously described is also applicable to lightweight concrete mixtures. Because the amount of water required in a lightweight concrete mixture is so dependent upon the sizes, shape, and texture of the lightweight aggregate and the fineness and amount of natural sand, if that is to be included, it is impossible to generalize on an estimate of mixing water. This must be determined by actual tests although a good estimate of it may perhaps be furnished by the vendor of the pozzolan or by the producer of the lightweight aggregate. Curves showing the relationship of cement, cement plus pozzolan, and cement plus pozzolan plus water-reducing or water-reducing-retarding admixture can then be developed. A further curve showing the optimum amount of pozzolan to be used in cement with the aggregates in question should also be drawn. With these two curves in hand it then becomes easy for the proportioner of lightweight concrete to use ACI 211.2-69 to determine a first trial mixture. An example of a family of curves developed for concrete containing an expanded shale lightweight aggregate and natural sand is shown in Figure 6-11. The relationship of fly ash to cement content with this particular combination of aggregate is shown in Figure 6-12.

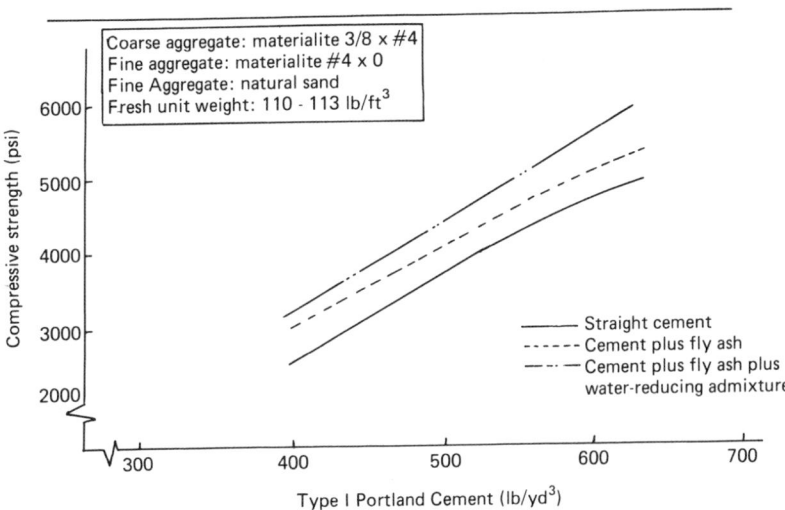

Figure 6-11 Average Compressive Strength of Laboratory-Made Air-Entrained Lightweight Concrete Cylinders at 28 Days.

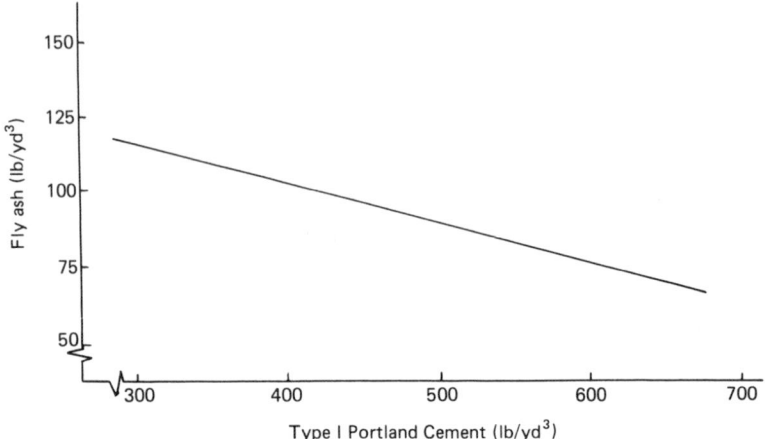

Figure 6-12   Optimum Amounts of Fly Ash at Varying Cement Contents with Materialite
Aggregate

SAMPLE COMPUTATIONS

Example 1.

Normal weight concrete is to be proportioned for a structure, part of which will be
exposed to frequent freezing and thawing. Structural demands require an average
compressive strength of 3500 psi for cylinders fabricated, cured, and tested in the
laboratory. It is expected that specimens from the field will average over 2500 psi.
Job conditions indicate that a slump of 4 in. $\pm \frac{1}{2}$ in. should be used and the coarse
aggregate will be crushed stone with a maximum size of one in. The dry rodded unit
weight of the coarse aggregate is 96.3 lb/ft^3 and the bulk specific gravity is 2.64. The
fine aggregate is natural sand of 2.66 specific gravity and fineness modulus of 2.70.
The designer has required that fly ash be used and its specific gravity is 2.44. The
concrete shall be air-entrained 5% $\pm \frac{1}{2}$%.

   (a) As indicated, the desired slump is 4 in.

   (b) The designer has required 5% entrained air plus fly ash with angular coarse
aggregate. Using Figure 6-2 it appears that the approximate amount of mixing water
to produce a 4 in. slump concrete with 1 in. maximum size aggregate is 285 lbs.

   (c) From Table 6-4 a water-cement ratio of 0.50 would be required for the
structure if cement alone were the cementing material. However, Table 6-3 estimates
that a water-cement ratio of 0.50 for air-entrained concrete will produce approximately
3800 psi compressive strength at 28 days. Although the designer has required only
3500 psi, the 3800 is required for durability, so this higher figure will govern and the
concrete should be designed accordingly.

   (d) Using Figure 6-7, entering at 3800 psi on the ordinate axis and proceeding
laterally to intersect the curve marked "Fly Ash Mixture" and then dropping down to
the abscissa axis, 390 lb./yd^3 of Type I portland cement is estimated as necessary to
fulfill the requirements.

   (e) To determine the optimum amount of fly ash for use in the mixture Figure
6-10 is used. Entering with 390 lb/yd^3 of Type I portland cement, proceeding upward

until intersecting the curve, then laterally to the ordinate scale, a value of 125 lb/yd³ of fly ash combines with the 390 lb/yd³ of Type I portland cement previously determined to give the total weight of the cementing material.

(f) The amount of coarse aggregate is estimated from Table 6-5. With a fineness modulus of the fine aggregate of 2.70 and one-in. maximum size of coarse aggregate, the table indicates that 0.68 ft³ of coarse aggregate on a dry rodded basis may be used in each cubic foot of concrete. Therefore, the coarse aggregate content is 27 × 0.68 = 18.36 ft³/yd³. Since it weighs 96.3 lb/ft³, the dry weight of coarse aggregate is 18.36 × 96.3 or 1768 lb/yd³.

(g) The quantities of cement, fly ash, coarse aggregate, entrained air, and water having been established, the following absolute volumes of ingredients are calculated:

Cement: $\dfrac{390}{3.15 \times 62.4} = 1.98 \text{ ft}^3$

Fly Ash: $\dfrac{125}{2.44 \times 62.4} = 0.82 \text{ ft}^3$

Coarse Aggregate: $\dfrac{1768}{2.64 \times 62.4} = 10.73 \text{ ft}^3$

Entrained Air: $0.05 \times 27.00 = 1.35 \text{ ft}^3$

Water: $\dfrac{285}{62.4} = 4.57 \text{ ft}^3$

Total solid volume of ingredients except sand = 19.45 ft³. Solid volume of sand required = 27 − 19.45 = 7.55 ft³. Required weight of dry sand = 7.55 × 2.66 × 62.4 = 1253 lb/yd³.

(h) We will not proceed further on this example as the method of making adjustments and compensations is well covered in ACI 211.1-70.

Example 2.

Concrete is required for a high-rise structure which will not be exposed to freezing and thawing, but because of the heavy reinforcement the designer has required that concrete shall be air-entrained 5% ±½%. Crushed stone aggregate shall be used with a top size of three-fourths inch and the slump shall be 3 in. ±½ in. The designer has further required that fly ash with an average specific gravity of 2.50 and a lignin-based water-reducer shall be used in the concrete. Structural requirements are such that laboratory-made specimens shall average 6000 psi at 28 days in compression.

(a) Slump and aggregate size and shape were given previously. Dry rodded weight of coarse aggregate is 100 lb/ft³ and its specific gravity is 2.67. The fineness modulus of sand is 2.80 and its specific gravity is 2.66.

(b) Since the designer has required air-entrained concrete with a lignin-based water-reducer and fly ash we proceed to Figure 6-4, which gives us an approximation of 260 lb/yd³ of water required for 3 in slump.

(c) Turning next to Figure 6-7, we enter at the 6000 psi ordinate and proceed horizontally to where we intersect the curve "Fly Ash + Water-Reducer Mixture". Moving downward to the abscissa scale we find that we require approximately 580 lb/yd³ of Type I portland cement to fulfill requirements.

(d) We then turn to Figure 6-10, where we obtain an estimated 85 lb/yd³ of fly ash which will be required for the job. Since the manufacturer of the water reducing admixture has recommended that 4.0 fluid ounces of the water reducer be used per 100 lbs of portland cement, we determine that we will require 23.2 fl oz of this admixture per cubic yard.

(e) The quantity of coarse aggregate is estimated from Table 6-5 to be 0.62 cubic feet on a dry rodded basis per cubic foot of concrete. For a cubic yard therefore, the coarse aggregate will be 16.74 ft³, and since the coarse aggregate weighs 100 lb per cubic foot the dry weight of coarse aggregate is 1674 lb/yd³.

(f) With quantities of cement, fly ash, coarse aggregate, water, and air known, the following calculations are made:

$$\text{Cement:} \qquad \frac{580}{3.15 \times 62.4} = 2.95 \text{ ft}^3$$

$$\text{Fly Ash:} \qquad \frac{85}{2.5 \times 62.4} = 0.55 \text{ ft}^3$$

$$\text{Coarse Aggregate:} \frac{1674}{2.67 \times 62.4} = 10.05 \text{ ft}^3$$

$$\text{Water:} \qquad \frac{260}{62.4} = 4.16 \text{ ft}^3$$

$$\text{Entrained Air:} \quad 0.05 \times 27.00 = 1.35 \text{ ft}^3$$

Total solid volume of ingredients except sand = 19.06 ft³. Solid volume of sand required = 27.00 − 19.06 = 7.94ft³. Required weight of dry sand = 7.94 × 2.66 × 62.4 = 1316 lb/yd³.

(g) Adjustments may be required and they can be performed as shown in ACI 211.1-70.

CONCLUSIONS

There are many methods which can be used to proportion concrete containing pozzolans. This method has given good results and is applicable to many different materials and situations. Certain preliminary work must be done before the method can be employed. It is of utmost importance that the type of pozzolan and other admixtures used are known and their characteristics well established before using the method. As is always the case, the proportioner must use good judgment and experience in applying the method.

REFERENCES

1. ACI Committee 212, "Guide for Use of Admixtures in Concrete", ACI Journal, Proceedings V. 68, No. 9, September 1971, pp. 646–676.

2. "Standard Specifications for Fly Ash and Raw or Calcined Natural Pozzolans For Use in Portland Cement Concrete", (ASTM C 618-71), *1971 Book of ASTM Standards,* Part 10, American Society for Testing and Materials, Philadelphia, Pennsylvania, pp. 345–349.

3. "Corps of Engineers Specifications For Pozzolan For Use in Portland-Cement Concrete", (CRD-C262-63), Handbook For Concrete and Cement, U.S. Army Engineer Waterways Experiment Station, Vicksburg, Mississippi, 1949, plus quarterly supplements.

4. "Specifications for Pozzolans for Use in Portland Cement Concrete", Federal Specifications SS-P-570b, General Services Administration, Washington, D.C.

5. "Admixtures in Concrete", Highway Research Board, Special Report 119, 1971.

6. LOVEWELL, C. E. and WASHA, G. W., "Proportioning Concrete Mixtures Using Fly Ash", ACI Journal, Proceedings V. 54, No. 12, June 1958, pp. 1093–1102.

7. CANNON, ROBERT W., "Proportioning Fly Ash Concrete Mixes for Strength and Economy", ACI Journal, Proceedings V. 65, No. 11, November 1968, pp. 969–979.

8. ACI Committee 211, "Recommended Practice for Selecting Proportions for Normal Weight Concrete", (ACI 211.1-70), American Concrete Institute, Detroit, Michigan, 1970.

9. WALLACE, GEORGE, B. and ORE, ELWOOD L., "Structural and Lean Mass Concrete As Affected by Water-Reducing, Set-Retarding Agents", ASTM Special Technical Publication No. 26, 1959, Philadelphia, Pa.

7

Performance Characteristics of Concrete

As indicated in Chapter 4, concrete has high compressive strength but low shear and tensile strength. It is porous, extensible, and fire-resistant, but can be damaged by intense heat. Under ordinary loading conditions, concrete is elastic, although it will creep under sustained, heavy loads. The shear and tensile strength of concrete can be increased by the addition of reinforcing bars or by prestressing with steel wires that are under tension. This fact allows much design flexibility, yet deterioration of concrete can pose significant problems. In the paragraphs that follow, the failure mechanism of concrete, the factors that effect deterioration of concrete, and those admixtures that may aid the concrete mixture will be discussed.

In discussions of concrete failure, cracking is usually the principal criterion upon which failure is based. However, not all cracking is an indication of concrete failure, as the concrete may still be able to perform a useful function. In determining whether or not the concrete structure has failed, considerable judgment must be exercised.

FAILURE MECHANISM OF CONCRETE

The failure mechanism of concrete has been under discussion for many years. Many investigators have implied that the failure mechanism is associated with internal microcracking. The formation and propagation of such microcracks have been studied indirectly by sonic velocity, acoustic methods, and by the observation of macrocracks on the surface of the models. Robinson and Hsu, Slate, Sturman, and Winter have directly observed the formation and propagation of microcracks by x-ray analysis. (Because of the limitations of the techniques employed, the detection of microcracks has been somewhat uncertain.)

Derucher also directly observed the formation and propagation of microcracks by utilizing the scanning electron microscope.

According to the research of Robinson and Hsu, Slate, Sturman, Winter, and Derucher, three types of microcracks were found to exist: bond, matrix, and aggregate microcracks. Further, bond microcracks (microcracks between the cement mortar matrix and aggregate particles) exist in the form of shrinkage microcracks (Figure 7-1) prior to the application of compressive stress fields.

Shrinkage Microcracks

The results obtained by these workers indicate the existence of shrinkage microcracks in concrete containing both rounded and angular aggregate. These initial shrinkage micro-cracks, which were in reality bond microcracks (microcracks between the cement mortar matrix and aggregate particles), exist for a variety of reasons. Sturman suggests that shrinkage microcracks may be formed by a variety of processes, including carbonation shrinkage, hydration shrinkage, segregation due to settlement, and drying shrinkage. All concrete, regardless of the measures or procedures used to eliminate shrinkage micro-cracks, will contain them.

Carbonation shrinkage occurs when any cement compound is stored in air and decomposed by carbon dioxide. The portland cement used in most construction projects is of sound character, so that carbonation shrinkage is unlikely to occur.

Hydration shrinkage occurs when the primary cement-paste volume decreases during hydration, resulting in the formation of microcracks. In most cases hydration shrinkage is responsible for a small amount of shrinkage microcracking in concrete. This type of microcracking is believed to be controlled by using an expansive cement.

The influence of *segregation due to settlement* on the formation of microcracks may be analyzed by applying Stokes' law to the viscous material first formed when sand, cement, and water are mixed to form mortar. Stokes found that for very small solid particles suspended in a viscous fluid, the steady-state or terminal velocity acquired by the larger and denser particles is greater than that of the smaller, less dense particles. Thus, in a sand-cement-water mixture, the large sand particles will settle first, the fines next, and the extremely fine, flocculated particles last. This leads to a condition in which there is a thin film of fluid adjacent to the aggregate which has an extremely high water-

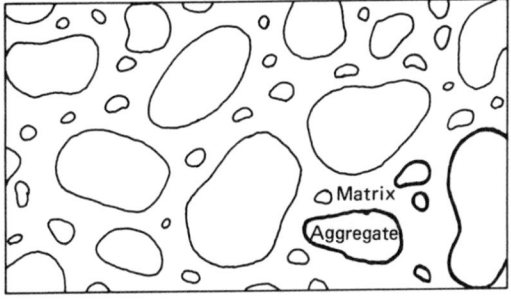

Figure 7-1 Shrinkage Microcracks

to-solids ratio. Eventually, this water is absorbed by the adjoining cement paste, which hydrates continuously, leaving a thin space on the aggregate. When this sedimentation occurs at the exposed horizontal surface of freshly poured concrete, it is referred to as bleeding. This phenomenon is likely to cause most of the shrinkage microcracks in concrete and is probably the leading reason for eventual concrete failure.

Drying shrinkage occurs in plastic concrete if the rate of evaporation exceeds 0.1 lb/ft² of surface area per hour. In other words, hydrostatic tension is present, resulting in shrinkage microcracks. This is a minor cause of shrinkage microcracks, as most concrete is cured in a water-saturated atmosphere.

Bond Microcracks

Bond microcracks (Figure 7-2) are extensions of shrinkage microcracks. As the compressive stress field increases, the shrinkage microcracks widen but do not propagate into the matrix. According to Derucher, this widening of the shrinkage microcracks occurs

Figure 7-2   Bond Microcracks

at 15 to 20 percent of the ultimate strength of concrete (f'_c). It should be kept in mind that the weakest point in concrete is the bond between the aggregate and the matrix.

Matrix Microcracks

Matrix microcracks (Figure 7-3) are microcracks that occur in the matrix which are extensions of the bond microcracks into the matrix. Again according to Derucher, the bond microcracks start to propagate at or around 20 percent of f'_c. They propagate to the

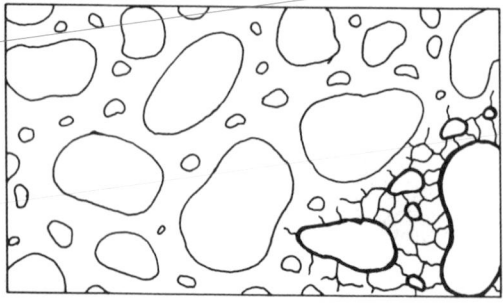

Figure 7-3   Matrix Microcracks

point where they begin to bridge one another at about 30 to 45 percent of the ultimate strength of concrete. At 75 percent of f'_c matrix microcracks start to bridge one another.

Aggregate Microcracks

Aggregate microcracks occur just before failure, at about 90 percent of f'_c; failure will definitely occur.

FACTORS THAT CAUSE DETERIORATION

The factors that cause deterioration may be grouped under the following areas: poor design details, construction deficiencies and operations, temperature variations, chemical attack, reactive aggregates (alkali-silica, alkali-carbonate, Kansas-Nebraska sand-gravel, and silicification of carbonate aggregate) and high-alkali cement, moisture absorption, wear and abrasion, shrinkage and flexure forces, collision damage, scouring, shock waves, overstress, fire damage, foundation movement, corrosion of reinforcing bars and prestressing wires, and aggregates that cause deterioration. The visual systems of deterioration consist of cracking, scaling, spalling, rust stains, surface disintegration, efflorescence and exudation, and vehicular damage. Each will be examined with the form of deterioration most responsible for that form of disintegration.

Poor Design Details

One of the poor design details that can cause concrete to crack is insufficient drainage. In bridge construction, scuppers may not be provided (or properly provided) with downspouts to keep the water discharge away from concrete surfaces, including caps and decks. The scuppers may be too small and, as a result, easily clogged. They should also be provided at low spots. Weep holes, if provided, may be too small, too few, or discharging over another concrete surface. Spalling may result when not enough space is provided at an expansion joint (Figure 7-4). Insufficient cover over rebars may cause corrosion of the rebars, or, in the case of prestressed concrete, of the prestressing wires. Of all these considerations, the need for sufficient expansion space cannot be overstated.

Construction Deficiencies

Construction deficiencies may result in concrete deterioration regardless of the care taken in the design procedure. Soft spots in the subgrade of a foundation will cause settlement, resulting in cracking (Figure 7-5). If formwork is removed between the time the concrete begins to harden and the specified time for formwork removal, cracks will probably occur. These cracks are especially dangerous, since they may occur internally, with no external manifestations. Water can collect in these cracks and cause spalling due to freezing and thawing or can cause corrosion of internal steel, which may eventually lead to cracking and spalling. If sufficient spacing does not exist between rebars in reinforced sections,

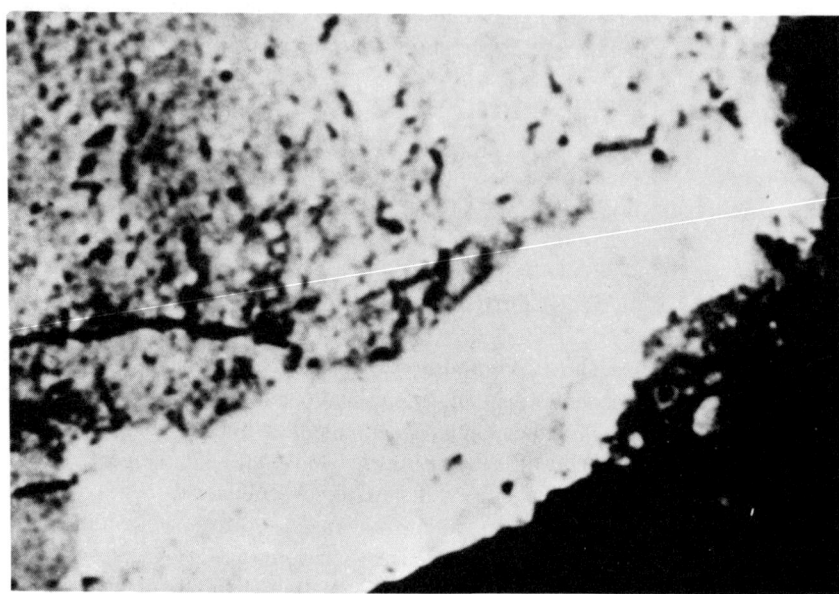

Figure 7-4   Spalling at Expansion Joint

Figure 7-5   Cracking Due to Foundation Settlement

voids may develop if the mix is not properly vibrated. These voids collect water, with end results similar to those discussed earlier for cracks that collect water. Excess vibration may cause segregation of the concrete mix, resulting in a weaker concrete than specified. The inclusion of clay or soft shale particles in the concrete mix will cause small holes to appear in the surface of the concrete as these particles dissolve. These tiny holes increase the porosity of the concrete and, as before, lead to cracking and spalling and possible corrosion of the internal steel (Figure 7-6).

Temperature Variations (Freezing and Thawing)

Freezing and thawing (temperature variation) is a common cause of concrete deterioration. Most ordinary concrete contains 1 or 2 percent air in the form of voids. This entrapped air makes the concrete very porous. Porous concrete absorbs water, and if allowed to freeze, the water will expand approximately 9 percent. In the process of expansion, unfrozen water is forced through small capillaries in the cement paste. Resistance to this movement of water through these small capillaries produces hydrostatic pressure. This

Figure 7-6 Surface Pitting Due to Included Clay

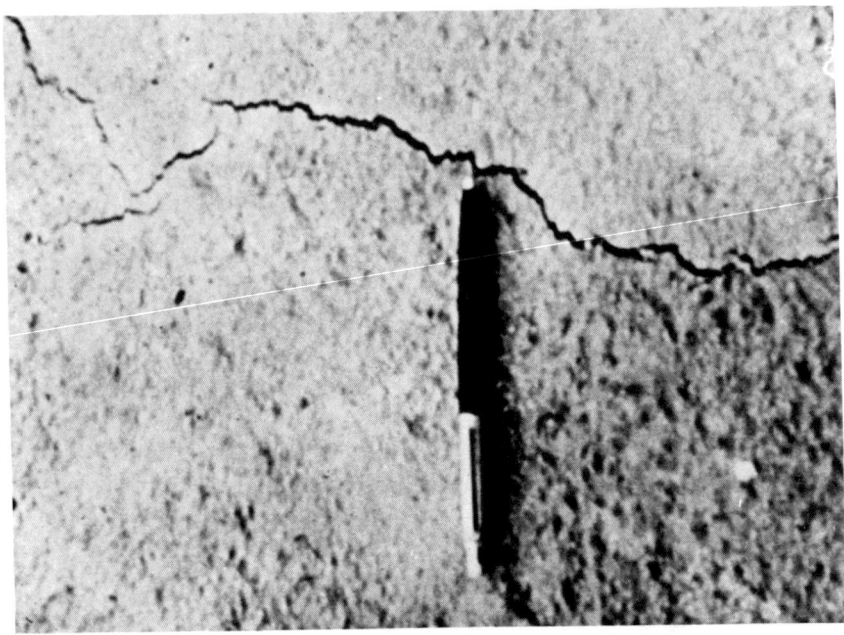

Figure 7-7   Freeze-Thaw Cracking

internal expansive pressure often produces cracking (Figure 7-7), spalling (Figure 7-8), or scaling.

Three factors influence the deteriorating effect of freezing and thawing:

1. Concrete permeability.
2. Ratio of freezable water to the air voids.
3. Rate of temperature decline upon freezing.

Factors 1 and 2 are influenced by the character and type of aggregate, the quality of the cement, mix procedures, placement, curing, and the number of air bubbles in the concrete. Factor 3 has little effect in practical field operations but plays a substantial role in laboratory testing. That is, the environment does not allow a temperature decline as fast as that which can be produced in the laboratory.

In most situations the effects of freezing and thawing can be eliminated by providing sound drainage under the concrete slab. Another method is the utilization of an air-entrained concrete, which allows for thousands of additional air voids. In this case, as water begins to freeze and expand, the unfrozen water is forced out of that space and into one that possibly has no water in it, thus minimizing the effect of hydrostatic pressure.

Temperature fluctuations may further affect concrete integrity if the coefficient of thermal expansion differs significantly from that of the mortar. Aggregates with lower coefficients may cause high tensile stresses, resulting in cracking and spalling. Further problems arise if the concrete section is restrained from expansion or contraction. The

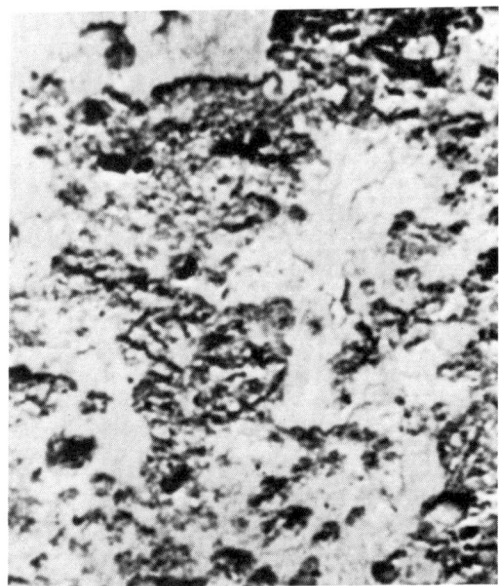

Figure 7-8   Surface Deterioration

internal forces set up under these circumstances are sufficient to result in cracking and spalling and eventual failure of the member.

Chemical Attack

Chemical attack of concrete may come from two sources. The use of salts or chemical deicing agents contributes to weathering through recrystallization, and salt may increase the water retention. The results are similar to the effects of freezing and thawing (cracking and spalling). Sodium and calcium chloride are the principal salts applied directly to the concrete pavement to keep ice and snow from the pavement. This result is most serious in concrete bridge decks, as they generally freeze before the adjacent roadway; hence, more salt is applied to the bridge deck. As a consequence, bridge decks deteriorate more quickly than the adjacent roadway. Several coats of linseed oil applied to the bridge deck help to prolong deck life.

Deicing agents generally result in scaling of the concrete surface. These agents lower the freezing point of water, which in effect increases the number of freeze-thaw cycles during a prolonged cold spell.

Further chemical attack may come from chemicals in the soil or water surrounding the concrete members. Ammonium and magnesium ions react with the calcium in the cement paste. Sodium, magnesium, and calcium sulfate react with the tricalcium aluminate in the cement paste, producing additional quantities of gypsum. Gypsum, you will recall, is added to the cement clinker to react with tricalcium aluminate in the portland cement to prevent a flash set (a situation in which the concrete gives the impression of being set but is not). In other words, gypsum acts as a retarder in preventing the concrete from setting too quickly, giving workers the time to mix, place, and work with the concrete.

However, if too much gypsum is interground with the cement clinker or allowed to form during the hydrolysis process, the concrete may not set for an indefinite period of time. Thus, if additional sulfate is introduced to the system during the hydration process, additional gypsum is produced and the process of sulfate attack occurs. Magnesium sulfate is one of the more important attacking agents. It reacts with lime [$Ca(OH)_2$] as follows:

$$Ca(OH)_2 + MgSO_4 \cdot 7H_2O \rightarrow CaSO_4 \cdot 2H_2O + Mg(OH)_2 + 5H_2O \qquad (7.1)$$

The calcium sulfate and the water combine to form gypsum, which then reacts with the tricalcium aluminate. Sulfate-resistant cement will restrict this type of action.

Reactive Aggregates and High-Alkali Cement

Reactive aggregates and high-alkali cements cause swelling, map cracking (Figure 7-9), and popouts (Figure 7-10) in portland cement concrete. The literature points out four types of reactions that may result due to reactive aggregates:

1. Alkali-silica reaction.
2. Alkali-carbonate reaction.
3. Kansas-Nebraska sand-gravel reaction.
4. Silicification of carbonate aggregate.

The alkali-silica reaction was first reported by Thomas E. Stanton in 1940. Stanton noticed that certain forms of silica and siliceous material may produce deterioration of concrete

Figure 7-9 Map Cracking

Figure 7-10   Popouts Due to Reactive Aggregate

structures by expansion and cracking of concrete through reaction with alkalies in cements in the form of sodium and potassium oxides. The expansion is caused by the formation of a gel-like material around the reactive aggregate. This material dries to a white amorphous substance around the aggregate (reactive rim). The reaction between the cement and aggregate may be explained as follows. During the initial mixing of the concrete and for several hours thereafter, the water in the concrete acquires alkalies (sodium oxide and potassium oxide) from the cement by preferential solubility, a solution being formed. As the hydration process continues, the calcium aluminates and the silicates of the cement paste extract water. Thus, less water is left in solution, resulting in an increase in the alkalies in the remaining water. If this results and if the cement is high in alkali, a rather caustic solution is formed. During curing this caustic solution reacts chemically with susceptible aggregates to form a silica-gel reaction rim around the aggregate particle. This rim has a great affinity for water, thus attracting water from the cement paste and reducing its viscosity. Osmotic pressures are set up by the drawing of the water from the cement paste, resulting in fracture of the cement paste close to the reaction rim. As the silica gel continues to grow along the fracture, cracks develop and extend until the entire concrete surface is a series of cracks. The cracks themselves may be detrimental to the concrete structure, but an even more serious problem results if the cracks allow water to enter and then go through various cycles of freezing and thawing. The most reactive aggregates have been found to contain opal, chalcedony, certain forms of chert, glass in felsites, rhyolite, basalts, cristobalite, and tridymite.

Three factors are necessary before the alkali-silica reaction can occur:

1. A reactive aggregate must be used.
2. Alkalies in the cement must be high.
3. The concrete must be partially or totally wet.

Obviously, to avoid the consequences of the alkali-silica reaction in concrete, three principles must be recognized:

1. Do not use reactive aggregates.
2. Do not use cement with a high alkali content.
3. If it becomes necessary to bring a reactive aggregate together with a high-alkali cement, use an admixture (such as a pozzolan) that will mitigate the deleterious consequences of the reaction.

To avoid the use of a reactive aggregate, it first becomes necessary to know which aggregates are reactive. If it is known that an aggregate has a satisfactory performance characteristic, continue to use it. If there is doubt or if the performance characteristics of a specific aggregate are not known, certain ASTM tests should be made on the aggregate. These tests entail a petrographic examination, a chemical test, and a proof test. ASTM C295 (Recommended Practice for Petrographic Examination of Aggregate for Concrete) provides specific information with reference to deleterious materials. Some reactive aggregates can be recognized readily, whereas others require more testing such as ASTM C289 [Test for Potential Reactivity of Aggregates (Chemical Method)] and/or ASTM C227 [Test for Potential Alkali Reactivity of Cement-Aggregate Combinations (Mortar-Bar Method)].

The second way of avoiding an alkali-silica reaction is to limit the alkali content in the cement to a value of 0.60 percent, as suggested by ASTM C150 (Specifications for Portland Cement). This value was arrived at empirically by correlation with field indications of distress.

The third method of avoiding an alkali-silica reaction is to use a pozzolanic or other admixture to mitigate the effects of alkali-aggregate reactions. This method of avoidance is covered by ASTM C441 (Test for the Effectiveness of Mineral Admixtures in Preventing Excess Expansion of Concrete Due to Alkali-Aggregate Reaction).

The alkali-carbonate reaction was first recognized by E. G. Swanson in 1957. Unlike the alkali-silica reaction, no visible gel, reaction product, or rim is apparent. In this reaction certain argillaceous (dolomitic limestones) aggregates react with the alkalies in the cement, which results in a destructive expansion. This expansion is referred to as *dedolomitization,* in which the dolomite is replaced by the formation of calcite and brucite. According to D. W. Hedley, the reaction is as follows:

$$CaMg(CO_3)_2 + 2MOH = Mg(OH)_2 + CaCO_3 + M_2CO_3 \qquad (7.2)$$

in which M represents potassium, sodium, or lithium. In concrete, the alkali carbonate produced by this reaction would react with hydration products of portland cement to regenerate alkali:

$$M_2CO_3 + Ca(OH)_2 \rightarrow 2MOH + CaCO_3 \qquad (7.3)$$

This process continues until all the dolomite has reacted or all the alkali has been used up.

Another important factor in the alkali-carbonate reaction is the texture of the aggregate. If the dolomite has a coarse grain, the reaction will not take place to any significant degree. For the reaction to occur, the dolomite must consist of fine-grained particles.

If it is known that an aggregate has a satisfactory performance characteristic, continue to use it. If there is doubt or if the performance characteristics of a specific aggregate are now known, certain ASTM tests should be made on the aggregate. These tests include ASTM C289 [Test for Potential Reactivity of Aggregates (Chemical Method)], ASTM C227 [Test for Potential Reactivity of Cement-Aggregate Combinations (Mortar-Bar Method)], and the draft form of another possible ASTM test (Test for Length Change of Concrete Due to Alkali-Carbonate Rock Reaction).

Concrete made with the sand-gravel aggregate (natural aggregate) found in the Kansas-Nebraska area as well as surrounding areas has been known to result in expansion, cracking, and eventually early destruction of the concrete. This reaction is known as the Kansas-Nebraska sand-gravel reaction. The problem is not a serious problem nationally, but where it does occur it is a major one. The cause of the reaction has not yet been determined. Many suspect that it could be an alkali-silica reaction or an as-yet-undefined chemical or physical phenomenon.

The final type of aggregate reaction is the silicification of carbonate aggregate. It has been observed in Iowa by J. Lemish, who proved that certain aggregates develop siliceous rims around the aggregate when used in portland cement concrete. This type of reaction is still under investigation by researchers.

Moisture Absorption

Moisture absorption will cause concrete to swell. Concrete cylinders 13 ft in diameter have grown as much as 6 in. in a marine environment. If restrained, the restraining material will break apart or the concrete will crack and spall. In addition, saltwater will tend to leach out the lime in the cement, leaving a powdery residue.

Wear and Abrasion

Wear and abrasion cause the surface of concrete to disintegrate over extended periods. In general, roadway pavements may become a problem if the aggregate is susceptible to polishing. In waterway structures made of concrete, wind and water-driven particles, particularly sand, play a significant role in abrasion near the mud line and above the high-water line. Piles may be damaged due to the rubbing action of vessels, resulting in scaling, swelling, and cracking at joint, and scarring.

Shrinking and Flexure Forces

Shrinking and flexure forces may set up tensile stresses that exceed the capacity of the concrete section. The end result is cracking. Setting shrinkage generally causes shallow surface cracking (Figure 7-11). Drying shrinkage may take place over extended periods

Figure 7-11   Shallow Surface Cracking

of time, producing tensile forces and eventually cracking. The time frame may amount to several years.

Collision Damage

Collision damage is generally considered to be an accidental occurrence. Any concrete structure that can be reached by a moving vehicle has suffered collision damage at some time. In most instances the vehicle is more severely damaged than the structure.

Scouring

Concrete surfaces in surf zones scarred by sand and silt (Figure 7-12) are referred to as scour. Ice flows in rivers and bays are also responsible for considerable damage to concrete piling and piers (Figure 7-13). Most damage of this type occurs between the low-and high-water marks.

Shock Waves

A shock wave can damage concrete due to the varying transmission rates through the aggregate, the paste, and the reinforcing steel. The shock waves then become partially additive, setting up conditions leading to cracking and spalling of the concrete mass. Concrete piles are also vulnerable to damage from shock waves while they are being driven.

Figure 7-12   Sand and Silt Scarring

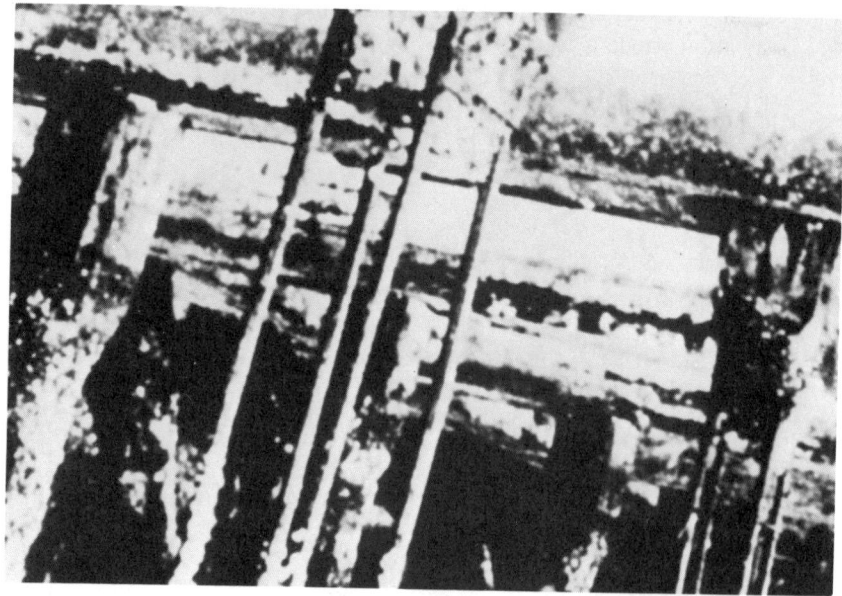

Figure 7-13   Total Removal of Concrete Matrix from Reinforcing Rods Due to Ice Flow

Overstressing

Overstressed concrete in a deck may exhibit longitudinal and lateral cracking. Over a bearing point there may be diagonal cracking at the end of a simple beam (Figure 7-14), vertical cracking running from the bottom at the center of a simple beam to the neutral axis, and vertical cracking from the top of the beam extending downward to the neutral axis for a beam that is continuous over the bearing area (Figure 7-15).

Fire Damage

Fire damage results from the extreme temperatures of a large fire. Temperatures above 570°F (300°C) will cause weakening in the cement paste and lead to cracking and spalling (Figure 7-16).

Foundation Movement

Foundation movement will cause serious cracking in concrete structures if it generates a sizable tensile stress in the concrete piers or abutments (vertical cracking predominates).

Corrosion

Corrosion of the internal reinforcing steel causes tremendous expansion pressures within the concrete. Fully corroded steel occupies seven times the volume of uncorroded steel, causing cracking and spalling as a result of the volume increase (Figure 7-17). The steel may also corrode if water finds its way into the concrete member.

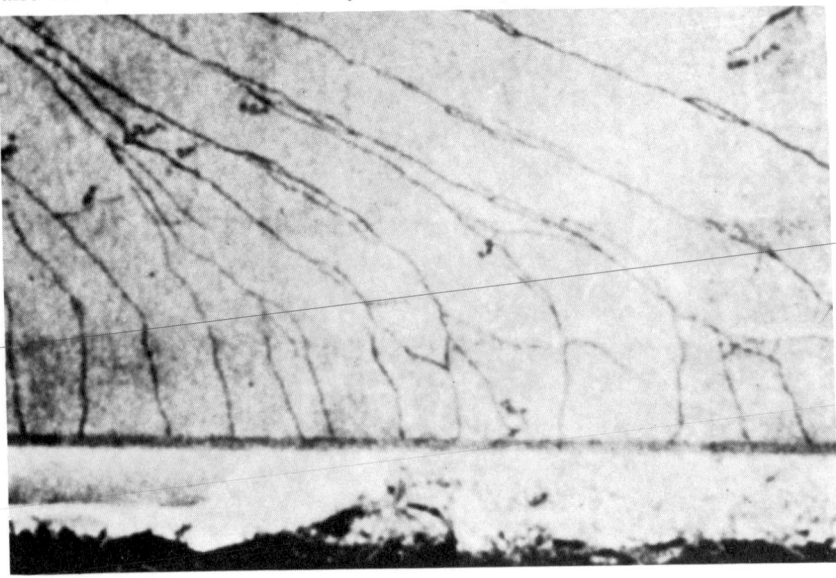

Figure 7-14   Diagonal Cracking

Figure 7-15   Concrete Member in Models Laboratory

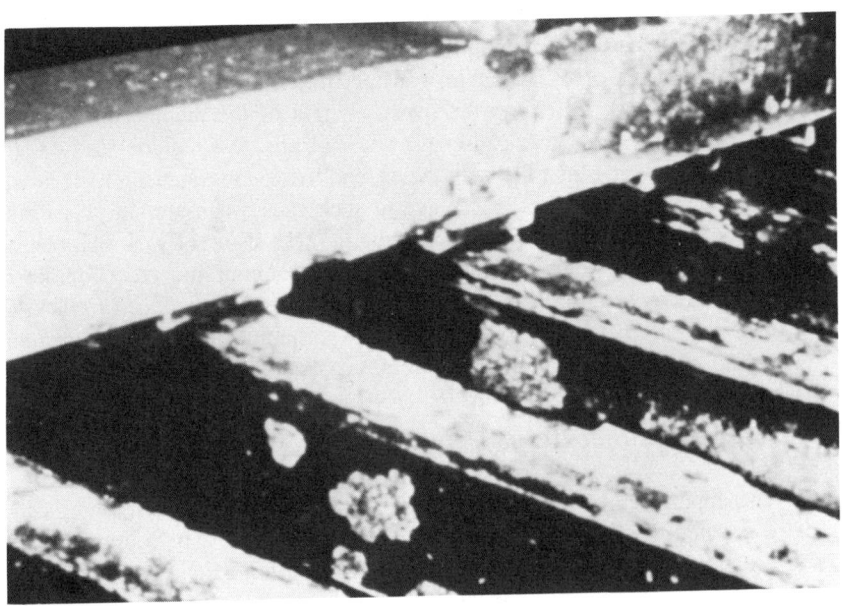

Figure 7-16   Cracking and Spalling Due to Fire

Figure 7-17 Piles Cracking

Corrosion of the prestressing wire combined with the high tensile stress required for prestressing may result in failure of a member due to stress corrosion. Failure of a significant number of prestressing wires in this manner will result in loss of tensile strength in the member and could lead to failure under heavy loading conditions. Further deterioration of the tensile strength of a member may result from creep of the prestressing steel; shrinkage of the concrete, causing relaxation in the prestressing wires; and creep of the concrete, shortening the overall length of the member and thereby relaxing the prestressing wires. Other causes for a loss of prestress include elastic deformation of the concrete, drawing in of the anchorages, and friction loss in the post-tensioning operations. The combined loss of prestress due to such causes can amount to as much as 25 to 35 percent of the member's design strength. The result of loss of prestress is cracking, particularly near the anchorages and in the compression face. Unlike cracks in high-tension areas of reinforced-concrete members, the appearance of cracks in a prestressed member may have a serious effect on its structural integrity. A prestressed member is usually under high compressive stress; consequently, cracking should not be expected.

Aggregates that Cause Deterioration

The importance of identifying aggregates that cause deterioration of concrete cannot be overstated. ASTM C88 (Soundness of Aggregates by Use of Sodium Sulfate or Magnesium Sulfate) covers the testing of aggregates to determine their resistance to disintegration by saturated solutions of sodium sulfate or magnesium sulfate. It further furnishes information helpful in judging the soundness of aggregates subjected to weathering action, particularly when adequate information is not available from service records of the material exposed to actual weathering conditions.

There are basically three classifications of distress in concrete due to unsound aggregate:

1. Pitting and popouts.
2. D-line deterioration.
3. Map cracking.

Pitting is the disintegration of weak, friable particles of aggregate as a result of frost action. It is a gradual deterioration near the surface of the deleterious particles. *Popouts* are caused by the rapid disruption of harder but saturated aggregate particles. Chert is the most common cause of popouts. The usual result of pitting and popouts is the poor appearance of the structure. However, the structure generally remains sound because of the insufficient quantity of the particles causing pitting of popouts. A considerable amount of unsound aggregate would be required to cause structural deficiencies. However, if water were to enter into crevices provided by pitting or popouts, severe damage could result if various cycles of freezing and thawing were encountered. If the unsound aggregate is sufficient in quantity, strength and appearance will suffer. In pavement, the concrete slab becomes susceptible to structural failure through heavy truckloads and blowups. Blowups are caused by the expansion that occurs within a pavement immediately after precipitation on a hot day.

D-line cracking is a result of a change in volume in coarse-aggregate particles, resulting in the failure of the aggregate-matrix bond. Further deterioration occurs as water enters the cracks. D-line cracks appear on first sight as very fine cracks along transverse joints in concrete pavements. As they progress, they appear along longitudinal joints and free edges.

Map cracking is the disintegration of concrete in which the cracking is random and appears over an entire surface. It is caused mainly by the property of unsound aggregates to expand while the cement mortar is shrinking. Freezing and thawing, alkali-silica, alkali-carbonate, and excess heat may cause map cracking.

ADMIXTURES

Admixtures are generally added to the basic ingredients that make up concrete in order to modify various properties of the concrete to make it more suitable for use. Admixtures are used only when the desired properties cannot be achieved by altering the design mix. ASTM C494 (Chemical Admixtures for Concrete) covers materials for use as chemical admixtures to be added to portland cement concrete mixtures in the field for the purpose or purposes indicated for the seven types as follows:

Type A. Water-reducing admixtures.
Type B. Retarding admixtures.

Type C. Accelerating admixtures.

Type D. Water-reducing and -retarding admixtures.

Type E. Water-reducing and -accelerating admixtures.

Type F. Water-reducing, high-range admixtures.

Type G. Water-reducing, high range, and retarding admixtures.

This specification stipulates tests for an admixture with suitable concreting materials or with cement, pozzolan, aggregates, and an air-entraining admixture proposed for specific work.

Water-reducing and -retarding Agents

There are five classes of water-reducing and -retarding agents:

1. Lignosulfonic acids and their salts.
2. Modifications and derivations of lignosulfonic acids and their salts.
3. Hydroxylated carboxylic acids and their salts.
4. Modifications and derivations of hydroxylated carboxylic acids and their salts.
5. Carbohydrates and modifications and derivatives thereof.

Water-reducing admixtures reduce the quantity of mixing water required to produce concrete of a given consistency. The most common water-reducing admixtures are of organic material. The water-reducing admixtures yield good strength advantages for a given cement content by reducing the water content of the concrete mix. The compressive strength increases 10 to 20 percent while the water content is reduced 5 to 15 percent. Water-reducing admixtures are effective with all types of portland cement, portland blast-furnace-slag cement, and portland-pozzolan cements.

Retarding admixtures increases the setting time for concrete, allowing a permissible period between batching and the final placement.

Accelerating Admixtures

An accelerating admixture accelerates the setting and early-strength development of concrete. Accomplishing this results in better scheduling of the workload, earlier removal of construction forms, and compensation effects of low temperature on early-strength development. The most common accelerator is calcium chloride ($CaCl_2$).

In addition to increasing strength, calcium chloride has the following secondary effects:

1. Increases workability by a small amount.
2. Increases air content and air-bubble size in air-entrained concrete.
3. Reduces bleeding.

4. Increases drying shrinkage slightly.

5. Lowers resistance to sulfate attack.

Water-reducing and -retarding Admixtures

Water-reducing and -retarding admixtures reduce the quantity of mixing water required to produce concrete of a given consistency and retard the setting of concrete.

Water-reducing and -accelerating Admixtures

Water-reducing and -accelerating admixtures reduce the quantity of mixing water required to produce concrete of a given consistency and accelerate the setting and early-strength development of concrete.

Air-entraining Agents and Admixtures

Admixtures may also be used to entrain air in the concrete. Air-entrained concrete is more plastic and workable than non-air-entrained concrete. It can be placed and compacted with less bleeding and less aggregate segregation. It improves both workability and durability, but reduces strength.

Air-entraining agents can also be interground with cement clinker at the mill to form air-entraining cements. The only advantage of air-entraining cements is the convenience of knowing that you do not have to include an admixture.

Air-entraining agents and admixtures are covered by ASTM standards. ASTM C233 (Air-Entraining Agents for Concrete) covers the testing of materials proposed for use as air-entraining admixtures to be added to concrete mixtures in the field. These tests are based on arbitrary stipulations permitting highly standardized testing in the laboratory and are not intended to simulate actual job conditions. ASTM C260 (Air-Entraining Admixtures for Concrete) covers materials proposed for use as air-entraining admixtures to be added to concrete mixtures in the field.

Other Admixtures

Of the large number of admixtures available for concrete, an important group falls into the category of finely divided mineral admixtures, which are divided into three classifications:

1. Those which are chemically inert.

2. Those which are pozzolanic.

3. Those which are cementitious.

The inert finely divided material includes quartz, limestone, bentonite, and hydrated lime. These finely divided materials are often substituted for portland cement in concrete to save on portland cement. To have the necessary workability and plasticity, most concrete must have large amounts of cement. Substituting finely divided material for portland cement saves a considerable amount of portland cement.

Pozzolanic or cementitious finely divided material contributes to the strength development of concrete and reduces the amount of portland cement required for a given strength. A pozzolan, as described by ASTM C618, is a siliceous or siliceous and aluminous material which itself has little or no cementitious value but which will, in finely divided form and in the presence of moisture, chemically react with calcium hydroxide at ordinary temperatures to form compounds that do have cementitious properties. Pozzolans include diatomaceous earth, opaline cherts, shales, volcanic ashes, fly ash, and clays. The most common pozzolan is fly ash, a finely divided residue that results from the combustion of ground or powdered coal. ASTM C618 (Fly Ash and Raw or Calcined Natural Pozzolan for Use as a Mineral Admixture in Portland Cement Concrete) covers fly ash and raw or calcined natural pozzolan for use as a mineral admixture in concrete where cementitious or pozzolanic action or both is desired; where other properties normally attributed to finely divided mineral admixtures may be desired; or where both objectives are to be achieved. These classifications exist under ASTM C618; Classes N, F, and C. Class N comprises raw or calcined natural pozzolans. They include diatomaceous earths; opaline cherts and shales; tuffs and volcanic ashes or pumices, any of which may or may not be processed by calcination; and various materials requiring calcination to induce satisfactory properties, such as some clays and shales.

Class F is fly ash normally produced from burning anthracite or bituminous coal which meets the applicable requirements for this class. This class of fly ash has pozzolanic properties.

Class C is fly ash normally produced from lignite or subbituminous coat that meets the applicable requirements for this class. This class of fly ash, in addition to having pozzolanic properties, has some cementitious properties. Some Class C fly ashes may have a lime content exceeding 10 percent.

PROBLEMS

7.1. Explain the failure mechanism of concrete according to Derucher.

7.2. Why does concrete shrink?

7.3. List the factors that cause deterioration of concrete.

7.4. What factors influence the deteriorating effect of freezing and thawing?

7.5. Explain the alkali-silica reaction.

7.6. Explain the alkali-carbonate reaction.

7.7. List and discuss three classifications of distress in concrete that result from the use of unsound aggregate.

7.8. List the five types of chemical admixtures outlined by the ASTM.

7.9. Calcium chloride is an accelerator in concrete. What is the primary advantage of using it? The secondary advantages?

7.10. List and discuss pozzolanic admixtures.

REFERENCES

BRANDTZAEG, A., "Study of the Failure of Concrete under Combined Compressive Stresses" *UIEES Bull. 185,* 1928.

BURG, O., "The Factors Controlling the Strength of Concrete," *Constr. Rev., 33,* No. 11 (1950), p. 19.

DERUCHER, K. N., "Applications of the Scanning Electron Microscope to Fracture Studies of Concrete," *Build. Environ., 13,* No. 2 (1978), pp. 135–141.

DERUCHER, K. N., and HEINS, C. P., *Bridge and Pier Protective Systems and Devices* Civil Engineering Series, Marcel Dekker Inc., New York, 1979.

HANSEN, T. C., "Microcracking of Concrete," *J. Am. Concr. Inst., 64,* No. 2 (1968), pp. 9–12.

HERMITE, J., "Present Day Ideas on Concrete Technology," *University of Illinois Bull. 18,* 1954, pp. 27–39.

HSU, T., SLATE, F., STURMAN, G., and WINTER, G., "Microcracking of Plain Concrete and the Shape of the Stress Strain Curve," *J. Am. Concr. Inst., 60,* No. 2 (1963), pp. 209–224.

JONES, R., "A Method of Studying the Formation of Cracks in a Material Subjected to Stress," *Br. J. Appl. Phys., 3* (1952), p. 329.

ROBINSON, J., "X-ray Analysis of Concrete Fracture," *J. Am. Concr. Inst., 27,* No. 4 (1955).

RUSCH, H., "Physical Problems in the Testing of Concrete," *Cement-Chalk, 12,* No. 1 (1959), pp. 1–9.

STANTON, T. E., "California Experience with the Expansion of Concrete through Reaction Between Cement and Aggregate," *J. Am. Concr. Inst., 39* (Jan. 1942).

STURMAN, G., "Shrinkage Microcracks in Concrete," Ph.D. dissertation, Cornell University, 1969.

SWENSON, E. F., "A Reactive Aggregate Undetected by ASTM Tests," *ASTM Bull. 226,* Dec. 1957, pp. 48–50.

Timber has been one of the basic materials of construction since the earliest days of humankind. Today it has been largely superseded by concrete and steel. However, the use of timber remains quite extensive.

In the United States 600 different tree species exist, of which 15 are utilized in the building trades (ash, birch, cedar, cypress, fir, gum, hemlock, hickory, maple, oak, pine, poplar, redwood, spruce, and walnut). A number of these include several varieties, each showing different characteristics. One variety of walnut, for example, might make a better building product than another.

Therefore, the engineer should have some knowledge of the classification of trees, as well as of their growth and structural ability, to understand the physical and mechanical properties of each.

CLASSES OF TREES

Exogens and Endogens

Botanists classify trees into two main categories: exogens and endogens. These terms refer to the pattern of growth of a particular species. *Exogenous trees* are those which grow diametrically, by adding new cells in a layer between the existing wood and the bark. Almost all wood that is of commercial use falls in this category. The exogenous growth pattern forms a characteristic transverse section with concentric annual rings which correspond to each year's growth. At the center of a cross section of a tree is pith, a dark area of small diameter that does not increase in size after the first year of growth. Encircling the outermost annual ring is a layer of microscopic thickness called the cambium ring, and it is here that new growth is formed. The outmost layer is known as the bark.

Endogenous trees add new living fiber to the old by allowing new fiber to intermingle with the old, thus producing growth both diametrically and longitudinally. Most endogens are fairly small plants, such as corn, sugar cane, palm, and bamboo. The latter two plants are used extensively throughout the Orient as a structural material.

Hardwoods and Softwoods

Exogenous trees are further classified as hardwoods or softwoods. These are somewhat misleading terms, since they refer not to a quality of the wood, but to the types of tree the wood comes from. Hardwoods comprise the broadleafed trees, mostly deciduous, and include the many varieties of oak, maple, ash, hickory, cherry, poplar, gum, walnut, elm, beech, birch, basswood, whitewood or tulip, and ironwood. These trees are in general slow-growing, forming heavy and hard wood that is used more extensively for furniture, interior finishing, and cabinetwork than for structural purposes. The softwoods are the conifers, such as pine, spruce, fir, hemlock, cedar, cypress, and redwood.

WOOD STRUCTURE

Many of the qualities that are unique to wood for use as a building material arise from the structure of wood. Wood is formed mostly of filamentous cells, fairly tubular in shape, varying in length from 1 to 3 mm and having a diameter that is about $1/100$ of their length. The shape of the cell in transverse section is approximately rectangular with rounded corners. Perforated pits in the cells make possible the flow of sap between cells.

Wood cells are composed primarily of the polymers cellulose and hemicellulose. These are similar to the sugar molecule, but form long chains that run in bundles helically. They are cemented together by a third polymer, lignin. Wood is typically 50 to 60 percent cellulose material and 20 to 35 percent lignin, with small amounts of other carboxydrates, such as resin and gum. It is the cellulose that gives wood its axial strength and elastic properties, and the lignin that is responsible for its compressive strength. These carboxydrates, although indigestible by man, are a favorite food for some other forms of life, such as termites, marine borers, and fungi, and measures must often be taken to protect wooden structures from these species.

Most fibers run parallel to the axis of the tree, and these cells, which are more or less empty in wood that is ready for use, act as a collection of empty tubes under stress. Since the tree develops from a sapling by adding new longitudinal cells in layers around the existing wood, differences in periods of growth give rings of cells with different characteristics. Growth is rapid in the spring and slows down in the summer, stopping altogether in the fall and winter. During the spring growth period, cells that are formed have relatively thin walls and an open texture, producing what is known as spring wood. The cross-sectional area of spring wood may form as little as 10 percent of the cell wall. The summer growth is slower and the wood formed is denser and stronger, with thicker cell walls. This summer wood may have cell walls constituting up to 90 percent of the cross-sectional area. The pattern of a layer of spring wood followed by increasingly dense wood that is formed over each year gives the wood its annual rings. These rings vary in size with the type of tree involved and the growth conditions during a particular year. A typical annual thickness is approximately $\frac{1}{10}$ in., with $\frac{1}{2}$ in. being about the maximum. In tropical trees, the growth rings correspond to the rainy and dry seasons.

Over a period of several years the wood ceases to function as part of the living tree and has only structural value for the tree. The living part is called *sapwood;* the older wood is *heartwood*. Heartwood is easily distinguished from sapwood because it is darker in color. It is more highly valued since it is more resistant to decay than sapwood, although the strength of each is about the same.

Also parallel to the axis of the tree and running the length of the tree are vessels formed by the fusion of chains of cells. These vessels are characteristic of hardwoods, and are sometimes visible in cross section without a microscope. Other groups of cells lie in radial arrangements, and these form what are known as radial, pith, or medullary rays. Medullary rays are present in all trees, but are especially prominent in woods such as oak. In the living tree they carry food from the inner bark to the cambrium layer and add strength in a radial direction.

The terminology of the tree is best illuminated by a cross section such as the one shown in Figure 8-1.

PHYSICAL CHARACTERISTICS OF WOOD

Season of Cutting; Slash and Rift Cutting

The first step in the manufacturing process of lumber is to select and fell the trees. This is normally done in the fall and winter when the flow of sap is minimal and destructive fungi and insects are least active. The logs are cut into standard-size boards at sawmills, as illustrated in Figure 8-2. The lumber must then be dried or seasoned.

The angle of the plane of the cut of the boards will have an effect on the resulting lumber. Wood that is quartersawed or *rift-cut* is cut parallel to the axis of the tree and radially across the annual rings. When the wood is cut tangential to the annual rings, it is called *slash-cut*. Rift-cut lumber generally warps less in the drying process and wears more evenly, but in the interest of practicality, more wood is slash-cut. A typical sawing pattern is shown in Figure 8-3.

Seasoning of Timber, Moisture Content

Wood must be seasoned before it is put to use so that its moisture content will become stabilized. Unseasoned wood readily takes up and retains moisture, and this moisture will adversely affect the wood in a number of ways. One is that changes in moisture content will produce dimensional changes in the timber. Also, a high moisture content diminishes

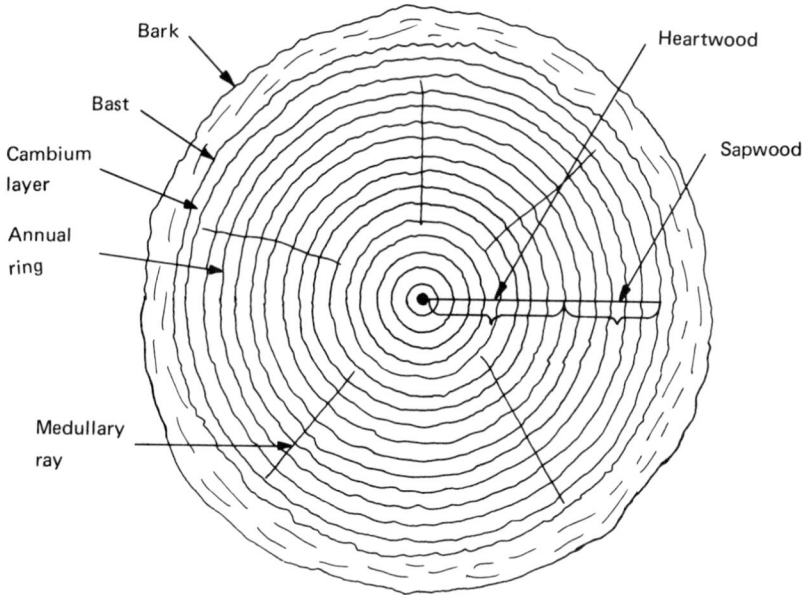

Figure 8-1 Tree Terminology

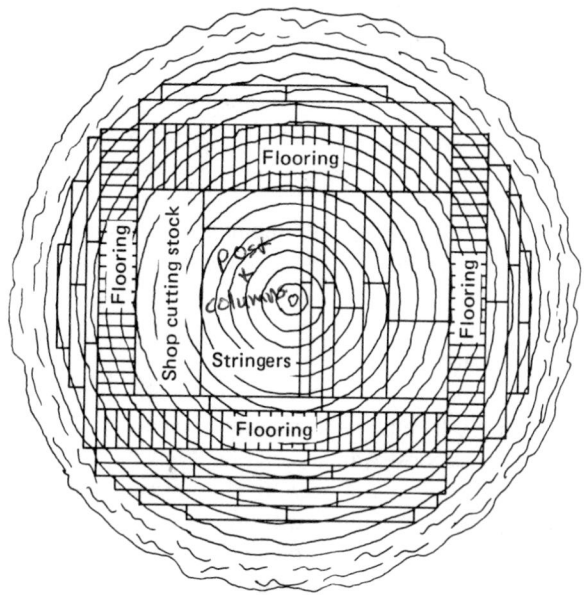

Figure 8-2 Log Cutting into Standard Size

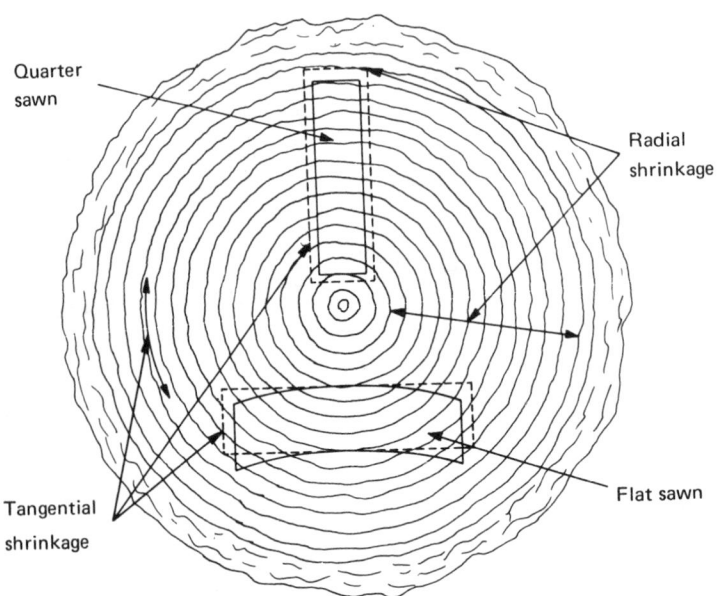

Figure 8-3 Sawing Pattern

the strength of the wood. Wood that is properly dried may be up to two and one-half times stronger than a comparable piece of green wood. Other problems, such as the presence of harmful fungi, occur when wood is cut. The moisture in wood is in two forms: free and combined. *Free moisture* is moisture outside the cell, and *combined moisture* is water that has been incorporated into the cell walls. In green wood, moisture will constitute from 25 to 200 percent of the oven-dry weight of the lumber, and most of this must be evaporated from the wood in the seasoning process.

The timber can be either air-dried or kiln-dried. As the wood dries, the free moisture is evaporated from the surface and the moisture from the interior of the porous lumber equalizes with the drier surface. When the moisture content stabilizes, the drying process is complete.

$$\text{moisture content (\%)} = \frac{\text{original amount of moist wood—oven-dry weight}}{\text{oven-dry weight}}$$

Air drying will reduce the moisture content to about 15 percent, a level that is acceptable for most construction purposes. Lumber that is to be used for purposes such as furniture, flooring, and cabinetwork is kiln-dried in a few days using temperatures of 155 to 180°F (68 to 82°C). A 1-in. pine board, for example, can be dried in 4 days. Hardwoods warp more easily, so the drying process is more gradual. They are normally air-dried for several months before they are kiln-dried. Temperatures used need to be lower, 100 to 120°F (38 to 49°C) for those woods that warp most easily. At the hotter kiln temperatures, a 1-in. hardwood board may take 6 to 10 days to dry.

All wood has the ability to reabsorb moisture from the atmosphere, depending on

[handwritten annotations at top: "homogeneous, isotropic — wood is not either." / "consistant makeup" / "is strength properties are the same in every direction"]

the humidity of its surroundings. This can cause a certain amount of dimensional instability, which may lead to problems in certain applications. Wood that has been subjected
to temperatures of 212°F (100°C) for a length of time during boiling, steaming, or soaking
will, with time, become less able to absorb humidity from the atmosphere.

Shrinkage, Warping, and Checking

[handwritten annotation: "Know what each are."]

Shrinkage will occur in the drying of wood when the combined moisture begins to be
evaporated from the cell walls. The contraction will be across the cell walls rather than
longitudinal, cells shrinking proportional to the thickness of their cell walls. The heavy-
celled summerwood thus shrinks far more than the thin-walled cells of springwood. In
a tangential direction, the bands of heavy summerwood force the lighter springwood to
contract with them, and for this reason there is a variation in the amount of shrinkage
that a piece of wood will exhibit in different directions. The usual limits in shrinkage
between a green condition and an over-dry condition are:

Volumetric shrinkage	7–21%
Longitudinal shrinkage	0.1–0.3%
Radial shrinkage	2–8%
Tangential shrinkage	4–14%

When shrinkage occurs unequally for any reason, warping results. If drying proceeds at
unequal rates over different portions of a piece of lumber, warping will occur. When
wood is cut tangentially, the dried piece will be convex because of unequal tangential
shrinkage. Irregularities in the grain of the wood will produce more pronounced irregularities in the dried boards.

    Checks are longitudinal cracks across the growth rings that can occur during drying
of timber. Unequal shrinkage produces strains within the wood, and checks can result.
They can be temporary, occurring when the outer portion of a piece of wood dries too
rapidly and contracts over the inner portion. These checks that appear in the outer layer
of the wood may close up and become imperceptible as the wood of the inner part of
the lumber dries and contracts. Other checks of a more serious nature can occur when
the difference in tangential shrinkage is too great relative to the radial shrinkage to be
accommodated. Large radial checks such as those sometimes seen in posts can then occur.

Defects in Lumber

Anything that adversely affects the strength, durability, or utility of a piece of wood is
a defect, and these may be within the wood itself or be produced by warping.

    Shakes are longitudinal cracks in the wood that follow the growth rings and develop
prior to the lumber's being cut. They are sometimes a result of heavy winds.

    Checks, as described above, are longitudinal splits across the growth rings resulting
from uneven drying.

Knots are formed at the base of branches where they extend into the wood of the tree. Only wood that comes from the base of the tree where there are no branches is free of knots. If the branch was dead, a *loose knot* is formed. A *spike knot* occurs when the cut is longitudinal to the branch. The effect of knots depends on the use of the wood and the location and size of the knots. They have little effect on wood in shear or in compression members. For beams in bending, however, a knot in the part of the beam subject to tension can significantly reduce its maximum load.

Pitch pockets are accumulations of resins in openings between the annual rings.

Bark pockets are formed when bark is wholly or partially encased in wood.

Waynes are areas where the lumber has been cut too close to the edge of the log and there is bark on the boards.

Compression wood is formed on the lower side of branches or leaning tree trunks. It is darker than normal wood, has a high lignin content, higher specific gravity, greater longitudinal shrinkage, and is not as tough as normal wood. The strength of compression wood is not predictable, and many failures in wood members have been found to be caused by compression wood. Tension wood, like compression wood, has higher specific

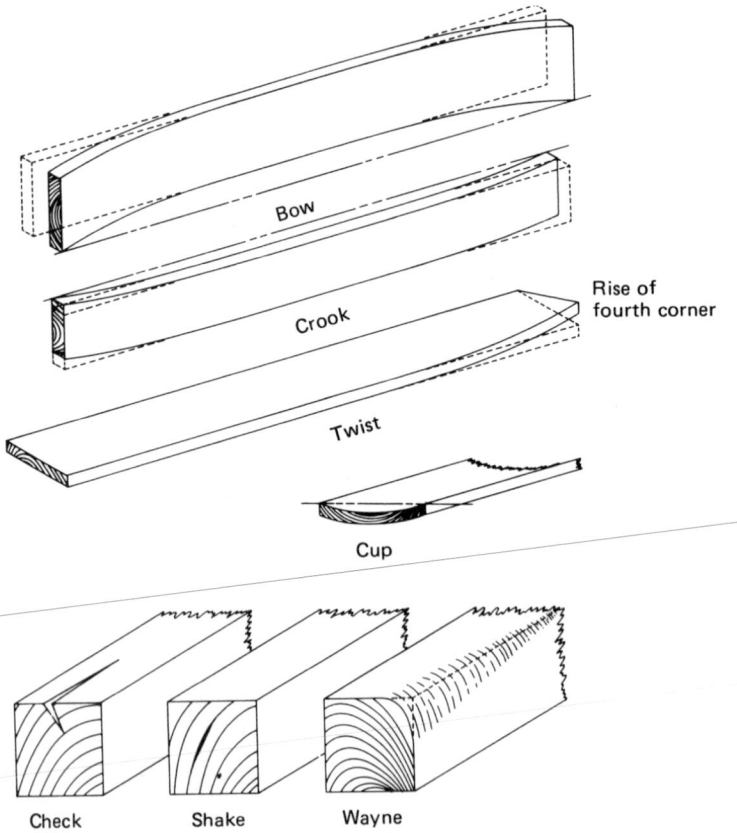

Figure 8-4　Unequal Shrinkage and Various Defects

gravity and greater longitudinal shrinkage. It is sometimes stronger and sometimes weaker than normal wood.

The effects of fungus or insect attack can be considered defects and are addressed later in the chapter.

Unequal shrinkage during the drying process produces warped lumber. The different types of warping are bow, crook, twist, and cup, as illustrated in Figure 8-4.

MECHANICAL PROPERTIES OF TIMBER

Strength of Wood

Because of the structure of wood, the fibers act like so many cylindrical tubes firmly bound together in the way that they withstand stress. Loads applied parallel to the grain are carried by the strongest of the fibers; in loads perpendicular to the grain, the weakest fibers have to take their share of the load.

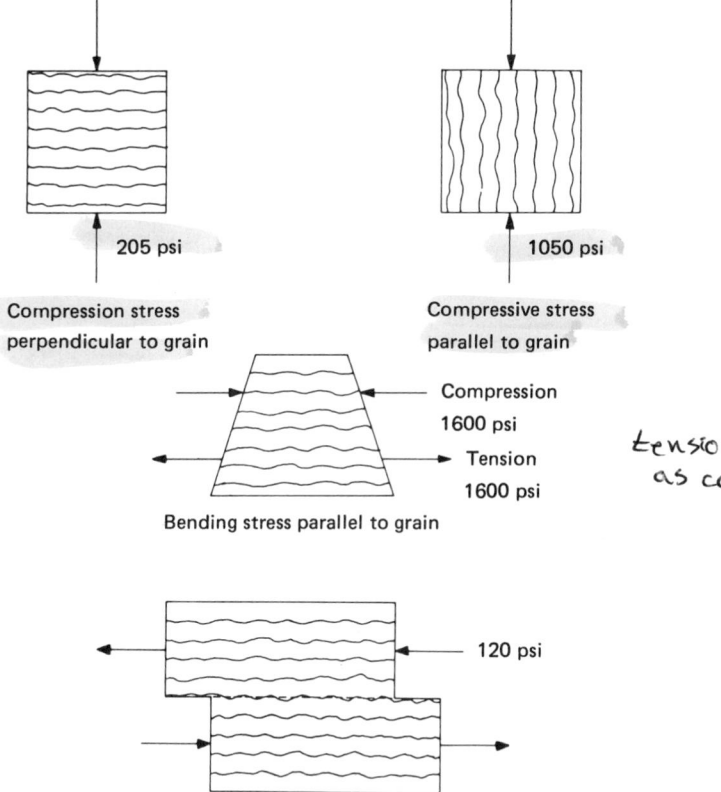

Tension, same as compression

Figure 8-5 Allowable Stresses for Douglas Fir

Tensile Strength
(Resistance to expansion)

Tensile strength in wood parallel to the grain is much higher (three times as much) than compressive strength. Thus, the limiting factor in tension members is usually compression or shear at the points of concentration. Thus far, no means have been developed to connect a tension member without it somehow compressing and eventually resulting in failure.

Tensile loads perpendicular to the grain of wood fibers cause the fibers to split apart. The tensile loads that can be carried are only one-tenth or less of the tensile load that can be withstood by the wood in a parallel direction with the grain. An example of the differences is shown in Figure 8-5, which shows Douglas fir and its appropriate values for various loading conditions.

Compression Strength

Compression loads parallel to the grain can be carried by the strongest fibers, whereas compression loads perpendicular to the grain are carried by both weak and strong fibers. Wood in compression parallel to the grain can carry three to four times the load that wood in compression perpendicular to the grain can carry.

Compression failure of wood perpendicular to the grain involves the complete crushing of the wood fiber. Compression failure of wood parallel to the grain involves the bending or buckling of the wood fibers.

Flexural Strength and Stiffness

For timber beams in flexure, the critical factors in evaluating the load that can be carried are the compressive strength parallel to the grain and the shear strength parallel to the grain. The resistance of shear strength parallel to the grain is very low. However, if the wood is free of defects, the initial failure will be compressive.

Stiffness also plays a role in timber beams, since deflection is usually limited to $1/_{360}$ of the span. Stiffness can be a measure of strength, in that stiffer timber beams are usually more dense.

Elastic Properties of Timber
(Restores to its original shape) plastic.

The modulus of elasticity in wood is high relative to its compressive strength, compared with other building materials. Wood does not, however, exhibit a well-defined yield point, and therefore the proportional limit is used to measure elastic strength. Therefore, for a specific load requirement and sufficient wood to carry such a load, there is a high degree of elastic strength. For this reason, wood members are considered to have good elastic properties.

Ductility -
brittle - breaking.
rigidity.

Factors Affecting Timber Strength

Many factors affect the strength of timber, such as direction of the wood fibers, moisture, weight, and rate of growth. The direction of the wood fibers is called the *grain*. The strength of timber depends so heavily on the direction of the loads with respect to the grain that it is an important consideration in design. Diagonal grain is measured by the tangent of the angle it makes with the cut edge of a piece of lumber. The diagonal grain or direction of the wood fibers will affect the strength ratios of lumber as shown in Table 8-1.

As indicated earlier, wood that is properly dried is far stronger than wood that has a high moisture content. When all the free water is evaporated and all the moisture is removed from the cell walls, the fiber saturation point is said to have been reached. The presence or absence of free moisture does not, in general, affect the strength properties. However, if drying occurs beyond the fiber saturation point, there will be an increase in strength. The increase in strength beyond the fiber saturation point may be two or three times the value of wet wood.

All other things being equal, the denser the wood, the stronger it will be. This is so true that a formula is based on it:

$$\text{modulus of rupture} = 26,200 \times (\text{specific gravity})^{1.25} \quad \text{(air-dry timber)}$$
$$\text{modulus of rupture} = 18,500 \times (\text{specific gravity})^{1.25} \quad \text{(green timber)}$$

These formulas were developed by the Forest Products Laboratory and are very accurate, regardless of species.

Finally, rate of growth greatly affects the strength of timber. In general, most timber used for construction will exhibit an optimal number of annual rings per inch of wood (Table 8-2). Usually, the greater number of rings per inch will be the stronger one.

Tabulation of the Mechanical Properties for Wood

Tables 8-3 and 8-4 show the mechanical properties for wood. The tables are based upon clear wood (wood that is free of defects) and a 100 percent strength ratio. The tables also have a factor of safety built into them for design purposes. Further, various grades

TABLE 8-1 STRENGTH RATIOS CORRESPONDING TO VARIOUS SLOPES OF GRAIN[a]

	Maximum Strength Ratio (%)	
Slope of Grain	*For Stress in Extreme Fiber in Bending (Beams and Stringers of Joists and Planks)*	*For Stress in Compression Parallel to Grain (Posts and Timbers)*
1 in 6	—	56
1 in 8	53	66
1 in 10	61	74
1 in 12	69	82
1 in 14	74	87
1 in 15	76	100
1 in 16	80	—
1 in 18	85	—
1 in 20	100	—

[a]ASTM D245-49T.

TABLE 8-2 GROWTH RINGS
FOR VARIOUS SPECIES

Species	Rings per Inch
Douglas fir	24
Shortleaf pine	12
Loblolly pine	6
Western hemlock	18
Redwood	30

of lumber have a strength ratio stamped on them, and by multiplying this number times that in the tables, basic stresses for that particular grade of lumber can be obtained.

DECAY, DURABILITY, AND PRESERVATION OF TIMBER

Mechanism of Decay

Various factors that cause damage to a wooden structure, the durability of the wooden structure, and its preservation will be discussed in detail. Molds, stains, and decay in timber are caused by fungi, microscopic plants that require organic material to live on. Reproduction occurs through thousands of small windblown particles called *spores*. Spores send out small arms that destroy timber through the action of enzymes. Fungi require a temperature of between 50 and 90°F (10 and 32°C), food, moisture, and air. Timber that is soaked or dried normally will not decay. Molds cause little direct staining because the color caused is largely superficial. The cottony or powdery surface growths range in color from white to black and are easily brushed or planed off the wood. Stains, on the other hand, cannot be removed by surface techniques. They appear as specks, spots, streaks, or patches of varying shades and colors, depending upon the organism that is infecting the timber. Although stains and molds should not be considered as stages of decay since the fungi do not attack the wood substance to any great degree, timber infected with mold and stain fungi has a greater capacity to absorb water and is therefore more susceptible to decay fungi. Decay fungi may attack any part of the timber, causing a fluffy surface condition indicative of decay or rot. Early stages of decay may show signs of discoloration or mushroom-type growths (Figure 8-6).

In some types the color differs only slightly from the normal color, giving the appearance of being water-soaked. Later stages of decay are easily recognized because of the definite change in color and physical properties of the timber (Figure 8-7). The surface becomes spongy, stringy, or crumbly, weak, and highly absorbent. It is further characterized by a lack of resonance when struck with a hammer. Brown, crumbly rot in a dry condition is known as *dry rot*. Serious decay problems are indicative of faulty design or construction or a lack of reasonable care in the handling of the timber. Principles that assure long service life and avoid decay hazards in construction include building with dry timber, using designs that will keep the wood dry and accelerate rain runoff, and using preservative-treated wood where the wood must be in contact with water.

TABLE 8-3 MECHANICAL PROPERTIES OF A FEW IMPORTANT WOODS GROWN IN THE UNITED STATES[a]

Common and Botanical Name	Weight (lb/ft³)		Shrinkage from Green to Oven-dried (% of green volume)	Modulus of Rupture (psi)		Modulus of Elasticity (1000 psi)	
	Green	Air-dried[b]		Green	Air-dried	Green	Air-dried
Hardwoods							
Ash, white (*Fraxinius* sp.)	48	41	12.8	9,500	14,600	1,410	1,680
Elm, American (*Ulmus americana*)	54	35	14.6	7,200	11,800	1,110	1,340
Hickory, true (*Carya* sp.)	63	51	17.9	11,300	19,700	1,570	2,190
Maple, red (*Acer rubrum*)	50	38	13.1	7,700	13,400	1,390	1,640
Maple, sugar (*Acer saccharum*)	56	44	14.9	9,400	15,800	1,550	1,830
Oak, red (*Quercus* sp.)	64	44	14.8	8,500	14,400	1,360	1,810
Oak, white (*Quercus* sp.)	63	47	16.0	8,100	13,900	1,200	1,620
Walnut, black (*Juglans nigra*)	58	38	11.3	9,500	14,600	1,420	1,680
Conifers							
Cedar, western red (*Thuja plicata*)	27	23	7.7	5,100	7,700	920	1,120
Cypress, bald (*Taxodium distichum*)	51	32	10.5	6,600	10,600	1,180	1,440
Douglas fir, coast type (*Pseudotsuga taxifolia*)	38	34	11.8	7,600	12,700	1,570	1,950
Fir, white (*Abies* sp.)	46	27	9.8	5,900	9,800	1,150	1,490
Hemlock, eastern (*Tsuga canadensis*)	50	28	9.7	6,400	8,900	1,070	1,200
Pine, longleaf (*Pinus palustris*)	55	41	12.2	8,700	14,700	1,600	1,990
Pine, shortleaf (*Pinus echinata*)	52	36	12.3	7,300	12,800	1,390	1,760
Pine, western white (*Pinus monticola*)	35	27	11.8	5,200	9,500	1,170	1,510
Redwood (old growth) (*Sequoia sempervirens*)	50	28	6.8	7,500	10,000	1,180	1,340
Spruce, Sitka (*Picea sitchensis*)	33	28	11.5	5,700	10,200	1,230	1,570
Spruce, eastern (*Picea* sp.)	34	28	12.6	5,600	10,100	1,120	1,450
Tamarack (*Larix laricina*)	47	37	13.6	7,200	11,600	1,240	1,640

(Continued)

TABLE 8-3 (Continued)

Common and Botanical Name	Maximum Crushing Strength Parallel to Grain (psi)		Compression Perpendicular to Grain, Proportional Limit (psi)		Maximum Shearing Strength, Parallel to Grain (psi)	
	Green	Air-dried	Green	Air-dried	Green	Air-dried
Hardwoods						
Ash, white (*Fraxinus* sp.)	4,060	7,280	860	1,510	1,350	1,920
Elm, American (*Ulmus americana*)	2,910	5,520	440	850	1,000	1,510
Hickory, true (*Carya* sp.)	4,570	8,970	1,080	2,310	1,360	2,130
Maple, red (*Acer rubrum*)	3,280	6,540	500	1,240	1,150	1,850
Maple, sugar (*Acer saccharum*)	4,020	7,830	800	1,810	1,460	2,330
Oak, red (*Quercus* sp.)	3,520	6,920	800	1,260	1,220	1,830
Oak, white (*Quercus* sp.)	3,520	7,040	850	1,410	1,270	1,890
Walnut, black (*Juglans nigra*)	4,300	7,580	600	1,250	1,220	1,370
Conifers						
Cedar, western red (*Thuja plicata*)	2,750	5,020	340	610	710	860
Cypress, bald (*Taxodium distichum*)	3,580	6,360	500	900	810	1,000
Douglas fir, coast type (*Pseudotsuga taxifolia*)	3,860	7,430	440	870	930	1,160
Fir, white (*Abies* sp.)	2,830	5,480	360	620	770	990
Hemlock, eastern (*Tsuga canadensis*)	3,080	5,410	440	800	850	1,060
Pine, longleaf (*Pinus palustris*)	4,300	8,440	590	1,190	1,040	1,500
Pine, shortleaf (*Pinus echinata*)	3,430	7,070	440	1,000	850	1,310
Pine, western white (*Pinus monticola*)	2,650	5,620	290	540	640	850
Redwood (old growth) (*Sequoia sempervirens*)	4,200	6,150	520	860	800	940
Spruce, Sitka (*Picea sitchensis*)	2,670	5,610	340	710	760	1,150
Spruce, eastern (*Picea* sp.)	2,600	5,620	290	590	710	1,070
Tamarack (*Larix laricina*)	3,480	7,100	480	990	860	1,280

[a]Compiled from R1903-10, June, 1952, Forest Products Laboratory, U.S. Dept. of Agriculture.

[b]Air-dried lumber contained 12% moisture.

TABLE 8-4 BASIC STRESSES FOR CLEAR MATERIAL[a,b]

Species	Extreme Fiber in Bending	Modulus of Elasticity	Compression Parallel to Grain[c] (L/d = 11 or less)	Compression Perpendicular to Grain	Maximum Horizontal Shear
Hardwoods					
Ash, commercial white	2,050	1,500,000	1,450	500	185
Elm, white	1,600	1,200,000	1,050	250	150
Hickory, true and pecan	2,800	1,800,000	2,000	600	205
Maple, sugar and black	2,200	1,600,000	1,600	500	185
Oak, commercial red and white	2,050	1,500,000	1,350	500	185
Conifers					
Cedar, western red	1,300	1,000,000	950	200	120
Cypress, southern	1,900	1,200,000	1,450	300	150
Douglas fir, coast region	2,200	1,600,000	1,450	320	130
Fir, commercial white	1,600	1,100,000	950	300	100
Hemlock, eastern	1,600	1,100,000	950	300	100
Pine, western white, eastern white, ponderosa, and sugar	1,300	1,000,000	1,000	250	120
Pine, Norway	1,600	1,200,000	1,050	220	120
Pine, southern yellow (longleaf or shortleaf)	2,200	1,600,000	1,450	320	160
Redwood	1,750	1,200,000	1,350	250	100
Spruce, red, white, and Sitka	1,600	1,200,000	1,050	250	120
Tamarack	1,750	1,300,000	1,350	300	140

[a]From Forest Products Laboratory, U.S. Dept. of Agriculture. See also ASTM D245-49T.

[b]All values in pounds per square inch and for material under long-time service conditions at maximum design load.

[c]L, unsupported length; d, least dimension of cross section.

Insect damage may result from infestation of the timber members with any of a number of insects, among them powder-post beetles, termites, and marine borers. Powder-post beetles are reddish brown to black, hard-shelled insects from $\frac{1}{8}$ to $\frac{1}{2}$ in. long. The life cycle of the beetle includes four distinct stages: egg, larva, transformation, and adult. The adult bores into the timber, producing a cylindrical tunnel just under the surface in which the eggs are laid. The larvae burrow through the wood, leaving tunnels packed with a fine powder $\frac{1}{16}$ to $\frac{1}{8}$ in. in diameter. Powder-post damage is indicated by this fine powder, either fallen from the timber or packed into tunnels within the wood and by the tunnels and holes in the timber. The beetles attack both sound and decayed wood but are not active in decayed wood that is water-soaked. They may also cause damage by transmitting destructive fungi from one site to another, thus spreading decay.

Termites resemble ants in size and general appearance and live in similarly organized colonies. Destruction is done by the workers only, not by the soldiers or winged sexual adults. Subterranean termites are responsible for most of the damage done in the United

Figure 8-6   Early Stages of Decay at Base of Timber Member

Figure 8-7   Advanced Stage of Decay

States. They are more prevalent in the southern states but are found in varying numbers throughout the states. Termites build dark, damp tunnels well below the surface of the ground, with some of these tunnels leading to the wood, which is the termites' source of food. Termites also require a constant water supply in order to survive.

Subterranean termites do not infest structures by being carried to the construction site. They must establish a colony in the soil before they are able to attack the timber. Tell-tale signs are the tunnels in the earth leading to unprotected timber and swarms of male and female winged adults in the early spring and fall (Figure 8-8). When termites successfully enter the wood, they make tunnels that follow the wood grain, leaving a shell of sound wood to conceal the tunnels (Figure 8-9). Methods of controlling termites include breaking the path from the timber to the ground, although the best method is to treat the timber with a preservative.

Wood-inhabiting termites are found in a narrow strip around the southern boundary of the United States. They are most common in southern California and southern Florida. They are fewer in number than the subterranean termites, do not multiply as fast, and do not cause as much damage. But because they can live without contact with the ground

Figure 8-8 Swarming Termites

and in either damp or dry wood, they are a definite problem and do considerable damage in the coastal states. They are carried to a building site in timber that has been infested prior to delivery, thereby making inspection prior to delivery a necessity. Full-length treatment with a good wood preservative is recommended.

Carpenter ants are usually found in stumps, trees, or logs, but are sometimes found in structural timbers. They range in color from brown to black and in size from large to small. Although carpenter ants are often confused with termites, they can be distinguished by comparison of the wings and thorax (waist) sizes of the insects. The carpenter ant has short wings and a narrow waist, whereas the termite has long wings and a thicker waist. Carpenter ants use wood as shelter, not food. They prefer naturally soft or decayed wood and construct tunnels that are very smooth and free of dust. Carpenter ants tunnel across the wood grain and cut small exterior openings for access to the food supplies (Figure 8-10). A large colony takes from 3 to 6 years to develop. Prevention of ant infestation can be accomplished by the use of preservatives along the length of the timber member.

By far the most damaging insect pest is the marine borer (Figure 8-11). Borers have been known to ruin piles and framing within a few months. No ocean is completely free of borers and, although some waters may be relatively free, the status of an area may change drastically within a relatively short period of time. The main point of attack is generally between the high tide level and the mud line. Submergence often hides tell-tale signs of infestation, so that the first sign of attack may be the failure of the structure. The borers that do the most damage are the mollusks, related to clams and oysters, and the crustaceans, related to lobsters and crabs.

The mollusk corers consist of the "shipworm" teredo, the "shipworm" bankia, and

Small home building is carpentry.

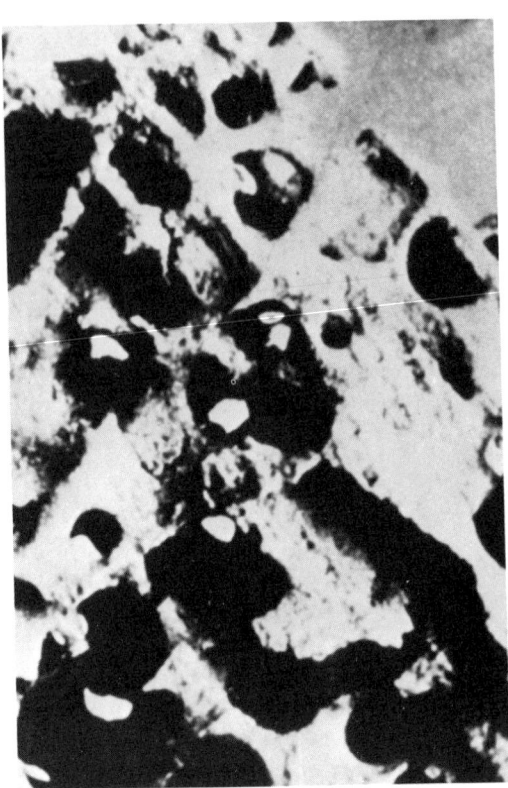

Figure 8-9   Characteristic Termite Tunnel   **Figure 8-10**   Carpenter Ant Tunnels

the Pholadidae. There are other types of shipworms throughout the world, but they all live and survive in much the same way, although size and environmental requirements may vary. The teredo has a wormlike, slimy gray body with two shells attached to the head which are used for boring. Two tubes that resemble a forked tail and normally remain outside the burrow are the only external indications that a shipworm is present. The shipworm can seal its entrance and thereby protect the burrow from intruders or foreign substances. The size of the teredo varies from $3/8$ to 1 in. in diameter and from 6 in. to 6 ft in length. The size of the bankia is generally larger than the teredo, but its other characteristics are the same. Both types bore tiny holes when they are young and grow to maturity inside the wood. Once the young shipworm has entered the wood, it normally turns down and expands its burrow to its full diameter. Extremely careful observation with a hand lens is required to detect the entrance. The only way to detect the extent of the damage is to cut the wood (Figure 8-12) or take borings by some other method. Because the first sign of marine borer infestation may be the failure of timber members (Figure 8-13), the shipworm should be considered extremely dangerous. Pholadidae resembles a clam with its body entirely enclosed in a two-part shell. It is of particular danger because it can bore holes up to $1\frac{1}{2}$ in. deep into the hardest timber.

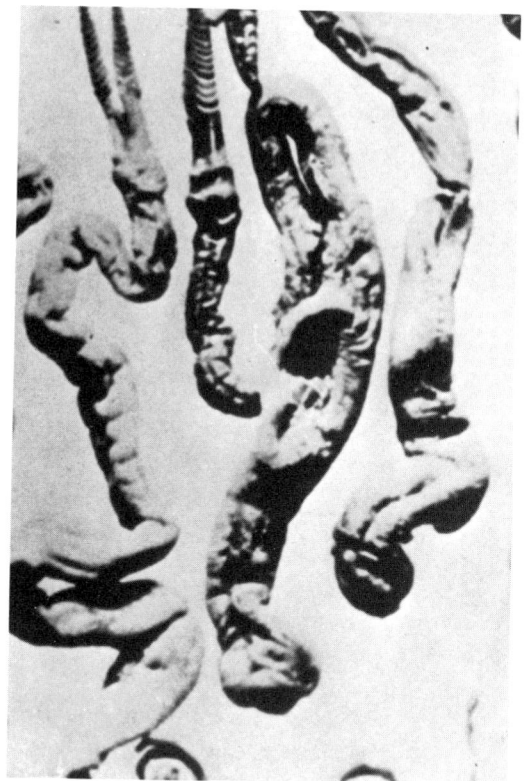

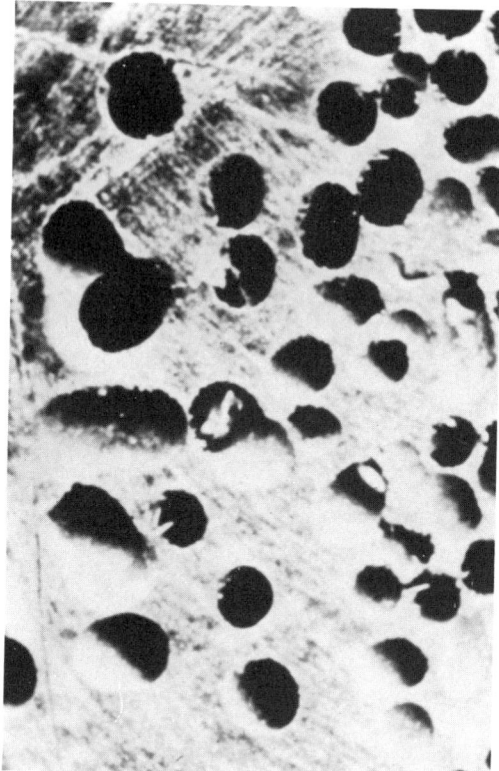

Figure 8-11 Marine Borers

Figure 8-12 Cross-Cut Timber Member Showing Extensive Damage

The most common crustacean borer is the limnoria or "wood louse." Its body is slipper-shaped, from $1/8$ to $1/4$ in. long and from $1/16$ to $1/8$ in. wide. It has a hard boring mouth, two sets of antennae, and seven sets of legs with sharp claws. The limnoria is able to roll itself into a ball, to swim, and to crawl. It will gnaw interlacing branching holes in the surface of the wood with as many as 200 to 300 holes per square inch. These holes follow the softer wooden rings and are 0.05 to 0.025 in. in diameter and seldom over $3/4$ in. long. As a result, the wood becomes a mass of thin walls between burrows which break away, exposing new areas to attack. In this manner the pile is slowly reduced in diameter (Figure 8-14). In soft woods such as pine and spruce, the diameter may be reduced as much as 2 in. per year. The chief area of attack is between the low-water level and the mud line, with occasional activity up to the high-water level. The limnoria does not seem to be affected by small environmental changes and may be found in salt or brackish water of any temperature, polluted or clean.

Complete protection from borer attack is essential. Metal armoring is falling into disuse, as is concrete casing. The very best method may prove to be jacketing creosoted piles. The most practical method involves heavy treatment with high-quality coal-tar

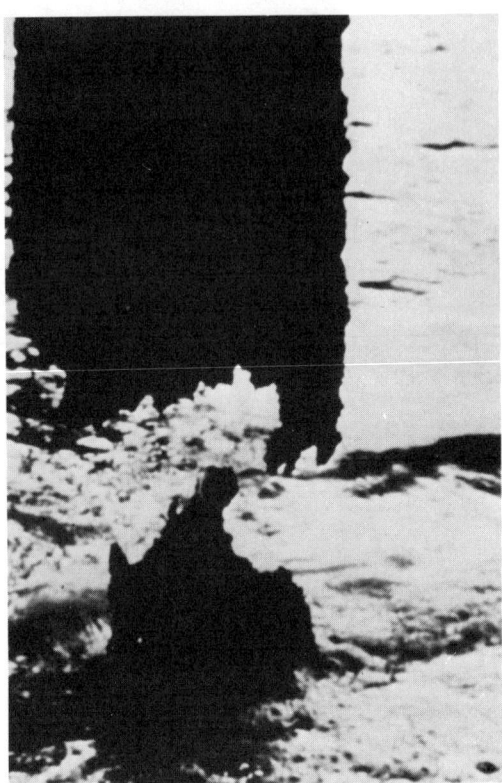

Figure 8-13 Insect Attack Resulting in Complete Failure

creosote by the full-cell method to the point of near saturation. Although shipworms will generally not attack a creosoted member, they may attack any area that has been damaged, and any untreated area. Limnoria attack the creosoted timber directly but at a decelerated rate. Other types of wood preservatives in use include a plastic outer wrap, which is successful as long as the plastic is not damaged, and cyanide treatment, which is messy to apply, leaches out with time, and may cause damage to the environment.

Fire damage to wood is a real problem on bridge and wharf structures. Treatment of wood with preservatives may protect the wood from fungi and insect attack, but it generally results in making the wood more susceptible to fire. Damage due to fire is readily apparent due to the charred appearance and burnt odor of fire-damaged wood.

Chemical damage to wood is very difficult to determine because it often resembles damage done by other factors. Fungi attack employs a chemical reaction between the enzymes particular to each fungus and the wood fiber. Fire damage involves the oxidation of the wood fiber, which is again a chemical process. It is therefore important to report any apparent damage to the timber as soon as possible.

Impact damage may occur in the event of a high-energy collision. Because wood has good impact characteristics, it may show only limited signs of external damage and must therefore be inspected carefully after an accident or suspected accident. The timbers may have a shattered appearance as opposed to the sagging appearance caused by overload.

Figure 8-14   Reduction of Water Line Due to Insect Attack

Compression failure will resemble wrinkled skin; tension failure looks as if the fibers have been pulled apart.

Another method of protecting timber is by injecting preservatives into the timber. There are two methods utilized: the pressure process and the nonpressure process. There are two types of *pressure processes:* the full cell and the empty cell. In the *full-cell process* it is intended that the cells remain filled with the preservative. The initial treatment removes most of the air and water before impregnation. The timber is then covered with preservatives under pressure. In the *empty-cell process,* the preservative merely forms a film over the cell walls. In the *nonpressure process* (hot and cold), the timber is immersed in the preservative in an open tank and heated to 200°F (93°C). The heat of the preserving fluid expands the air within the cells. The cooling of the bath on the immersing of the timber in a cold medium causes a contraction of the air still remaining in the cells, which tends to produce infiltration by the preservative.

PROBLEMS

8.1. List the two principal categories of trees and describe each.

8.2. How do hardwood trees differ from softwood trees?

8.3. Discuss the role of the following polymers: cellulose, hemicellulose, and lignin.

8.4. Explain the differences between slash cut and rift cut.

8.5. What role does moisture content play in timber used for structural purposes?

8.6. What are some defects in lumber? Describe each.

8.7. How does compressive strength parallel to the grain differ from compressive strength perpendicular to the grain?

8.8. Explain the mechanism of decay in lumber.

8.9. How do marine borers damage timber?

8.10. Explain the methods utilized for protecting timber.

REFERENCES

American Society for Testing Materials, *Book of Standards*, Part 22, 1979.

BROWN, H. P., PANSHIN, A. J., and FORSAITH, C. C., *Textbook of Wood Technology*, McGraw-Hill Book Company, New York, 1949.

BRUST, A. W., and BERKLEY, E. E., "The Distributions and Variations of Certain Strength and Elastic Properties of Clear Southern Yellow Pine Wood," *Proc. ASTM, 35*, Part II (1935), pp. 643–693.

CLAPP, W. F., "Recent Increases in Marine Borer Activity," *Civil Eng., 7*, No. 12 (Dec. 1937), pp. 836–838; and *17*, No. 6 (June 1947), pp. 324–327.

"Defects in Timber Caused by Insects," *U.S. Dept. Agr. Farmers Bull. 1490*, 1927.

DESCH, H. E., *Timber, Its Structure and Properties*, 2nd ed., Macmillan Publishing Co., Inc., New York, 1947.

DIETZ, A. G. H., *Materials of Construction, Wood, Plastics, Fabrics*, D. Van Nostrand Company, New York, 1949, pp. 1–189.

DIETZ, A. G. H., and GRINSFELDER, A., "Fatigue Tests on Compressed and Laminated Wood," *ASTM Bull. 129*, Aug. 1944.

ELMENDORF, A., "The Uses and Properties of Water-resistant Plywood," *Proc. ASTM, 20*, Part II (1920), p. 324.

Forest Products Laboratory, *Wood Handbook*, rev. ed., Superintendent of Documents, Government Printing Office, Washington, D.C., 1940.

FREAS, A. D., "Studies of the Strength of Glued Laminated Wood Construction," *ASTM Bull. 70*, Dec. 1950; and *Bull. 73*, Apr. 1953.

GAY, C. M., and PARKER, H., *Materials and Methods of Architectural Construction*, 2nd ed., John Wiley & Sons, Inc., New York, 1943, pp. 39–50, pp. 304–371.

HANSEN, H. J., *Timber Engineers Handbook*, John Wiley & Sons, Inc., New York, 1948.

HOLTMAN, D. F., *Wood Construction*, McGraw-Hill Book Company, New York, 1929.

HUNT, G. M., and GARRATT, G. A., *Wood Preservation*, McGraw-Hill Book Company, New York, 1938.

KOFOID, C. A., *Termites and Termite Control*, University of California Press, Berkeley, Calif., 1934.

KUERRZI, E. Q., "Testing of Sandwich Construction at the Forest Products Laboratory," *ASTM Bull. 164*, Feb. 1950.

LEWIS, W. C., "Fatigue of Wood and Glued Wood Construction," *Proc. ASTM, 46* (1946), p. 814.

MARKWARDT, L. J., and WILSON, T. R. C., "Strength and Related Properties of Woods Grown in the United States," *U.S. Dept. Agr., Tech. Bull. 479*, 1935.

MILLS, A. P., HAYWORD, H. W., and RADER, L. F., *Materials of Construction*, 6th ed., John Wiley & Sons, Inc., New York, 1955, pp. 508–543.

"Preventing Damage by Termites and White Ants," *U.S. Dept. Agr. Farmers Bull. 1972*, 1927.

TRUAX, T. R., "The Gluing of Wood," *U.S. Dept. Agr. Bull. 1500*, 1929.

United States Coast Guard, *The State of the Art: Bridge Protective Systems and Devices*, A Report by the U.S.C.G. Bridge Modification Branch, R. T. Mancill, Editor, 1979.

9

Asphalt Cements

as opposed to Portland Cement

Asphalt materials have been utilized since 3500 B.C. in building and road construction. Their main uses have been as adhesives, waterproofing agents, and as mortars for brick walls. These early asphalt materials were native asphalts. *Native asphalts* occur when petroleum rises to the earth's crust and the volatile oils are evaporated. These native asphalts were found in pools and asphalt lakes. One of the more well-known deposits of native asphalts is the "Trinidad Lake" deposit on the island of Trinidad off the north coast of South America. Prior to the development of the processes for producing asphalt cement from crude petroleum products, native asphalts were the only sources of supply for early pavement projects. The first asphaltic pavement was built in 1869 in London, England. A year later construction of road and street pavements began in the United States in Newark, New Jersey.

With the invention of the automobile, which required smooth all-weather pavements, the demand for asphaltic products for pavements grew. Thus, the processes for producing asphalt cement from crude petroleum products grew, which led to the development of modern asphalt cement.

ASPHALT CEMENTS

Asphalt cement, according to ASTM D8 (Materials for Roads and Pavements), is fluxed or unfluxed asphalt specially prepared as to quality and consistency for direct use in the manufacture of bituminous pavements, and having a penetration at 77°F (25°C) of between 5 and 300, under a load of 0.2 lb (100 g) applied for 5 seconds. Asphalt cements fall within a broad category known as bitumens. *Bitumen,* according to ASTM D8, is a class of black or dark-colored (solid, semisolid, or viscous) cementitious substances, natural or manufactured, composed principally of high-molecular-weight hydrocarbons, of which asphalts, tars, pitches, and asphaltites are typical.

Bitumen by definition is soluble in carbon disulfide. The hydrocarbons that make up bitumen can generally be made up of the following:

1. Asphaltenes.
2. Resins.
3. Oils.

Asphaltenes are large, high-molecular-weight hydrocarbon fractions precipitated from asphalt by a designated paraffinic naphtha solvent at a specified solvent-asphalt ratio. Asphaltenes have a carbon-to-hydrogen ratio of 0.8. Asphaltenes constitute the body of the asphalt. *Resins* are hydrocarbon molecules with a carbon-to-hydrogen ratio of more than 0.6 but less than 0.8. Resins affect the adhesiveness and ductility properties of asphalt. *Oils* are hydrocarbon molecules with a carbon-to-hydrogen ratio of less than 0.6. Oils influence the viscosity and flow of the asphalt.

Ductility and adhesiveness are two properties that make asphalt cement attractive as a highway material. Oxidation of an asphalt cement causes a loss of ductility and adhesiveness, resulting in an asphalt cement that is harder and less ductile and adhesive. Oxidation results in the creation of more asphaltenes at the expense of resins. Thus, oxidation is a serious problem.

Production and Distillation

Asphalt cement is a valuable by-product obtained when petroleums are processed to obtain gasoline, kerosene, fuel oil, motor oil (diesel and lubricating), and other asphalt products. There are three types of petroleum found in the earth's crust:

1. Asphaltic-base crude oils.
2. Paraffin-base crude oils.
3. Mixed-base crude oils.

Asphalt cement is easily obtained from asphaltic-base crude oils by a straight-run distillation process. Bituminous products may be obtained from paraffin-base crude oils by a destructive distillation process involving chemical changes. These bituminous materials should not be classified as asphalt. Asphalt cement may also be obtained from mixed-base crude oils but the process is complicated.

All distillation of asphalt-base petroleum is fractional. During the distillation of petroleum several fractions are separated from the petroleum as given in Table 9-1.

Figure 9-1 shows a flowchart of the refining process necessary for the production of asphalt cement. When the process is controlled to prevent overheating and eventual chemical changes, the asphalt cement that remains as a residue is a straight-run asphalt. In the first operation the petroleum is pumped through the tube heater, in which the petroleum is placed under pressure and heated to a temperature of about 550°F (288°C).

TABLE 9-1 FRACTIONS OF PETROLEUM

Fraction	Product Type	Boiling-Point Range (°F)
Light distillate	Gasoline	100–400
Medium distillate	Kerosene	350–575
Heavy distillate	Diesel oil	425–700
Very heavy	Lubricating oil	Over 650
Residue	Asphalt	

This crude product, released through the bottom of the tube heater, enters an atmospheric fractionating column (tower distillation). Upon being exposed to atmospheric pressure within the column the more volatile fractions rise to the top of the column, where the vapors pass through condensers and coolers. Traps are arranged in the column at different levels where each trap is just below the boiling point of the liquid at which it is to collect. Once the vapor passes through the appropriate traps, the vapors are condensed and cooled. These condensed vapors are referred to as distillates (light, medium, and heavy). After further processing, gasoline, kerosene, and diesel oil result. The product that settles down in the atmospheric fractionating column is *residuum,* sometimes called *hot topped crude.*

The crude then enters a second fractionation column. This column allows steam to be introduced at the bottom of the fractionation column in such a way as to become mixed with the hot topped crude. The boiling point of the various fractions becomes a combination of the boiling points of the oils being vaporized plus water. A second procedure (which is an aid for the use of lower temperatures) is the application of a partial vacuum in the fractionation column. The greater the vacuum applied, the lower the boiling point of the fractions to be separated from the crude. By controlling the steam, the temperature, and the partial vacuum, the quality of asphalt cement is maintained. The residual that results from the second fractionation column is asphalt cement. The volatiles are light vacuum distillate, nonvolatile oils, and heavy vacuum distillate.

Air is sometimes used to improve an asphalt material. The air is blown through an asphalt stock at temperatures of 400 to 500°F (205 to 288°C). A reaction results in which the oxygen from the air combines with the hydrogen in the hydrocarbon molecule to form water, which is emitted as steam. As a result, polymerization of the hydrocarbons takes place to form heavier and harder materials. This chemical reaction may be speeded up by the addition of ferric chloride or phosphorous pentoxide. The main advantage of air blowing is that the final product is less susceptible to temperature fluctuations than an unheated asphalt.

METHODS OF TESTING

Most present day standards were developed in the early 1900s. Very few new tests or standards have been developed in recent years. Equipment has changed, becoming electric, and better control can be accomplished during the testing period of the product. We shall look at some of these standards in greater detail.

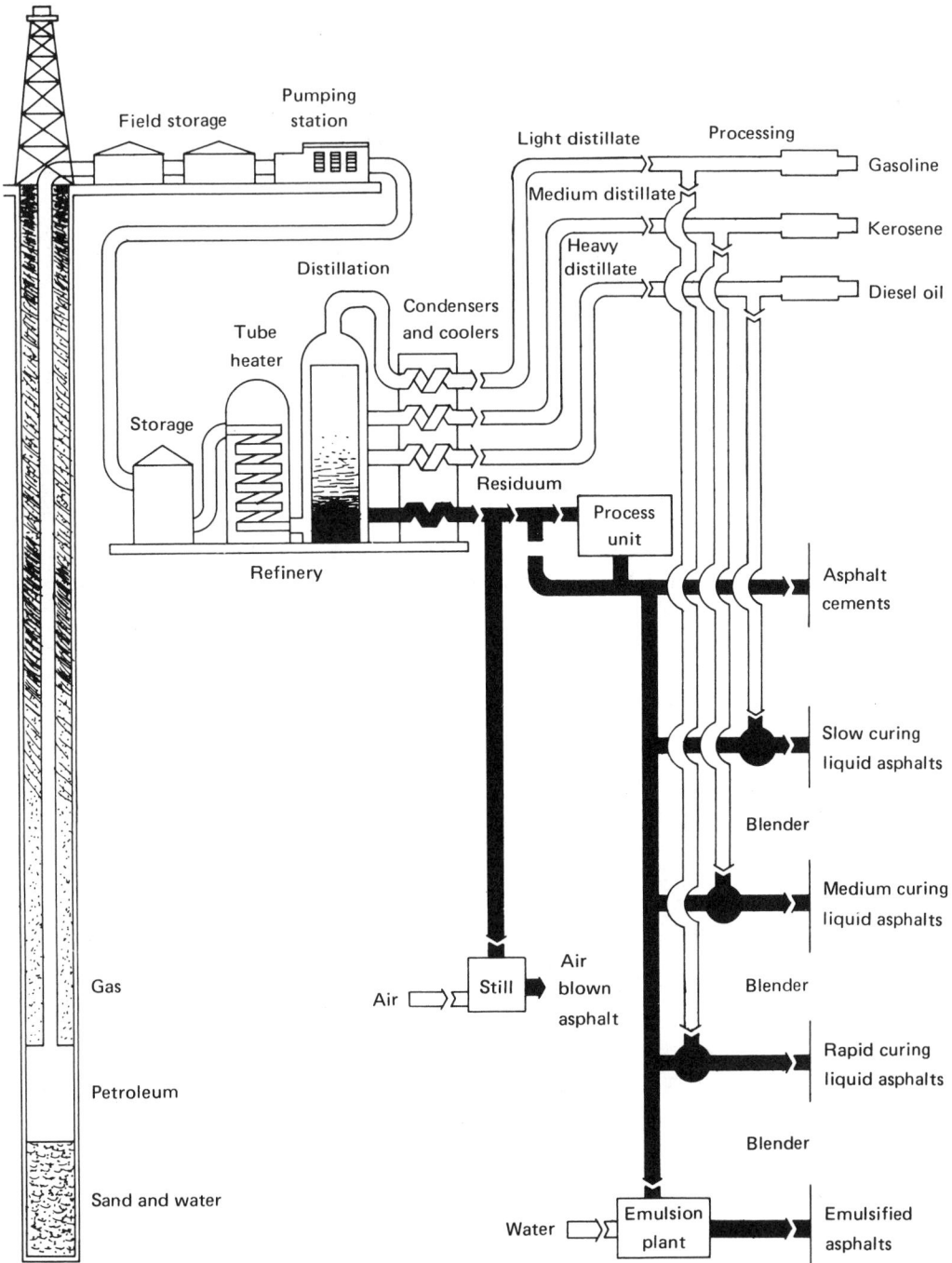

Figure 9-1 Flowchart of the Refining Process for Asphalt Cement (Courtesy of Asphalt Institute)

Penetration of Bituminous Materials

ASTM D5 (Penetration of Bituminous Materials) covers determination of the penetration of semisolid and solid bituminous materials. In other words, it measures the hardness or softness of the material. It is a consistency test for bituminous material expressed as the distance in tenths of a millimeter that a standard needle vertically penetrates a sample of the material under known conditions of loading, time, and temperature.

The procedure for the standard test, which is implied unless other conditions are stated, is for pressure to be applied to a load of 100 grams for a period of 5 sec at a temperature of 77°F (25°C).

In running the test at least three determinations should be made on the surface of the sample at points not less than 10 mm from the side of the container and not less than 10 mm apart.

The loss on weight with today's manufactured products is quite low, and thus the test has lost much of its meaning. The test back in the early 1900s was extremely valuable in characterizing early steam-refined asphalts, which had low flash points and high losses on heating.

Lower penetration grades are generally needed in warmer climates to avoid softening in the summer. Obviously, higher penetration grades will be used in colder climates so that excessive brittleness does not occur during cold winter weather.

The penetration test has provided long experience and service records associated with asphalts classified in this matter. However, the test is empirical and many engineers would like to replace it with ASTM D2171 (Viscosity of Asphalts by Vacuum Capillary Viscometer)

Specific Gravity

Know what it is.

ASTM D70 (Specific Gravity and Density of Semi-solid Bituminous Materials) covers the determination of the specific gravity and density of semisolid bituminous materials, asphalt cements, and soft tars pitches by use of a pycnometer.

In running this test, the sample is heated until it can be poured. The material is placed in a pycnometer. The asphalt volume is determined by taking the difference between the total volume of the bottle and the volume of water required to complete the filling. From this information the specific gravity can be expressed as the ratio of the weight of a given volume of the material at 77°F (25°C) or at 60°F (15.6°C) to that of an equal volume of water at the same temperature. The following formula may be used to calculate the specific gravity:

$$\text{specific gravity} = \frac{C - A}{(B - A) - (D - C)}$$

where A = weight of pycnometer (plus stopper)
       B = weight of pycnometer filled with water
       C = weight of pycnometer partially filled with asphalt
       D = weight of pycnometer plus asphalt plus water

The density is determined by multiplying the specific gravity by the density of water at test temperature in desired units where the specific gravity is calculated as in the given equation.

The specific gravity and density is reported to the nearest third decimal place at 77°F (25°C) or 60°F (15.6°C).

Specific gravity is usually not a requirement of asphalt specifications but is of value in the design of bituminous mixtures as well as for determining costs.

Ductility

ASTM D113 (Ductility of Bituminous Materials) covers the procedure for determining ductility. The ductility of a bituminous material is measured by the distance to which it will elongate before breaking when two ends of a briquette specimen of the material are pulled apart at a specified speed and temperature.

The results of the ductility test are controversial. The test is believed to measure the adhesiveness and elasticity of the asphalt. The property of adhesiveness is most important, since asphalt cement is used to bind stone, sand, and filler to make bituminous concrete. Also of concern is the temperature at which the product is run, 77°F (25°C). It is believed that this temperature is too high and that a much lower temperature should be used, because lack of ductility or brittleness, is more serious in cold weather.

Float Test

ASTM D139 (Float Test for Bituminous Materials) is a consistency test used for materials that are too soft to undergo the standard penetration test and too hard for use with the Saybolt-Furol viscosity test.

The test is performed by filling a brass collar with the asphalt or asphalt product to be tested, cooling the holder with its contents to 41°F (5°C), screwing the collar to a float, and floating the apparatus in water at a specified temperature. The float-test results are reported as the time in seconds from placing the float in water until the water breaks through the material in the collar.

Sampling Bituminous Materials

ASTM D140 (Sampling Bituminous Materials) covers the method used to sample bituminous materials at points of manufacture, storage, or delivery. The purpose of this procedure is to determine the true nature and condition of the material. Samples are taken as a representation of the bulk of the material and to ascertain the maximum variation in characteristics that the material possesses. This procedure covers sampling of semisolid or uncrushed solid material and of crushed or powdered material at place of manufacture, from containers, tankcars, vehicle tanks, distributor trucks, recirculating storage tanks, tankers, barges, pipelines, drums, and barrels.

Viscosity

Viscosity of asphalt materials can be determined by one of two methods: ASTM D2170 [Kinematic Viscosity of Asphalt (Bitumens)] and ASTM D2171 (Viscosity of Asphalts by Vacuum Capillary Viscometer).

ASTM D2170 [Kinematic Viscosity of Asphalt (Bitumens)] covers determination of the kinematic viscosity of liquid asphalts (bitumens), road oils, and distillation residues of liquid asphalts (bitumens), all at 140°F (60°C), and of asphalt cements at 275°F (135°C) in the range of 6 to 100,000 centistokes. Kinematic viscosity is the ratio of the viscosity to the density of a liquid. It is a measure of the resistance to flow of a liquid under gravity. The SI unit of kinematic viscosity is m^2/s; for practical use, a submultiple (mm^2/s) is more convenient. The centistoke (cSt) is 1 mm^2/s and is customarily used. *Viscosity* as referred to herein is the ratio between the applied shear stress and the rate of shear, called the *coefficient of dynamic viscosity*. This coefficient is a measure of the resistance to flow of a liquid. The SI unit of viscosity is the pascal-second; for practical use, a submultiple (mPa•s) is more convenient. The centipoise, 1 mPa•s, is customarily used. Density as referred to herein is the mass per unit volume of liquid. The cgs unit of density is 1 g/cm^3 and the SI unit of density is 1 kg/m^3.

The method includes measuring the time for a fixed volume of the liquid to flow through the capillary of a calibrated glass capillary viscometer under an accurately reproducible head and at a closely controlled temperature. The kinematic viscosity is then calculated by multiplying the efflux time in seconds by the viscometer calibration factor.

The results of this test can be used to calculate viscosity when the density of the test material at the test temperature is known or can be determined.

The second method, ASTM D2171 (Viscosity of Asphalts by Vacuum Capillary Viscometer), covers procedures for the determination of viscosity of asphalt (bitumen) by vacuum capillary viscometers at 140°F (60°C). It is applicable to materials having viscosities in the range from 0.036 to over 200,000 poises (P).

In this procedure the time is measured for a fixed volume of the liquid to be drawn up through a capillary tube by means of vacuum, under closely controlled conditions of vacuum and temperature. The viscosity, in poise, is calculated by multiplying the flow time in seconds by the viscometer calibration factor.

There has been a strong movement to change the system of grading asphalt cements from penetration units at 77°F (25°C) to absolute viscosity units at 140°F (160°C), in accordance with ASTM D2171. Many researchers feel that this temperature represents a critical service temperature in asphalt pavements and that its use tends to minimize viscosity differences at higher and lower temperatures among asphalts of the same penetration grade having different temperature susceptibilities. However, opponents indicate that the 140°F (60°C) temperature is too high—that the asphalt begins to behave as a Newtonian liquid. In any event, some manufacturers are now utilizing both classification techniques. It is doubtful that the penetration method of classification will be dropped entirely for several years to come, as this would result in the loss of many years of accumulated field experience.

MODIFICATIONS OF ASPHALT CEMENTS

Asphalt cements may be modified into several different products to make it easier for distribution and use. Asphalt cement is generally hard and relatively solid. It must therefore be heated or treated before it can be mixed with aggregates to produce an asphaltic concrete (pavement). Asphalt cement can be made into a liquid at lower temperatures by mixing it with a volatile oil, or it can be emulsified with water to produce liquids at normal temperatures. When the volatile oils evaporate or the emulsion breaks, a hard stable asphalt cement remains which with the aggregate particles make up the pavement.

Liquid Asphalts (Cutbacks)

When volatile oils are mixed with asphalt cement to make a liquid product, the product is referred to as a *cutback*. The purpose of the cutback is to allow relatively easy placement of the asphalt product without the use of high temperatures. After the material has been placed, the product reverts to its natural penetration value through the evaporation of the volatile oils. In general, there are three types of liquid asphalt in the cutback category:

1. Rapid-curing (RC), (ASTM D-2028)
2. Medium-curing (MC), (ASTM D-2027)
3. Slow-curing (SC; road oils), (ASTM D-2026)

Rapid-curing (RC) cutbacks are made by diluting gasoline or naphtha with asphalt cement. The four grades of RC cutbacks range from RC 70 to RC 3000 depending on the amount of gasoline or naphtha used. Careful control must be maintained when utilizing gasoline or naphtha to keep the flash point of the material above 80°F (27°C). The percentage of the total volatile oils driven off at 500°F (260°C) varies from 70 percent for RC 70 to 25 percent for RC 3000. Obviously, the viscosity increases from RC 70 to RC 3000.

Medium-curing (MC) cutbacks are similar to rapid-curing cutbacks, with the exception that kerosene is used rather than gasoline or naphtha to liquefy the asphalt cement. The asphalt cement used to make medium-curing cutbacks should have a penetration of 70 to 250 at 77°F (25°C). There are five grades of MC cutbacks which range from MC 30 to MC 3000.

Comparing rapid-curing cutbacks with medium-curing cutbacks, it can be shown that rapid-curing cutbacks have a harder base asphalt and that the gasoline or naphtha will evaporate at lower temperatures, resulting in a material that is believed to cure rapidly. The medium-curing cutbacks have a softer base asphalt and a less volatile solvent (kerosene), and as a result will cure much more slowly than will a rapid-curing cutback.

Therefore, if one is constructing a pavement in a northern region, one might well choose the medium-curing cutback, as it has a softer base, and after curing the material would be less brittle in the winter and subject to less cracking. In a southern region, one might well choose the rapid-curing cutback, to obtain a harder asphalt base. This harder base may well provide a much more stable pavement under the hot sun.

Many state specifications for rapid-curing and medium-curing cutbacks are based upon viscosity tests run at 140°F (60°C). In this system RC and MC asphalt cutbacks have the following grades: 70, 250, 800, and 3000 for RC and 30, 70, 250, 800, and 3000 for MC. These numbers refer to minimum allowable range of viscosity as determined by the kinematic viscosity test, in units of centistokes (Table 9-2).

Cutback liquid asphalts are governed by ASTM D2027 [Cutback Asphalt (Medium-Curing Type)] and ASTM D2028 [Cutback Asphalt (Rapid-Curing Type)].

Road Oils

Slow-curing liquid asphalts (*road oils*) were originally manufactured by a straight-run distillation process and were really liquid asphalt cements. In other words, they were manufactured like asphalt cements, but the distillation process was cut off earlier and many of the volatile oils remained as part of the asphalt. Today, slow-curing (SC) road oils are cutback asphalts. However, they are not cutback with gasoline, naphtha, or kerosene but are fluxed with nonvolatile oils.

The grading system for SCs is the same as that for RC and MC materials. Their main applications are for dust control (dust binding). *Dust binding* is a light application of bituminous material for the express purpose of laying and bonding loose dust. Slow-curing road oils are governed by ASTM D2026 [Cutback Asphalt (Slow-Curing Type)].

Asphalt Emulsions

An *asphalt emulsion* is a suspension of minute globules of water or an aqueous solution in a liquid bituminous material. It is a mixture of asphalt cement and water in which water is the continuous phase and asphalt cement is the dispersed phase. Asphalt emulsions are another way of liquefying asphalt cements. The two most commonly used types of emulsified asphalts are anionic emulsions and cationic emulsions. *Anionic emulsions* are a type of emulsion such that a particular emulsifying agent establishes a predominance of negative charges on the discontinuous phase. *Cationic emulsions* are a type of emulsion

TABLE 9-2   CUTBACK GRADES

Viscosity	Grade (MC)	Grade (RC)
30–60	MC-30	
70–140	MC-70	RC-70
250–500	MC-250	RC-250
800–1600	MC-800	RC-800
3000–6000	MC-3000	RC-3000

such that a particular emulsifying agent establishes a predominance of positive charges on the discontinuous phase. Originally, all asphalt emulsions were the anionic type. The type of asphalt emulsion is determined by the kind of emulsifying agent or soap used. An anionic emulsion works well with aggregates that have a positive charge (thus, opposites attract). Cationic emulsions work well with all types of aggregate. Cationic emulsions can extend the paving season, as they can better handle cold weather and are not damaged by sudden rain.

The setting time of each of the three general grades of asphalt emulsions (rapid setting, medium setting and slow setting) is dependent upon the breaking time of the emulsion and the evaporation of the water. Each type of emulsion has two grades, 1 and 2. There also exists an MS-2h grade, which is suitable for cold and hot plant mix as well as four HFMS grades. Most types are used cold, although in special cases they may be used hot.

The advantages of asphalt emulsions are as follows:

1. Can be used with cold or hot aggregate.
2. Can be used with aggregate that is dry, damp, or wet.
3. Eliminates the fire and toxicity hazards of cutback liquid asphalt.

Testing of Asphalt Cements Emulsions

Most of the ASTM test procedures previously discussed for asphalt cements may be used for emulsions. However, ASTM D244 (Emulsified Asphalt) covers specific testing for emulsified asphalts. The method covers the compositions, consistency, stability, and examination of residue of asphalt emulsions composed principally of a semisolid or liquid asphaltic base, water, and an emulsifying agent. The composition test includes water content, residue by distillation, identification of oil distillation by microdistillation, residue by evaporation, and particle charge of emulsified asphalts. The consistency test is the Saybolt viscosity test. The stability test includes demulsibility, settlement, cement mixing, sieve test, coating, miscibility with water, modified miscibility with water, freezing, coating ability and water resistance, and storage stability of asphalt emulsions. The final test is examination of residue.

ROAD TARS

Road tars were the most common paving materials until the development of the petroleum industry. With the widespread use of gasoline for automobile consumption, the asphalt by-product of the refining process has become so readily available that it has put tars out of use. However, in England, road tars are still in use because of the large bituminous coal industry.

Manufacture

Tars are brown or black bituminous material, liquid or semisolid in consistency, in which the predominating constituents are bitumens obtained as condensates in the destructive distillation of coal, petroleum, oil shale, wood, or other organic materials, and which yield substantial quantities of pitch when distilled. *Road tars* are the product of straight-run distillation of crude tars. The two most common methods of production are the coke-oven process and the water-gas method.

In the *coke-oven method*, bituminous coal is processed to produce coke with crude coke-oven tar as a by-product. Approximately 10 tons of bituminous coal are heated to 2500°F (1370°C) in a brick-lined oven. The crude tar is part of the volatile product that is removed in the heating process, the residue being coke. With this method the properties of the tar vary with the makeup of the coal, the kind of oven, the temperature, the length of time the temperature is applied, and the pressure of the system. When the vapors are removed from the oven, condensed, and cooled, crude tar is formed. The crude tar is then placed through a straight-run distillation process to form tar.

Although most of the road tars are produced by destructive distillation of bituminous coal in the coke-oven tar procedure, a very small amount is produced by the *water-gas operation* of cracking petroleum. In this method, the crude tar is a by-product of the second stage of producing heating gas. Initially, steam is passed over a bed of incandescent coke which decomposes the steam; carbon monoxide and hydrogen gases are formed, which are known as *water gas*. This water-gas tar is too low in Btu for heating purposes, and hydrocarbon gases must be added for enrichment. These enriching hydrocarbon gases result from the passing of petroleum oils over hot firebrick in a carburetor. The high temperature in this process causes cracking (destructive distillation) of the petroleum oils, and in addition to the hydrocarbon gases, which enrich the water gas, a heavy residue of hydrocarbons is condensed to form crude tar. The tar is referred to as water-gas tar if a light petroleum fraction is used, and as residuum tar if a heavier residue is used.

In the manufacture of tar, the crude tar is stored until the water settles out. At this point the crude tar is transported to the refining plant, where the small amount of water that remains is heated out. In the tar refining process, many other products are produced besides road tars. The crude tar is treated by fractional distillation in much the same manner as petroleum is treated to produce asphalt. The viscosity of tars is produced by straight-run distillation. The distillation is carried to that extent required for the grade desired. Road tars are graded from RT 1 up to RT 12. RT 1 is a light fraction of tar and RT 12 is as hard in consistency as the No. 200 penetration asphalt. Cutback tars also exist in two grades, RTCB 5 and RTCB 6. These two grades are made by using light or middle oils with RT 10, 11, or 12 tar.

RT 1 is used for dust control. RT 2 and 3 are light prime coats. RT 4 is for prime coats and sometimes surface treatments. RT 5, 6, and 7 are used for surface treatments and road mixes. RT 8 and 9 are for use as surface treatments, seal coats, and road mixes. RT 10 and 11 are used in seal coats, tar concrete, and hot repairs. RT 12 is used in penetration macadam work, tar concrete, and hot repairs.

The RT 1-6 and RTCB 5-6 can be used at temperatures up to 150°F (66°C). RT 7 and above may be used at higher temperatures.

Methods of Test

Road tars are tested in much the same manner as asphalt cements, but the tests differ somewhat, primarily because the tests were developed at different times. In general, the tars are many years older.

PROPORTIONING ASPHALTIC MIXES

The most suitable asphaltic concrete is one that produces a stable, durable, flexible, and skid-resistant pavement at a minimum cost with adequate aggregate. It is not possible to optimize all four properties; thus, compromises result. We will now look at the compromises. One was made in the design of bituminous-concrete mix design.

Properties

The term *stability* is related to strength and refers to the ability of a pavement to resist deformation under application of loads. The stability depends on the distribution of loads by point-to-point contact of the aggregate particles. The forces of this distribution are affected by aggregate interlocking developed among aggregate particles and the cohesiveness supplied by the asphalt cement.

In maximizing stability, the aggregate would have to be of crushed angular particle shape with a rough surface texture. It should also be hard and dense, its grade approaching the Fuller curve. Further, there should be just enough asphalt to coat the aggregate particles such that all particles benefit from the adhesiveness.

Rounded particles would result in low stability and load distribution. Rounded particles tend to slide over one another, whereas angular particles interlock with one another. If soft aggregate particles were used, they would break and wear under the impact of vehicular loads and reduce stability greatly.

Durability refers to the resistance of the pavement to disintegration under traffic loads. In other words, the pavement will remain smooth and serviceable during summer heat and will not crack or ravel during the winter cold. It will further resist the forces of freezing and thawing. In general, the greater protection of the aggregate provided by the asphalt cement, the more durable the pavement. Thus, in maximizing durability, one would not only wish to coat all the aggregate particles with asphalt but also to fill the voids within it. In this matter, the aggregate would be immune to the forces of freezing and thawing, for no water could enter the mix. Further stripping would not occur, because the water could not get between the aggregate and the asphalt. In addition, oxygen would not penetrate the pavement; thus, oxidation cannot take place and durability and adhesiveness would last longer.

In maximizing durability, stability and skid resistance are compromised. In a thick coating of asphalt over the aggregate particles, the aggregates tend to float in the asphalt. The result is that no interlocking of the aggregate particles takes place, and stability is lost.

As traffic tends to compact a pavement in which there is more asphalt than that needed to cover the aggregate, skid resistance is compromised. As the pavement compacts, bleeding of the asphalt occurs (asphalt comes to the surface), making a very slippery pavement.

Flexibility refers to the ability of a pavement to withstand deflections and bending without cracking. To maximize flexibility, one would use an open-graded aggregate mixture. However, in this case, stability is compromised. The stability might be improved by increasing the pavement thickness or by increasing such flexibility decreases.

Skid resistance is a form of stability, and there are two categories that cause slippery pavements:

1. Pavement bleeding.
2. Aggregate polishing.

In *pavement bleeding* too few voids are left in the mix and compaction occurs under traffic loads on a hot day, forcing asphalt to the surface of the roadway. This covers the exposed rough aggregate and results in a slippery pavement. With voids in pavements as low as $1/2$ or 1 percent, slippery pavements will not occur; however, most specifications require 2 or 3 percent air voids to prevent surface bleeding.

Aggregate polishing results after a pavement receives continuous wear from traffic and the surface aggregates polish, forming a slippery pavement. One method to prevent aggregate polishing is to use relatively hard aggregates. Another is to use a mixture of varying aggregates of different hardnesses. In this way, as one type of aggregate polishes, the other, harder aggregates do not, so that the road does not totally polish.

Mix Design

A good asphalt pavement is one that compromises among stability, durability, flexibility, and skid resistance for a given gradation of aggregate with just enough asphalt to cover the aggregate particles for good adhesive properties. In proportioning asphaltic concrete, seven steps should be followed:

1. Evaluate and select the aggregates to be used.
2. Select the aggregate gradation.
3. Make trial mixes with various amounts of asphalt cement.
4. Measure the relative stability of each mix.
5. Compute the void content of each compacted mix.

6. Compute the voids in the mineral aggregate and voids in the aggregate filled with asphalt.
7. Select the optimum design.

In step 1, the contractor is allowed to select the most convenient and economical aggregate source that meets the general specifications. These specifications usually demand an aggregate of the highest quality that is economically feasible.

For step 2, aggregate gradation, many highway and government agencies have specifications that define aggregate gradings for various types of mixes and sizes of aggregates. For reference, the Asphalt Institute publishes various manuals that give minimum recommendations on grading and sizes.

Step 3, the making of the trial mixes with different amounts of asphalt cements (2 to 6 percent), is probably the most important step. As previously discussed, the optimum mix should be one in which a compromise is made among stability, durability, flexibility, and skid resistance with sufficient asphalt to coat the aggregate particles of a given size and gradation.

Step 4, which allows for the measurement of stability, may utilize either of two laboratory tests that measure the deformation under load of compacted laboratory specimens. These tests are both governed by ASTM specifications and are as follows:

1. Marshall, ASTM D1559.
2. Hveem Stabilometer, ASTM D1560.

The most common test of these two is the Marshall test, ASTM D1559.

In step 5, the void content of the compacted specimen is computed. The percentage of voids in a compacted asphaltic concrete influences the stability and can be used as a specification to prevent pavement bleeding and thus slippery roadways.

The void content is computed by comparing the volume of the compacted mix with the volume occupied by each of the ingredients. The volume of any amount of material can be computed by knowing its specific gravity or density.

$$\text{specific gravity} = \frac{\text{density of the material}}{\text{density of water}} \tag{9.1}$$

$$\frac{\text{solid volume}}{\text{of each ingredient}} = \frac{\text{weight of the ingredient in the specimen}}{\text{specific gravity of the ingredient}} \tag{9.2}$$

$$\text{percent voids} = 100 \, \frac{V_S - V_1}{V_S} \tag{9.3}$$

where V_S = volume of specimen
V_1 = combined solid volumes of all ingredients in the specimen

In step 6, we compute the voids in the mineral aggregate. When one discusses the total voids within the compacted aggregate particles, it becomes clear that asphalt occupies most of the void space. As indicated, if the voids are completely filled with asphalt, the pavement loses its stability. Therefore, it is useful to compute the voids in the mineral aggregate (VMA) and the percentage of voids filled with asphalt. The VMA and the voids filled with asphalt are usually controlled by specifications. The VMA may be computed as follows:

$$\text{VMA} = \text{volume of specimen} - \text{volume occupied by the aggregate} \quad (9.4a)$$

Therefore,

$$\text{VMA} = V_S - V_{agg} \quad (9.4b)$$

$$\text{percent VMA filled with asphalt} = 100 \, \frac{V_a}{\text{VMA}} \quad (9.5)$$

where V_a is the volume of asphalt in the aggregate voids.

The final step, 7, is to select the optimum asphalt content according to the design specifications.

Example 9.1:

Calculate the percent voids and VMA in a compacted mixture with a unit weight of 144 lb/ft^3 with the following criteria:

Ingredients	Percent of Total Aggregate	Effective Specific Gravity
Coarse aggregate	70	2.70
Fine aggregate	25	2.65
Filler aggregate	5	2.60

Percent asphalt = 5 with an effective specific gravity = 1.00

Solution:

Weight		**Volume**
0	Air	$0.06 = 1 - 0.94$
$0.05 \times 144 = 7.2$	Asphalt	$0.12 = \dfrac{7.2}{1 \times 62.4}$
$0.047 \times 144 = 6.7$	Filler	$0.04 = \dfrac{6.8}{2.60 \times 62.4}$
$0.24 \times 144 = 34.1$	Fine aggregate	$0.21 = \dfrac{34.1}{2.65 \times 62.4}$
$0.67 \times 144 = \underline{95.9}$	Coarse aggregate	$\underline{0.57} = \dfrac{95.9}{2.70 \times 62.4}$
Σ 144 lbs.		Σ 1 ft.^{3}

1. Assume a 1-ft³ volume and determine the total weight of the mix.
2. Compute the percent of aggregate based upon 95 percent of total ingredients.
3. Determine the weights of each ingredient.
4. Determine the volume of each ingredient by eq. (9.2).
5. Use eq. (9.3) to determine the volume of the air.
6. Use eq. (9.4b) to determine the VMA.

Answer: 6 percent air and 18 percent VMA.

In Example 9.1 the term "effective specific gravity" was used, and needs an explanation. The specific gravity of aggregates is an important consideration in the determination of void contents in compacted bituminous materials. There are three types of specific gravity values used in the field: apparent, bulk, and effective.

The *apparent specific gravity* is the ratio of the weight of an aggregate particle to the weight of a volume of water equal to the volume of solid aggregate and pores impermeable to water. The diagram below illustrates the term of apparent specific gravity:

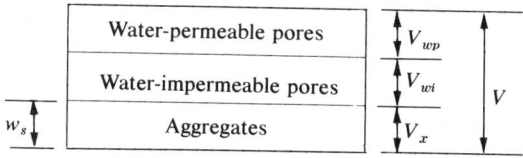

The apparent specific gravity is as follows:

$$\text{apparent specific gravity} = \frac{W_s}{(V_s + V_{wi})\gamma_w} \tag{9.6}$$

where W_s = weight of the aggregate particles (dry)
V_s = volume of solid aggregate
V_{wi} = volume of water-impermeable pores
γ_w = unit of water
V_{wp} = volume of water-permeable pores

ASTM C127 and C128 give the procedure for the determination of the specific gravity (apparent) for coarse aggregate and fine aggregate, respectively.

The *bulk specific gravity* is the ratio of the weight of aggregate particles to the weight of a volume of water equal to the volume of solid aggregate, pores impermeable to water, and pores permeable to water. In viewing the previous diagram, the bulk specific gravity is as follows:

$$\text{bulk specific gravity} = \frac{W_s}{(V_{wp} + V_{wi} + V_s)\gamma_w} \quad \text{or} \quad \frac{W_s}{V_w} \tag{9.7}$$

where V is the total volume of aggregate. The same ASTM procedures, ASTM C127 and C128, may be used to determine the bulk specific gravity.

In the determination of the air-void content in bituminous concrete mixtures, the use of the apparent or bulk specific gravity is incorrect. If the apparent specific gravity is used, it is incorrect because this assumes that the permeable voids are filled with bitumen to the same extent as with water. On the other hand, if the bulk specific gravity is used, the asphalt is assumed not to penetrate into the permeable voids. Therefore, effective specific gravity is used.

The *effective specific gravity* is the ratio of the weight of aggregate particles to the weight of a volume of water equal to the volume of solid aggregate and pores impermeable to asphalt. Again using a diagram, the effective specific gravity is illustrated.

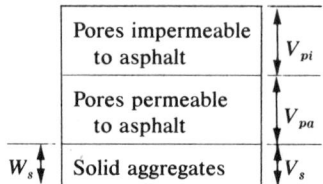

$$\text{effective specific gravity} = \frac{W_s}{(V_s + V_{pi})\gamma_w} \qquad (9.8)$$

where V_{pi} = pores impermeable to asphalt
$\quad\ V_{pa}$ = pores permeable to asphalt

Generally speaking, an average effective specific gravity can be obtained as follows:

$$\text{effective specific gravity} = \frac{\text{apparent SG} + \text{bulk SG}}{2} \qquad (9.9)$$

Marshall Mix Design

The Marshall mix design procedure is probably the most widely used method of bituminous mix design. It is governed by ASTM C1559 (Resistance to Plastic Flow of Bituminous Mixtures Using Marshall Apparatus). This method covers the measurement of the resistance to plastic flow of cylindrical specimens of bituminous paving mixtures loaded on the lateral surface by means of the Marshall apparatus. This method is for use with mixtures containing asphalt cement, asphalt cutback or tar, and aggregate up to 1 in. (2.54 cm) maximum size.

In this procedure, a 4-in. (10.16-cm) diameter by 2.5-in. (6.35-cm)-high specimen is prepared by compacting in a mold with a compaction hammer that weighs 10 lb (4.54 kg) and has a free fall of 18 in. (45.72 cm). Depending upon the expected design traffic, either 35, 50, or 75 blows of the hammer are applied to each side of the specimen. After 24 hours of curing, the density and voids are determined and the specimen is heated to 140°F (60°C) for the Marshall stability and flow tests. The specimen is situated in a cylindrical-shaped half-split breaking head and the specimen is loaded at a rate of 2 in./min (5.08 cm/min). The maximum load registered, in pounds (kilograms) is referred

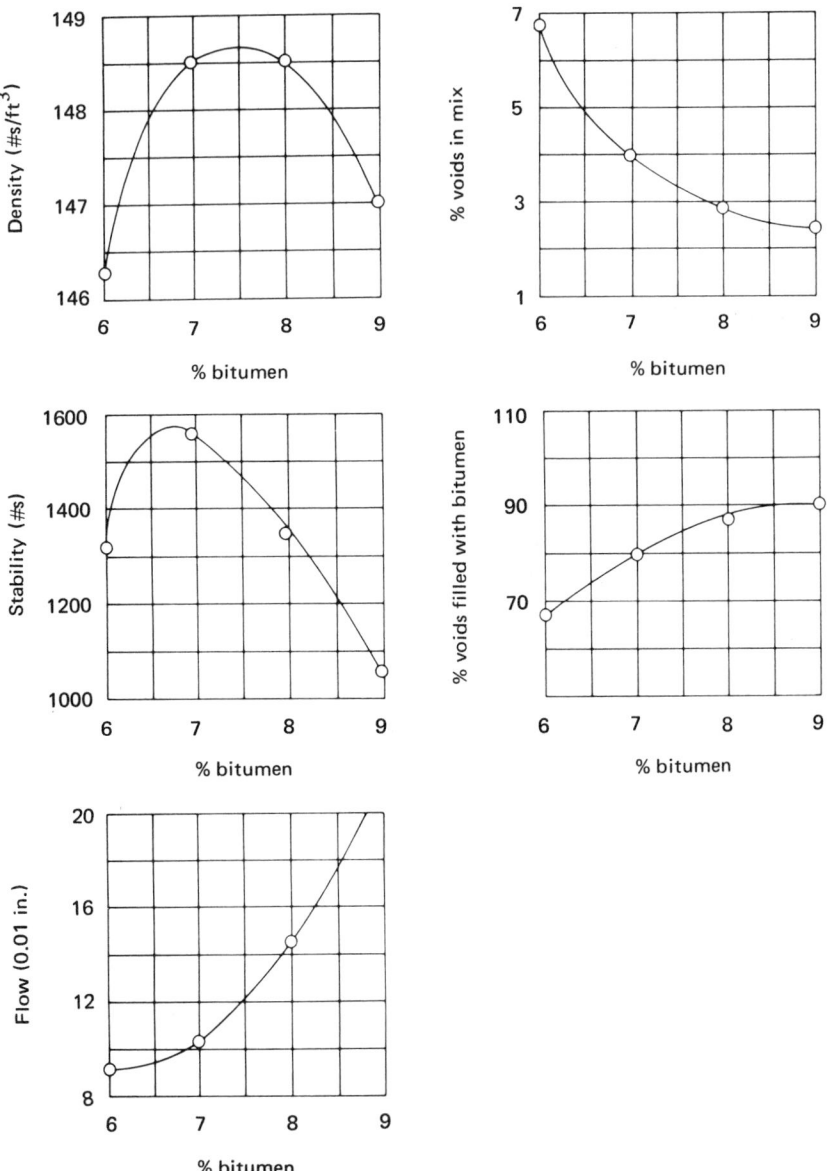

Figure 9-2 Typical Test Data for Marshall Mix

TABLE 9-3 MARSHALL DESIGN CRITERIA

Traffic Category:	Heavy and Very Heavy		Medium		Light	
Number of Compaction Flows Each End of Specimen:	75		50		35	
Test Property	Min.	Max.	Min.	Max.	Min.	Max.
Stability, all mixtures	750	—	500	—	500	—
Flow, all mixtures	8	16	8	18	8	20
Percent air voids						
Surfacing or leveling	3	5	3	5	3	5
Sand or stone sheet	3	5	3	5	3	5
Sand asphalt	5	8	5	8	5	8
Binder or base	3	8	3	8	3	8
Percent voids in mineral aggregate	Varies with particle size					

to as the Marshall stability of the specimen. The amount of movement between no load and maximum load is referred to as *strain* and is measured in units of 0.01 in.; it is generally referred to as the *flow*.

Figure 9-2 illustrates test property curves for hot mix-design data by the Marshall method. As shown, five curves are presented: unit weight, percent air voids, Marshall stability, percent voids filled with bitumen, and flow. The stability value obtained for each test specimen is modified if the thickness is greater or less than a height of 2 in. (5.08 cm).

Table 9-3 illustrates the Marshall design criteria. This information is based upon the Asphalt Institute Marshall design criteria. These criteria are applied to the curves and a suitable asphalt percentage is determined. The most common procedure for doing this is to take the most desirable asphalt percentages for stability, unit weight, and percent voids and to average them. This average value should satisfy the required criteria.

PROBLEMS

9.1. Define asphalt cement.

9.2. List and explain the purpose of the hydrocarbons that make up bitumen.

9.3. What is the best type of petroleum found in the earth's crust? Why?

9.4. Explain the distillation process of crude oil.

9.5. Discuss the specific ASTM standards for asphalt cements. Give the purpose and procedure for each test.

9.6. Explain liquid asphalts and discuss in detail.

9.7. Describe the manufacturing process for road tars.

9.8. List and describe the four essential properties of a good bituminous mix design.

9.9. What are the seven basic steps in proportioning asphaltic concrete? Explain.

9.10. Calculate the percent voids and VMA in a compacted mixture with a unit weight of 135 lb/ft^3 with the following criteria:

Ingredient	Percent of Total Aggregate	Effective Specific Gravity
Coarse aggregate	76	2.67
Fine aggregate	20	2.65
Filler aggregate	4	2.55

Percent asphalt = 6 with an effective specific gravity = 1.00

REFERENCES

Asphalt as a Material, *Asphalt Inst. Inform. Ser. 93,* 1965.

Asphalt Institute, *Asphalt Paving Manual,* 1962.

Asphalt Institute, "Brief Introduction to Asphalt," *Manual MS-5,* 1974.

BROOME, D. C., "Native Bitumens," in Arnold J. Holberg, *Bituminous Materials: Asphalts, Tars and Pitches,* vol. 2, part 1, Interscience Publishers, John Wiley & Sons, Inc., New York, 1965.

GOETZ, W. H., and WOOD, L. E., "Bituminous Materials and Mixtures," in K. B. Woods, *Highway Engineering Handbook,* sec. 18, McGraw-Hill Book Company, New York, 1960.

KREBS, R. D., and WALKER, R. D., *Highway Materials,* McGraw-Hill Book Company, New York, 1971.

WHITEHURST, E. A., and GOODWIN, W. A., *Bituminous Materials and Bitumen Aggregate Mixes,* Pitman Publishing Corp., New York, 1958.

This test will be on Monday.

10

The Metallic State

To understand the macroscopic properties of metals and alloys in any detail, it is necessary to have some knowledge of the metallic state at the atomic level. All of the large-scale effects metals exhibit under stress may be accurately explained by the interactions occurring microscopically. Indeed, it is knowledge of the physics involved between individual atoms and groups of atoms that allow predictions of large-scale behavior to be made. The degree to which one can correlate actual test results with known atomic structure will mark the extent to which one can make correct decisions in the selection of metals for a given design. Such ability is only possible through an understanding of the manner in which metallic properties derive from the nature of the metallic state. To facilitate this understanding, we will begin with a discussion of the crystalline structure of metals.

THE CRYSTALLINE NATURE OF METALS

The Metallic Bond

In a pure metal or alloy, each atom is independent of its neighbors; that is, there are no definable molecules. The valence electrons of each atom exist in an unconfined state, free to move randomly among the other atoms at very high velocity. This freedom of movement is due to the nearly equal energy levels existing on the valence shells of the atoms in the lattice. Thus, electrons may move from one atom to the next without disrupting the electronic equilibrium of the structure. The aggregation of all free valence electrons is termed the *electron cloud*, a singular property of the metallic state. As we shall see, the freedom of electrons from individual atoms is responsible for the unique elastic behavior of metals.

214

When the metal atoms give up their valence electrons to the electron cloud in metallic bonding, they become positive ions, held in position through their attraction to the electron cloud and their mutual repulsion. For the structure to exist in equilibrium, these forces must balance in such a way that the resultant is zero. At some specific atomic spacing a level of lowest energy is reached for which the internal forces cancel each other. The structure will seek to return to this equilibrium position if displaced, giving rise to the elastic behavior of metals in tension and compression. Figure 10-1 is a graph of force versus distance for two ions. The points above the horizontal axis correspond to repulsive force, those below represent the action of attraction, and the point where the curve crosses the horizontal axis is the equilibrium spacing or mean interatomic distance. If the structure is placed in tension (i.e., an attempt is made to increase the atomic spacing), the attractive force predominates, resisting the deformation and restoring the equilibrium position after load is removed. Similarly, when compressive forces are ap-

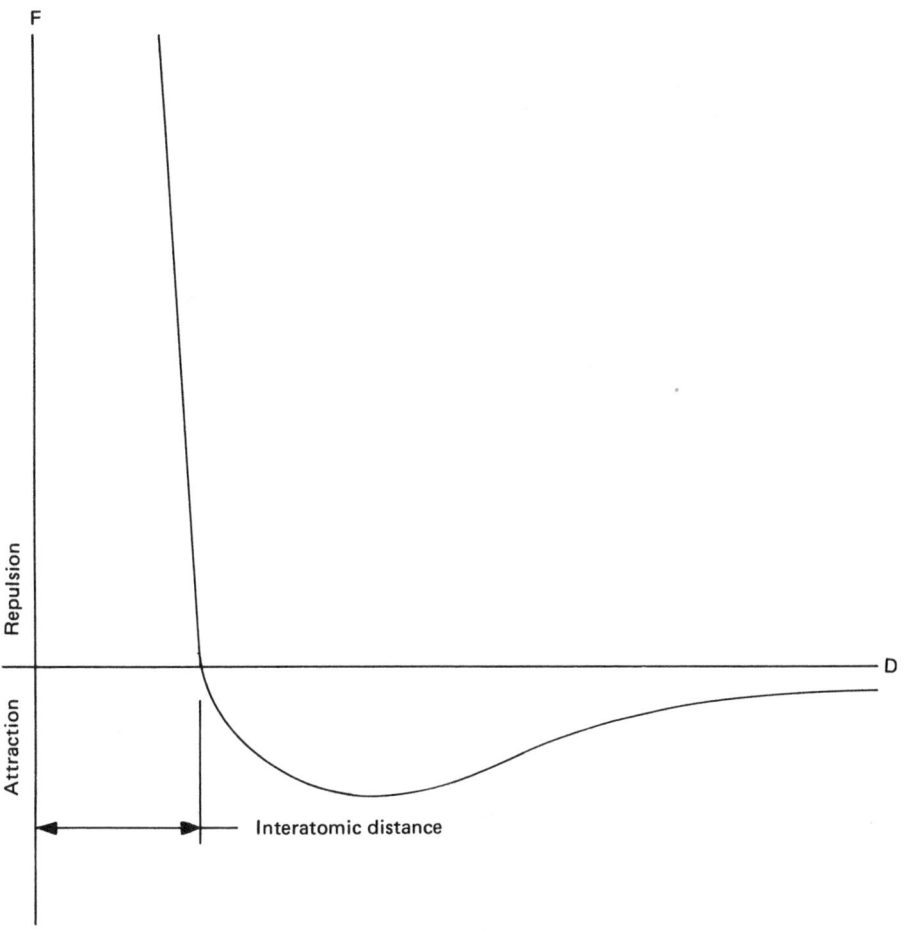

Figure 10-1 Force Versus Distance

plied, the ions are pushed closer together and the repulsive force due to their like charges resists the deformation, again restoring the lattice to its equilibrium spacing when load is removed.

For equilibrium to exist throughout the structure (in the absence of any external or internal forces), the interatomic distance must be the same for all atoms in the lattice. Each metal or alloy is associated with a unique interatomic distance which allows individual metals or alloys to be identified through their characteristic x-ray diffraction pattern. This pattern is a result of the crystal structure and atomic spacing, thus possessing the same uniqueness as the interatomic distance of an individual metal.

Unit Cells

The absence of any bonding requirement beyond a uniform spacing results in a very closely packed structure resembling spheres stacked in a box. The majority of metals have as lattice structure one of the following three types, illustrated in Figure 10-2: body-centered cubic (BCC), face-centered cubic (FCC), or hexagonal close-packed (HCP). These geometrical figures, called *unit cells,* represent the smallest configuration of atoms that preserves the characteristic arrangement of the lattice. Table 10-1 lists some of the important metallic elements with their interatomic spacing and unit cell configuration in order of increasing interatomic distance.

Allotropic Behavior

Inspection of Table 10-1 will reveal several elements that have more than one crystalline configuration. The different crystal formations characteristic of an element are called its allotropic forms or *allotropes*. The property itself is termed *polymorphism* and occurs due to the almost identical interatomic distances existing in the allotropes of elements that exhibit this behavior. Examination of the interatomic spacing corresponding to the allotropes of the polymorphous elements cobalt, iron, and titanium listed in Table 10-1 will reveal this similarity.

The interatomic spacing, and hence the crystal structure of an allotrope, is determined by the energy level of the lattice. Iron, for example, has a FCC structure with an interatomic distance of 2.58 angstroms (Å) at temperatures above 3038°F (1670°C). This is a higher energy state than that of the BCC structure. It occurs at lower temperatures with the closer spacing of 2.48 Å.

Grain Formation and Growth

The majority of metals used in industry are polycrystalline; they are composed of a great many small crystals, called *grains,* each of which is made up of metallic ions arranged in the particular space lattice characteristic of that metal. Grain size is determined by the rate of cooling from the liquid state. As the temperature of the liquid metal falls below its melting point, a phase change begins to occur; low-energy atoms precipitate out of the liquid state first, forming the nuclei upon which further growth takes place. The rate at which nuclei precipitate and growth progresses is a direct function of the cooling rate of the melt. If cooling is slow, few nuclei precipitate and growth is slow. This tends to produce a matrix of large, well-developed crystals. If cooling is rapid, as when the metal is quenched (immersed in a medium such as soil, water, or brine at much lower temperature), many nuclei form almost instantaneously and the structure produced is very fine grained.

Regardless of the rate of cooling, individual grains form in the dendritic pattern illustrated in Figure 10-3, with growth occurring predominantly at the ends of the dendrites. Eventually, the growth of an individual dendrite progresses to the point that other dendrites restrict its expansion. When this happens, growth continues on the existing arms until all internal liquid is frozen. This is a random process, determined by the charge positions that nuclei occupy when they precipitate. Thus, the final shape of a grain is irregular and bears little resemblance to its highly ordered internal structure.

The space separating individual grains, typically one or two atoms wide, is a region of higher energy (less order) than the grains themselves. It is a transition zone from the

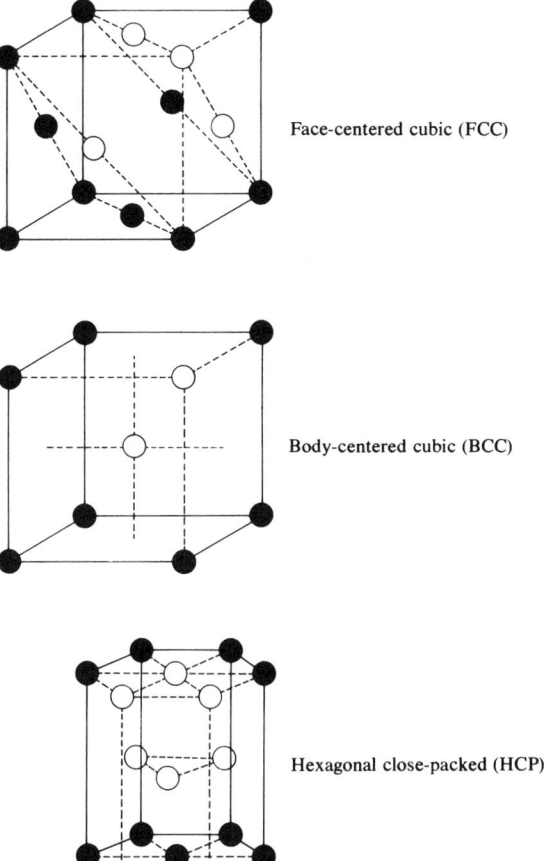

Face-centered cubic (FCC)

Body-centered cubic (BCC)

Hexagonal close-packed (HCP)

Figure 10-2 Lattice Structure

TABLE 10-1   INTERATOMIC SPACING

Element	Unit Cell	Interatomic Spacing (Å)
Beryllium	HCP	2.23
Iron	BCC	2.48
	FCC	2.58
Nickel	FCC	2.49
Chromium	BCC	2.50
Cobalt	HCP	2.506
	FCC	2.511
Copper	FCC	2.56
Vanadium	BCC	2.63
Zinc	BCC	2.66
Molybdenum	BCC	2.73
Tungsten	BCC	2.74
Platinum	FCC	2.78
Aluminum	FCC	2.86
Tantalum	BCC	2.86
Gold	FCC	2.88
Titanium	HCP	2.89
	BCC	2.89
Silver	FCC	2.89
Magnesium	HCP	3.20
Lead	FCC	3.50

ordered crystalline direction of one grain to the differing orientation of adjacent grains. We shall see that grain boundaries play a significant role in influencing the characteristic behavior of a metal when stresses exceed its elastic limit. A diagram of a grain boundary is shown in Figure 10-4.

Alloying

There are certain additional elements, predominantly other metals, which will combine with a pure metal in such a way that the resulting solid solution, called an *alloy*, is also metallic. This is possible only if the alloying element lies within specific boundaries of size and valence. The size factor divides alloys into two distinct groups: substitutional and interstitial alloys. For simplicity, we consider only binary alloys, those composed of two constituent elements. Alloys composed of three or more elements are very common. However, the complexity of dealing with them increases rapidly as their number of constituent elements increases. We limit the scope of our examination accordingly.

In the *substitutional* alloy, the alloying element replaces the primary element in random positions in its lattice. To accomplish this without destroying the structure, the alloying element must have an atomic radii within 15 percent of that of the primary element. If the radial difference is greater than this, the energy required to place the alloying element in position is more than the structure can accommodate and still retain its crystalline form.

As the name implies, an *interstitial* alloying element fits into the spaces or interstices

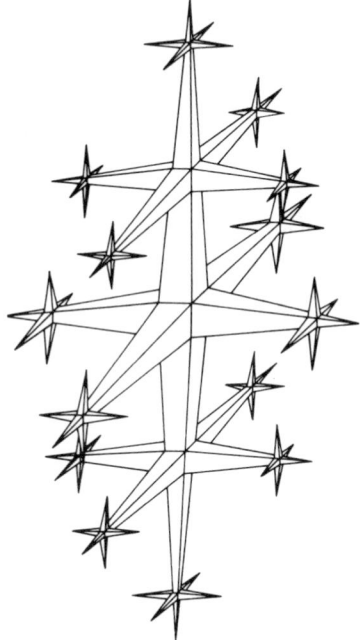

Figure 10-3 Density Pattern

of the primary element's lattice. There is a definite geometrical restriction on the size of the interstitial atom; to fit into the lattice spaces, the alloying element's radius must be less than approximately six-tenths that of the primary element. This corresponds to the maximum-size sphere that will fit into the spaces existing in a close-packed lattice of larger spheres. The transition metals are the main group of elements which act as primary

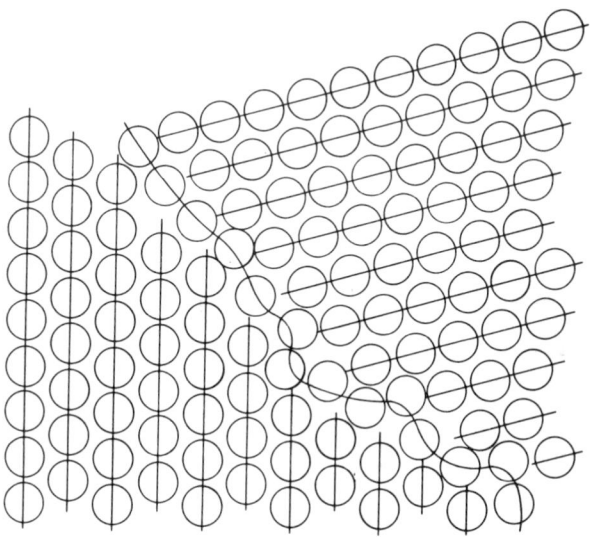

Figure 10-4 Grain Boundary

elements in interstitial alloys. There are only four elements small enough to meet the size requirement for interstitial alloying to occur when the primary elements are of this group. They are hydrogen, boron, carbon, and nitrogen and are termed *metalloid elements*.

Table 10-2 is a list of the metallic elements in a form known as the *electromotive series*. As one moves down the series, elements contain increasingly more complete valence shells and thus tend to lose electrons less easily than do the elements above them.

For two metallic elements to form a substitutional alloy over a wide range of composition, they must lie near each other in the series; that is, they must be of similar valence. If their valences differ by a sufficient amount, they will tend to exchange electrons and form a compound rather than a metallic bond when their relative proportions approach the necessary ratio. This is also true of interstitial alloys where the metalloid elements tend to form ionic compounds with nontransition metals which are relatively much more electronegative.

Alloys containing more than two elements are often both substitutional and interstitial. An important class of alloys in this category is the nickel steels; nickel and carbon are substitutional and interstitial elements, respectively, in iron, the primary element.

Phase Diagram

The *phase diagram* is a very useful tool in illustrating the characteristics of an alloy in relation to its temperature and composition. Also called an *equilibrium diagram*, this is a graph of the phases existing in an alloy for any combination of temperature and composition. Figure 10-5 is a typical phase diagram for substitutional binary alloys. In the diagram, relative percentages of one of the two constituent elements are plotted on the horizontal axis with the vertical scale corresponding to temperature. Any two distinct phases are separated by a region in which both exist simultaneously. The curves separating the different regions are the specific points of temperature and composition where one phase or combination of phases can exist in equilibrium with another. These are the loci of saturated equilibrium states.

Three different phases are shown in Figure 10-5: α, β, and L, where α and β are solid phases and L is a liquid phase. Also shown are the three two-phase regions separating

TABLE 10-2 METALLIC
ELEMENTS

Lithium	Cobalt
Potassium	Nickel
Sodium	Tin
Barium	Lead
Calcium	Antimony
Magnesium	Bismuth
Aluminum	Copper
Manganese	Mercury
Zinc	Silver
Chromium	Platinum
Iron	Gold

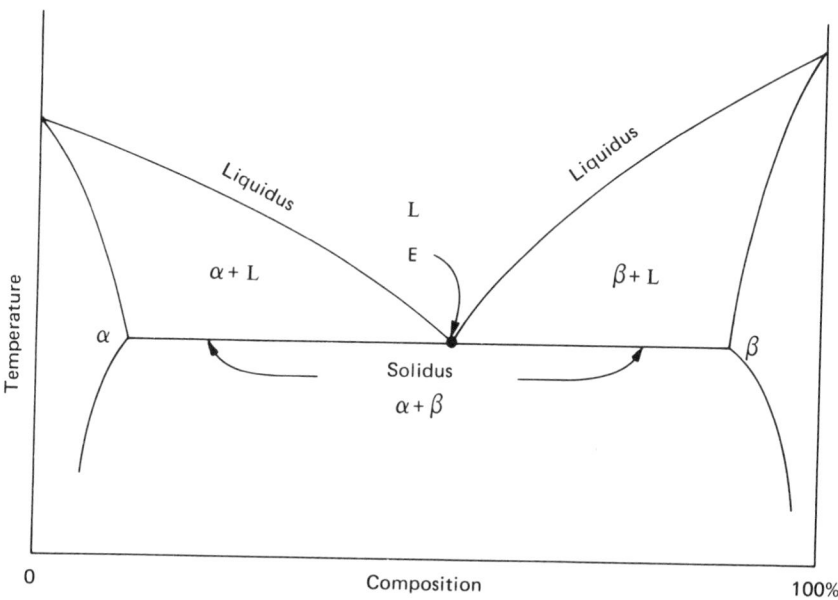

Figure 10-5 Phase Diagram of Binary Alloy

the single-phase regions, $\alpha + \beta$, $\alpha + L$, and $\beta + L$. The liquidus and the solidus are the boundaries between the liquid and solid phases, respectively, and the mixed regions. They converge to point E, the intersection of three regions, which describes the singular temperature and composition at which three phases may exist simultaneously. Such a point, which has a single liquid phase above two solid phases, is called a *eutectic;* one with a single solid phase above two other solid phases is termed a *eutectoid.* Many binary alloys exhibit phase relationships similar to Figure 10-5, with the sizes of the regions and position of the eutectic varying substantially from one alloy to the next. There are also binary alloys with a completely different form from that of Figure 10-5, and eutectoids. An example of an alloy with a phase diagram similar to Figure 10-5 is the copper-tin system. One differing markedly from this is the aluminum-nickel binary system.

For alloys with three or more elements, graphic illustration of phase relationships becomes increasingly difficult. The ternary or three-element alloys, for example, require a three-dimensional graph for complete description. Two-dimensional plots of the system may be made by taking a horizontal cross section of the three-dimensional graph, resulting in the phase relationships of the three constituents for varying compositions at a constant temperature.

MECHANICAL BEHAVIOR OF METALS

In this section we discuss the mechanical behavior of metals. Included in our discussion will be the resistance of metals to failure caused by slip, fatigue, impact, flow or creep, stress rupture, and corrosion and wear.

Slip (Inelastic Action)

Slip is a phenomenon of movement along gliding planes of crystals due to the application of stress beyond the material's elastic limit. When a material undergoes slip, one will observe slip lines or slip bends (fine lines) across the faces of a number of the crystalline grains of the overstressed metal.

Slip is an inelastic action that does not occur to any appreciable extent in any of the ordinary materials of construction until a fairly well defined limiting stress has been applied; and if slip occurs, it does not continue indefinitely, under load, but ceases after a short time.

The most satisfactory explanation of the fact that under a load above the elastic strength of a metal slip soon ceases, and of the observed increase of elastic strength after such overload, was furnished by the slip interference theory of Jeffries and Archer. As slip takes place, the crystalline grains of metal are fragmented into thin plates and the surfaces of the sliding plates tend to "dig into" each other, causing increasing resistance as slip proceeds. In addition to this, adjacent grains, in general, have slip planes lying in different directions, and as slip proceeds, the slip in one grain tends to hinder the slip in the adjacent grain.

Moreover, in metals composed of two or more different kinds of crystalline grains, the stronger grains act as "keys," tending to stop or to hinder slip in weaker grains. Slip interference theory further holds that a metal may contain definite, very small grain sizes which will be more effective as brakes on slip than larger, more widely scattered grains, and also more effective than extremely small grains, grains so small that slip can proceed right by them without much deviation from its direct path. In the case of steel, such an "optimum" size of grain may be produced by heat treatment.

Fatigue

For structural parts (axles, machine parts, shafts, etc.) under infrequent variation of load, the outstanding danger of failure is failure by inelastic distortion (slip).

Examination under the microscope and the scanning electron microscope of the behavior of the crystalline constituents of metals under repeated stress has shown that failure of crystals is caused by a succession of shear slips on parallel planes of vast strength. The failure of the piece as a whole is due to the successive failures of individual crystals and the development of cracks. Failure very seldom occurs at crystalline boundaries.

Fatigue was once thought of as a phenomenon of the cohesion between crystalline constituents of the metal subjected to many repetitions or reversals under comparatively low stresses. However, the failure of a material under repeated stress is a process of gradual or progressive fracture of the crystals themselves. Thus, the term "fatigue" is improperly used and the appropriate term should be "progressive fracture."

In a general way, the difference in the behavior of material under repeated loads and under a single load is shown by the tendency toward gradual cracking of the material under repeated loads. Under a single application of load the material of a structure either withstands the load or fails. Under a repeated load, the load may be applied many thousands of times and the material may withstand the load for a while, and then fail by the gradual spread of cracks in the material. Under repeated loads, local strains (which would be of little importance in a single load) may form a nucleus for damage which gradually spreads until the whole member fails. Slip in crystals is likely to occur at points where high localized stresses are set up. These stresses depend upon the distribution of stress between crystals, which in turn depends upon the homogeneity of the structure. Thus, flaws and cracks in the internal structure tend to cause internal stresses, which were formed by cooling, heat treatment, or mechanical working of the metal. These stresses would not normally affect the static structural strength but tend to produce slip in crystals when the metal is subjected to repeated stresses.

Impact

Impact is the resistance of a material to failure due to brittleness under service conditions in a structure. In selecting materials for members in a structure that must resist impact, two factors must be considered:

1. Total stress allowable.
2. Total strain allowable.

Resistance to impact is a function of both of these factors and both must be taken into consideration in design.

For materials that must withstand heavy accidental impact without actual rupture, toughness of the material is of primary concern. Toughness of a material may be measured by the area under the stress-strain curve of the material. According to Moore, a striking illustration of the resistance of materials to rupture under impact is furnished by comparing the action of oak with that of cast iron. Under static load cast iron is about three times as strong as oak, but the strain that oak will stand before rupture is about nine times the strain that cast iron will stand. The area under the stress-strain diagram for oak is about three times that for cast iron, and under impact loads oak requires about three times as much energy for fracture as cast iron.

The ability of a material to resist impact without a permanent distortion is measured by the area under the stress-strain diagram up to the elastic limit: in other words, the area under the straight-line portion of the stress-strain curve. If elastic resistance to impact is desired, a material with a high elastic strength or a low modulus of elasticity should be used.

Flow or Creep

Creep is the very slow flow of a material at elevated temperatures under sustained stress. The rate of flow of a given metal depends on the magnitude of the stress and the temperature. Lead and zinc, for example, will creep under normal temperatures at relatively low stresses. However, most metals require a relatively high temperature before they will begin to flow under low stresses. Creep may continue for an indefinite amount of time under a sustained load tending to distort the material and eventually cause failure of the material by rupture. Unlike repeated loads (fatigue), creep affects the entire body of the material under stress instead of producing a localized rupture.

Stress Rupture

With the development of jet propulsion and aircraft gas turbines, greater attention has been placed on metals to withstand stresses at high temperature. In the aircraft industry this temperature may be as high as 982°C (1800°F). Various laboratory tests have been developed and proven reliable in indicating the performance of the metal at high temperatures under stress. These laboratory tests eliminate time-consuming engine tests. In determining the material's life under continued high temperature and loading, stress necessary to cause rupture at a given time and temperature is the property of greatest importance. This property is sometimes referred to as *stress rupture*.

In jet engines and aircraft gas turbines operations, severe local overheating and overstressing may occur in a very short time. Thus, designs must provide for short-term strengths of alloys at temperatures somewhat higher than expected operating levels as well as sufficient stress rupture strengths at the anticipated operating temperature levels.

Corrosion and Wear

Each year millions of dollars worth of damage to iron and steel structures is caused by corrosion. Further, millions of dollars are spent to replace worn-out parts of machine structures. Corrosion and wear damage take place gradually. They seldom cause sudden or dramatic structural disasters, as parts damaged by corrosion or wear can be replaced or repaired before failure occurs.

Most metals associated with construction materials come in contact with water which contains dissolved oxygen or with moist air and enter into solution readily. The rate of solution is usually retarded by a film of hydrogen forming on the metal or by coating the metal with a protective coating. However, oxygen will combine with the hydrogen and over a period of time will strip it away from the metal, and thus further corrosion will result.

Five classifications of corrosion for metals exist:

1. Atmospheric.
2. Water immersion.
3. Soil.
4. Chemicals other than water.
5. Electrolytic.

In atmospheric corrosion a large excess of oxygen is available and the rate of corrosion is largely determined by the quantity of moisture in the air and the length of time in contact with the metal.

When metals are immersed in water, the amount of oxygen dissolved in the water is an important factor. If the water does not contain any dissolved oxygen, the metal will not corrode. If the water is acidic, the corrosion rate is increased, whereas water that is alkaline has very little corrosion activity unless the solution is highly concentrated.

In soil corrosion and in corrosion by chemicals other than water, the most important item is the ingredient coming in contact with the iron or steel.

Corrosion by electrolysis due to stray currents from power circuits may be disastrous, but in nearly all cases it can be prevented by suitable electrical precautions.

The most common protective coating against corrosion for iron and steel is paint. The paint coating is usually mechanically weak and it cracks and wears out. Thus, to do a satisfactory job, the paint must be renewed every 2 or 3 years. Before the structure is painted, it should first be cleaned and the rust removed.

If the structure is to be immersed in water or if it comes in contact with water, paint provides little protection. Thus, the portion that is in contact with water might require a coating of asphalt or coal tar to protect it.

Another excellent method of preventing corrosion is to encase the iron or steel in concrete. Although concrete is porous, it will provide adequate protection for years. However, if the concrete becomes cracked, it loses most of its protecting ability and should be replaced if possible, or patched.

Metals under stress, especially those beyond their elastic strength, corrode more rapidly than do unstressed metals.

In nearly all cases the failure of materials by mechanical wear under abrasion occurs gradually, the progress of wear is evident, and the failure is not a definitely defined event but one whose occurrence is a matter of judgment on the part of the user of the material. Failure by wear rarely leads to disaster, and usually involves repair or replacement of a part.

TESTING AND EVALUATION OF METALS

Tests are conducted on materials of construction in order to determine their quality and their suitability for specific uses in machines and structures. It is necessary for the producer, consumer, and the general public to have tests for the determination of quantitative properties of materials such that the material may be properly selected, specified, and designed. Tests are further needed to duplicate materials and to check upon the uniformity of different shipments.

Testing and evaluation of the many various metals and alloys requires hundreds of tests and specifications. The ASTM specifications dealing with metals and their alloys comprise several volumes. ASTM, Section 3, which covers various tests for steel products, is a primary reference for civil and highway engineers.

The most common mechanical tests (for metals and metallic products) are listed in ASTM A370 as follows:

1. ASTM E8: Tension Testing of Metallic Materials.
2. ASTM E10: Test for Brinell Hardness of Metallic Materials.

3. ASTM E18: Test for Rockwell Hardness and Rockwell Superficial Hardness of Metallic Materials.

4. ASTM E23: Notched-Bar Impact Testing of Metallic Materials.

ASTM E8: Tension Testing of Metallic Materials

Most commercial specifications for metals have requirements for physical properties as determined by the tensile strength test. The tension test is one of the most important tests for determining the structural and mechanical properties of metals. These properties include the ultimate strength (tensile), ductility (elongation and reduction of area), modulus of elasticity, yield point, offset yield strength, proportional limit, and the elastic and inelastic range, among others.

The *tensile test* is performed by gripping the opposite ends of a test specimen called a "coupon" and pulling it apart. Various standard ASTM coupons are shown in Figure 10-6. Standard test specimens are obtained from sheared, blanked, sawed, trepanned, or oxygen-cut materials that are being tested. In all cases, special attention should be taken to ensure the removal by machining of all distorted cold-worked or heat-affected areas. Test specimens should be machined such that they have a reduced cross section at the midpoint to localize the zone of fracture. In addition, the specimen should be gaged—marked such that the percent elongation may be determined.

When loading a metal specimen, the rate of loading is usually unimportant but should not exceed 690 MPa/min. Most modern testing machines are equipped with electronic strain gages; otherwise, extensometers or foil gages (in conjunction with strain indicators and switching and balancing units) may be employed. When the coupon is fastened in the machine and the strain gage is fastened, the test begins. As the specimen is pulled apart, a load-strain (stress-strain) diagram may be plotted (automatically if electronic strain gages are used) as shown in Figure 10-7. From this diagram the various physical and/or mechanical properties may be obtained.

In a tensile test it is customary to give the strength of a material in terms of unit stress or internal force per unit of area. Also, the point at which yielding starts is expressed as unit stress.

Further, in a tensile test, strain is measured as unit strain. *Unit strain* in any direction is the deformation per unit of length in that direction. If electronic strain gages are employed in a tensile test, the plot on the stress-strain diagram is a function of the force per unit area versus the unit strain.

However, in some cases the unit strain cannot be obtained directly (as in the use of an extensometer), so the total strain or deformation is measured. *Deformation* in any direction is the total change in the dimension of a member in that direction. When the loading is such that the unit strain is constant over the length of a member, it may be computed by dividing the deformation by the original length of the member (gage length). Thus, unit strain equals total strain divided by the gage length.

Tensile Strength   The *tensile strength* of metals is the maximum axial load (ultimate load) observed in a tension test divided by the original cross-sectional area. The strength increases and reaches a maximum in mild steel, after extensive elongation and

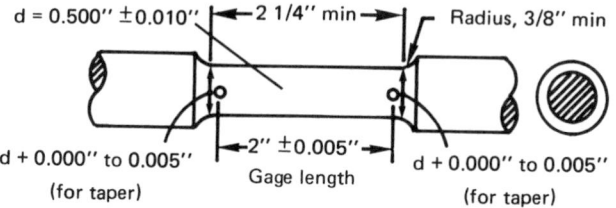

(a) Standard round specimen with 2" gage length

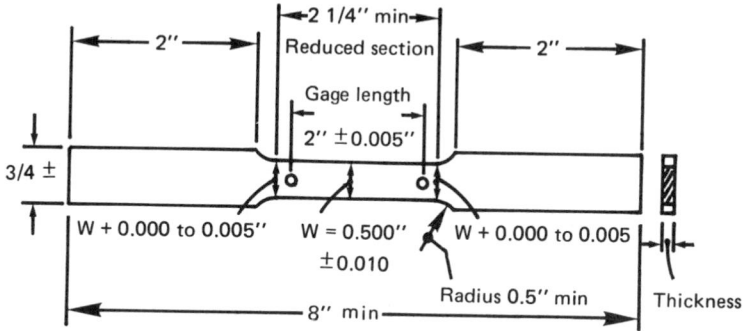

(b) Standard rectangular specimen with 2"
gage length for testing metals in form of
plate, sheet, etc. having thickness from
0.005" to 5/8".

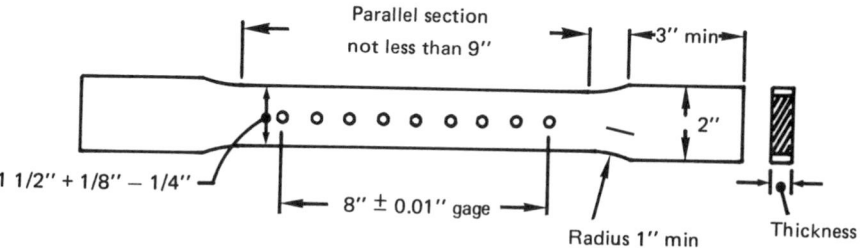

Figure 10-6 Various Coupons (Courtesy of ASTM)

necking. As indicated, it is characterized by the beginning of necking down, a decrease
in cross-sectional area of the specimen, or local instability. The tensile strength is the
ultimate strength expressed in units of pounds per square inch (Newtons per square meters)
and as shown in Figure 10-7 is 68,000 psi.

In design, one cannot base the working stress on the ultimate strength because
excessive strain cannot be tolerated. Thus, one must find the ultimate unit stress produced
by the design loads in the member and reduce it by a factor of safety such that one works
with an allowable unit stress.

Ductility *Ductility* is the ability of a material to undergo large deformations
without fracture. Thus, the material will deform in the inelastic (plastic) range. Ductility

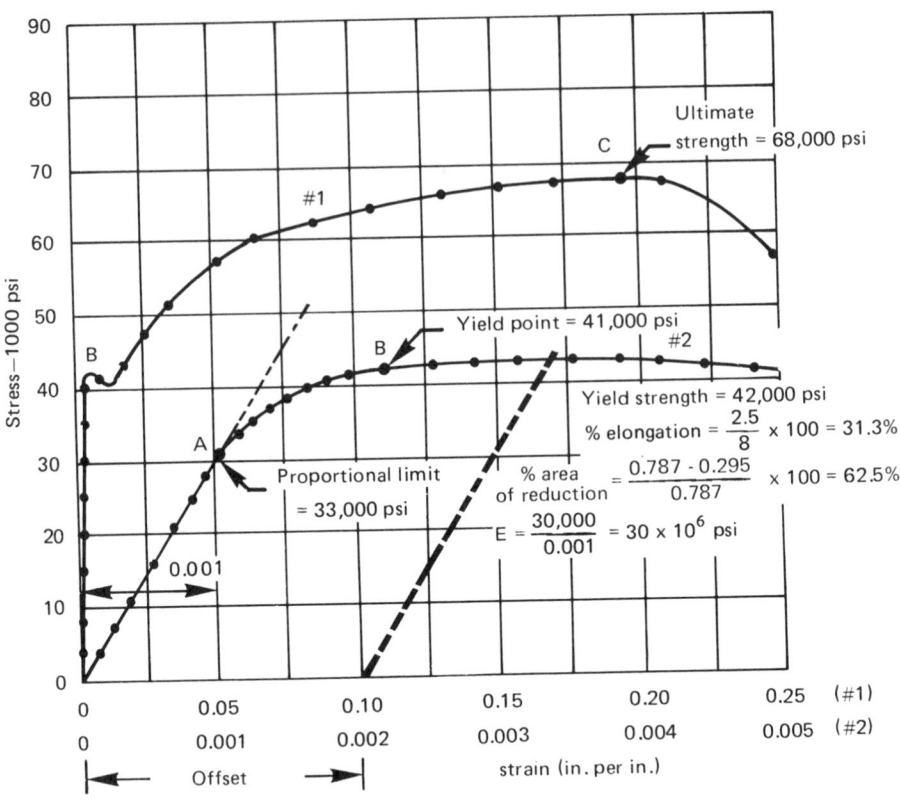

Figure 10-7   Stress–Strain Diagram

is measured by the elongation and reduction of area in a tension test and is expressed as a percentage:

$$\text{percent elongation} = \frac{\text{final length} - \text{original length}}{\text{original length}} \times 100 \qquad (10.1)$$

and

$$\frac{\text{percent reduction}}{\text{in area}} = \frac{\text{original area} - \text{area after fracture}}{\text{original area}} \times 100 \qquad (10.2)$$

For our example (as given in Figure 10-7 the percent elongation is 31.3 and the percent reduction in area is 62.5.

Modulus of Elasticity   The *modulus of elasticity (E)* is given by the slope of the straight-line portion of the stress-strain curve. It is a measure of the inherent rigidity or stiffness of a material. For given geometric configuration, a material with a large *E* deforms less under the same stress.

The modulus of elasticity (Young's modulus) is the ratio of unit stress to unit strain in the elastic range of the stress-strain curve as follows:

$$E = \frac{\text{stress (psi)}}{\text{strain (in./in.)}} \qquad (10.3)$$

or in SI units,

$$E = \frac{\text{stress (MPa)}}{\text{strain (cm/cm)}} \qquad (10.4)$$

In Figure 10-7 the modulus of elasticity is 30,000,000 psi.

Yield Point   At the termination of the linear portion of the stress-strain curve, some materials (such as a low-carbon steel) develop a yield point. The *yield point* is the first load at which there is a marked increase in strain without an increase in stress. This behavior may be a consequence of inertia due to the effects of the testing machine and the deformation characteristics of the test specimen. The yield point is sometimes taken as the proportional limit and elastic limit, which is an incorrect practice. Most metals do not have a yield point, and thus an offset method is utilized. In Figure 10-7 the yield point would correspond to 41,000 psi.

Offset Yield Strength   The *offset yield strength* is defined as the stress corresponding to a permanent deformation, usually 0.10 or 0.20 percent (0.001 or 0.002 in./in.). The offset method is usually used with materials that have a definite straight-line portion to their stress-strain curve. One measures the corresponding offset percentage on the stress-strain curve and projects upward a straight line parallel with the straight-line portion of the stress-strain curve. Where the line intersects the stress-strain curve, the value is read off as the offset yield strength. In Figure 10-7 this value would be 42,000 psi.

In situations when the stress-strain curve does not exhibit a straight-line portion, the secant modulus or tangent modulus method may be used.

Proportional Limit   The *proportional limit* is the greatest stress that a material is capable of without deviating from the law of proportionality of stress to strain (Hooke's law). *Hooke's law* is defined as

$$f = E\varepsilon \qquad (10.5)$$

where f = unit stress
   ε = unit strain
   E = Young's modulus of elasticity

Metals are elastic within the proportional limit and thus the proportional limit has significance in the elastic stability of columns and shells. In our example, as shown by Figure 10-7, the proportional limit is 33,000 psi.

Elastic Limit and Inelastic Limit   The *elastic limit* is the largest unit stress that can be developed without a permanent set remaining after the load is removed. In most cases the elastic limit is difficult to determine, and many materials do not have a well-defined proportional limit, or any at all; thus, the offset yield strength is used to measure the beginning of plastic deformation (inelastic limit).

Modulus of Rigidity   The *modulus of rigidity* or the shearing modulus of elasticity, as it is sometimes called, is defined as

$$G = \frac{v}{\gamma} \tag{10.6}$$

where G = modulus of rigidity
  v = unit shearing stress
  γ = unit shearing strain

The modulus of rigidity may also be rewritten and related to the modulus of elasticity by

$$G = \frac{E}{2(1 + \mu)} \tag{10.7}$$

where μ is a constant known as *Poisson's ratio* and for structural steel equals 0.3.

Modulus of Toughness   The *toughness* of a material is the ability of a material to absorb large amounts of energy. The *modulus of toughness* can be related to the area under the entire stress-strain curve, and it depends on both strength and ductility. Because of the difficulty of determining toughness analytically, toughness is often measured by the energy required to fracture a specimen, usually notched and sometimes at low temperatures, in an impact test such as the Charpy or the Izod.

Modulus of Resilience   The *resilience* of a material is that property of an elastic body by which energy can be stored up in the body by loads applied to it and given up in recovering its original shape when the loads are removed. Thus, the *modulus of resilience* is equal to the area under the straight-line portion of the stress-strain curve (a triangle).

ASTM E10: Brinell Hardness of Metallic Materials

The *hardness* of metals is usually determined by measuring the resistance to penetration of a ball, cone, or pyramid. The results of a hardness test can be directly related to the tensile strength of a material. Thus, it provides a quick check on the tensile strength of a material without going through a time-consuming tensile test.

The *Brinell hardness method* is based upon determining the resistance offered to indentation by a hard ball of specific diameter that is subjected to a given pressure. The pressure used in testing steel is usually 6600 lb (3000 kg) and the diameter of the ball

TABLE 10-3 RELATIONSHIP OF BRINELL AND ROCKWELL HARDNESS NUMBERS TO TENSILE STRENGTH

Brinell Indentation Diameter (mm)	Brinell Hardness Number		Rockwell Hardness Number		Rockwell Superficial Hardness Number, Superficial Diamond Penetrator			(Approximate) Tensile Strength (MPa)
	Standard Ball	Tungsten Carbide Ball	B Scale	C Scale	15− N Scale	30− N Scale	45− N Scale	
2.50		601		57.3	89.0	75.1	63.5	2262
2.60		555		54.7	87.8	72.7	60.6	2055
2.70		514		52.1	86.5	70.3	47.6	1890
2.80		477		49.5	85.3	68.2	54.5	1738
2.90		444		47.1	84.0	65.8	51.5	1586
3.00	415	415		44.5	82.8	63.5	48.4	1462
3.10	388	388		41.8	81.4	61.1	45.3	1331
3.20	363	363		39.1	80.0	58.7	42.0	1220
3.30	341	341		36.6	78.6	56.4	39.1	1131
3.40	321	321		34.3	77.3	54.3	36.4	1055
3.50	302	302		32.1	76.1	52.2	33.8	1007
3.60	285	285		29.9	75.0	50.3	31.2	952
3.70	269	269		27.6	73.7	48.3	28.5	897
3.80	255	255		25.4	72.5	46.2	26.0	855
3.90	241	241	100.0	22.8	70.9	43.9	22.8	800
4.00	229	229	98.2	20.5	69.7	41.9	20.1	766
4.10	217	217	96.4					710
4.20	207	207	94.6					682
4.30	197	197	92.8					648
4.40	187	187	90.7					621
4.50	179	179	89.0					607
4.60	170	170	86.8					579
4.70	163	163	85.0					566
4.80	156	156	82.9					552
4.90	149	149	80.8					503
5.00	143	143	78.7					490
5.10	137	137	76.4					462
5.20	131	131	74.0					448
5.30	126	126	72.0					434
5.40	121	121	69.0					414
5.50	116	116	67.6					400
5.60	111	111	65.7					386

is 0.4 in. (10 mm). When softer metals are utilized, a pressure of 1100 lb (500 kg) is used. The Brinnell hardness number can be computed by the following formula:

$$BH = \frac{2P}{\pi D(D - \sqrt{D^2 - d^2})}$$ (10.8)

where P = pressure, kg
D = diameter of the steel ball, mm
d = average diameter of indentation, mm

As is obvious from the equation, the smaller the indentation, the greater the Brinell hardness number.

ASTM E18: Test for Rockwell Hardness

ASTM E18 specifies the test for the determination of the Rockwell hardness and Rockwell superficial hardness of metallic materials. The *Rockwell hardness method* employs either a ball or a diamond cone in a precision testing instrument that is designed to measure depth of penetration accurately. Two superimposed impressions are made. The depth to which the major load drives the ball or cone below that depth to which the minor load has previously driven it is a measure of the hardness. For the harder steels, greater accuracy is obtained by use of a diamond cone.

Rockwell superficial hardness machines are used for the testing of very thin shells or thin surface layers of materials. In this case the minor load is 6.6 lb (3 kg) and the major load varies from 33 to 99 lb (15 to 45 kg).

In all cases the Rockwell hardness number is read directly from the scales on the machine. The relationship of the Brinell and Rockwell hardness numbers to tensile strength is shown in Table 10-3.

ASTM E23: Notched-Bar Impact Testing of Metallic Materials

Impact tests are performed primarily for two reasons:

1. To determine the ability of the material to resist impact under service conditions.
2. To determine the quality of the metal from a metallurgical standpoint.

As previously indicated, two types of impact tests are usually performed, the Charpy and the Izod tests. Both apply a dynamic load by use of a pendulum that has enough kinetic energy to rupture a specimen in its path.

PROBLEMS

10.1. Explain the crystalline nature of metals.

10.2. What is meant by allotropic behavior?

10.3. Why is grain formation and growth important in metals?

10.4. Discuss the phenomenon of slip.

10.5. Why is "fatigue" an improper term as it applies to metals?

10.6. Explain the process of flow or creep.

10.7. What are the five classifications of corrosion? Discuss.

10.8. Explain by use of your own diagram the tension testing of steel. Show all physical and mechanical properties and explain each.

10.9. Explain the two most common hardness tests.

10.10. Using the library and ASTM standards, explain the Charpy impact test.

REFERENCES

BRICK, R. M., GORDEN, R. B., and PHILLIPS, A., *The Structure and Properties of Alloys*, McGraw-Hill Book Company, New York, 1964.

CORDON, W. A., *Properties, Evaluation, and Control of Engineering Materials*, McGraw-Hill Book Company, New York, 1979.

COTTRELL, A. H., *An Introduction to Metallurgy*, Edward Arnold (Publishers) Ltd., London, 1967.

DAVIS, H. E., TROXELL, G. E., and WISKOCIL, C. T., *The Testing and Inspection of Engineering Materials*, 3rd ed., McGraw-Hill Book Company, New York, 1964.

DIETER, G. E., *Mechanical Metallurgy*, McGraw-Hill Book Company, New York, 1961.

DOAN, G. E., and MAHLA, E. M., *The Principles of Physical Metallurgy*, McGraw-Hill Book Company, New York, 1941.

GILLET, H. W., *The Behavior of Engineering Metals*, John Wiley & Sons, Inc., New York, 1951.

JASTRZEBSKI, Z. D., *The Nature and Properties of Engineering Materials*, John Wiley & Sons, Inc., New York, 1959.

JEFFRIES, Z., and ARCHER, R. S., *The Science of Metals*, McGraw-Hill Book Company, New York, 1924.

MOORE, H. F., *Materials of Engineering*, 7th ed. McGraw-Hill Book Company, New York, 1947.

SEITZ, F. B., *The Physics of Metals*, McGraw-Hill Book Company, New York, 1943.

VAN VLACK, L. H., *Elements of Material Science*, 2nd ed., Addison-Wesley Publishing Company, Inc., Reading, Mass., 1964.

11

Ferrous Metals

In general, metals can be classified into two major groups: ferrous and nonferrous. A *ferrous metal* is one in which the principal element is iron, as in cast iron, wrought iron, and steel. A *nonferrous metal* is one in which the principal element is not iron, as in copper, tin, lead, nickel, aluminum, and refractory metals.

In this chapter we discuss ferrous metals. Special attention will be placed on the production techniques of steel.

GENERAL

Sources of Metals

In general, over 45 metals of industrial importance are found within the earth's crust. With the exception of aluminum, iron, magnesium, and titanium, which occur in appreciable percentages within the earth's crust, all other metals comprise less than 1 percent of the earth's crust. Thus, most metals occur in the form of ore, in which the metal has to be extracted. An ore is usually referred to as a *mineral,* which is a chemical compound or mechanical mixture. The material associated with the ore which has no commercial use is referred to as *gangue.* Basically, six classifications of ore exist:

1. Native metals.
2. Oxides.
3. Sulfides.
4. Carbonates.

5. Chlorides.
6. Silicates.

The native metals consist of copper and precious metals. Oxides are the most important ore source, in that iron, aluminum, and copper can be extracted from them. Sulfides include ores of copper, lead, zinc, and nickel. Carbonates include ores of iron, copper, and zinc. The chlorides include ores of magnesium, and the silicates include ores of copper, zinc, and beryllium.

Production of Metals

Four operations are required for the production of most metals:

1. Mining the ore.
2. Preparing the ore.
3. Extracting the metal from the ore.
4. Refining the metal.

In the mining operation, the methods of open-pit borrowing and underground mining are both utilized. The most famous examples of an open-pit operation are the iron mines in the Mesabi Range in Minnesota. An example of underground mines are the copper mines in upper Michigan. Most underground mines are of the room-and-pillar type for working horizontal veins or of the stepping type for working vertical veins by cutting a series of steps.

In the preparation process the ore is crushed and large quantities of gangue are removed by a heavy-media-separation method. In some cases, the preparation of the ore may involve roasting or calcining. In roasting, the ore of sulfide is heated to remove the sulfur and in calcining the carbonate ores are heated to remove carbon dioxide and water.

The extraction of the metal from the ore is accomplished through chemical processes. These chemical processes reduce the compounds, such as oxides, by releasing the oxygen from chemical combinations and thus freeing the metal.

Basically, three types of processes of extraction are used:

1. Pyrometallurgy.
2. Electrometallurgy.
3. Hydrometallurgy.

In the *pyrometallurgy* process (generally referred to as *smelting*) the ore is heated in a furnace producing a molten solution, from which the metal can be obtained by chemical separation. The blast furnace or reverberatory furnace is used in this process.

In the *electrometallurgy* process metals are obtained from ores by electrical processes utilizing an electric furnace or an electrolytic process.

Hydrometallurgy or *leaching* involves subjecting the ore to an aqueous solution from which the metal is dissolved and recovered.

As a result of the extraction process, the metals will contain impurities, which must be removed by a refining process. If the metal was extracted by the pyrometallurgy process, the most common method of refining is by oxidizing the impurities in a furnace (steel from pig iron). However, other methods are utilized, such as liquidation (tin), distillation (zinc), electrolysis (copper), and the addition of a chemical reagent (manganese to molten steel).

FERROUS METALS

Ferrous metals comprise three general classes of materials of construction:

1. Cast iron. *pipes*
2. Wrought iron. *decorative.*
3. Steel. *building*

All of these classes are produced by the reduction of iron ores to pig iron and the subsequent treatment of the pig iron to various metallurgical processes. Both cast iron and wrought iron have fallen in production with the advent of steel, as steel tends to exhibit better engineering properties than do cast and wrought iron. The application of steel and steel alloys is so widespread it has been estimated that there are over a million uses. In construction, steel has three principal uses:

1. Structural steel. *super structure.*
2. Reinforcing steel. *foundation*
3. Forms and pans.

Classification of Iron and Steel

Iron products may be grouped under six headings:

1. Pig iron.
2. Cast iron.
3. Malleable cast iron.
4. Wrought iron.
5. Ingot iron.
6. Steel.

Pig iron is obtained by reducing the iron ore in a blast furnace. This is accomplished by charging alternate layers of iron, ore, coke, and limestone in a continuously operating *blast furnace*. Blasts of hot air are forced up through the charge to accelerate the combustion of coke while raising the temperature sufficiently to reduce the iron ore to molten iron. The limestone is a flux which unites with impurities in the iron ore to form slag. The blast furnace accomplishes three functions:

1. Reduction of iron ore.
2. Absorption of carbon.
3. Separation of impurities.

The amount of carbon present in pig iron is usually greater than 2.5 percent but less than 4.5 percent. The iron may be cast into bars, referred to as *pigs*.

Cast iron is pig iron remelted after being cast into pigs or about to be cast in final form. It does not differ from pig iron in composition and it is not in a malleable form.

Malleable cast iron is cast iron that has undergone special annealing treatment after casting and has been made malleable or semimalleable.

Wrought iron is a form of iron that contains slag, is initially malleable but normally possesses little to no carbon, and will harden quickly when rapidly cooled.

Ingot iron is a form of iron (or a low-carbon steel) that has been cast from a molten condition.

Steel is an iron–carbon alloy which is cast from a molten mass whose composition is such that it is malleable in some temperature range. Carbon steel is steel that has a carbon content of less than 2 percent and generally of less than 1.5 percent; its properties are dependent on the amount of carbon it contains. Alloy steels are steels in which the properties are due to elements other than carbon.

MANUFACTURE OF STEEL

As previously stated, the first process in the manufacture of steel is the reduction of iron ore to pig iron by use of a blast furnace. This is followed by the removal of impurities, and four principal methods are used to refine the pig iron and scrap metal:

1. Open-hearth furnace.
2. Bessemer furnace.
3. Electric furnace.
4. Basic oxygen furnace.

In this section we discuss in detail the blast furnace, the open-hearth furnace, the Bessemer converter, the electric furnace, and the basic oxygen furnace. Most of the discussion is taken directly from the *Steel Products Manual* of the American Iron and Steel Institute.

Blast Furnace

Blast furnaces (Figure 11-1) are so named because of the continuous blast of air required to bring about the necessary heat and chemical reactions in the raw materials in the stack. As much as 4.5 tons of air may be needed to make 1 ton of pig iron.

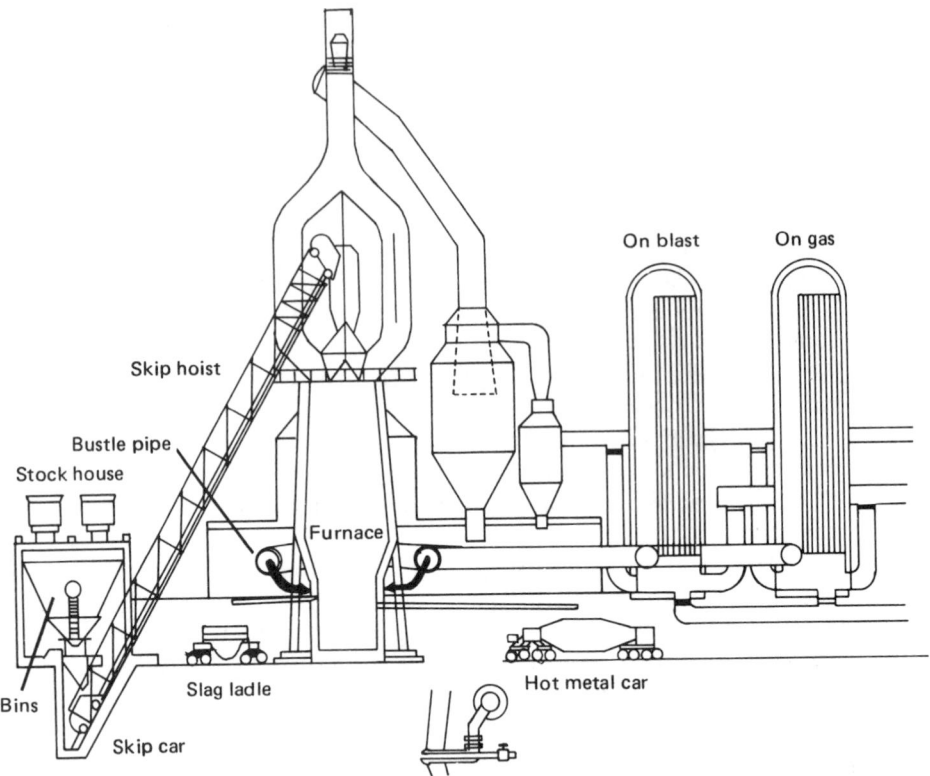

Figure 11-1   Blast Furnace (Courtesy of AISI)

Most blast furnaces, and all of the more modern installations, are served by *turboblowers*. These enormously powerful steam engines are designed to receive steam at 700 psi (492,100 kg/m²) at a temperature of 750°F (399°C).

The blowers force air from the atmosphere through piping into *stoves*, which are the most prominent auxiliaries serving a blast furnace. Each furnace must have at least two stoves and may have three.

Essentially, a stove consists of two parts: a *combustion chamber*, which is a vertical passageway wherein cleaned blast furnace gas is burned, and *brick checkerwork,* which contains many small passageways for heating the air as it passes over the hot masonry.

The design of the stoves varies in complexity, but it is important to remember that each stove serves two alternating functions. A stove will receive cold air from the blowers and pass it on through the heated brick checkerwork to the blast furnace; when the atmospheric air from the blower has cooled the brick checkerwork, another stove will take over the function and hot air from the top of the blast furnace will burn in the combustion chamber to reheat the brick checkerwork in the furnace stove.

Modern stoves for large furnaces are somewhat like farm silos in appearance (cylindrical in shape with a domed top). They may be up to 28 ft (8.5 m) in diameter and about 120 ft (36 m) high from the bottom to the top of the dome. Depending upon the type of brick checkerwork used, the stoves may contain upwards of 250,000 ft² (22,500 in.²) of heating surface.

It is common for steel columns to support steel grids inside the stove, and these grids bear the brick checkerwork. Insulation between the brick and the steel shells prevents distortion of the stove.

Special valves control the flow of air from the blower to the heated stove, which is referred to as being "on blast." The stove or stoves in the process of having their checkerwork heated are referred to as being "on gas." In a three-stove arrangement serving one blast furnace, each stove is on gas twice as long as it is on blast.

The air from an on-blast stove speeds toward the furnace at a temperature that ranges from 1400 to 2100°F (760 to 1150°C). Air enters a bustle pipe made of heavy steel plate which completely encircles the stack of the blast furnace. From this bustle pipe smaller pipes extend at an angle downward and enter the stack of the furnace. These entry pipes, called *tuyeres,* enter the furnace at a level considerably above the hearth, in which lies a pool of molten iron at approximately 2700°F (1482°C) with the layer of impurities, called *slag,* floating on top. In modern practice it is not uncommon for fuels, such as oil or a coal sledge, to be injected through the tuyeres along with the air for higher output.

The term "pig iron" may seem obscure to modern city dwellers, but its origin was perfectly clear in an earlier agricultural age when small blast furnaces largely confined their services to their own communities. In those days molten iron from a furnace ran down a channel prepared in the ground and flowed from the channel into small holes dug on either side of it. To neighboring farmers the arrangement resembled a familiar sign in their lives (newborn pigs, sucklings). The central channel was known as a "sow."

Today's typical pig casting machines feature twin conveyor lines of shallow molds. Molten iron is poured from a ladle or hot metal car into a short channel which divides to feed the molds passing on the conveyor lines at a controlled rate so that there is little or no spilling. Iron in the molds is cooled with water. By the time the mold is overturned at the end of the conveyor line, the iron is solid and each piece is called a *pig.* It probably weighs 40 lb (18 kg) or more and simply drops into a railway car. There it is further cooled by spraying. Samples are taken and transported to a laboratory for analysis. Finally, the carload of cool iron is taken to a numbered pile and stored until customers require it.

Most of the iron from pig casting machines is made to rigid specifications which vary widely according to end use. The generic name for this product is *merchant pig iron,* because it is intended to be sold as an end product, although some small percentage of it may be used in steel making.

The customers for merchant pig iron are foundries that cast pipe for water and waste, for oil burner parts, and for automobile engine blocks, to name a few of a hundred uses. Among others are the dutch ovens used in many household kitchens, intricately shaped pipe fittings, heavy bases of machine tools, and major elements in a magnet developed at the Department of Energy's Argonne Laboratories.

A merchant pig iron producer may have several hundred piles of different grades of iron, each manufactured to meet the specifications of a given customer. A merchant pig iron producer may blend a dozen or more grades of iron ore to arrive at a desired chemical composition. Thus, the different products may be considered as iron alloys.

Most merchant pig iron furnaces are considerably smaller than the machines producing iron for steelmaking purposes. Even so, it is worthy of note that while, for example, 30,000 ft³ (900 m³) of solids weighing just under 1000 tons is being processed in a furnace stack, the chemical composition of the pigs is being controlled within fractions of 1 percent relative to carbon, silicon, sulfur, phosphorus, and manganese.

Foundries buy and melt scrap to supplement merchant pig iron in making their products. Some of them also buy imported iron. In a recent year about 20 percent of the available supply of merchant pig iron in the United States was of foreign origin. But domestic pig iron remains the standard for quality control in making castings because foundries are dependent on the chemistry of the iron they buy from various producers in the United States.

Serious changes in chemistry occur when too much poorly graded scrap and iron are consumed. The precise specifications to which American merchant iron is made provide a standard by which foundries can control the quality of their product.

Open-Hearth Furnace

Open-hearth furnaces (Figure 11-2) are so named because limestone, scrap steel, and molten iron are charged into a shallow steelmaking area called a "hearth" and are exposed, or "open," to the sweep of flames that emanate alternately from opposite ends of the

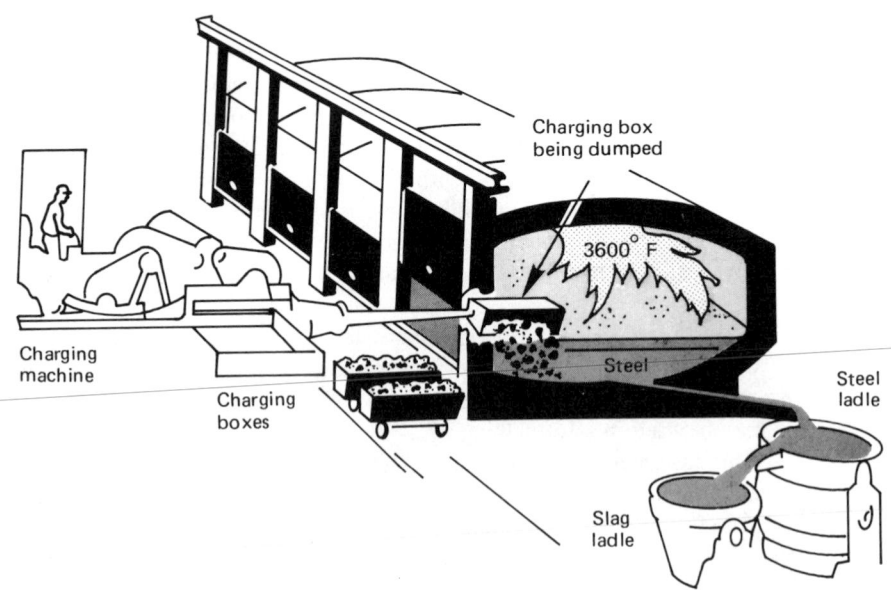

Figure 11-2   Open-Hearth Furnace (Courtesy of AISI)

furnace. The first open hearths for steelmaking were built in the United States during the late 1800s, and they have since been developed to a high degree of sophistication. In the beginning, as now, one of their most attractive features concerns the use of steel scrap.

Theoretically, an open-hearth furnace can operate using either blast furnace iron alone or steel scrap alone, but most are designed to operate using both in approximately equal proportions. These proportions vary according to such economic factors as the price of scrap or the availability of molten iron. For more than half a century open-hearth furnaces were the outstanding high-tonnage producers of steel for America's expanding industry and are still major factors in the steel industry. A furnace that will produce a fairly typical 350 tons (318,000 kg) of steel in 5 to 8 hours may be about 90 ft (27 m) long and 30 ft (9 m) wide. It is very likely to be one of several furnaces contained in a single building referred to as an *open-hearth shop*. These very large furnaces are installed close together so that they may be serviced by the same units of highly specialized auxiliary equipment.

Because open-hearth shops are so large, it is difficult to obtain meaningful photographs. The arrangement can be likened to that of a split-level house with the entrance on the top floor and with a bottom level exit and heating system.

The charging side of the furnace is on the upper level and faces a charging floor full of highly specialized equipment. The most notable piece of equipment is a charging machine, which is electrically operated on a very broad gage track. Between the door side of the furnace and the charging machine is another track system. Special cars containing boxes of raw materials are run into the open-hearth shop on this track and are parked in position near the furnace doors. The charging machines are equipped to pick up the boxes of raw materials from their buggies and to thrust them in through the temporarily opened doors of the furnace, empty them by overturning them and then put them back on to their buggies. The empty charging boxes are then hauled away and refilled in preparation for the next charge.

In the rafters above the tracks for raw material, buggies, and charging machines are enormously strong overhead cranes, also running on tracks extending the full length of the shop. This crane system is used in many ways, but primarily to carry ladles of molten iron from a blast furnace or a mixer to the open hearths. Special troughs are wheeled into place in front of a furnace door and the huge ladles of molten iron are tilted so that a stream of the white-hot metal pours from the lip of the ladle through the trough and into the furnace. These paragraphs are devoted primarily to the layout of the shop; the actual processes of operation will be described later.

The only other major elements on the charging-floor side of the open-hearth furnace are the control areas, which contain panels permitting operators to control most of the processes at a safe distance from the heat and moving equipment.

The dish-shaped hearth of an open-hearth steelmaking furnace is quite shallow. The walls containing the charging doors are vertical and each unit is covered by an arched refractory roof. On the side of the furnace opposite the charging floor is a *taphole*, which is so arranged that molten steel can rush by gravity through a spout into a large ladle on the pouring floor or pit side of the open-hearth shop, which is at a considerably lower level than the charging floor.

A very considerable portion of an open-hearth facility is not visible at all. Brick *checker chambers* are located at both ends of each furnace below the level of the charging floor. The bricks in these chambers are arranged to leave a great number of passages through which the hot waste gases from the furnace pass and heat the brickwork prior to going through the gas cleaning equipment. Later, the flow of gas is reversed, and the atmospheric air supporting combustion in the furnace passes through the heated bricks and is heated on its way to the hearth.

Today, practically all open-hearth furnaces have been converted to use oxygen. The gas is fed into the open hearth through the roof by means of retractable lances. The use of gaseous oxygen in open hearth increases temperatures and thereby speeds up the melting process.

Each open-hearth shop is supervised by a melter supervisor. This person is in charge of all furnaces in the shop and their crews. He or she operates the controls, directs any repairs to the furnace in operation, and reports any variations from standard procedure. There are many variables in open-hearth steelmaking practice, the most significant being the ratio of pig iron to scrap. For the purpose of this book, the rather common "fifty-fifty" practice is chosen for description. In this practice the limits of molten pig iron are usually 45 to 55 percent and the charge may include relatively small amounts of solid pig iron or iron scrap generated within the plant. Limestone and some ore agglomerates are also charged.

Assuming that the raw steel has been tapped from an open-hearth furnace, the first step in making a new heat of steel is to examine the interior of the furnace carefully to determine if any damage has been done to the hearth or roof. If so, a special patching gun may be wheeled into position and the damage patched by literally shooting a refractory material into the holes or cracks.

As the furnace is being prepared, buggies containing boxes of raw materials are rolled into position next to the furnace. The long-armed charging machine picks up boxes of limestone and fairly light steel scrap. One by one these boxes are thrust through the furnace doors and dumped. The flame of burning fuel oil, tar, or gases shoots from one end of the furnace across the solid materials in the hearth, partially melting them. (Most of the combustion gases are collected at the other end of the furnace and used to heat the checker brick.) Molten iron is then poured into the furnace.

All of this is much more complex than it sounds, because timing is extremely important. It is necessary to have the solid materials at such a temperature and degree of oxidation that, on the one hand, the molten pig iron will not be chilled by the scrap and, on the other hand, the oxidation of the metalloids of the pig iron will not be delayed by insufficient oxygen support from the oxidized scrap. The present-day usage of oxygen roof lances has greatly aided this critical phase of charging.

The chemical reactions for making steel in an open-hearth furnace relate to the removal of carbon, manganese, phosphorus, sulfur, and silicon from the metallic bath. First, silicon and manganese are oxidized and become part of the slag, which is a layer of molten limestone and other materials floating on top of the molten steel. Next, the oxidation of carbon speeds up, creating carbon monoxide gas, which causes agitation as it leaves the metal. Eventually, phosphorus and sulfur are transferred to the slag.

The agitation of the metal bath caused by carbon monoxide is called the *ore boil*. The more violent turbulence caused by calcination of limestone is called the *lime boil*. After both "boils" have subsided, the refining phase of open-hearth steelmaking begins, with the aim of lowering the phosphorus and sulfur contents to meet end-product specifications, controlling the carbon content, and generally bringing the composition and grade of the steel to predetermined levels.

The manufacture of steel in open-hearth furnaces takes much more time per ton than it does by the basic oxygen process, but it affords very good control of chemical analysis and is capable, in some instances, of making very large batches—sometimes 600 tons (540,000 kg) of steel of a given analysis in a single heat.

When a heat of steel is ready to be tapped, a member of the furnace crew, working from the tapping side, breaks out the clay plug and refractory that has closed the taphole since before the furnace was charged. The operation may involve the use of a special explosive charge and is undertaken following the best safety practice.

The highest level of the taphole is located at the lowest part of the hearth and slopes down to a spout. Most of the steel surges out of the furnace into a large ladle before any of the slag floating on top of the molten bath appears. The late appearance of slag permits alloy and recarburized and deoxidized materials to be added to the steel in the ladle. The ladle is so placed beneath the spout that the stream of molten steel is given a swirling motion which tends to mix and make more homogeneous the metal and the additions.

The capacity of the ladle is matched to the amount of steel in the furnace. When the slag comes through the taphole, a layer of it is permitted to lie on top of the molten metal as a covering while the remainder overflows through a special notch into a *slag thimble*, a small vessel placed next to the steel ladle.

Some very large open-hearth furnaces are provided with two tapholes, therefore requiring two ladles on the pit-side floor.

Bessemer Process

In the Bessemer process (Figure 11-3) the impurities in pig iron are removed by oxidation, by finely divided air currents blowing through a bath of molten iron contained in a vessel called a *converter*.

The Bessemer converter consists of a heavy steel pear-shaped shell with refractory brick. It is supported by two trunnions upon which it can rotate 180°. The upper portion can be either concentric or eccentric. The bottom of the converter is pierced with a large number of small holes, called tuyeres, through which the air blast is forced by means of a blower from the windbox up through the molten metal, oxidizing the impurities.

Converters generally have a capacity of 1 to 40 tons, with the average around 25 tons. For a small converter of 15-ton capacity, the clear-mouth opening is 2 to 2.5 ft (0.6 to 0.75 m) in diameter with the inside cylindrical diameter at the bottom opening 8 ft (2.4 m) and the height of the converter approximately 15 ft (4.5 m).

The lining of the converter is usually 12 to 15 in. (0.3 to 0.33 m) thick and is made of a refractory material of strongly acidic character, with silica being the primary constituent. This lining lasts for several months before it has to be replaced.

The bottom is lined with 24 to 30 in. (0.6 to 0.75 m) of damp siliceous material bound together with clay in which the tuyere bricks are set. Each tuyere brick is about 30 in. long and has about 10 blast holes, each approximately 0.25 in. (0.6 cm) in diameter. Tuyere bricks have a useful life of about 1 month.

In the operation of the Bessemer process, the converter is tilted to a horizontal position to receive a charge. The blast is turned on after charging and before righting, to prevent the metal from entering the tuyere.

As soon as the blow is on, the silicon and manganese begin to burn to form oxides and thus are reduced to traces before the oxidation of the carbon becomes appreciable. As soon as the carbon is burned out, the converter is turned down, the blasts are shut off, and the recarburizer is added.

This method of making steel is of little importance today because of the advent of better production facilities.

Electric Furnace

Electric arc furnaces (Figure 11-4) have a long history of producing alloy, stainless, tool, and other specialty steels. More recently, operators have also learned to make larger heats of carbon steels in these furnaces. Therefore, the electric steelmaking process is presently becoming a high-tonnage producer.

Electric arc furnaces are shallow steel cylinders lined with refractory brick. In the first half of the twentieth century most electric furnaces were loaded, or changed, with scrap iron and steel through a door in the side of the cylinder. However, today, the entire roof of an electric furnace is mounted on cantilevered steel beams so that it can be lifted and swung to one side. Thus, electric furnaces are now charged in one operation from buckets or other containers brought in by overhead cranes.

The roof of an electric furnace is pierced so that three carbon or graphite electrodes can be lowered into the furnace, and these electrodes give the electric arc furnace its name, because, in operation, the current arcs from one electrode to the metallic charge and then from the charge to the next electrode. This provides intense heat.

Opposite one of the doors is a tapping spout through which the molten steel is poured into a ladle. The entire furnace is mounted on "rockers" so that it can be tilted to permit the molten steel to emerge through the spout.

From this description and the accompanying diagram it can be seen that the charge for electric furnace steelmaking is entirely solid. The principal metallic charge is steel scrap. When a complex specialty steel is to be made, the scrap is carefully selected so that no unwanted elements are present. Pig iron is charged into some electric furnaces and, more recently, prereduced iron ore, which may contain up to 98 percent of the element iron. Such ore, under these conditions, is handled as a primary metallic source.

One of the major factors in making electric furnaces economically competitive as high-tonnage producers of high-carbon steel is the increased size of the units. Most experienced operators remember when a furnace capable of producing 50 tons of steel was considered large. Today, some furnaces average 300 tons per heat.

A relatively few years ago new electric arc furnaces customarily operated on 25,000 to 35,000 kilovolt-amperes compared with up to 80,000 kVA for today's units. Electric

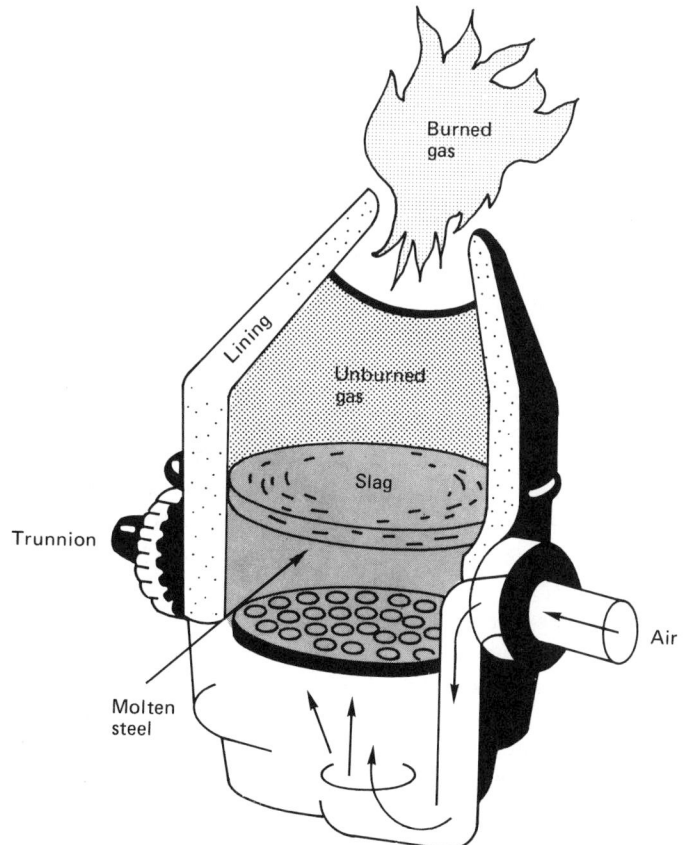

Figure 11-3 Bessemer Process (Courtesy of AISI)

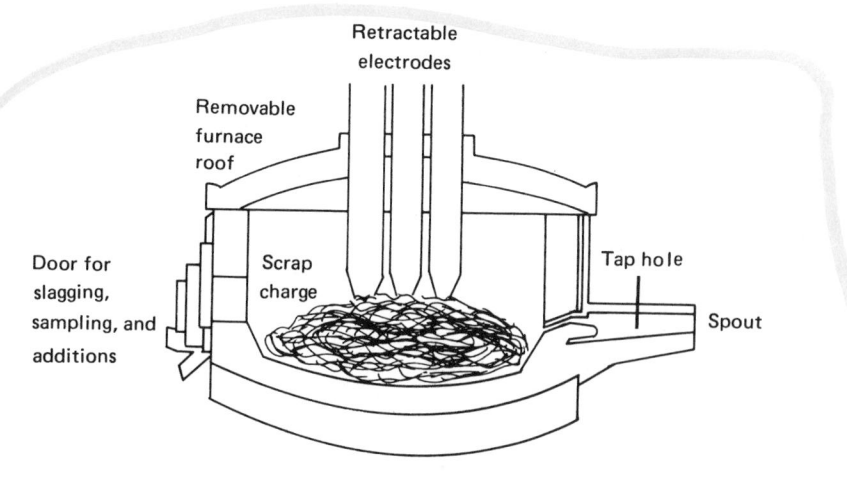

Figure 11-4 Electric Arc Furnace

furnace steelmakers figure their electricity costs on the basis of kilowatt-hours per ton of raw steel. As power input and furnace size increase, the time required to produce a heat of steel decreases. Therefore, the kilowatt-hours per ton also decrease. As indicated, this economy has thrust electric arc furnaces into the high-tonnage production picture, particularly in areas where molten iron is not readily available.

The charge to most electric furnaces is largely scrap with small amounts of burned lime and mill scale. The scrap is segregated (separated) into stockpiles of identified grades, a procedure that is necessary for several reasons. Primary among these is the matter of economics. It would be wasteful to use valuable alloy steel scrap as a material for making common grades of carbon steel. The most economical way of operating an electric furnace is to assure, in advance, that only the elements desired for a special heat are introduced to the furnace. This means that the preliminary operations in collecting, sorting, and preparing scrap are extremely important. Steelmakers are very concerned with the recycling aspect of environmental quality control. However, this does not mean that any classification of steel scrap can be put back into a furnace without further treatment, and this is the case with electric arc furnaces. The fragmentation of old automobiles and the recycling of "tin cans" and other steel-base containers cannot make up too large a portion of the charge. These items contain contaminants such as copper and tin which might spoil a new heat of steel.

Once the sorted scrap has been placed into the hearth of the furnace, ferroalloys and other elements that are not easily oxidized may be charged in the furnace prior to melting down. If the metallic charge is too low in carbon, additions of coke or of scrap electrodes are included with the scrap. In all instances, the amount of carbon charged into the electric arc furnace is higher than is required to produce the desired chemical analysis in the finished steel.

The top of the furnace is now swung into position, the electrodes are lowered through the roof, and the electric power is turned on. The heat resulting from the arcing between electrodes and metal begins to melt the solid scrap. The electrodes, in some cases 2 ft thick and 24 ft long, turn white-hot to a considerable distance above the arc. As the charge melts, they may rise and fall vertically, a process called *searching*.

At a given point, iron ore is added to reduce the carbon content and cause a "bubbling" or boiling action. This is one of the most important factors in the production of high-quality steel.

Limestone and flux are charged on top of the molten bath. Through a chemical interaction, impurities in the steel rise in the molten slag, which floats on top of the melt. When the desired product is a carbon steel, a *single-slag* practice is used. This means that the furnace is tilted slightly and much of the slag is raked off through the charge door. In the *double-slag* method, an oxidizing slag is first formed, then raked off, and an additional slag is formed for reducing purposes.

The direct use of oxygen gas is extremely important in modern practice. It is of great value in the rapid removal of carbon from the bath.

In the double-slag method, the original slag is removed from the surface of the bath by cutting off the power to the electrodes, back-tilting the furnace slightly, and then pouring the slag out through the charge door into a slag thimble. The original slag must be removed thoroughly to prevent delay in making up the second slag. This might cause the reversion of some elements from slag to metal. Once the first slag is removed, additional lime and iron ore can be added.

The steel should not be held under the second slag any longer than is absolutely necessary. As soon as the results of the last analysis are reported, additions are made to the bath to adjust the carbon and alloying element content. When all of the additions are in solution, ferrosilicon may be added and aluminum thrust through the slag into the bath to control the grain size of the steel.

When the chemical composition of the steel meets specifications, the furnace (roof, electrodes, and all) is tilted forward so that molten metal may pour out the taphole through the spout. The slag comes after the steel and serves as an insulating blanket during tapping into a ladle.

Basic Oxygen Process

The basic oxygen steelmaking process (Figure 11-5) uses as its principal raw material molten pig iron from a blast furnace. The other source of metal is scrap. Lime, rather than limestone, is the fluxing agent. As the name implies, heat is provided by the use of oxygen.

Although the basic oxygen steel production in the United States was first reported by the American Iron and Steel Institute in 1954, the modern technique was pioneered in Austria shortly after World War II, and the concept of using oxygen in a pneumatic steelmaking process originated with Sir Henry Bessemer in the mid-1800s, at the time when he was developing the process now bearing his name. Bessemer recognized that

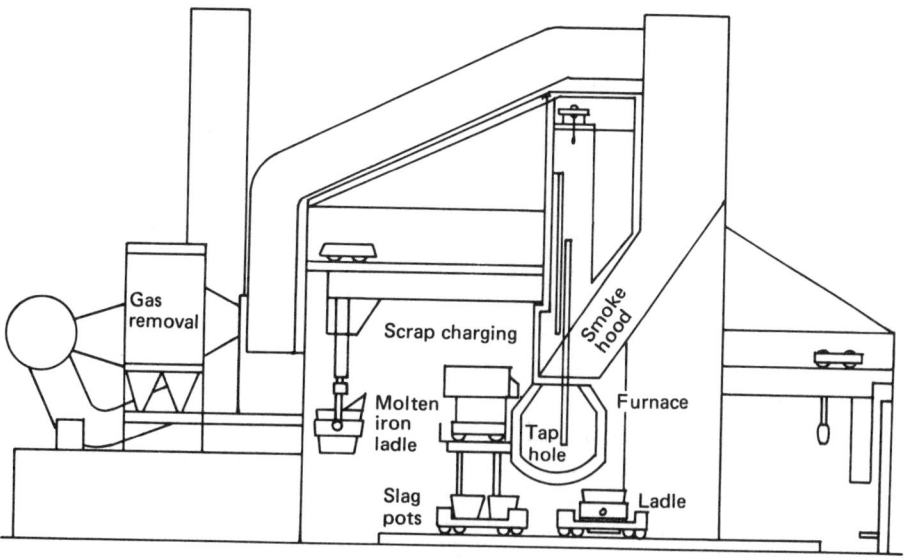

Figure 11-5   Basic Oxygen Steelmaking Process

blowing pure oxygen into his converter would be advantageous, but the technology for the bulk production of oxygen had not yet been conceived.

The first European basic oxygen furnaces were too small to be of commercial value in this country. Much of the developmental work in adapting the processes to big furnaces, containing over 300 tons of molten metal, was done in the United States.

The basic oxygen furnace is a steel shell lined with refractory materials. The body of the furnace is cylindrical, the bottom is slightly cupped, and the top is shaped like a truncated cone with its open base set on top of the cylinder. The narrow part of the truncated cone is open to receive raw materials and a jet of oxygen and also to permit dirty waste gases from the steelmaking process to escape into extensive air treatment facilities. The entire furnace is supported on horizontal trunnions so that it can be tilted.

Usually, these furnaces are installed in pairs so that one of them can be making steel while the other is being filled with raw materials. A basic oxygen furnace (BOF) may produce batches of over 300 tons in 45 minutes as against 5 to 8 hours for the older open-hearth process. Most grades of steel can now be made in BOFs, although this was not true in their early history. Even today, many foreign BOFs concentrate on making only a few uncomplicated high-tonnage steels, whereas in the United States it is the practice to produce a wide range of compositions.

The first step for making a heat of steel in a BOF is to tilt the furnace and charge it with steel scrap. This is done by swinging the furnace on its trunnions through an arc so that the open top can be reached by a charger which runs on rails at an appropriate level. The charger looks like a combination small railroad car and dump truck. The boxlike part of the charger is filled with carefully graded steel scrap and is carried by the railcar to a position where it can dump its contents into the tilted furnace. Pistons, or jointed arms, then lift and tilt the box so that the material slides out the open end into the preheated furnace. The furnace is rotated forward to distribute the scrap over the furnace bottom.

Immediately following the scrap charge, an overhead crane presents a ladle of molten iron from a blast furnace or from a holding device called a *mixer*. The iron, accounting for 65 to 80 percent of the charge, is also poured into the top of the tilted furnace.

As soon as the furnace with its charge of scrap and molten iron is in a vertical position, the oxygen lance is lowered and the oxygen is turned on to a flow of up to 6000 ft^3 (170 m^3) per minute at a pressure of up to 160 psi (480 kg/cm^2). The tip of the water-cooled lance may be about 6 ft above the metal surface when it is locked into place.

In a very short period of time, ignition causes increased heat and provides correct conditions for adding lime, fluorspar, and sometimes scale via a retractable chute to the metallic charge.

From that point on, the blowing procedure is uninterrupted. Oxygen combines with carbon and other unwanted elements, eliminating those impurities from the molten charge and converting it to steel. The lime and fluorspar help to carry off the impurities as a floating layer of slag on top of the metal, which is now entirely molten. In this function, lime is usually consumed at a rate of about 150 lb (680 kg) per ton of raw steel produced.

A BOF shop must be designed so that it is possible for the oxygen lance to be lowered into the vessel through the open top while that aperture is also hooded by the open end of the ductwork which carries away the waste gases. These gases are conducted to air treatment facilities, which are among the most complex and effective of the many pollution control systems employed by the iron and steel industry. In fact, the gas cleaning equipment is so complex that it may account for one-third or more of the overall structure of a BOF shop.

Experienced operators of BOFs can identify the completion of the steelmaking process by a variety of signs, including a decrease in the flame, a change in the wound level, a reading of the amount of oxygen used as indicated by the composition of gases emitted from the furnace, and a consideration of the blowing time for iron of a given composition.

When the batch of steel is complete, the clamps on the lance are released and the lance is retracted through the hood. The furnace is then rotated back toward the charging floor until the slag floating on top of the metal is even with the lip at the top of the furnace. Special equipment is then used to determine the temperature of the bath, which may be 3000°F (1649°C) for many of the more common types of steel. Once a correct temperature reading has been obtained, it may be necessary to test for carbon content. When both temperature and carbon content are acceptable, the furnace is rotated away from the charging platform past the vertical position and onto the opposite tilt. In this position the taphole, set into the cone-shaped portion of the furnace top, is aimed at a ladle which receives the molten steel. The slag, which floats on top of the steel, is caused to stay above the taphole by the progressive tilt of the furnace. Alloys are added to the ladle of steel, often by chutes extended from above the teeming floor.

The ladle into which the metallic contents of the furnace have been teemed is usually mounted on a railcar which is removed to a position where an overhead crane can lift it. The overhead crane carries the ladle of molten steel to a position where it can be poured into ingot molds or into a strand casting machine for solidification.

Meanwhile, the molten slag remaining in the furnace is emptied into a receptacle which is carried away for disposal. Except for possible minor refractory lining repair, the BOF is then ready to be recharged immediately. At approximately the same time, the other furnace in a pair should be ready to begin the steelmaking process.

STRUCTURE OF IRON AND STEEL

Carbon Steel

Carbon steel is an alloy of iron and carbon. The carbon atoms actually replace or enter into solution among the lattice structure of the iron atoms and limit the slip planes in the lattice structure. The amount of carbon within the lattice determines the properties of the steel.

Alloys containing less than 0.008 percent carbon are classed as irons. Steel is an iron–carbon alloy in which the carbon content is less than 2.0 percent. These steel products, including structural steel and reinforcing steel, can be rolled and molded into a shape. However, as the carbon content goes above 2.0 percent, the material becomes increasingly hard and brittle. Thus, cast iron has a carbon content above 2.0 percent. *High-strength steels* are alloys containing less than 0.8 percent carbon (the eutectoid composition) and are sometimes referred to as *hypoeutectoid steels*. *Structural steels* are alloys containing less than 2.0 percent but more than 0.8 percent carbon and are referred to as *hypereutectoid steels*. Wrought iron is a combination of iron and slag.

Phase Diagrams

A typical phase diagram of iron–carbon is shown in Figure 11-6. This equilibrium diagram is plotted for amounts of carbon up to 5 percent, which is sufficient for practical purposes. The diagram at times may be extended up to 6.67 percent of carbon, as this corresponds to 100 percent cementite. *Cementite,* a compound of iron and carbon, is a high-carbon steel that exists in a stable phase as iron carbide (Fe_3C). Cementite contains 6.67 percent carbon and 93.33 percent iron and is very hard and brittle. Other terms of importance shown on the phase diagram of Figure 11-6 are austenite, eutectic, eutectoid, ferrite, graphite, and pearlite. These terms are used to define the various stages or phases of iron–carbon alloys. *Austenite* is gamma iron with carbon in solution. The *eutectic* of the carbon–steel alloy is the combination that melts at the lowest temperature. Thus, according to Figure 11-6, the eutectic at 4.3 percent carbon melts at 2060°F (1130°C) and will contain a solid solution of austenite and a solid cementite. Below the 1330°F (723°C) mark, the solution changes to cementite and pearlite. The *eutectoid* in an equilibrium diagram for a solid solution is the point at which the solution on cooling is converted to a mixture of solids. Pearlite is a eutectoid and changes to a solid at 1330°F (723°C). *Ferrite* is iron that has not combined with carbon in pig iron as steel. This fact allows the steel to be cold-worked. *Graphite* (black lead) occurs as small flakes of carbon that become mixed with the steel. *Pearlite* is a lamellar aggregate of ferrite and cementite often occurring in carbon steels and in cast irons.

Properties of Cementite, Ferrite, and Pearlite

Cementite is a hard and very brittle material with a Brinell hardness of 650 and a diamond hardness of 760. Ferrite is a soft and ductile material with a Brinell hardness of 90 and a diamond hardness of 170. It has a 40 percent elongation in 2-in. (5.08-cm) specimen. Pearlite is harder and less ductile than ferrite but is softer and less brittle than cementite. Pearlite has a Brinell hardness of 275 and a diamond hardness of 300. It has an elongation of 15 percent in a 2-in. (5.08-cm) specimen.

Impurities in Steel

The principal impurities in steel are silicon, phosphorus, sulfur, and manganese.

The amount of silicon in structural steel is less than 1 percent and forms a solid solution with iron. This small amount of silicon increases both the ultimate strength and

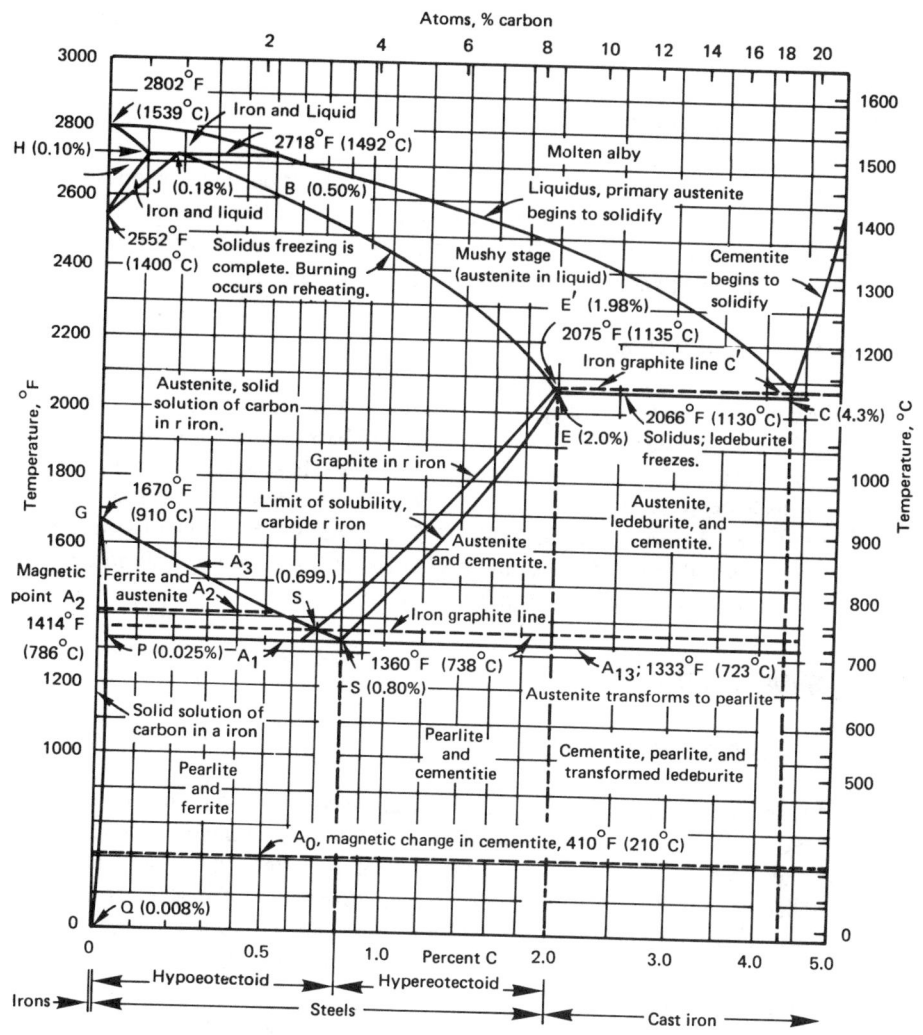

Figure 11-6 Phase Diagram of Iron–Carbon

the elastic limit of steel with no appreciable change in its ductility. Silicon may further prevent the solution of carbon in iron.

The phosphorus in steel is in the form of iron phosphide (Fe_3P). For low-grade structural steel the amount of phosphorus is about 0.1 percent and decreasing to 0.05 percent for high-grade structural steel. Tool steel is approximately 0.02 percent phosphorus.

Sulfur in steel combines with the iron to form iron sulfate (FeS). This compound has a low melting point and segregation may take place.

Manganese has an affinity for sulfur and combines with such as well as with other impurities to form slag. In other words, manganese acts like a cleanser. Manganese is used to harden steels.

Structural Steel

Table 11-1 illustrates the uses of structural steels. It becomes obvious that the steel must have strength, toughness, and, above all, durability. The requirements for structural steel are presented in Table 11-2. In most cases the maximum percent of carbon is less than 0.27, but most structural steels average 0.2 percent.

Reinforcing Steel

As was explained in Chapter 4, concrete exhibits great compressive strength but little tensile or flexural strength. Thus, deformed bars of structural steel are embedded in the concrete to take up the tensile or flexural forces. These deformed bars have been developed in such a way as to force the concrete between the deformations such that failure in shear will occur before slippage. Table 11-3 lists the ASTM standard-size reinforcing bars as to their weights and dimensions. ASTM has also standardized reinforcing steel according to its yield point, as shown in Table 11-4.

HEAT TREATMENTS

Hardening or Quenching

Whenever a solid solution, such as steel, decomposes due to a falling temperature into the eutectoid, the decomposition may be more or less completed, depending on the cooling rate. This process is utilized in the hardening of steel. If the steel is cooled slowly, the changes just discussed will take place; however, if the steel is cooled too quickly, decomposition into the eutectoid will be prevented and a structure called *martensite* is produced rather than steel. Martensite is a hard structure with little ductility (a necessary property in steels).

The successful hardening of steel may be achieved by the application of three general principles:

1. Steel should always be annealed before hardening, to remove forging or cooling stains.
2. Heating for hardening should be slow.
3. Steel should be quenched on a rising, not on a falling temperature.

Quenching media vary, but are basically of three types:

1. Brine for maximum hardness.
2. Water for rapid cooling of the common steels.

3. Oils (light, medium, or heavy) for use with common steel parts of irregular shapes or for alloy steels.

All hardened steel is in a state of strain, and steel pieces with sharp angles or grooves sometimes crack immediately after hardening. For this reason, tempering must follow the quenching operation as soon as possible.

Tempering

Tempering of steel is defined as the process of reheating a hardened steel to a definite temperature below the critical temperature, holding it at that point for a time, and cooling it, usually by quenching for the purpose of obtaining toughness and ductility in the steel.

Annealing

Annealing has basically the opposite objective of hardening. *Annealing* has the process of heating a metal above the critical temperature range, holding it at that temperature for the proper period of time, and then slowly cooling. During the cooling process, pearlite, ferrite, and/or cementite form. The objectives of annealing are:

1. To refine the grain.
2. To soften the steel to meet definite specifications.
3. To remove internal stresses caused by quenching, forging, and cold working.
4. To change ductility, toughness, electrical, and magnetic properties.
5. To remove gases.

PHYSICAL PROPERTIES OF STEELS

In general, and as previously mentioned, three principal factors influence the strength, ductility, and elastic properties of steel:

1. The carbon content.
2. The percentages of silicon, sulfur, phosphorus, manganese, and other alloying elements.
3. The heat treatment and mechanical working.

Carbon

The various properties of different grades of steel are due more to variations in the carbon content of the steel than to any other single factor. Carbon acts as both a hardener and a strengthener, but at the same time it reduces the ductility.

TABLE 11-1 STRUCTURAL STEEL[a]

ASTM Designation	Product	Use
A36	Carbon-steel shapes, plates, and bars	Welded, riveted, and bolted construction; bridges, buildings, towers, and general structural purposes
A53	Welded or seamless pipe, black or galvanized	Welded, riveted, and bolted construction; primary use in buildings, particularly columns and truss members
A242	High-strength, low-alloy shapes, plates, and bars	Welded, riveted, and bolted construction; bridges, buildings, and general structural purposes; atmospheric-corrosion resistance about four times that of carbon steel; a weathering steel
A245	Carbon-steel sheets, cold- or hot-rolled	Cold-formed structural members for buildings, especially standardized buildings; welded, cold-riveted, bolted, and metal-screw construction
A374	High-strength, low-alloy, cold-rolled sheets and strip	Cold-formed structural members for buildings, especially standardized buildings; welded, cold-riveted, bolted, and metal-screw construction
A440	High-strength shapes, plates, and bars	Riveted or bolted construction; bridges, buildings, towers, and other structures; atmospheric-corrosion resistance double that of carbon steel
A441	High-strength, low-alloy manganese–vanadium steel shapes, plates, and bars	Welded, riveted, or bolted construction but intended primarily for welded construction; bridges, buildings, and other structures; atmospheric-corrosion resistance double that of carbon steel
A446	Zinc-coated (galvanized) sheets in coils or cut lengths	Cold-formed structural members for buildings, especially standardized buildings; welded, cold-riveted, bolted, and metal-screw construction

[Handwritten annotations: "most commonly used" pointing to A36 and A53; near A36: "36,000 psi steel strength (ultimate)"]

A500	Cold-formed welded or seamless tubing in round, square, rectangular, or special shapes	Welded, riveted, or bolted construction; bridges, buildings, and general structural purposes
A501	Hot-formed welded or seamless tubing in round, square, rectangular, or special shapes	Welded, riveted, or bolted construction; bridges, buildings, and general structural purposes
A514	Quenched and tempered plates of high yield strength	Intended primarily for welded bridges and other structures; welding technique must not affect properties of the plate, especially in heat-affected zone
A529	Carbon-steel plates and bars to ½ in. thick	Buildings, especially standardized buildings; welded, riveted, or bolted construction
A570	Hot-rolled carbon-steel sheets and strip in coils or cut lengths	Cold-formed structural members for buildings, especially standardized buildings; welded, cold-riveted, bolted, and metal-screw construction
A572	High-strength, low-alloy columbian–vanadium steel shapes, plates, sheet piling, and bars	Welded, riveted, or bolted construction of buildings in all grades; welded bridges in grades 42, 45, and 50 only
A588	High-strength, low-alloy steel shapes, plates, and bars	Intended primarily for welded bridges and buildings; atmospheric-corrosion resistance about four times that of carbon steel; a weathering steel
A606	High-strength, low-alloy hot- and cold-rolled sheet and strip	Intended for structural and miscellaneous purposes where savings in weight or added durability are important

[a]After E. H. Gaylord, Jr., and C. N. Gaylord, *Design of Steel Structures*, McGraw-Hill Book Company, New York, 1972.

TABLE 11-2 ASTM STRUCTURAL STEEL REQUIREMENTS

Product:	Shapes[a]	Plates					Bars			
Thickness [in. (mm)]:	All	To ¾ (19), incl.	Over ¾ to 1½ (19 to 38), incl.	Over 1½ to 2½ (38 to 64), incl.	Over 2½ to 4 (64 to 102), incl.	Over 4 (102)	To ¾ (19), incl.	Over ¾ to 1½ (19 to 38), incl.	Over 1½ to 4 (102), incl.	Over 4 (102)
Carbon, maximum (%)	0.26	0.25	0.25	0.26	0.27	0.29	0.26	0.27	0.28	0.29
Manganese, (%)	—	—	0.80–1.20	0.80–1.20	0.85–0.20	0.85–0.20	—	0.60–0.90	0.60–0.90	0.60–0.90
Phosphorus, max. (%)	0.04	0.04	0.04	0.04	0.04	0.04	0.04	0.04	0.04	0.04
Sulfur, max. (%)	0.05	0.05	0.05	0.05	0.05	0.05	0.05	0.05	0.05	0.05
Silicon (%)	—	—	0.15–0.30	0.15–0.30	0.15–0.30	0.15–0.30	—	—	—	—
Copper when copper steel is specified, min. (%)	0.20	0.20	0.20	0.20	0.20	0.20	0.20	0.20	0.20	0.20

Minimum Tensile Properties of Structural Steels

ASTM Designation	Yield (ksi)	Strength (ksi)	Elongation (8 in. Unless Noted) (%)
Carbon steels			
A36	36	58–80	20
A529	42	60–85	19

High strength steels			
A242, A440, A441			
To ¾ in. thick	50	70	18
Over ¾ in. to 1½ in.	46	67	19
Over 1½. in. to 4 in.	42	63	16
A572			
Grade 42, to 4 in. incl.	42	60	20
Grade 45, to 1½ in. incl.	45	60	19
Grade 50, to 1½ in. incl.	50	65	18
Grade 55, to 1½ in. incl.	55	70	17
Grade 60, to 1 in. incl.	60	75	16
Grade 65, to ½ in. incl.	65	80	15
A588			
To 4 in. thick	50	70	19–21[b]
Over 4 in. to 5 in.	46	67	19–21[b]
Over 5 in. to 8 in.	42	63	19–21[b]
Quenched and tempered steels			
A514			
To 2½ in. thick	100	115–135	18[b]
Over 2½ in. to 4 in.	90	105–135	17[b]

[a]Manganese content of 0.85–1.35% and silicon content of 0.15–0.30% is required for shapes over 426 lb/ft.
[b]In 2 in.

TABLE 11-3 ASTM STANDARD-SIZE REINFORCING BARS[a]

| Bar Designation Number[c] | Nominal Weight (kg/m) | Nominal Dimensions[b] | | | Deformation Requirements (mm) | | |
		Diameter (mm)	Cross-Sectional Area (cm²)	Perimeter (mm)	Maximum Average Spacing	Minimum Average Height	Maximum Gap (Chord 12½% of Nominal Perimeter)
3	0.560	9.52	0.71	29.9	6.7	0.38	3.5
4	0.994	12.70	1.29	39.9	8.9	0.51	4.9
5	1.552	15.88	2.00	49.9	11.1	0.71	6.1
6	2.235	19.05	2.84	59.8	13.3	0.96	7.3
7	3.042	22.22	3.87	69.8	15.5	1.11	8.5
8	3.973	25.40	5.10	79.8	17.8	1.27	9.7
9	5.059	28.65	6.45	90.00	20.1	1.42	10.9
10	6.403	32.26	8.19	101.4	22.6	1.62	11.4
11	7.906	35.81	10.06	112.5	25.1	1.80	13.6
14[d]	11.384	43.00	14.52	135.1	30.1	2.16	16.5
18[d]	20.238	57.33	25.81	180.1	40.1	2.59	21.9

[a]Based on ASTM 6615, 6616, 6617.

[b]The nominal dimensions of a deformed bar are equivalent to those of a plain round bar having the same weight per foot as the deformed bar.

[c]Bar numbers are based on the number of eighths of an inch included in the nominal diameter of the bars.

[d]Available in billet steel only.

Silicon, Sulfur, Phosphorus, and Manganese

The effect of silicon on strength and ductility in ordinary proportions (less than 0.2 percent) is very slight. If the silicon content is increased to 0.3 or 0.4 percent, the elastic limit and ultimate strength of the steel are raised without reducing the ductility. This is a procedure used for steel castings.

Sulfur within ordinary limits (0.02 to 0.10 percent) has no appreciable effect upon the strength or ductility of steels. It does, however, have a very injurious effect upon the properties of the hot metal, lessening its malleability and weldability, thus causing difficulty in rolling, called "red-shortness."

Phosphorus is the most undesirable impurity found in steels. It is detrimental to toughness and shock-resistance properties, and often detrimental to ductility under static load.

Manganese improves the strength of plain carbon steels. If the manganese content is less than 0.3 percent, the steel will be impregnated with oxides that are injurious to the steel. With a manganese content of between 0.3 and 1.0 percent, the beneficial effect depends upon the amount of carbon content. As the manganese content rises above 1.5 percent, the metal becomes brittle and worthless.

TABLE 11-4 KINDS AND GRADES OF REINFORCING BARS

Type of Steel and ASTM Specification Number	Grade Designation	Size Numbers	Yield Point Minimum (MPa)	Tensile Strength Minimum (MPa)	Elongation in 203-mm Minimum (%)	Diameter Bend Test Pin[a]
Billet steel	40	3	276	483	11	4d
A615		4, 5			12	4d
		6			12	5d
		7			11	5d
		8			10	5d
		9			9	5d
		10			8	5d
		11			7	5d
		14, 18			—	None
	60	3, 4, 5	415	621	9	4d
		6			9	5d
		7, 8			8	6d
		9, 10, 11			7	8d
		14, 18			7	None
Rail Steel	50	3	345	550	6	6d
A616		4, 5, 6			7	6d
		7			6	6d
		8			5	6d
		9, 10, 11[b]			5	8d
	60	3	415	620	6	6d
		4, 5, 6			6	6d
		7			5	6d
		8			4.5	6d
		9, 10, 11[b]			4.5	8d
Axle steel	40	3	275	480	11	4d
A617		4, 5			12	4d
		6			12	5d
		7			11	5d
		8			10	5d
		9			9	5d
		10			8	5d
		11			7	5d
	60	3, 4, 5			8	4d
		6			8	5d
		7			7	6d
		8	415	620	8	6d
		9, 10, 11			7	8d

[a]d, diameter of specimen.

[b]Number 11, 90° bend; all others 180°.

Effect of Heat Treatment upon Physical Properties

The effects of various heat treatments upon the mechanical properties of wrought or rolled carbon steels of various compositions are shown in Table 11-5.

Tensile Strength

The tensile strength and properties of various carbon steels as given by the ASTM are shown in Table 11-6. The modulus of elasticity of all grades varies from 28,000,000 to 30,000,000 psi (1.9×10^{11} to 2.1×10^{11} N/m^2).

Structural Steel

The chemical compositions for structural steel used in bridges and buildings as given by the ASTM are listed in Table 11-7. Notice that the carbon content is not specified. The reason for this is that the manufacturer is permitted to vary the carbon percentage to meet the tensile strength requirements of Table 11-6.

Torsional Shear

The strength of steel in torsional shear is shown in Table 11-8. The modulus of elasticity for shear is about 12,000,000 psi (8.2×10^7 N/m^2).

ALLOY STEELS

Alloy steels are steels that owe their distinctive properties to elements other than carbon. Common alloys include chromium, nickel, manganese, molybdenum, silicon, copper, vanadium, and tungsten.

These alloys can be classified into two groups: those which combine with the carbon to form carbides, such as nickel, silicon, and copper, and those which do not combine with carbon to form carbides, such as manganese, chromium, tungsten, molybdenum, and vanadium.

Alloys are added to steel for three principal reasons:

1. To increase hardness.
2. To increase the strength.
3. To add special properites, such as
   a. Toughness.
   b. Improved magnetic and electrical properties.
   c. Corrosion resistance.
   d. Machinability.

TABLE 11-5 MECHANICAL PROPERTIES OF HEAT-TREATED WROUGHT OR ROLLED CARBON STEELS[a]

Composition (%)			Yield Point (psi)	Tensile Strength (psi)	Elongation in 2 in. (%)	Reduction of Area (%)	Brinell Hardness
Carbon	Manganese	Heat Treatment					
0.14	0.45	Hot-rolled	45,000	59,500	37.5	67.0	112
		Annealed	31,000	54,500	39.5	67.0	107
		Quenched in water	—	90,000	21.0	67.0	170
		Quenched in oil	56,500	71,500	34.0	75.5	134
0.32	0.5	Hot-rolled	49,500	75,500	30.0	51.9	144
		Annealed	41,000	70,000	30.5	51.9	131
		Quenched in water	—	135,000	8.0	16.9	255
		Quenched in oil	67,500	101,000	23.5	62.3	207
		Quenched in oil, tempered at 650°F	61,500	84,000	30.0	71.4	163
0.46	0.40	Hot-rolled	52,500	86,500	22.5	30.7	160
		Annealed	48,000	79,500	28.5	46.2	153
		Quenched in water	—	220,000	1.0	0.0	600
		Quenched in oil	87,500	126,500	20.5	51.9	225
		Quenched in oil, tempered at 560°F	81,500	111,500	24.0	57.2	—
		Quenched in oil, tempered at 650°F	73,000	98,000	25.5	59.8	192
0.57	0.65	Hot-rolled	57,000	106,500	19.0	27.4	220
		Annealed	50,000	95,000	25.0	40.3	183
		Quenched in water	—	215,000	0.0	0.0	578
		Quenched in oil	105,000	152,000	16.5	40.3	311
		Quenched in oil, tempered at 460°F	97,500	145,000	16.0	46.2	293
		Quenched in oil, tempered at 650°F	79,500	113,000	24.0	62.3	228
0.71	0.67	Hot-rolled	66,000	128,000	15.0	20.5	240
		Annealed	46,500	111,500	16.5	24.0	217
		Quenched in oil	100,000	184,500	1.5	0.0	364
		Quenched in oil, tempered at 460°F	115,500	177,000	10.0	34.0	340
		Quenched in oil, tempered at 560°F	106,000	148,500	17.0	43.0	311
		Quenched in oil, tempered at 650°F	91,000	125,500	19.5	57.2	269

[a]Data from the American Society for Steel Treating.

TABLE 11-6 TENSILE PROPERTIES OF VARIOUS STEELS

Kind and Use of Steel	Tensile Strength (psi)	Yield Point, Min. (psi)	Elongation Min. in 8 in. (%)	Elongation Min. in 2 in. (%)	Reduction of Area, Min. (%)
Structural steel for bridges and buildings					
Structural	60,000–72,000	33,000	21	22	
Rivet	52,000–62,000	28,000	24		
Structural steel for ships					
Structural	58,000–71,000	32,000	21	22	
Rivet	55,000–65,000	30,000	23		
Carbon steel forgings for locomotives and cars					
Annealed or normalized	75,000 min.	37,500		20	33
Normalized quenched, and tempered	115,000 min.	75,000		16	35
Carbon steel bolting material, heat-treated					
Grade BO	100,000 min.	75,000		16	45
Boiler and firebox steel					
Flange	55,000–65,000	0.5 u.t.s.[a]	1,500,000/u.t.s.	1,750,000/u.t.s	
Firebox, grade A	55,000–65,000	0.5 u.t.s.	1,550,000/u.t.s.	1,750,000/u.t.s	
Firebox, grade B	48,000–58,000	0.5 u.t.s.	1,550,000/u.t.s.	1,750,000/u.t.s	
Boiler-rivet steel					
Grade A	45,000–55,000	0.5 u.t.s.	1,500,000/u.t.s.		
Grade B	58,000–68,000	0.5 u.t.s.	1,500,000/u.t.s.		
Billet-steel-concrete reinforcing bars					
Plain bars					
Structural grade	55,000–70,000	33,000	1,400,000/u.t.s.		
Intermediate grade	70,000–90,000	40,000	1,300,000/u.t.s.		
Hard grade	80,000 min.	50,000	1,100,000/u.t.s.		
Deformed bars					
Structural grade	55,000–75,000	33,000	1,200,000/u.t.s.		
Intermediate grade	70,000–90,000	40,000	1,100,000/u.t.s.		
Hard grade	80,000 min.	50,000	1,000,000/u.t.s.		
Concrete reinforcing bars from rerolled steel rails					
Plain bars	80,000 min.	50,000	1,000,100/u.t.s.		
Deformed bars	80,000 min.	50,000	1,000,000/u.t.s.		

[a]u.t.s., ultimate tensile strength.

TABLE 11-7 CHEMICAL COMPOSITION OF STRUCTURAL STEEL FOR BRIDGES AND BUILDINGS, ASTM SPECIFICATIONS

Element	Ladle Analysis, Max. (%)	Check Analysis, Max. (%)
Open-Hearth and Electric-Furnace Structural and Structural-Rivet Steel		
Phosphorus		
Acid process	0.06	0.075
Basic process	0.04	0.05
Sulfur	0.05	0.063
Copper (when specified)	0.020	0.18
Acid Bessemer Structural Steel[a]		
Phosphorus	0.11	0.138

[a]Not permitted in bridges or in building members subject to dynamic loads.

Chromium

Chromium is primarily a hardening agent and is generally added to steel in amounts of 0.70 to 1.20 percent, with a variation in carbon content of 0.17 to 0.55 percent. Its value is due principally to its property of combining intense hardness after quenching with very high strength and elastic limit. Thus, it is well suited to withstand abrasion, cutting, or shock. It does lack ductility, but this is unimportant in view of its high elastic limit. Chromium steels corrode less rapidly than do carbon steels.

Nickel–Chromium

Nickel–chromium steels, when properly heat-treated, have a very high tensile strength and elastic limit, with considerable toughness and ductility. The nickel content is usually 3.5 percent, with a carbon content ranging from 0.15 to 0.50 percent.

One very important property of nickel–chromium steels is that by adding aluminum, cobalt, copper, manganese, silicon, silver, or tungsten, stainless steel results.

TABLE 11-8 STRENGTH OF STEEL IN TORSIONAL SHEAR

Class of Steel	Computed Extreme Fiber Stress (psi)	Shearing Modulus of Elasticity (psi)
Mild Bessemer	64,200	11,320,000
Medium Bessemer	68,300	11,570,000
Hard Bessemer	74,000	11,700,000
Cold-rolled	79,900	11,950,000

Manganese

As previously indicated, manganese is present in all steels as a result of the manufacturing process. When the manganese is 1.0 percent or greater in solution with steel, it is considered an alloy. Manganese will add hardness to steel if used within the proper range.

Molybdenum

Molybdenum provides strength and hardness in steel. It inhibits grain growth on heating as a result of its slow solubility of austenite. When in solution in the austenite, it decreases the cooling rate and, therefore, increases the depth of hardening.

Silicon

Silicon is added to carbon steel for the purpose of deoxidizing. For this reason, silicon may be added in amounts of up to 0.25 percent. Silicon does not form carbides but does dissolve in the ferrite up to about 15 percent. Silicon decreases hysteresis and eddy-current losses, and thus is valuable for electrical machinery.

Vanadium

Vanadium is a powerful element for alloying in steel. It forms stable carbides and improves the hardenability of steels. Vanadium promotes a fine-grained structure and promotes hardness at high temperatures. The amount of vanadium present is 0.10 to 0.30 percent when used.

Copper

Copper increases the yield strength, tensile strength, and hardness of steel. However, ductility may be decreased by about 2 percent. The most important use of copper is to increase the resistance of steel to atmospheric corrosion.

Tungsten

Tungsten increases the strength, hardness, and toughness of steel. After moderately rapid cooling from high temperatures, tungsten steel exhibits remarkable hardness, which is still retained upon heating to temperatures considerably above the ordinary tempering heats of carbon steels. It is this property of tungsten that makes it a valuable alloy, in conjunction with chromium or manganese, for the production of high-speed tool steel.

NONFERROUS METALS

In this section, various nonferrous metals will be listed with only a brief statement; specific details will be omitted.

Basically, three groups of nonferrous metals exist. In the first group, those of

Test on this material Next Monday.

greatest industrial importance, are aluminum, copper, lead, magnesium, nickel, tin, and zinc. The second group includes antimony, bismuth, cadmium, mercury, and titanium. The third and final group, important in that they are used to form alloy steels, includes chromium, cobalt, molybdenum, tungsten, and vanadium.

The nonferrous alloys of the greatest importance are alloys of copper with tin (bronzes), alloys of copper with zinc (brasses), and alloys of aluminum, magnesium, nickel, and titanium.

PROBLEMS

11.1. Explain the purpose and use of the blast furnace.

11.2. Discuss four manufacturing processes of steel.

11.3. Why is carbon so important in steel?

11.4. Define austenite, cementite, eutectic, eutectoid, ferrite, graphite, and pearlite.

11.5. Discuss the various heat-treatment processes.

11.6. What is tempering?

11.7. What factors influence the strength, ductility, and elastic properties of steel?

11.8. Discuss the importance of silicon, sulfur, phosphorus, and manganese in steel.

11.9. List and discuss the various alloy steels.

11.10. Discuss briefly the use of nonferrous metals that are alloyed with steel.

REFERENCES

American Iron and Steel Institute, *The Making of Steel,* Washington, D.C., 1970.

American Society for Metals, *Metals Handbook,* Cleveland, Ohio, 1948.

American Society for Testing and Materials, "Specifications for Structural Steel," *ASTM A-36, C1020.*

BULLENS, D. K., *Steel and Its Heat Treatment,* 5th ed., 3 vols., John Wiley & Sons, Inc., New York, 1948. See especially vol. 1, *Principles,* and vol. 2, *Tools, Processes, and Control.*

DIGGES, T. G., "Effect of Carbon on the Hardening of High Purity Iron Carbon Alloys," *Trans. Am. Soc. Met.,* 25 (1938).

GAYLORD, E. H., JR., and GAYLORD, C. N., *Design of Steel Structures,* McGraw-Hill Book Company, New York, 1972.

LIPSON, H., and PARKER, A. M. B., "The Structure of Martensite," *J. Iron Steel Inst. (Lond.), 149* (1944).

SISCO, F. T., *Modern Metallurgy for Engineers,* 2nd ed., Pitman Publishing Corp., New York, 1948.

TEICHERT, E. J., *Ferrous Metallurgy,* 2nd ed., McGraw-Hill Book Company, New York, 1944.

12

Plastics

A textbook on materials for civil and highway engineers would not be complete without a chapter on plastics, as plastics comprise an important group of materials of construction. Many plastic substances have combinations of properties that cannot be duplicated by other materials. In general, plastics exhibit a number of outstanding characteristics:

1. Lightness in weight (generally half as light as aluminum).
2. High dielectric strength (electrical insulation).
3. Low heat conductivity (heat insulation).
4. Special properties toward lights (colorability).
5. Extremely resistant toward chemicals.
6. Metal inserts may be molded into the plastic (since plastics are inert toward such materials).
7. Many high-quality products can be developed by using a lathe, sawing, punching, and drilling.

CLASSIFICATION

A *plastic* is a polymeric material (usually organic) of high molecular weight which can be shaped by flow. The term usually refers to the final product, with fillers, plasticizers, pigments, and stabilizers included (as opposed to the resin, the homogeneous polymeric starting material). Most plastics are synthetic organic compounds which derive their coherence and strength from large chain-linked macromolecules formed from one or more single molecules (monomers) to a macromolecule (polymer) by a chemical reaction called *polymerization.* Examples include polyvinyl chloride, polyethylene, and urea–formaldehyde. Most organic plastics contain a *binder,* which imparts the plastic properties to the composition. In addition, many plastics contain a *filler,* an inert extender that adds hardness, strength, and other desirable properties. In general, organic plastics can be divided into three general classifications:

1. Thermoplastics.
2. Thermosetting plastics.
3. Chemically setting plastics.

Table 12-1 shows the classification of organic plastics.

Thermoplastics

Thermoplastics are organic plastics, either natural or synthetic, which remain permanently soft at elevated temperatures. Upon cooling, they again become hard. These materials can be shaped and reshaped any number of times by repeated heating and cooling. Natural thermoplastics include asphalts, bitumen, pitches, and resin, to name some of the most familiar.

Thermosetting Plastics

Thermosetting plastics are organic plastics that were originally soft or soften at once upon heating, but upon further heating, they harden permanently. Thermosetting plastics are hardened by chemical changes due to heat, a catalyst, or to both. Thermosetting plastics remain hardened without cooling and do not soften appreciably when reheated. The most common thermosetting plastic is polyester.

Chemically Setting Plastics

Chemically setting plastics are those that harden by the addition of a suitable chemical to the composition just before molding or by subsequent chemical treatment following fabrication.

TYPES OF PLASTICS

Polymerization and Condensation

Polymerization involves unsaturated molecules that contain double or triple bonds between carbon atoms which are weaker than single bonds. Unsaturated molecules are unstable and they react in such a way as to break the multiple bond. The mechanism by which polymerization takes place is grouped into two categories:

1. Addition polymerides.
2. Condensation polymerides.

TABLE 12-1 CLASSIFICATION OF ORGANIC PLASTICS

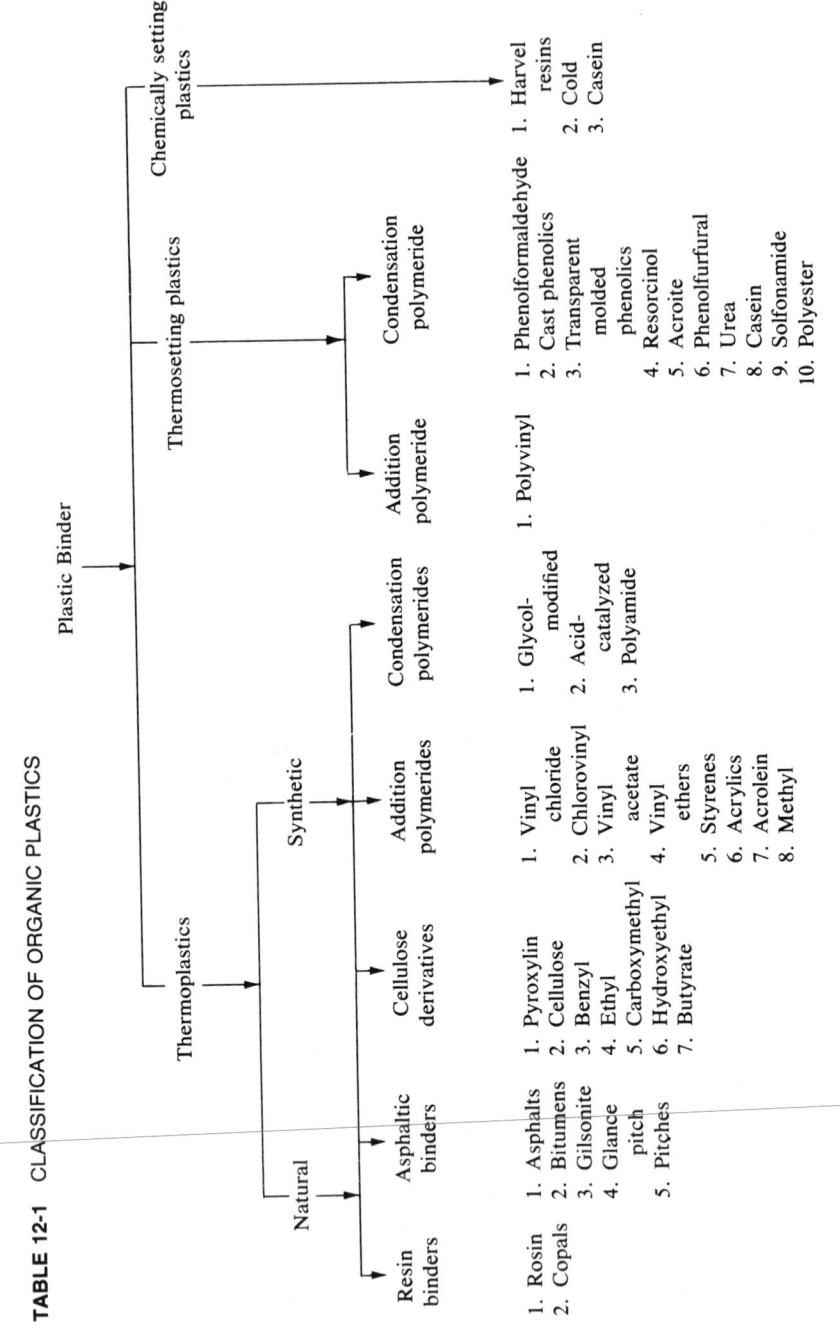

Plastic Binder

Thermoplastics — Thermosetting plastics — Chemically setting plastics

Natural — Synthetic

Resin binders

1. Rosin
2. Copals

Asphaltic binders

1. Asphalts
2. Bitumens
3. Gilsonite
4. Glance pitch
5. Pitches

Cellulose derivatives

1. Pyroxylin
2. Cellulose
3. Benzyl
4. Ethyl
5. Carboxymethyl
6. Hydroxyethyl
7. Butyrate

Addition polymerides

1. Vinyl chloride
2. Chlorovinyl
3. Vinyl acetate
4. Vinyl ethers
5. Styrenes
6. Acrylics
7. Acrolein
8. Methyl

Condensation polymerides

1. Glycol-modified
2. Acid-catalyzed
3. Polyamide

Addition polymeride

1. Polyvinyl

Condensation polymeride

1. Phenolformaldehyde
2. Cast phenolics
3. Transparent molded phenolics
4. Resorcinol
5. Acroite
6. Phenolfurfural
7. Urea
8. Casein
9. Solfonamide
10. Polyester

Chemically setting plastics

1. Harvel resins
2. Cold
3. Casein

268

Addition polymerides are mixtures of polymers that have been formed by addition of like molecules. The molecules are added to increase the average molecular size and weight. This reaction occurs by the breaking of double bonds between atoms in monomers and the forming of two single bonds in their place.

Condensation polymerides are formed by chemical reactions in which two or more different molecules combine with the separation of water or other simple substances in the formation of resins. Condensation polymerides are produced by a by-product as well as the growing polymer molecules. The nonpolymerizable molecule is usually water or some simple molecule.

Copolymers are produced by mixing monomers and then polymerizing. In other words, copolymers are produced by the simultaneous polymerization of two or more chemically different monomers.

Thermoplastics

Properties and uses of common plastics of the thermoplastic type are shown in Table 12-2. Thermoplastic types of binders are divided into two groups: natural and synthetic. Of the *natural* type two categories exist: resin and asphaltic binders. The resin binders at this point in materials of construction are of little importance. The asphaltic binders were thoroughly discussed in Chapter 9.

The *synthetic* group consists of three categories: cellulose derivatives, addition polymerides, and condensation polymerides. *Cellulose* plastic is the term employed for plastics made from derivatives of cellulose, such as pyroxylin, cellulose acetate, benzyl cellulose, ethyl cellulose, carboxymethyl cellulose, hydroxyethyl cellulose, and cellulose acetate butyrate. Cellulose plastics are true thermoplastics and exhibit the greatest toughness and resilience of any of the plastics. They are used for objects having thin-walled sections where other plastics may be too brittle.

Addition and condensation polymerides are grouped together and referred to as *noncellulose* type. This group consist of vinyl chloride, chlorovinyl chloride, vinyl acetate, vinyl ethers, styrene, acrylics, acrolein, methyl methacrylate, cumarone–indene, glycol-modified alkyds, acid-catalyzed phenol–formaldehyde, and polyamide (nylon). The noncellulose plastics are all odorless, tasteless, nontoxic, and transparent. These plastics are strong, tough, and chemically inert with little water absorption. The noncellulose plastics are suited for electrical insulation or cable coatings, safety glass, nylon, and Saran (covering to prevent corrosion and/or chemical attack on pipes and tubing).

Thermosetting Plastics

Thermosetting plastics can be divided into two groups: addition polymerides and condensation polymerides. Addition polymerides consist of any polyvinyl types of resins; condensation polymerides, the largest group, consist of phenolformaldehyde, cast phenolics, transparent molded phenolics, resorcinol–formaldehyde, acroite, phenolfurfural, urea, casein–formaldehyde, sulfonamide resins, and polyesters.

Plastics of the thermosetting group have excellent mechanical and electrical prop-

TABLE 12-2 PROPERTIES AND USES OF COMMON PLASTICS (THERMOPLASTICS)

Property	Shellac	Polyethylene	Polymono-chloro-trifluoro-ethylene	Vinylidene Chloride Molding	Polystyrene	Methyl methacrylate, Cast	Polyamide (Nylon) Molding	Cellulose Acetate Molding	Cellulose Nitrate (Pyroxylin)
Injection molding pressure (1000 psi)	1.00–1.20	8–15	20–60	10–30	10–30	—	10–25	8–32	8–32
Specific gravity	1.1–2.7	0.92	2.10	1.65–1.72	1.05–1.07	1.18–1.20	1.14	1.27–1.37	1.35–1.40
Tensile strength (1000 psi)	0.9–2.0	1.5–1.8	5.7	3–5	5–9	6–7	7–9	1.9–8.5	7–8
Elongation (% in 2 in.)	—	50–400	28–36	20–250	1–3.6	2–7	40–100	6–50	40–45
Modulus of elasticity in tension (100,000 psi)	5–6	0.19	1.9	0.5–0.8	4–6	3.5–5	2.6–4	0.86–4.0	1.9–2.2
Compressive strength, (1000 psi)	10–17	—	32–80	7.5–8.5	11.5–16	11–19	7.2–13	13–36	22–35
Impacts strength, Izod test on ½ × ½-in. notched bar (ft-lb./in. width of notch)	2.6–2.9	Less than 16	3.6	0.3–1.0	0.26–0.50	0.4–0.5	1.0	0.4–5.2	5–7
Hardness, Rockwell[a]	—	R11	R110–115	M50–65	M65–90	M90–100	M111–118	R50–125	R95–115

Highest usable temperature continuous (°F)	150–190	212	390	160–200	150–205	140–200	270–300	140–220	140
Thermal conductivity 10^{-4} cal/(sec)(cm³)(°C)	—	8	1.4	3	2.4–3.3	4–6	5.2–5.5	4–8	3.1–5.5
Thermal expansion (10^{-5} in./in. °C)	—	16–18	4.5–7.0	19	6–8	9	10–15	8–16	8–12
Dielectric strength, short time, ⅛ in. thickness (V/mil)	200–600	400	2500	350	500–700	450–500	385–470	250–365	300–600
Water absorption 24 hr, ⅛ in. thick (%)	0–0.1	Less than 0.01	0.00	0–0.1	0.03–0.05	0.3–0.4	0.4–1.5	1.9–6.5	1.0–2.0
Effect of strong acids	Deteriorated	Attacked by oxidizing acids	None	Highly resistant	Attacked by oxidizing acids	Attacked by oxidizing acids	Attacked	Decomposed	Decomposed
Color possibilities	Limited	Unlimited	Unlimited	Extensive	Unlimited	Unlimited	Unlimited	Unlimited	Unlimited
Common uses	Phonograph records, electrical insulation	Bottle stoppers, flexible bottles, wire insulation, textiles, tablewear	Filter disks, insulators, gaskets	Screening, chemical tubing, auto seat covers	Electrical insulators, battery boxes, lenses, toys, boxes	Windows, furniture, dentures, picture frames	Bearings, cups, fabrics, bristles	Fountain pens, tools, toys, spectacles, packaging	Packaging, foils, glazing, materials, photographic film

[a]Rockwell scales: M, ¼-in.-diameter ball, 100-kg major load; R, ½-in.-diameter ball; 60-kg major load.

erties and are highly resistant to heat (Table 12-3). These plastics are also highly resistant to water, oil, alkalies, and acids. They also exhibit very little shrinkage. Thus, they are excellent where high first degree of precision is required. In this group the phenol–formaldehyde resins are probably the most important resins. The phenol–formaldehyde group can be subdivided into four general classes:

1. Cellulose-filled compositions.
2. Mineral-filled compositions.
3. Molding sheet.
4. Impact-resistant materials.

The *cellulose-filled composition* utilizes wood flour as the filler and the final product has high dielectric strength, mechanical strength, and it is very light. Uses have been found in the aerospace and automotive industries.

The *mineral-filled composition* utilizes asbestos as the filler. Thus, articles made of these plastics resist chemicals and heat. Uses are for insulation of high-voltage transmissions.

Molding sheets are made by impregnating paper with phenolic resins. These sheets can be softened by heat and become brittle in the cold weather. These sheets are used for molding applications where the material must be able to give.

In the final group, *impact-resistant materials,* the plastic is made by impregnating paper and fabric fillers, built up in layers. This product is called laminated plastic.

Chemically Setting Plastics

In the chemically setting group three categories of groups exist: harvel resins, cold-molded plastics, and casein. These plastics are generally used where resistance to heat and arcing are of primary importance. Thus, they are utilized for electrical insulating parts.

MANUFACTURE OF ORGANIC PLASTICS

In general, four main steps are required in the manufacturing of articles made from organic plastics:

1. The production of intermediate materials (chemical) from the raw materials of coal, petroleum, and cotton.
2. The manufacture of synthetic resins from the above.
3. The preparation of molding powders, fillers, rods, and sheets from step 2.
4. Molding the articles from the powders, fillers, rods, and sheets.

Methods of Forming and Fabricating Plastics

Most of the methods utilized in the forming and fabrication of plastic have their counterpart in the processing of other materials of engineering. Examples include the following:

1. Casting.
2. Compression molding.
3. Injection molding.
4. Transfer molding.
5. Extruding.
6. Blowing.
7. Laminating.

Casting is the simplest molding method available. Casting utilizes low-cost lead–antimony dies (molds) where the melted resins are poured into open molds and cured. The cast resins are baked in an oven at about 167°F (75°C) for several days, until a permanent set is obtained. Both thermoplastics and thermosetting plastics may be cast.

Compression molding is the most widespread molding operation for thermosetting plastics. Compression molding requires the use of pressure on the molding compound placed in the mold of heated platens of the press. Thermoplastics may be compression-molded, but the mold must be cooled after each molding before the article is removed from the press. This is time-consuming and uneconomical, so other methods are utilized for thermoplastic plastics.

Injection molding is one of the most widely used and most rapid methods of producing articles of intricate shape. Injection molding consists of forcing softened plastic materials into a closed mold maintained at a temperature below the softening point of the compression. This method is widely used for thermoplastic materials. The thermoplastic material is fed into a pressure chamber and a plunger compresses the material and forces it into a heating chamber, where it is progressively heated to a uniform temperature. The thermoplastic material emerges through a nozzle in a thoroughly softened condition and is then forced into a cold mold, where pressure causes it to take the shape of the cavity. The plastic is cooled, the mold is opened, and the part removed.

Transfer molding is used for thermosetting materials when the article is to include delicate metal inserts for any reason. In this method the molding compound is heated under pressure until it is soft enough to flow. The compound is then placed into a cavity in which the part is formed and allowed to cool.

Extruding is basically the opposite of injection molding. Extruding of thermoplastic plastics is a widely used practice for producing rods, tubes, and other cylindrical shapes.

Blowing of thermoplastics is a process by which hollow objects are made. It is accomplished in much the same manner as the blowing of glass.

Laminating is used to produce hard boards or sheets of resin-impregnated papers, wood veneers, and/or fabrics. Laminates are made into stock sizes of flat sheets under

TABLE 12-3 PROPERTIES AND USES OF COMMON PLASTICS (THERMOSETTING)

Property	Phenol–Formaldehyde Resin		Urea–Formaldehyde, α-Cellulose, Molded	Melamine– Formaldehyde, Asbestos, Paper or Fabric Laminate	Polyester, Glass Fiber, Mat, Laminate	Silicone Glass Fabric Laminate	Cold-molded[a] Cement Binder Asbestos-filled	Hard Rubber,[b] No Filler
	Macerated Cotton Fabric or Cord Filler, Molded	Mechanical Grade, No Filler, Cast						
Compression molding pressure (1000 psi)	2.00–8.00	0	2.00–8.00	1.00–1.80	0.01–0.15	1.00–2.00	1.00–10.00	1.20–1.80
Specific gravity	1.34–1.47	1.25–1.30	1.45–1.55	1.75–1.85	1.5–1.8	1.6–1.8	1.6–2.2	1.4
Tensile strength (1000 psi)	2–9	4–7	6–13	6.5–12	10–20	10–25	1.6–2.2	8–10
Elongation (% in 2 in.)	0.4–0.6	Very small	0.5–1.0	Very small	Very small	Very small	Very small	5–7.5
Modulus of elasticity in tension (100,000 psi)	9–13	5–7	12–15	16–39	10–19	20	—	3.0
Compressive strength (1000 psi)	15–30	15–20	25–35	27–50	30–50	35–46	16	8–12
Impact strength, Izod test on ½ × ½ in. notched bar (ft-lb/in. width of notch)	1–8	0.3–0.4	0.24–0.36	0.7–5.0	11–25	5–22	0.4	0.5
Hardness, Rockwell[c]	M110–120	M70–110	M115–120	M110–115	M90–100	M100	M75–95	HR95
Highest usable temperature, continuous (°F)	250	250	170	225–245	300–400	400–480	900–1300	—

Thermal conductivity [10^{-4} cal/(sec) (cm³) (°C)]	4–7	3–5	7–10	10–17	8–12	3.5	—	2.9
Thermal expansion (10^{-5} in./in. °C)	1–4	8–11	2.5–4.5	2.0–4.8	1.0–3.0	0.5	—	7.7
Dielectric strength, short time ⅛ in. thickness (V/mil)	200–400	—	300–400	40–150	250–400	200–480	45	470
Water absorption, 24 hr, ⅛ in. thick (%)	0.04–1.8	0.2–0.4	0.4–0.8	1–5	0.3–1.0	0.2–0.7	0.5–15	0.02
Effect of strong acids	Decomposed by oxidizing acids	Decomposed by oxidizing acids	Decomposed	Decomposed	Some attack	Very slight	Decomposed	Attacked by oxidizing acids
Color possibilities	Limited	Limited	Unlimited	Limited	Unlimited	Limited	Gray and black	Limited
Common uses	Serving trays, radio cabinets, electrical parts	Punches and dies	Tablewear, electrical controls, housings	Aircraft structural parts, high-strength electrical parts, stove switches	Aviation and automotive structures, decorative applications	High-temperature-resisting electrical insulation	Arc-shield terminal insulators, electric heater elements	Beakers, funnels, etc., for chemicals, combs

[a] The cement binder is not strictly a thermosetting material but is set by chemical combination with water from steam (hydration).

[b] Hard rubber is not usually classified as thermosetting but as vulcanizing.

[c] Rockwell scales: M, ¼-in.-diameter ball, 100-kg major load; HR (hard rubber), ¼-in.-diameter ball, 60-kg major load.

TABLE 12-4 MECHANICAL PROPERTIES OF MOLDED COMPOSITIONS

Type of Material	Tensil Strength (psi)	Impact Strength Notched Bar, (ft-lb./in.)	Modulus of Rupture (psi)	Modulus of Elasticity (psi)	Specific Gravity	Heat Distortion (°C)	Hardness	Coefficient of Linear Expansion per °C
Phenolic laminated, paper base	6,000–13,000	0.8–2.4C[a]	13,000–20,000	1.0–2.0×10^6	1.34–1.55	100–140	85–125R[b]	20–50×10^{-6}
Phenolic laminated, canvas base	8,000–12,000	1.6–10.4C	12,000–19,000	1.0–2.0×10^6	1.34–1.55	100–140	95–115R	30–70×10^{-6}
Phenolic molded, woodflour filled	6,000–11,000	0.26–0.50C	8,000–15,000	0.8–1.5×10^6	1.25–1.52	120–140	95–120R	35–80×10^{-6}
Phenolic molded, cellulose filled	6,000–11,000	0.40–0.80C	8,000–15,000	0.8–1.5×10^6	1.32–1.48	120–140	95–115R	35–80×10^{-6}
Phenolid molded, fabric filled	6,000–8,000	0.80–6.0C	8,000–13,000	0.8–1.5×10^6	1.35–1.40	120–140	90–115R	35–80×10^{-6}
Phenolic molded, mineral filled	5,000–9,000	0.26–1.0C	8,000–18,000	1.0–5.0×10^6	1.70–2.05	120–150	100–120R	25–50×10^{-6}
Cast phenolics	3,000–7,000	0.30–0.50C	3,000–14,000	0.25–0.75×10^6	1.26–1.70	40–80	70–110R	70–160×10^{-6}
Phenolic molded, transparent	8,000	—	1,600		1.27	107	—	—
Furfuryl-phenol molded, woodflour filled	5,000–12,000	1.0–6.5I[c]	10,000–16,000	—	1.3–1.4	131	35–40B[d]	30×10^{-6}
Furfuryl-phenol molded, asbestos filled	4,000–12,000	1.0–6.0I	8,000–14,000	—	1.6–2.0	136	44–46B	20×10^{-6}

276

Furfuryl-phenol molded, fabric filled	5,000–10,000	20–39I	10,000–16,000	—	1.3–1.4	—	30–35B	—
Furfuryl-phenol laminated, paper base	10,000–20,000	5.0–20I	20,000–30,000	—	1.3–1.4	—	—	—
Furfuryl-phenol laminated, cloth base	9,000–12,000	10.0–50I	—	—	1.3–1.4	—	—	—
Urea molded	9,000–12,000	0.28–0.36C	10,000–14,000	$1.2–1.9 \times 10^6$	1.45–1.55	95–130	110–125R	$65–75 \times 10^{-6}$
Polystyrene molded	5,000–7,000	0.40–0.60C	6,000–8,000	$0.40–0.60 \times 10^6$	1.05–1.07	75–80	82–92R	80×10^{-6}
Acrylate molded	4,000–8,000	0.3–4.0C	9,000–16,000	$0.4–0.6 \times 10^6$	1.18–1.19	51–60	18–20B	
Methyl methacrylate molded	9,000–12,000	0.2–3.0C	12,000–14,000	—	1.18–1.20	60–135	17–20B	$70–90 \times 10^{-6}$
Cellulose nitrate molded	4,900–8,500	10–11.5I	5,000–8,000	$0.2–0.4 \times 10^6$	1.35–1.60	71–91	7–12B	$120–160 \times 10^{-6}$
Cellulose acetate molded	4,000–5,000	1.7–3.0C	5,000–7,000	$0.20–0.40 \times 10^6$	1.27–1.63	50–80	85–120R	$140–160 \times 10^{-6}$
Ethyl cellulose	5,000–9,000	3.2–9.6I	—	0.3×10^6	1.14	100–130	—	—
Shellac	900–2,000	—	—	—	1.1–2.7	66–90	—	—
Cold-molded composition	700–1,700	1.5–4.5	3,500–7,800	—	1.9–2.12	182–260	—	—
Hard rubber	1,500–10,000	0.5I	—	0.33×10^6	1.12–1.80	—	31B	80×10^{-6}

[a] C, Charpy.
[b] R, Rockwell.
[c] I, Izod.
[d] B, Brinell.

TABLE 12-5 ELECTRICAL PROPERTIES OF MOLDED COMPOSITIONS

Type of Material	Power Factor			Dielectric Constant		
	60 cycles/sec	1000 cycles/sec	10^6 cycles/sec	60 cycles/sec	1000 cycles/sec	10^6 cycles/sec
Phenolic laminated, paper base	0.02–0.15	0.02–0.10	0.02–0.06	4.5–6.5	4.5–6.0	4.0–5.5
Phenolic laminated, canvas base	0.05–0.30	0.05–0.20	0.04–0.10	5.0–9.0	4.5–8.0	4.0–6.0
Phenolic molded, woodflour filled	0.02–0.30	0.02–0.15	0.035–0.08	4.5–10.0	4.5–10.0	4.0–8.0
Phenolic molded, cellulose filled	0.05–0.30	0.04–0.15	0.04–0.10	4.5–10.0	4.5–10.0	4.0–8.0
Phenolic molded, fabric filled	0.05–0.30	0.04–0.15	0.04–0.10	4.5–15	4.5–15	4.0–10
Phenolic molded, mineral filled	0.020–0.50	0.01–0.20	0.005–0.10	4.5–20	4.5–15	4.0–10
Cast phenolics	0.070–0.50	0.030–0.30	0.04–0.13	7.0–30	6.0–25	5.5–15
Phenolic molded, transparent	0.06	0.04	0.019	5.5	5.0	4.5
Furfuryl-phenol molded, woodflour filled	—	0.04–0.15	0.01–0.06	—	4.0–8.0	6–7.5
Furfuryl-phenol molded, asbestos filled	—	0.1–0.15	0.06–0.15	—	4.5–20	5–18
Furfuryl-phenol molded, fabric filled	—	0.08–0.2	0.05–0.08	—	4.5–6	6
Furfuryl-phenol laminated, paper base	—	—	0.15–0.48	—	4.5	—
Furfuryl-phenol laminated, cloth base	—	—	0.41–0.64	—	4.5	—
Urea molded	0.050–0.13	0.035–0.07	0.035–0.040	8.0–10	8.0–9.0	6.9–7.5
Polystyrene molded	0.0001–0.0002	0.0001–0.0002	Under 0.0002	2.6	2.6	2.6
Acrylate molded	0.05–0.06	—	—	3.4–3.6	—	—
Methyl methacrylate, molded	0.06–0.08	—	0.02	4.0–4.4	—	2.8
Cellulose nitrate, molded	0.062–0.149	—	0.07–0.09	6.7–7.3	—	6.2
Cellulose acetate, molded	0.03–0.05	0.035–0.07	0.035–0.07	4.5–7.0	4.5–6.5	4.0–4.5
Ethyl cellulose, molded	0.03–0.06	0.025	—	2.6–2.9	3.9	—
Shellac, molded filled	0.004–0.018P[a]	—	—	3–4	—	—
Cold-molded composition	0.2	—	0.07	15.0	—	6.0
Hard rubber	—	—	0.003–0.008	2.8	2.9–3.0	3.0

[a]P, Paper filled.

Type of Material	Loss Factor			Resistivity (megohm-cm)	Dielectric Strength Step Test (V/mil)
	60 cycles/sec	1000 cycles/sec	10^6 cycles/sec		
Phenolic laminated, paper base	0.1–1.0	0.1–0.6	0.10–0.30	$2 \times 10^6 - 1 \times 10^8$	400–600
Phenolic laminated, canvas base	0.25–2.0	0.15–1.0	0.15–1.0	$3 \times 10^5 - 4 \times 10^7$	250–400
Phenolic molded, woodflour filled	0.10–3.0	0.20–1.5	0.15–0.80	$10^4 - 10^6$	200–300
Phenolic molded, cellulose filled	0.25–3.0	0.20–1.5	0.15–1.0	$10^4 - 10^5$	200–300
Phenolic molded, fabric filled	0.25–4.5	0.20–2.0	0.15–1.0	$10^3 - 10^5$	200–300
Phenolic molded, mineral filled	0.10–4.0	0.045–3.0	0.020–1.0	$10^3 - 10^8$	200–375
Cast phenolics	0.70–16.0	0.20–4.0	0.20–2.0	$10^4 - 10^7$	120–300
Phenolic molded, transparent	—	—	—	7.5×10^5	325
Furfuryl-phenol molded, woodflour filled	—	—	—	$10^{10} - 10^{12}$	400–600
Furfuryl-phenol molded, asbestos filled	—	—	—	$10^9 - 10^{11}$	200–500
Furfuryl-phenol molded, fabric filled	—	—	—	—	200–500
Furfuryl-phenol laminated, paper base	—	—	—	10^{10}	900–1800
Furfury-phenol laminated, cloth base	—	—	—	—	300–700
Urea molded	0.40–1.2	0.28–0.65	0.24–0.30	$1 - 10 \times 10^6$	275–325
Polystyrene molded	0.00026–0.00053	0.00026–0.00053	Under 0.0005	Over 10^{10}	500–525
Acrylate molded	—	—	—	1.0×10^{15}	500
Methyl methacrylate, molded	—	—	—	1.0×10^{15}	480
Cellulose nitrate, molded	—	—	—	$2 - 30 \times 10^{10}$	660–780
Cellulose acetate, molded	0.13–0.30	1.15–0.40	0.15–0.40	$10^6 - 10^8$	300–350
Ethyl cellulose, molded	—	—	—	—	1500
Shellac, molded filled	0.016–0.16P	—	—	$10^8 - 10^{16}$	200–600
Cold-molded composition	—	—	—	1.3×10^{12}	85–100
Hard rubber	—	—	—	$10^{12} - 10^{15}$	250–1000

279

TABLE 12-6 STRENGTH-WEIGHT RELATIONSHIPS FOR PLASTIC AND SOME COMMON STRUCTURAL MATERIALS[a]

Material	Specific Gravity	Tensile Strength (psi)	Tensile Strength / Specific Gravity	Tensile Strength / (Specific Gravity)^2
Chrome-vanadium steel	7.85	164,000	20,800	2,700
Structural steel	7.83	65,000	8,300	1,060
Cast iron, gray	7.0	35,000	5,000	720
Titanium alloy	4.7	145,000	31,000	6,600
Aluminum alloy	2.8	56,000	20,000	7,200
Magnesium alloy	1.81	44,000	24,300	13,400
Glass fabric laminate	1.9	45,000	23,600	12,500
Asbestos cloth laminate	1.7	9,000	5,300	3,100
Paper laminate	1.33	20,000	15,000	11,300
Cellulose acetate	1.3	5,000	3,900	3,000
Methyl methacrylate	1.18	8,500	7,200	6,100
Polystrene	1.06	5,500	5,200	4,900
Polyethylene	0.92	1,300	1,400	1,500
Sitka spruce	0.40	17,000	42,500	106,000

[a]The values given here are average values. Different compositions, heat treatments, etc., of a given material may differ widely from the values shown here.

medium to high pressure in a press. The process involves building up individual resin-coated fabric or veneer over a form, inserting the assembly into a large rubber bag (called bag-molding process), withdrawing the air from the bag, and subjecting the bag and contents to pressure in an autoclave.

PROPERTIES OF ORGANIC PLASTICS

Table 12-4 shows the mechanical properties of molded compositions. Table 12-5 shows the electrical properties of molded compositions which most plastics are used for. Table 12-6 illustrates the strength–weight relationships for plastics and some common structural materials.

PROBLEMS

12.1. List seven outstanding characteristics of plastics.

12.2. Define organic plastic.

12.3. List the three general classifications of organic plastics.

12.4. Describe a thermoplastic plastic and list several.

12.5. Describe a thermosetting plastic and list several.

12.6. Describe a chemically setting plastic and list several.

12.7. Explain the difference between polymerization and condensation.

12.8. Describe the procedure for manufacturing organic plastics.

12.9. List and describe the methods of forming and fabricating plastics.

12.10. List the mechanical properties of organic plastics and discuss how they compare to other materials of construction.

REFERENCES

American Society for Testing Materials, *Symposium on Plastics*, Philadelphia, 1938.

GILMORE, G., and SPENCER, R., "Injection Molding Process," *Modern Plastics*, No. 27 (Apr. 1950), p. 143.

MACTAGGORT, E. F., and CHAMBERS, H. H., *Plastic and Building*, Sir Isaac Pitman & Sons Ltd., London, 1951.

MILLS, A. P., HAYWARD, H. W., and RADER, L. F., *Materials of Construction*, John Wiley & Sons, Inc., New York, 1955.

MOORE, H. F., and MOORE, M. B., *Materials of Engineering*, McGraw-Hill Book Company, New York, 1953.

Appendices

Ⱥ§ṬM Designation: D 422 – 63 (Reapproved 1972)[ε1]

Standard Method for
PARTICLE-SIZE ANALYSIS OF SOILS[1]

This standard is issued under the fixed designation D 422; the number immediately following the designation indicates the year of original adoption or, in the case of revision, the year of last revision. A number in parentheses indicates the year of last reapproval. A superscript epsilon (ε) indicates an editorial change since the last revision or reapproval.

[ε1] NOTE—Section 2 was added editorially and subsequent sections renumbered in July 1984.

1. Scope

1.1 This method covers the quantitative determination of the distribution of particle sizes in soils. The distribution of particle sizes larger than 75 µm (retained on the No. 200 sieve) is determined by sieving, while the distribution of particle sizes smaller than 75 µm is determined by a sedimentation process, using a hydrometer to secure the necessary data (Notes 1 and 2).

NOTE 1—Separation may be made on the No. 4 (4.75-mm), No. 40 (425-µm), or No. 200 (75-µm) sieve instead of the No. 10. For whatever sieve used, the size shall be indicated in the report.

NOTE 2—Two types of dispersion devices are provided: (*1*) a high-speed mechanical stirrer, and (*2*) air dispersion. Extensive investigations indicate that air-dispersion devices produce a more positive dispersion of plastic soils below the 20-µm size and appreciably less degradation on all sizes when used with sandy soils. Because of the definite advantages favoring air dispersion, its use is recommended. The results from the two types of devices differ in magnitude, depending upon soil type, leading to marked differences in particle size distribution, especially for sizes finer than 20 µm.

2. Applicable Documents

2.1 *ASTM Standards:*

D 421 Practice for Dry Preparation of Soil Samples for Particle-Size Analysis and Determination of Soil Constants[2]

E 11 Specification for Wire-Cloth Sieves for Testing Purposes[3]

E 100 Specification for ASTM Hydrometers[4]

3. Apparatus

3.1 *Balances*—A balance sensitive to 0.01 g for weighing the material passing a No. 10 (2.00-mm) sieve, and a balance sensitive to 0.1 % of the mass of the sample to be weighed for weighing the material retained on a No. 10 sieve.

3.2 *Stirring Apparatus*—Either apparatus A or B may be used.

3.2.1 Apparatus A shall consist of a mechanically operated stirring device in which a suitably mounted electric motor turns a vertical shaft at a speed of not less than 10 000 rpm without load. The shaft shall be equipped with a replaceable stirring paddle made of metal, plastic, or hard rubber, as shown in Fig. 1. The shaft shall be of such length that the stirring paddle will operate not less than ¾ in. (19.0 mm) nor more than 1½ in. (38.1 mm) above the bottom of the dispersion cup. A special dispersion cup conforming to either of the designs shown in Fig. 2 shall be provided to hold the sample while it is being dispersed.

3.2.2 Apparatus B shall consist of an air-jet dispersion cup[5] (Note 3) conforming to the general details shown in Fig. 3 (Notes 4 and 5).

NOTE 3—The amount of air required by an air-jet dispersion cup is of the order of 2 ft³/min; some small air compressors are not capable of supplying sufficient air to operate a cup.

NOTE 4—Another air-type dispersion device, known as a dispersion tube, developed by Chu and Davidson at Iowa State College, has been shown to give

[1] This method is under the jurisdiction of ASTM Committee D-18 on Soil and Rock and is the direct responsibility of Subcommittee D18.03 on Texture, Plasticity, and Density Characteristics of Soils.

Current edition approved Nov. 21, 1963. Originally published 1935. Replaces D 422 – 62.

[2] *Annual Book of ASTM Standards*, Vol 04.08.

[3] *Annual Book of ASTM Standards*, Vol 14.02.

[4] *Annual Book of ASTM Standards*, Vol 14.01.

[5] Detailed working drawings for this cup are available at a nominal cost from the American Society for Testing and Materials, 1916 Race St., Philadelphia, PA 19103. Order Adjunct No. 12-404220-00.

results equivalent to those secured by the air-jet dispersion cups. When it is used, soaking of the sample can be done in the sedimentation cylinder, thus eliminating the need for transferring the slurry. When the air-dispersion tube is used, it shall be so indicated in the report.

NOTE 5—Water may condense in air lines when not in use. This water must be removed, either by using a water trap on the air line, or by blowing the water out of the line before using any of the air for dispersion purposes.

3.3 *Hydrometer*—An ASTM hydrometer, graduated to read in either specific gravity of the suspension or grams per litre of suspension, and conforming to the requirements for hydrometers 151H or 152H in Specifications E 100. Dimensions of both hydrometers are the same, the scale being the only item of difference.

3.4 *Sedimentation Cylinder*—A glass cylinder essentially 18 in. (457 mm) in height and 2½ in. (63.5 mm) in diameter, and marked for a volume of 1000 mL. The inside diameter shall be such that the 1000-mL mark is 36 ± 2 cm from the bottom on the inside.

3.5 *Thermometer*—A thermometer accurate to 1°F (0.5°C).

3.6 *Sieves*—A series of sieves, of square-mesh woven-wire cloth, conforming to the requirements of Specification E 11. A full set of sieves includes the following (Note 6):

3-in. (75-mm)	No. 10 (2.00-mm)
2-in. (50-mm)	No. 20 (850-μm)
1½-in. (37.5-mm)	No. 40 (425-μm)
1-in. (25.0-mm)	No. 60 (250-μm)
¾-in. (19.0-mm)	No. 140 (106-μm)
⅜-in. (9.5-mm)	No. 200 (75-μm)
No. 4 (4.75-mm)	

NOTE 6—A set of sieves giving uniform spacing of points for the graph, as required in Section 17, may be used if desired. This set consists of the following sieves:

3-in. (75-mm)	No. 16 (1.18-mm)
1½-in. (37.5-mm)	No. 30 (600-μm)
¾-in. (19.0-mm)	No. 50 (300-μm)
⅜-in. (9.5-mm)	No. 100 (150-μm)
No. 4 (4.75-mm)	No. 200 (75-μm)
No. 8 (2.36-mm)	

3.7 *Water Bath or Constant-Temperature Room*—A water bath or constant-temperature room for maintaining the soil suspension at a constant temperature during the hydrometer analysis. A satisfactory water tank is an insulated tank that maintains the temperature of the suspension at a convenient constant temperature at or near 68°F (20°C). Such a device is illustrated in Fig. 4. In cases where the work is performed in a room at an automatically controlled constant temperature, the water bath is not necessary.

3.8 *Beaker*—A beaker of 250-mL capacity.

3.9 *Timing Device*—A watch or clock with a second hand.

4. Dispersing Agent

4.1 A solution of sodium hexametaphosphate (sometimes called sodium metaphosphate) shall be used in distilled or demineralized water, at the rate of 40 g of sodium hexametaphosphate/litre of solution (Note 7).

NOTE 7—Solutions of this salt, if acidic, slowly revert or hydrolyze back to the orthophosphate form with a resultant decrease in dispersive action. Solutions should be prepared frequently (at least once a month) or adjusted to pH of 8 or 9 by means of sodium carbonate. Bottles containing solutions should have the date of preparation marked on them.

4.2 All water used shall be either distilled or demineralized water. The water for a hydrometer test shall be brought to the temperature that is expected to prevail during the hydrometer test. For example, if the sedimentation cylinder is to be placed in the water bath, the distilled or demineralized water to be used shall be brought to the temperature of the controlled water bath; or, if the sedimentation cylinder is used in a room with controlled temperature, the water for the test shall be at the temperature of the room. The basic temperature for the hydrometer test is 68°F (20°C). Small variations of temperature do not introduce differences that are of practical significance and do not prevent the use of corrections derived as prescribed.

5. Test Sample

5.1 Prepare the test sample for mechanical analysis as outlined in Practice D 421. During the preparation procedure the sample is divided into two portions. One portion contains only particles retained on the No. 10 (2.00-mm) sieve while the other portion contains only particles passing the No. 10 sieve. The mass of air-dried soil selected for purpose of tests, as prescribed in Practice D 421, shall be sufficient to yield quantities for mechanical analysis as follows:

5.1.1 The size of the portion retained on the No. 10 sieve shall depend on the maximum size of particle, according to the following schedule:

Nominal Diameter of Largest Particles, in. (mm)	Approximate Minimum Mass of Portion, g
⅜ (9.5)	500
¾ (19.0)	1000

Nominal Diameter of Largest Particles, in. (mm)	Approximate Minimum Mass of Portion, g
1 (25.4)	2000
1½ (38.1)	3000
2 (50.8)	4000
3 (76.2)	5000

5.1.2 The size of the portion passing the No. 10 sieve shall be approximately 115 g for sandy soils and approximately 65 g for silt and clay soils.

5.2 Provision is made in Section 5 of Practice D 421 for weighing of the air-dry soil selected for purpose of tests, the separation of the soil on the No. 10 sieve by dry-sieving and washing, and the weighing of the washed and dried fraction retained on the No. 10 sieve. From these two masses the percentages retained and passing the No. 10 sieve can be calculated in accordance with 12.1.

NOTE 8—A check on the mass values and the thoroughness of pulverization of the clods may be secured by weighing the portion passing the No. 10 sieve and adding this value to the mass of the washed and oven-dried portion retained on the No. 10 sieve.

SIEVE ANALYSIS OF PORTION RETAINED ON NO. 10 (2.00-mm) SIEVE

6. Procedure

6.1 Separate the portion retained on the No. 10 (2.00-mm) sieve into a series of fractions using the 3-in. (75-mm), 2-in. (50-mm), 1½-in. (37.5-mm), 1-in. (25.0-mm), ¾-in. (19.0-mm), ⅜-in. (9.5-mm), No. 4 (4.75-mm), and No. 10 sieves, or as many as may be needed depending on the sample, or upon the specifications for the material under test.

6.2 Conduct the sieving operation by means of a lateral and vertical motion of the sieve, accompanied by a jarring action in order to keep the sample moving continuously over the surface of the sieve. In no case turn or manipulate fragments in the sample through the sieve by hand. Continue sieving until not more than 1 mass % of the residue on a sieve passes that sieve during 1 min of sieving. When mechanical sieving is used, test the thoroughness of sieving by using the hand method of sieving as described above.

6.3 Determine the mass of each fraction on a balance conforming to the requirements of 3.1. At the end of weighing, the sum of the masses retained on all the sieves used should equal closely the original mass of the quantity sieved.

HYDROMETER AND SIEVE ANALYSIS OF PORTION PASSING THE NO. 10 (2.00-mm) SIEVE

7. Determination of Composite Correction for Hydrometer Reading

7.1 Equations for percentages of soil remaining in suspension, as given in 14.3, are based on the use of distilled or demineralized water. A dispersing agent is used in the water, however, and the specific gravity of the resulting liquid is appreciably greater than that of distilled or demineralized water.

7.1.1 Both soil hydrometers are calibrated at 68°F (20°C), and variations in temperature from this standard temperature produce inaccuracies in the actual hydrometer readings. The amount of the inaccuracy increases as the variation from the standard temperature increases.

7.1.2 Hydrometers are graduated by the manufacturer to be read at the bottom of the meniscus formed by the liquid on the stem. Since it is not possible to secure readings of soil suspensions at the bottom of the meniscus, readings must be taken at the top and a correction applied.

7.1.3 The net amount of the corrections for the three items enumerated is designated as the composite correction, and may be determined experimentally.

7.2 For convenience, a graph or table of composite corrections for a series of 1° temperature differences for the range of expected test temperatures may be prepared and used as needed. Measurement of the composite corrections may be made at two temperatures spanning the range of expected test temperatures, and corrections for the intermediate temperatures calculated assuming a straight-line relationship between the two observed values.

7.3 Prepare 1000 mL of liquid composed of distilled or demineralized water and dispersing agent in the same proportion as will prevail in the sedimentation (hydrometer) test. Place the liquid in a sedimentation cyclinder and the cylinder in the constant-temperature water bath, set for one of the two temperatures to be used. When the temperature of the liquid becomes constant, insert the hydrometer, and, after a short interval to permit the hydrometer to come to the temperature of the liquid, read the hydrometer at the top of the meniscus formed on the stem. For hydrometer 151H the composite correction is the difference between this reading and one; for hydrometer 152H it is the difference between the reading and zero. Bring the liquid and the hydrometer to the other temperature to be used, and secure the composite correction as before.

8. Hygroscopic Moisture

8.1 When the sample is weighed for the hydrometer test, weigh out an auxiliary portion of from 10 to 15 g in a small metal or glass container, dry the sample to a constant mass in an oven at 230 ± 9°F (110 ± 5°C), and weigh again. Record the masses.

9. Dispersion of Soil Sample

9.1 When the soil is mostly of the clay and silt sizes, weigh out a sample of air-dry soil of approximately 50 g. When the soil is mostly sand the sample should be approximately 100 g.

9.2 Place the sample in the 250-mL beaker and cover with 125 mL of sodium hexametaphosphate solution (40 g/L). Stir until the soil is thoroughly wetted. Allow to soak for at least 16 h.

9.3 At the end of the soaking period, disperse the sample further, using either stirring apparatus A or B. If stirring apparatus A is used, transfer the soil - water slurry from the beaker into the special dispersion cup shown in Fig. 2, washing any residue from the beaker into the cup with distilled or demineralized water (Note 9). Add distilled or demineralized water, if necessary, so that the cup is more than half full. Stir for a period of 1 min.

NOTE 9—A large size syringe is a convenient device for handling the water in the washing operation. Other devices include the wash-water bottle and a hose with nozzle connected to a pressurized distilled water tank.

9.4 If stirring apparatus B (Fig. 3) is used, remove the cover cap and connect the cup to a compressed air supply by means of a rubber hose. A air gage must be on the line between the cup and the control valve. Open the control valve so that the gage indicates 1 psi (7 kPa) pressure (Note 10). Transfer the soil - water slurry from the beaker to the air-jet dispersion cup by washing with distilled or demineralized water. Add distilled or demineralized water, if necessary, so that the total volume in the cup is 250 mL, but no more.

NOTE 10—The initial air pressure of 1 psi is required to prevent the soil - water mixture from entering the air-jet chamber when the mixture is transferred to the dispersion cup.

9.5 Place the cover cap on the cup and open the air control valve until the gage pressure is 20 psi (140 kPa). Disperse the soil according to the following schedule:

Plasticity Index	Dispersion Period, min
Under 5	5
6 to 20	10
Over 20	15

Soils containing large percentages of mica need be dispersed for only 1 min. After the dispersion period, reduce the gage pressure to 1 psi preparatory to transfer of soil - water slurry to the sedimentation cylinder.

10. Hydrometer Test

10.1 Immediately after dispersion, transfer the soil - water slurry to the glass sedimentation cylinder, and add distilled or demineralized water until the total volume is 1000 mL.

10.2 Using the palm of the hand over the open end of the cylinder (or a rubber stopper in the open end), turn the cylinder upside down and back for a period of 1 min to complete the agitation of the slurry (Note 11). At the end of 1 min set the cylinder in a convenient location and take hydrometer readings at the following intervals of time (measured from the beginning of sedimentation), or as many as may be needed, depending on the sample or the specification for the material under test: 2, 5, 15, 30, 60, 250, and 1440 min. If the controlled water bath is used, the sedimentation cylinder should be placed in the bath between the 2- and 5-min readings.

NOTE 11—The number of turns during this minute should be approximately 60, counting the turn upside down and back as two turns. Any soil remaining in the bottom of the cylinder during the first few turns should be loosened by vigorous shaking of the cylinder while it is in the inverted position.

10.3 When it is desired to take a hydrometer reading, carefully insert the hydrometer about 20 to 25 s before the reading is due to approximately the depth it will have when the reading is taken. As soon as the reading is taken, carefully remove the hydrometer and place it with a spinning motion in a graduate of clean distilled or demineralized water.

NOTE 12—It is important to remove the hydrometer immediately after each reading. Readings shall be taken at the top of the meniscus formed by the suspension around the stem, since it is not possible to secure readings at the bottom of the meniscus.

10.4 After each reading, take the temperature of the suspension by inserting the thermometer into the suspension.

11. Sieve Analysis

11.1 After taking the final hydrometer reading, transfer the suspension to a No. 200 (75-µm) sieve and wash with tap water until the wash water is clear. Transfer the material on the No. 200 sieve to a suitable container, dry in an oven at 230 ± 9°F (110 ± 5°C) and make a sieve analysis of the portion retained, using as many sieves as desired, or required for the material, or upon the specification of the material under test.

CALCULATIONS AND REPORT

12. Sieve Analysis Values for the Portion Coarser than the No. 10 (2.00-mm) Sieve

12.1 Calculate the percentage passing the No. 10 sieve by dividing the mass passing the No. 10 sieve by the mass of soil originally split on the No. 10 sieve, and multiplying the result by 100. To obtain the mass passing the No. 10 sieve, subtract the mass retained on the No. 10 sieve from the original mass.

12.2 To secure the total mass of soil passing the No. 4 (4.75-mm) sieve, add to the mass of the material passing the No. 10 sieve the mass of the fraction passing the No. 4 sieve and retained on the No. 10 sieve. To secure the total mass of soil passing the ⅜-in. (9.5-mm) sieve, add to the total mass of soil passing the No. 4 sieve, the mass of the fraction passing the ⅜-in. sieve and retained on the No. 4 sieve. For the remaining sieves, continue the calculations in the same manner.

12.3 To determine the total percentage passing for each sieve, divide the total mass passing (see 12.2) by the total mass of sample and multiply the result by 100.

13. Hygroscopic Moisture Correction Factor

13.1 The hydroscopic moisture correction factor is the ratio between the mass of the oven-dried sample and the air-dry mass before drying. It is a number less than one, except when there is no hygroscopic moisture.

14. Percentages of Soil in Suspension

14.1 Calculate the oven-dry mass of soil used in the hydrometer analysis by multiplying the air-dry mass by the hygroscopic moisture correc-

tion factor.

14.2 Calculate the mass of a total sample represented by the mass of soil used in the hydrometer test, by dividing the oven-dry mass used by the percentage passing the No. 10 (2.00-mm) sieve, and multiplying the result by 100. This value is the weight W in the equation for percentage remaining in suspension.

14.3 The percentage of soil remaining in suspension at the level at which the hydrometer is measuring the density of the suspension may be calculated as follows (Note 13): For hydrometer 151H:

$$P = [(100\,000/W) \times G/(G - G_1)](R - G_1)$$

NOTE 13—The bracketed portion of the equation for hydrometer 151H is constant for a series of readings and may be calculated first and then multiplied by the portion in the parentheses.

For hydrometer 152H:

$$P = (Ra/W) \times 100$$

where:

a = correction faction to be applied to the reading of hydrometer 152H. (Values shown on the scale are computed using a specific gravity of 2.65. Correction factors are given in Table 1),

P = percentage of soil remaining in suspension at the level at which the hydrometer measures the density of the suspension,

R = hydrometer reading with composite correction applied (Section 7),

W = oven-dry mass of soil in a total test sample represented by mass of soil dispersed (see 14.2), g,

G = specific gravity of the soil particles, and

G_1 = specific gravity of the liquid in which soil particles are suspended. Use numerical value of one in both instances in the equation. In the first instance any possible variation produces no significant effect, and in the second instance, the composite correction for R is based on a value of one for G_1.

15. Diameter of Soil Particles

15.1 The diameter of a particle corresponding to the percentage indicated by a given hydrometer reading shall be calculated according to Stokes' law (Note 14), on the basis that a particle of this diameter was at the surface of the suspension at the beginning of sedimentation and had settled to the level at which the hydrometer is measuring the density of the suspension. According to Stokes' law:

$$D = \sqrt{[30n/980(G - G_1)]} \times L/T$$

where:

D = diameter of particle, mm,

n = coefficient of viscosity of the suspending medium (in this case water) in poises (varies with changes in temperature of the suspending medium),

L = distance from the surface of the suspension to the level at which the density of the suspension is being measured, cm. (For a given hydrometer and sedimentation cylinder, values vary according to the hydrometer readings. This distance is known as effective depth (Table 2)),

T = interval of time from beginning of sedimentation to the taking of the reading, min,

G = specific gravity of soil particles, and

G_1 = specific gravity (relative density) of suspending medium (value may be used as 1.000 for all practical purposes).

NOTE 14—Since Stokes' law considers the terminal velocity of a single sphere falling in an infinity of liquid, the sizes calculated represent the diameter of spheres that would fall at the same rate as the soil particles.

15.2 For convenience in calculations the above equation may be written as follows:

$$D = K\sqrt{L/T}$$

where:

K = constant depending on the temperature of the suspension and the specific gravity of the soil particles. Values of K for a range of temperatures and specific gravities are given in Table 3. The value of K does not change for a series of readings constituting a test, while values of L and T do vary.

15.3 Values of D may be computed with sufficient accuracy, using an ordinary 10-in. slide rule.

NOTE 15—The value of L is divided by T using the A- and B-scales, the square root being indicated on the D-scale. Without ascertaining the value of the square root it may be multiplied by K, using either the C- or CI-scale.

16. Sieve Analysis Values for Portion Finer than No. 10 (2.00-mm) Sieve

16.1 Calculation of percentages passing the various sieves used in sieving the portion of the sample from the hydrometer test involves several steps. The first step is to calculate the mass of the fraction that would have been retained on the No. 10 sieve had it not been removed. This mass is equal to the total percentage retained on the No. 10 sieve (100 minus total percentage passing) times the mass of the total sample represented by the mass of soil used (as calculated in 14.2), and the result divided by 100.

16.2 Calculate next the total mass passing the No. 200 sieve. Add together the fractional masses retained on all the sieves, including the No. 10 sieve, and subtract this sum from the mass of the total sample (as calculated in 14.2).

16.3 Calculate next the total masses passing each of the other sieves, in a manner similar to that given in 12.2.

16.4 Calculate last the total percentages passing by dividing the total mass passing (as calculated in 16.3) by the total mass of sample (as calculated in 14.2), and multiply the result by 100.

17. Graph

17.1 When the hydrometer analysis is performed, a graph of the test results shall be made, plotting the diameters of the particles on a logarithmic scale as the abscissa and the percentages smaller than the corresponding diameters to an arithmetic scale as the ordinate. When the hydrometer analysis is not made on a portion of the soil, the preparation of the graph is optional, since values may be secured directly from tabulated data.

18. Report

18.1 The report shall include the following:

18.1.1 Maximum size of particles,

18.1.2 Percentage passing (or retained on) each sieve, which may be tabulated or presented by plotting on a graph (Note 16),

18.1.3 Description of sand and gravel particles:

18.1.3.1 Shape—rounded or angular,

18.1.3.2 Hardness—hard and durable, soft, or weathered and friable,

18.1.4 Specific gravity, if unusually high or low,

18.1.5 Any difficulty in dispersing the fraction passing the No. 10 (2.00-mm) sieve, indicating any change in type and amount of dispersing agent, and

18.1.6 The dispersion device used and the length of the dispersion period.

290

NOTE 16—This tabulation of graph represents the gradation of the sample tested. If particles larger than those contained in the sample were removed before testing, the report shall so state giving the amount and maximum size.

18.2 For materials tested for compliance with definite specifications, the fractions called for in such specifications shall be reported. The fractions smaller than the No. 10 sieve shall be read from the graph.

18.3 For materials for which compliance with definite specifications is not indicated and when the soil is composed almost entirely of particles passing the No. 4 (4.75-mm) sieve, the results read from the graph may be reported as follows:

(1) Gravel, passing 3-in. and retained on
No. 4 sieve ..... %
(2) Sand, passing No. 4 sieve and retained on No. 200 sieve ..... %
  (a) Coarse sand, passing No. 4 sieve and retained on No. 10 sieve ..... %
  (b) Medium sand, passing No. 10 sieve and retained on No. 40 sieve ..... %
  (c) Fine sand, passing No. 40 sieve and retained on No. 200 sieve ..... %
(3) Silt size, 0.074 to 0.005 mm ..... %

(4) Clay size, smaller than 0.005 mm ..... %
Colloids, smaller than 0.001 mm ..... %

18.4 For materials for which compliance with definite specifications is not indicated and when the soil contains material retained on the No. 4 sieve sufficient to require a sieve analysis on that portion, the results may be reported as follows (Note 17):

SIEVE ANALYSIS

Sieve Size	Percentage Passing
3-in.	....
2-in.	....
1½-in.	....
1-in.	....
¾-in.	....
⅜-in.	....
No. 4 (4.75-mm)	....
No. 10 (2.00-mm)	....
No. 40 (425-µm)	....
No. 200 (75-µm)	....

HYDROMETER ANALYSIS

0.074 mm	....
0.005 mm	....
0.001 mm	....

NOTE 17—No. 8 (2.36-mm) and No. 50 (300-µm) sieves may be substituted for No. 10 and No. 40 sieves.

TABLE 1   Values of Correction Factor, α, for Different Specific Gravities of Soil Particles[A]

Specific Gravity	Correction Factor[A]
2.95	0.94
2.90	0.95
2.85	0.96
2.80	0.97
2.75	0.98
2.70	0.99
2.65	1.00
2.60	1.01
2.55	1.02
2.50	1.03
2.45	1.05

[A] For use in equation for percentage of soil remaining in suspension when using Hydrometer 152H.

TABLE 2   Values of Effective Depth Based on Hydrometer and Sedimentation Cylinder of Specified Sizes[A]

Hydrometer 151H		Hydrometer 152H			
Actual Hydrometer Reading	Effective Depth, L, cm	Actual Hydrometer Reading	Effective Depth, L, cm	Actual Hydrometer Reading	Effective Depth, L, cm
1.000	16.3	0	16.3	31	11.2
1.001	16.0	1	16.1	32	11.1
1.002	15.8	2	16.0	33	10.9
1.003	15.5	3	15.8	34	10.7
1.004	15.2	4	15.6	35	10.6
1.005	15.0	5	15.5		
1.006	14.7	6	15.3	36	10.4
1.007	14.4	7	15.2	37	10.2
1.008	14.2	8	15.0	38	10.1
1.009	13.9	9	14.8	39	9.9
1.010	13.7	10	14.7	40	9.7
1.011	13.4	11	14.5	41	9,6
1.012	13.1	12	14.3	42	9.4
1.013	12.9	13	14.2	43	9.2
1.014	12.6	14	14.0	44	9.1
1.015	12.3	15	13.8	45	8.9
1.016	12.1	16	13.7	46	8.8
1.017	11.8	17	13.5	47	8.6
1.018	11.5	18	13.3	48	8.4
1.019	11.3	19	13.2	49	8.3
1.020	11.0	20	13.0	50	8.1
1.021	10.7	21	12.9	51	7.9
1.022	10.5	22	12.7	52	7.8
1.023	10.2	23	12.5	53	7.6
1.024	10.0	24	12.4	54	7.4
1.025	9.7	25	12.2	55	7.3
1.026	9.4	26	12.0	56	7.1
1.027	9.2	27	11.9	57	7.0
1.028	8.9	28	11.7	58	6.8
1.029	8.6	29	11.5	59	6.6
1.030	8.4	30	11.4	60	6.5

TABLE 2   Continued

Hydrometer 151H		Hydrometer 152 H	
Actual Hydrometer Reading	Effective Depth, L, cm	Actual Hydrometer Reading	Effective Depth, L, cm
1.031	8.1		
1.032	7.8		
1.033	7.6		
1.034	7.3		
1.035	7.0		
1.036	6.8		
1.037	6.5		
1.038	6.2		

[A] Values of effective depth are calculated from the equation:

$$L = L_1 + \tfrac{1}{2}\,[L_2 - (V_B/A)]$$

where:
L = effective depth, cm,
L_1 = distance along the stem of the hydrometer from the top of the bulb to the mark for a hydrometer reading, cm,
L_2 = overall length of the hydrometer bulb, cm,
V_B = volume of hydrometer bulb, cm^3, and
A = cross-sectional area of sedimentation cylinder, cm^2
Values used in calculating the values in Table 2 are as follows:
For both hydrometers, 151H and 152H:
L_2 = 14.0 cm
V_B = 67.0 cm^3
A = 27.8 cm^2
For hydrometer 151H:
L_1 = 10.5 cm for a reading of 1.000
 = 2.3 cm for a reading of 1.031
For hydrometer 152H:
L_1 = 10.5 cm for a reading of 0 g/litre
 = 2.3 cm for a reading of 50 g/litre

TABLE 3 Values of *K* for Use in Equation for Computing Diameter of Particle in Hydrometer Analysis

Temperature, °C	Specific Gravity of Soil Particles								
	2.45	2.50	2.55	2.60	2.65	2.70	2.75	2.80	2.85
16	0.01510	0.01505	0.01481	0.01457	0.01435	0.01414	0.01394	0.01374	0.01356
17	0.01511	0.01486	0.01462	0.01439	0.01417	0.01396	0.01376	0.01356	0.01338
18	0.01492	0.01467	0.01443	0.01421	0.01399	0.01378	0.01359	0.01339	0.01321
19	0.01474	0.01449	0.01425	0.01403	0.01382	0.01361	0.01342	0.1323	0.01305
20	0.01456	0.01431	0.01408	0.01386	0.01365	0.01344	0.01325	0.01307	0.01289
21	0.01438	0.01414	0.01391	0.01369	0.01348	0.01328	0.01309	0.01291	0.01273
22	0.01421	0.01397	0.01374	0.01353	0.01332	0.01312	0.01294	0.01276	0.01258
23	0.01404	0.01381	0.01358	0.01337	0.01317	0.01297	0.01279	0.01261	0.01243
24	0.01388	0.01365	0.01342	0.01321	0.01301	0.01282	0.01264	0.01246	0.01229
25	0.01372	0.01349	0.01327	0.01306	0.01286	0.01267	0.01249	0.01232	0.01215
26	0.01357	0.01334	0.01312	0.01291	0.01272	0.01253	0.01235	0.01218	0.01201
27	0.01342	0.01319	0.01297	0.01277	0.01258	0.01239	0.01221	0.01204	0.01188
28	0.01327	0.01304	0.01283	0.01264	0.01244	0.01255	0.01208	0.01191	0.01175
29	0.01312	0.01290	0.01269	0.01249	0.01230	0.01212	0.01195	0.01178	0.01162
30	0.01298	0.01276	0.01256	0.01236	0.01217	0.01199	0.01182	0.01165	0.01149

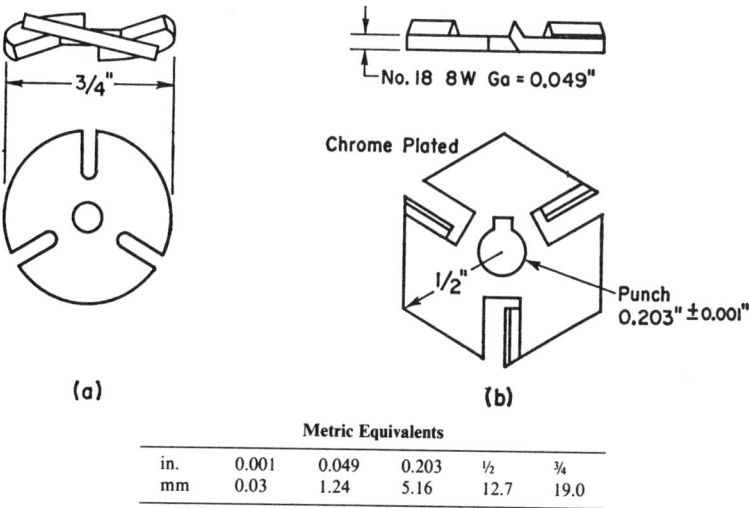

(a)

No. 18 8W Ga = 0.049"

Chrome Plated

1/2"

Punch 0.203" ±0.001"

(b)

Metric Equivalents

in.	0.001	0.049	0.203	½	¾
mm	0.03	1.24	5.16	12.7	19.0

FIG. 1 Detail of Stirring Paddles

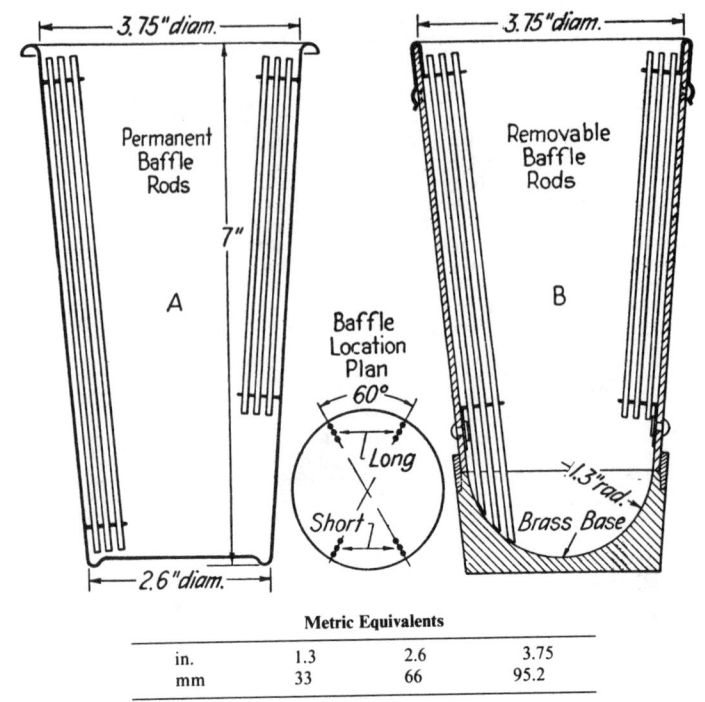

Metric Equivalents

in.	1.3	2.6	3.75
mm	33	66	95.2

FIG. 2   Dispersion Cups of Apparatus

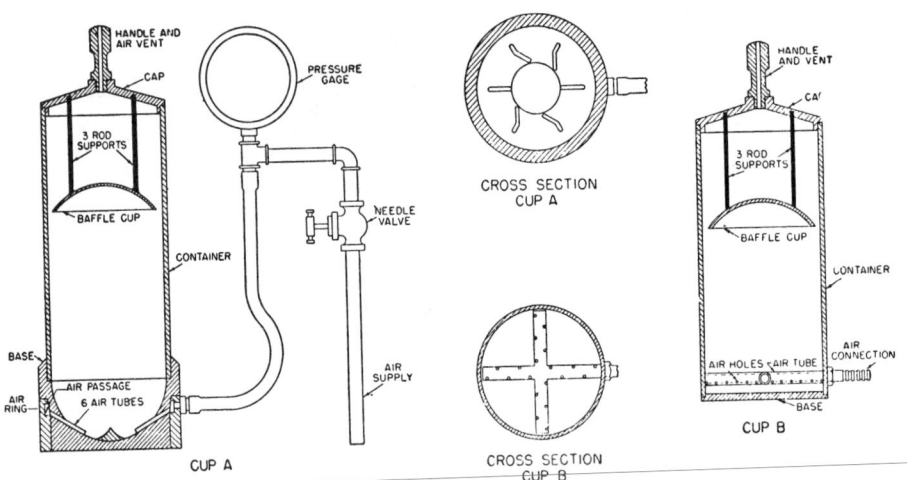

FIG. 3   Air-Jet Dispersion Cups of Apparatus B

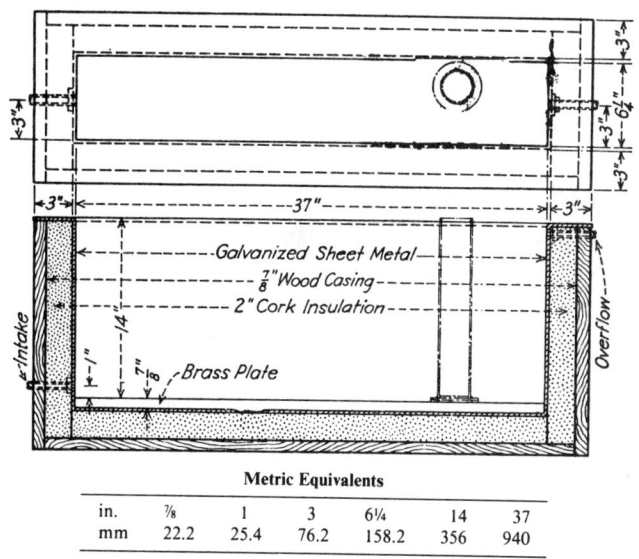

Metric Equivalents

in.	⅞	1	3	6¼	14	37
mm	22.2	25.4	76.2	158.2	356	940

FIG. 4 Insulated Water Bath

Designation: D 4318 – 84

Standard Test Method for
LIQUID LIMIT, PLASTIC LIMIT, AND PLASTICITY INDEX OF SOILS[1]

This standard is issued under the fixed designation D 4318; the number immediately following the designation indicates the year of original adoption or, in the case of revision, the year of last revision. A number in parentheses indicates the year of last reapproval. A superscript epsilon (ε) indicates an editorial change since the last revision or reapproval.

This test method has been approved for use by agencies of the Department of Defense and for listing in the DoD Index of Specifications and Standards.

1. Scope

1.1 This test method covers the determination of the liquid limit, plastic limit, and the plasticity index of soils as defined in Section 3.

1.1.1 Two procedures for preparing test specimens and two procedures for performing the liquid limit are provided as follows:

A   Multipoint test using a wet preparation procedure, described in Sections 10.1, 11, and 12.

B   Multipoint test using a dry preparation procedure, described in Sections 10.2, 11, and 12.

C   One-point test using a wet preparation procedure, described in Sections 13, 14, and 15.

D   One-point test using a dry preparation procedure, described in Sections 13, 14, and 15.

The procedure to be used shall be specified by the requesting authority. If no procedure is specified, Procedure A shall be used.

NOTE 1—Prior to the adoption of this test method, a curved grooving tool was specified as part of the apparatus for performing the liquid limit test. The curved tool is not considered to be as accurate as the flat tool described in 6.2 since it does not control the depth of the soil in the liquid limit cup. However, there are some data which indicate that typically the liquid limit is slightly increased when the flat tool is used instead of the curved tool.

1.1.2 The plastic limit test procedure is described in Sections 16, 17, and 18. The plastic limit test is performed on material prepared for the liquid limit test. In effect, there are two procedures for preparing test specimens for the plastic limit test.

1.1.3 The procedure for calculating the plasticity index is given in Section 19.

1.2 The liquid limit and plastic limit of soils (along with the shrinkage limit) are often collectively referred to as the Atterberg limits in recognition of their formation by Swedish soil scientist, A. Atterberg. These limits distinguish the boundaries of the several consistency states of plastic soils.

1.3 As used in this test method, soil is any natural aggregation of mineral or organic materials, mixtures of such materials, or artificial mixtures of aggregates and natural mineral and organic particles.

1.4 The multipoint liquid limit procedure is somewhat more time consuming than the one-point procedure when both are performed by experienced operators. However, the one-point procedure requires the operator to judge when the test specimen is approximately at its liquid limit. In cases where this is not done reliably, the multipoint procedure is as fast as the one-point procedure and provides additional precision due to the information obtained from additional trials. It is particularly recommended that the multipoint procedure be used by inexperienced operators.

1.5 The correlations on which the calculations of the one-point procedure are based may not be valid for certain soils, such as organic soils or

---

[1] This test method is under the jurisdiction of ASTM Committee D-18 on Soil and Rock and is the direct responsibility of Subcommittee D18.03 on Texture, Plasticity and Density Characteristics of Soils.

Current edition approved Oct. 26, 1984. Published December 1984. Originally published as D 4318 – 83. Last previous edition D 4318 – 83[r1].

soils from a marine environment. The liquid limit of these soils should therefore be determined by the multipoint procedure (Procedure A).

1.6 The liquid and plastic limits of many soils that have been allowed to dry before testing may be considerably different from values obtained on undried samples. If the liquid and plastic limits of soils are used to correlate or estimate the engineering behavior of soils in their natural moist state, samples should not be permitted to dry before testing unless data on dried samples are specifically desired.

1.7 The composition and concentration of soluble salts in a soil affect the values of the liquid and plastic limits as well as the water content values of soils (see Method D 2216). Special consideration should therefore be given to soils from a marine environment or other sources where high soluble salt concentrations may be present. The degree to which the salts present in these soils are diluted or concentrated must be given consideration if meaningful results are to be obtained.

1.8 Since the tests described herein are performed only on that portion of a soil which passes the 425-μm (No. 40) sieve, the relative contribution of this portion of the soil to the properties of the sample as a whole must be considered when using these tests to evaluate the properties of a soil.

1.9 The values stated in acceptable metric units are to be regarded as the standard. The values given in parentheses are for information only.

1.10 *This standard may involve hazardous materials, operations, and equipment. This standard does not purport to address all of the safety problems associated with its use. It is the responsibility of whoever uses this standard to consult and establish appropriate safety and health practices and determine the applicability of regulatory limitations prior to use.*

2. Applicable Documents

2.1 *ASTM Standards:*
C 702 Methods for Reducing Field Samples of Aggregate to Testing Size[2]
D 75 Practice for Sampling Aggregates[3]
D 420 Recommended Practice for Investigating and Sampling Soil and Rock for Engineering Purposes[4]

D 653 Terms and Symbols Relating to Soil and Rock[4]
D 1241 Specification for Materials for Soil-Aggregate Subbase, Base, and Surface Courses[4]
D 2216 Method for Laboratory Determination of Water (Moisture) Content of Soil, Rock, and Soil-Aggregate Mixtures[4]
D 2240 Test Method for Rubber Property—Durometer Hardness[5]
D 2487 Test Method for Classification of Soils for Engineering Purposes[4]
D 2488 Practice for Description and Identification of Soils (Visual-Manual Procedure)[4]
D 3282 Practice for Classification of Soils and Soil-Aggregate Mixtures for Highway Construction Purposes[4]
E 11 Specification for Wire-Cloth Sieves for Testing Purposes[6]
E 319 Methods of Testing Single-Arm Balances[6]
E 898 Method of Testing Top-Loading, Direct-Reading Laboratory Scales and Balances[6]

3. Definitions

3.1 *Atterberg limits*—originally, seven "limits of consistency" of fine-grained soils were defined by Albert Atterberg. In current engineering usage, the term usually refers only to the liquid limit, plastic limit, and in some references, the shrinkage limit.

3.2 *consistency*—the relative ease with which a soil can be deformed.

3.3 *liquid limit (LL)*—the water content, in percent, of a soil at the arbitrarily defined boundary between the liquid and plastic states. This water content is defined as the water content at which a pat of soil placed in a standard cup and cut by a groove of standard dimensions will flow together at the base of the groove for a distance of 13 mm (½ in.) when subjected to 25 shocks from the cup being dropped 10 mm in a standard liquid limit apparatus operated at a rate of 2 shocks per second.

[2] *Annual Book of ASTM Standards*, Vol 04.02.
[3] *Annual Book of ASTM Standards*, Vols 04.02, 04.03, and 04.08.
[4] *Annual Book of ASTM Standards*, Vol 04.08.
[5] *Annual Book of ASTM Standards*, Vol 09.01.
[6] *Annual Book of ASTM Standards*, Vol 14.02.

NOTE 2—The undrained shear strength of soil at the liquid limit is considered to be 2 ±0.2 kPa (0.28 psi).

3.4 *plastic limit (PL)*—the water content, in percent, of a soil at the boundary between the plastic and brittle states. The water content at this boundary is the water content at which a soil can no longer be deformed by rolling into 3.2 mm (⅛ in.) in diameter threads without crumbling.

3.5 *plastic soil*—a soil which has a range of water content over which it exhibits plasticity and which will retain its shape on drying.

3.6 *plasticity index (PI)*—the range of water content over which a soil behaves plastically. Numerically, it is the difference between the liquid limit and the plastic limit.

3.7 *liquidity index*—the ratio, expressed as a percentage, of (*1*) the natural water content of a soil minus its plastic limit, to (*2*) its plasticity index.

3.8 *activity number (A)*—the ratio of (*1*) the plasticity index of a soil to (*2*) the percent by weight of particles having an equivalent diameter smaller than 0.002 mm.

4. Summary of Method

4.1 The sample is processed to remove any material retained on a 425-µm (No. 40) sieve. The liquid limit is determined by performing trials in which a portion of the sample is spread in a brass cup, divided in two by a grooving tool, and then allowed to flow together from the shocks caused by repeatedly dropping the cup in a standard mechanical device. The multipoint liquid limit, Procedures A and B, requires three or more trials over a range of water contents to be performed and the data from the trials plotted or calculated to make a relationship from which the liquid limit is determined. The one-point liquid limit, Procedures C and D, uses the data from two trials at one water content multiplied by a correction factor to determine the liquid limit.

4.2 The plastic limit is determined by alternately pressing together and rolling into a 3.2 mm (⅛ in.) diameter thread a small portion of plastic soil until its water content is reduced to a point at which the thread crumbles and is no longer able to be pressed together and rerolled. The water content of the soil at this stage is reported as the plastic limit.

4.3 The plasticity index is calculated as the difference between the liquid limit and the plastic limit.

5. Significance and Use

5.1 This test method is used as an integral part of several engineering classification systems to characterize the fine-grained fractions of soils (see Test Method D 2487 and Practice D 3282) and to specify the fine-grained fraction of construction materials (see Specification D 1241). The liquid limit, plastic limit, and plasticity index of soils are also used extensively, either individually or together with other soil properties to correlate with engineering behavior such as compressibility, permeability, compactibility, shrink-swell, and shear strength.

5.2 The liquid and plastic limits of a soil can be used with the natural water content of the soil to express its relative consistency or liquidity index and can be used with the percentage finer than 2-µm size to determine its activity number.

5.3 The one-point liquid limit procedure is frequently used for routine classification purposes. When greater precision is required, as when used for the acceptance of a material or for correlation with other test data, the multipoint procedure should be used.

5.4 These methods are sometimes used to evaluate the weathering characteristics of clay-shale materials. When subjected to repeated wetting and drying cycles, the liquid limits of these materials tend to increase. The amount of increase is considered to be a measure of a shale's susceptibility to weathering.

5.5 The liquid limit of a soil containing substantial amounts of organic matter decreases dramatically when the soil is oven-dried before testing. Comparison of the liquid limit of a sample before and after oven-drying can therefore be used as a qualitative measure of organic matter content of a soil.

6. Apparatus

6.1 *Liquid Limit Device*—A mechanical device consisting of a brass cup suspended from a carriage designed to control its drop onto a hard rubber base. A drawing showing the essential features of the device and the critical dimensions is given in Fig. 1. The design of the device may vary provided that the essential functions are preserved. The device may be operated either by a hand crank or by an electric motor.

6.1.1 *Base*—The base shall be hard rubber having a D Durometer hardness of 80 to 90, and a resilience such that an 8-mm (⁵⁄₁₆-in.) diameter polished steel ball, when dropped from a height of 25 cm (9.84 in.) will have an average rebound of at least 80 % but no more than 90 %. The tests shall be conducted on the finished base with feet attached.

6.1.2 *Feet*—The base shall be supported by rubber feet designed to provide isolation of the base from the work surface and having an A Durometer hardness no greater than 60 as measured on the finished feet attached to the base.

6.1.3 *Cup*—The cup shall be brass and have a weight, including cup hanger, of 185 to 215 g.

6.1.4 *Cam*—The cam shall raise the cup smoothly and continuously to its maximum height, over a distance of at least 180° of cam rotation. The preferred cam motion is a uniformly accelerated lift curve. The design of the cam and follower combination shall be such that there is no upward or downward velocity of the cup when the cam follower leaves the cam.

NOTE 3—The cam and follower design in Fig. 1 is for uniformly accelerated (parabolic) motion after contact and assures that the cup has no velocity at drop off. Other cam designs also provide this feature and may be used. However, if the cam-follower lift pattern is not known, zero velocity at drop off can be assured by carefully filing or machining the cam and follower so that the cup height remains constant over the last 20 to 45° of cam rotation.

6.1.5 *Carriage*—The cup carriage shall be constructed in a way that allows convenient but secure adjustment of the height of drop of the cup to 10 mm (0.394 in.). The cup hanger shall be attached to the carriage by means of a pin which allows removal of the cup and cup hanger for cleaning and inspection.

6.1.6 *Optional Motor Drive*—As an alternative to the hand crank shown in Fig. 1, the device may be equipped with a motor to turn the cam. Such a motor must turn the cam at 2 ±0.1 revolutions per second, and must be isolated from the rest of the device by rubber mounts or in some other way that prevents vibration from the motor being transmitted to the rest of the apparatus. It must be equipped with an ON-OFF switch and a means of conveniently positioning the cam for height of drop adjustments. The results obtained using a motor-driven device must not differ from those obtained using a manually operated device.

6.2 *Flat Grooving Tool*—A grooving tool having dimensions shown in Fig. 2. The tool shall be made of plastic or noncorroding metal. The design of the tool may vary as long as the essential dimensions are maintained. The tool may, but need not, incorporate the gage for adjusting the height of drop of the liquid limit device.

6.3 *Gage*—A metal gage block for adjusting the height of drop of the cup, having the dimensions shown in Fig. 3. The design of the tool may vary provided the gage will rest securely on the base without being susceptible to rocking, and the edge which contacts the cup during adjustment is straight, at least 10 mm (³⁄₈ in.) wide, and without bevel or radius.

6.4 *Containers*—Small corrosion-resistant containers with snug-fitting lids for water content specimens. Aluminum or stainless steel cans 2.5 cm (1 in.) high by 5 cm (2 in.) in diameter are appropriate.

6.5 *Balance*—A balance readable to at least 0.01 g and having an accuracy of 0.03 g within three standard deviations within the range of use. Within any 15-g range, a difference between readings shall be accurate within 0.01 g (Notes 4 and 5).

NOTE 4—See Methods E 898 and E 319 for an explanation of terms relating to balance performance.

NOTE 5—For frequent use, a top-loading type balance with automatic load indication, readable to 0.01 g, and having an index of precision (standard deviation) of 0.003 or better is most suitable for this method. However, nonautomatic indicating equal-arm analytical balances and some small equal arm top pan balances having readabilities and sensitivities of 0.002 g or better provide the required accuracy when used with a weight set of ASTM Class 4 (National Bureau of Standards Class P) or better. Ordinary commercial and classroom type balances such as beam balances are not suitable for this method.

6.6 *Storage Container*—A container in which to store the prepared soil specimen that will not contaminate the specimen in any way, and which prevents moisture loss. A porcelain, glass, or plastic dish about 11.4 cm (4½ in.) in diameter and a plastic bag large enough to enclose the dish and be folded over is adequate.

6.7 *Ground Glass Plate*—A ground glass plate at least 30 cm (12 in.) square by 1 cm (³⁄₈ in.) thick for mixing soil and rolling plastic limit threads.

6.8 *Spatula*—A spatula or pill knife having a

blade about 2 cm (¾ in.) wide by about 10 cm (4 in.) long. In addition, a spatula having a blade about 2.5 cm (1 in.) wide and 15 cm (6 in.) long has been found useful for initial mixing of samples.

6.9 *Sieve*—A 20.3 cm (8 in.) diameter, 425-μm (No. 40) sieve conforming to the requirements of Specification E 11 and having a rim at least 5 cm (2 in.) above the mesh. A 2-mm (No. 10) sieve meeting the same requirements may also be needed.

6.10 *Wash Bottle*, or similar container for adding controlled amounts of water to soil and washing fines from coarse particles.

6.11 *Drying Oven*—A thermostatically controlled oven, preferably of the forced-draft type, capable of continuously maintaining a temperature of 110 ±5°C throughout the drying chamber. The oven shall be equipped with a thermometer of suitable range and accuracy for monitoring oven temperature.

6.12 *Washing Pan*—A round, flat-bottomed pan at least 7.6 cm (3 in.) deep, slightly larger at the bottom than a 20.3-cm (8-in.) diameter sieve.

6.13 *Rod* (optional)—A metal or plastic rod or tube 3.2 mm (⅛ in.) in diameter and about 10 cm (4 in.) long for judging the size of plastic limit threads.

7. Materials

7.1 A supply of distilled or demineralized water.

8. Sampling

8.1 Samples may be taken from any location that satisfies testing needs. However, Methods C 702, and Practice D 75, and Recommended Practice D 420 should be used as guides for selecting and preserving samples from various types of sampling operations. Samples which will be prepared using the wet preparation procedure, 10.1, must be kept at their natural water content prior to preparation.

8.2 Where sampling operations have preserved the natural stratification of a sample, the various strata must be kept separated and tests performed on the particular stratum of interest with as little contamination as possible from other strata. Where a mixture of materials will be used in construction, combine the various components in such proportions that the resultant sample represents the actual construction case.

8.3 Where data from this test method are to be used for correlation with other laboratory or field test data, use the same material as used for these tests where possible.

8.4 Obtain a representative portion from the total sample sufficient to provide 150 to 200 g of material passing the 425-μm (No. 40) sieve. Free flowing samples may be reduced by the methods of quartering or splitting. Cohesive samples shall be mixed thoroughly in a pan with a spatula, or scoop and a representative portion scooped from the total mass by making one or more sweeps with a scoop through the mixed mass.

9. Calibration of Apparatus

9.1 *Inspection of Wear:*

9.1.1 *Liquid Limit Device*—Determine that the liquid limit device is clean and in good working order. The following specific points should be checked:

9.1.1.1 *Wear of Base*—The spot on the base where the cup makes contact should be worn no greater than 10 mm (⅜ in.) in diameter. If the wear spot is greater than this, the base can be machined to remove the worn spot provided the resurfacing does not make the base thinner than specified in 6.1 and the other dimensional relationships are maintained.

9.1.1.2 *Wear of Cup*—The cup must be replaced when the grooving tool has worn a depression in the cup 0.1 mm (0.004 in.) deep or when the edge of the cup has been reduced to half its original thickness. Verify that the cup is firmly attached to the cup hanger.

9.1.1.3 *Wear of Cup Hanger*—Verify that the cup hanger pivot does not bind and is not worn to an extent that allows more than 3-mm (⅛-in.) side-to-side movement of the lowest point on the rim.

9.1.1.4 *Wear of Cam*—The cam shall not be worn to an extent that the cup drops before the cup hanger (cam follower) loses contact with the cam.

9.1.2 *Grooving Tools*—Inspect grooving tools for wear on a frequent and regular basis. The rapidity of wear depends on the material from which the tool is made and the types of soils being tested. Sandy soils cause rapid wear of grooving tools; therefore, when testing these materials, tools should be inspected more frequently than for other soils. Any tool with a tip width greater than 2.1 mm must not be used. The depth of the tip of the grooving tool must be 7.9 to 8.1 mm.

Note 6—The width of the tip of grooving tools is conveniently checked using a pocket-sized measuring magnifier equipped with a millimetre scale. Magnifiers of this type are available from most laboratory supply companies. The depth of the tip of grooving tools can be checked using the depth measuring feature of vernier calipers.

9.2 *Adjustment of Height of Drop*—Adjust the height of drop of the cup so that the point on the cup that comes in contact with the base rises to a height of 10 ±0.2 mm. See Fig. 4 for proper location of the gage relative to the cup during adjustment.

Note 7—A convenient procedure for adjusting the height of drop is as follows: place a piece of masking tape across the outside bottom of the cup parallel with the axis of the cup hanger pivot. The edge of the tape away from the cup hanger should bisect the spot on the cup that contacts the base. For new cups, placing a piece of carbon paper on the base and allowing the cup to drop several times will mark the contact spot. Attach the cup to the device and turn the crank until the cup is raised to its maximum height. Slide the height gage under the cup from the front, and observe whether the gage contacts the cup or the tape. See Fig. 4. If the tape and cup are both contacted, the height of drop is approximately correct. If not, adjust the cup until simultaneous contact is made. Check adjustment by turning the crank at 2 revolutions per second while holding the gage in position against the tape and cup. If a ringing or clicking sound is heard without the cup rising from the gage, the adjustment is correct. If no ringing is heard or if the cup rises from the gage, readjust the height of drop. If the cup rocks on the gage during this checking operation, the cam follower pivot is excessively worn and the worn parts should be replaced. Always remove tape after completion of adjustment operation.

MULTIPOINT LIQUID LIMIT—PROCEDURES A AND B

10. Preparation of Test Specimens

10.1 *Wet Preparation*—Except where the dry method of specimen preparation is specified (10.2), prepare specimens for test as described in the following sections.

10.1.1 *Samples Passing the 425-μm (No. 40) Sieve*—When by visual and manual procedures it is determined that the sample has little or no material retained on a 425-μm (No. 40) sieve, prepare a specimen of 150 to 200 g by mixing thoroughly with distilled or demineralized water on the glass plate using the spatula. If desired, soak soil in a storage dish with small amount of water to soften the soil before the start of mixing.

Adjust the water content of the soil to bring it to a consistency that would require 25 to 35 blows of the liquid limit device to close the groove (Note 8). If, during mixing, a small percentage of material is encountered that would be retained on a 425-μm (No. 40) sieve, remove these particles by hand, if possible. If it is impractical to remove the coarser material by hand, remove small percentages (less than about 15 %) of coarser material by working the specimen through a 425-μm (No. 40) sieve using a piece of rubber sheeting, rubber stopper, or other convenient device provided the operation does not distort the sieve or degrade material that would be retained if the washing method described in 10.1.2 were used. If larger percentages of coarse material are encountered during mixing, or it is considered impractical to remove the coarser material by the methods just described, wash the sample as described in 10.1.2. When the coarse particles found during mixing are concretions, shells, or other fragile particles, do not crush these particles to make them pass a 425-μm (No. 40) sieve, but remove by hand or by washing. Place the mixed soil in the storage dish, cover to prevent loss of moisture, and allow to stand for at least 16 h (overnight). After the standing period and immediately before starting the test, thoroughly remix the soil.

Note 8—The time taken to adequately mix a soil will vary greatly, depending on the plasticity and initial water content. Initial mixing times of more than 30 — in may be needed for stiff, fat clays.

10.1.2 *Samples Containing Material Retained on a 425-μm (No. 40) Sieve:*
10.1.2.1 Select a sufficient quantity of soil at natural water content to provide 150 to 200 g of material passing the 425-μm (No. 40) sieve. Place in a pan or dish and add sufficient water to cover the soil. Allow to soak until all lumps have softened and the fines no longer adhere to the surfaces of the corase particles (Note 9).

Note 9—In some cases, the cations of salts present in tap water will exchange with the natural cations in the soil and significantly alter the test results should tap water be used in the soaking and washing operations. Unless it is known that such cations are not present in the tap water, distilled or demineralized water should be used. As a general rule, water containing more than 100 mg/L of dissolved solids should not be used for washing operations.

10.1.2.2 When the sample contains a large percentage of material retained on the 425-μm

(No. 40) sieve, perform the following washing operation in increments, washing no more than 0.5 kg (1 lb) of material at one time. Place the 425-μm (No. 40) sieve in the bottom of the clean pan. Pour the soil water mixture onto the sieve. If gravel or coarse sand particles are present, rinse as many of these as possible with small quantities of water from a wash bottle, and discard. Alternatively, pour the soil water mixture over a 2-mm (No. 10) sieve nested atop the 425-μm (No. 40) sieve, rinse the fine material through and remove the 2-mm (No. 10) sieve. After washing and removing as much of the coarser material as possible, add sufficient water to the pan to bring the level to about 13 mm (½ in.) above the surface of the 425-μm (No. 40) sieve. Agitate the slurry by stirring with the fingers while raising and lowering the sieve in the pan and swirling the suspension so that fine material is washed from the coarser particles. Disaggregate fine soil lumps that have not slaked by gently rubbing them over the sieve with the fingertips. Complete the washing operation by raising the sieve above the water surface and rinsing the material retained with a small amount of clean water. Discard material retained on the 425-μm (No. 40) sieve.

10.1.2.3 Reduce the water content of the material passing the 425-μm (No. 40) sieve until it approaches the liquid limit. Reduction of water content may be accomplished by one or a combination of the following methods: (a) exposing the air currents at ordinary room temperature, (b) exposing to warm air currents from a source such as an electric hair dryer, (c) filtering in a Buckner funnel or using filter candles, (d) decanting clear water from surface of suspension, or (e) draining in a colander or plaster of paris dish lined with high retentivity, high wet-strength filter paper.[7] If a plaster of paris dish is used, take care that the dish never becomes sufficiently saturated that it fails to actively absorb water into its surface. Thoroughly dry dishes between uses. During evaporation and cooling, stir the sample often enough to prevent overdrying of the fringes and soil pinnacles on the surface of the mixture. For soil samples containing soluble salts, use a method of water reduction such as a or b that will not eliminate the soluble salts from the test specimen.

10.1.2.4 Thoroughly mix the material passing the 425-μm (No. 40) sieve on the glass plate using the spatula. Adjust the water content of the mixture, if necessary, by adding small increments of distilled or demineralized water or by allowing the mixture to dry at room temperature while mixing on the glass plate. The soil should be at a water content that will result in closure of the groove in 25 to 35 blows. Return the mixed soil to the mixing dish, cover to prevent loss of moisture, and allow to stand for at least 16 h. After the standing period, and immediately before starting the test, remix the soil thoroughly.

10.2 *Dry Preparation:*

10.2.1 Select sufficient soil to provide 150 to 200 g of material passing the 425-μm (No. 40) sieve after processing. Dry the sample at room temperature or in an oven at a temperature not exceeding 60°C until the soil clods will pulverize readily. Disaggregation is expedited if the sample is not allowed to completely dry. However, the soil should have a dry appearance when pulverized. Pulverize the sample in a mortar with a rubber tipped pestal or in some other way that does not cause breakdown of individual grains. When the coarse particles found during pulverization are concretions, shells, or other fragile particles, do not crush these particles to make them pass a 425-μm (No. 40) sieve, but remove by hand or other suitable means, such as washing.

10.2.2 Separate the sample on a 425-μm (No. 40) sieve, shaking the sieve by hand to assure thorough separation of the finer fraction. Return the material retained on the 425-μm (No. 40) sieve to the pulverizing apparatus and repeat the pulverizing and sieving operations as many times as necessary to assure that all finer material has been disaggregated and material retained on the 425-μm (No. 40) sieve consists only of individual sand or gravel grains.

10.2.3 Place material remaining on the 425-μm (No. 40) sieve after the final pulverizing operations in a dish and soak in a small amount of water. Stir the soil water mixture and pour over the 425-μm (No. 40) sieve, catching the water and any suspended fines in the washing pan. Pour this suspension into a dish containing the dry soil previously sieved through the 425-μm (No. 40) sieve. Discard material retained on the 425-μm (No. 40) sieve.

10.2.4 Adjust the water content as necessary by drying as described in 10.1.2.3 or by mixing on the glass plate, using the spatula while adding increments of distilled or demineralized water,

_____
[7] S and S 595 filter paper, available in 32-cm circles, has proven satisfactory.

302

until the soil is at a water content that will result in closure of the groove in 25 to 35 blows.

10.2.5 Put soil in the storage dish, cover to prevent loss of moisture and allow to stand for at least 16 h. After the standing period, and immediately before starting the test, thoroughly remix the soil (Note 8).

11. Procedure

11.1 Place a portion of the prepared soil in the cup of the liquid limit device at the point where the cup rests on the base, squeeze it down, and spread it into the cup to a depth of about 10 mm at its deepest point, tapering to form an approximately horizontal surface. Take care to eliminate air bubbles from the soil pat but form the pat with as few strokes as possible. Heap the unused soil on the glass plate and cover with the inverted storage dish or a wet towel.

11.2 Form a groove in the soil pat by drawing the tool, beveled edge forward, through the soil on a line joining the highest point to the lowest point on the rim of the cup. When cutting the groove, hold the grooving tool against the surface of the cup and draw in an arc, maintaining the tool perpendicular to the surface of the cup throughout its movement. See Fig. 5. In soils where a groove cannot be made in one stroke without tearing the soil, cut the groove with several strokes of the grooving tool. Alternatively, cut the groove to slightly less than required dimensions with a spatula and use the grooving tool to bring the groove to final dimensions. Exercise extreme care to prevent sliding the soil pat relative to the surface of the cup.

11.3 Verify that no crumbs of soil are present on the base or the underside of the cup. Lift and drop the cup by turning the crank at a rate of 1.9 to 2.1 drops per second until the two halves of the soil pat come in contact at the bottom of the groove along a distance of 13 mm (½ in.). See Fig. 6.

NOTE 10—Use the end of the grooving tool, Fig. 2, or a scale to verify that the groove has closed 13 mm (½ in.).

11.4 Verify that an air bubble has not caused premature closing of the groove by observing that both sides of the groove have flowed together with approximately the same shape. If a bubble has caused premature closing of the groove, reform the soil in the cup, adding a small amount of soil to make up for that lost in the grooving

operation and repeat 11.1 to 11.3. If the soil slides on the surface of the cup, repeat 11.1 through 11.3 at a higher water content. If, after several trials at successively higher water contents, the soil pat continues to slide in the cup or if the number of blows required to close the groove is always less than 25, record that the liquid limit could not be determined, and report the soil as nonplastic without performing the plastic limit test.

11.5 Record the number of drops, N, required to close the groove. Remove a slice of soil approximately the width of the spatula, extending from edge to edge of the soil cake at right angles to the groove and including that portion of the groove in which the soil flowed together, place in a weighed container, and cover.

11.6 Return the soil remaining in the cup to the glass plate. Wash and dry the cup and grooving tool and reattach the cup to the carriage in preparation for the next trial.

11.7 Remix the entire soil specimen on the glass plate adding distilled water to increase the water content of the soil and decrease the number of blows required to close the groove. Repeat 11.1 through 11.6 for at least two additional trials producing successively lower numbers of blows to close the groove. One of the trials shall be for a closure requiring 25 to 35 blows, one for closure between 20 and 30 blows, and one trial for a closure requiring 15 to 25 blows.

11.8 Determine the water content, W_N, of the soil specimen from each trial in accordance with Method D 2216. Make all weighings on the same balance. Initial weighings should be performed immediately after completion of the test. If the test is to be interrupted for more than about 15 min, the specimens already obtained should be weighed at the time of the interruption.

12. Calculations

12.1 Plot the relationship between the water content, W_N, and the corresponding number of drops, N, of the cup on a semilogarithmic graph with the water content as ordinates on the arithmetical scale, and the number of drops as abscissas on the logarithmic scale. Draw the best straight line through the three or more plotted points.

12.2 Take the water content corresponding to the intersection of the line with the 25-drop abscissa as the liquid limit of the soil. Computational methods may be substituted for the graphical method for fitting a straight line to the data and determining the liquid limit.

ONE-POINT LIQUID LIMIT—PROCEDURES C AND D

13. Preparation of Test Specimens

13.1 Prepare the specimen in the same manner as described in Section 10, except that at mixing, adjust the water content to a consistency requiring 20 to 30 drops of the liquid limit cup to close the groove.

14. Procedure

14.1 Proceed as described in 11.1 through 11.5 except that the number of blows required to close the groove shall be 20 to 30. If less than 20 or more than 30 blows are required, adjust the water content of the soil and repeat the procedure.

14.2 Immediately after removing a water content specimen as described in 11.5, reform the soil in the cup, adding a small amount of soil to make up for that lost in the grooving and water content sampling operations. Repeat 11.2 through 11.5, and, if the second closing of the groove requires the same number of drops or no more than two drops difference, secure another water content specimen. Otherwise, remix the entire specimen and repeat.

NOTE 11—Excessive drying or inadequate mixing will cause the number of blows to vary.

14.3 Determine water contents of specimens as described in 11.8.

15. Calculations

15.1 Determine the liquid limit for each water content specimen using one of the following equations:

$$LL = W_N\left(\frac{N}{25}\right)^{0.121} \text{ or}$$

$$LL = K(W_N)$$

where:

N = the number of blows causing closure of the groove at water content,

W_N = water content, and

K = a factor given in Table 1.

The liquid limit is the average of the two trial liquid limit values.

15.2 If the difference between the two trial liquid limit values is greater than one percentage point, repeat the test.

PLASTIC LIMIT

16. Preparation of Test Specimen

16.1 Select a 20-g portion of soil from the material prepared for the liquid limit test, either after the second mixing before the test, or from the soil remaining after completion of the test. Reduce the water content of the soil to a consistency at which it can be rolled without sticking to the hands by spreading and mixing continuously on the glass plate. The drying process may be accelerated by exposing the soil to the air current from an electric fan, or by blotting with paper that does not add any fiber to the soil, such as hard surface paper toweling or high wet strength filter paper.

17. Procedure

17.1 From the 20-g mass, select a portion of 1.5 to 2.0 g. Form the test specimen into an ellipsoidal mass. Roll this mass between the palm or fingers and the ground-glass plate with just sufficient pressure to roll the mass into a thread of uniform diameter throughout its length (Note 12). The thread shall be further deformed on each stroke so that its diameter is continuously reduced and its length extended until the diameter reaches 3.2 ±0.5 mm (0.125 ±.020 in.), taking no more than 2 min (Note 13). The amount of hand or finger pressure required will vary greatly, according to the soil. Fragile soils of low plasticity are best rolled under the outer edge of the palm or at the base of the thumb.

NOTE 12—A normal rate of rolling for most soils should be 80 to 90 strokes per minute, counting a stroke as one complete motion of the hand forward and back to the starting position. This rate of rolling may have to be decreased for very fragile soils.

NOTE 13—A 3.2-mm (⅛-in.) diameter rod or tube is useful for frequent comparison with the soil thread to ascertain when the thread has reached the proper diameter, especially for inexperienced operators.

17.1.1 When the diameter of the thread becomes 3.2 mm, break the thread into several pieces. Squeeze the pieces together, knead between the thumb and first finger of each hand, reform into an ellipsoidal mass, and reroll. Continue this alternate rolling to a thread 3.2 mm in diameter, gathering together, kneading and rerolling, until the thread crumbles under the pressure required for rolling and the soil can no longer be rolled into a 3.2-mm diameter thread (See Fig. 7). It has no significance if the thread breaks into threads of shorter length. Roll each of these shorter threads to 3.2 mm in diameter. The only requirement for continuing the test is that they are able to be reformed into an ellipsoidal mass and rolled out again. The operator shall at no time attempt to produce failure at exactly 3.2 mm diameter by allowing the thread to reach 3.2 mm, then reducing the rate of rolling or the hand pressure, or both, while continuing the rolling without further deformation until the thread falls apart. It is permissible, however, to reduce the total amount of deformation for feebly plastic soils by making the initial diameter of the ellipsoidal mass nearer to the required 3.2-mm final diameter. If crumbling occurs when the thread has a diameter greater than 3.2 mm, this shall be considered a satisfactory end point, provided the soil has been previously rolled into a thread 3.2 mm in diameter. Crumbling of the thread will manifest itself differently with the various types of soil. Some soils fall apart in numerous small aggregations of particles, others may form an outside tubular layer that starts splitting at both ends. The splitting progresses toward the middle, and finally, the thread falls apart in many small platy particles. Fat clay soils require much pressure to deform the thread, particularly as they approach the plastic limit. With these soils, the thread breaks into a series of barrel-shaped segments about 3.2 to 9.5 mm (⅛ to ⅜ in.) in length.

17.2 Gather the portions of the crumbled thread together and place in a weighed container. Immediately cover the container.

17.3 Select another 1.5 to 2.0 g portion of soil from the original 20-g specimen and repeat the operations described in 17.1 and 17.2 until the container has at least 6 g of soil.

17.4 Repeat 17.1 through 17.3 to make another container holding at least 6 g of soil. Determine the water content, in percent, of the soil contained in the containers in accordance with Method D 2216. Make all weighings on the same balance.

NOTE 14—The intent of performing two plastic limit trials is to verify the consistency of the test results. It is acceptable practice to perform only one plastic limit trial when the consistency in the test results can be confirmed by other means.

18. Calculations

18.1 Compute the average of the two water contents. If the difference between the two water contents is greater than two percentage points, repeat the test. The plastic limit is the average of the two water contents.

PLASTICITY INDEX

19. Calculations

19.1 Calculate the plasticity index as follows:

$$PI = LL - PL$$

where:
LL = the liquid limit,
PL = the plastic limit.

Both LL and PL are whole numbers. If either the liquid limit or plastic limit could not be determined, or if the plastic limit is equal to or greater than the liquid limit, report the soil as nonplastic, NP.

20. Report

20.1 Report the following information:

20.1.1 Sample identifying information,

20.1.2 Any special specimen selection process used, such as removal of sand lenses from undisturbed sample,

20.1.3 Report sample as airdried if the sample was airdried before or during preparation,

20.1.4 Liquid limit, plastic limit, and plasticity index to the nearest whole number and omitting the percent designation. If the liquid limit or plastic limit tests could not be performed, or if the plastic limit is equal to or greater than the liquid limit, report the soil as nonplastic, NP,

20.1.5 An estimate of the percentage of sample retained on the 425-μm (No. 40) sieve, and

20.1.6 Procedure by which liquid limit was performed, if it differs from the multipoint method.

21. Precision and Bias

21.1 No interlaboratory testing program has as yet been conducted using this test method to determine multilaboratory precision.

21.2 The within laboratory precision of the results of tests performed by different operators at one laboratory on two soils using Procedure A for the liquid limit is shown in Table 2.

TABLE 1 Factors for Obtaining Liquid Limit from Water Content and Number of Drops Causing Closure of Groove

N (Number of Drops)	K (Factor for Liquid Limit)
20	0.974
21	0.979
22	0.985
23	0.990
24	0.995
25	1.000
26	1.005
27	1.009
28	1.014
29	1.018
30	1.022

TABLE 2 Within Laboratory Precision for Liquid Limit

	Average Value, $\bar{x}$	Standard Deviation, s
Soil A:		
PL	21.9	1.07
LL	27.9	1.07
Soil B:		
PL	20.1	1.21
LL	32.6	0.98

DIMENSIONS

LETTER	A△	B△	C△	E△	F	G	H	J△	K△	L△	M△
MM	54 ± 0.5	2 ± 0.1	27 ± 0.5	56 ± 2.0	32	10	16	60 ± 1.0	50 ± 2.0	150 ± 2.0	125 ± 2.0

LETTER	N	P	R	T	U△	V	W	Z
MM	24	28	24	45	47 ± 1.0	3.8	13	6.5

△ ESSENTIAL DIMENSIONS

CAM ANGLE DEGREES	CAM RADIUS
0	0.742 R
30	0.753 R
60	0.764 R
90	0.773 R
120	0.784 R
150	0.796 R
180	0.818 R
210	0.854 R
240	0.901 R
270	0.945 R
300	0.974 R
330	0.995 R
360	1.000 R

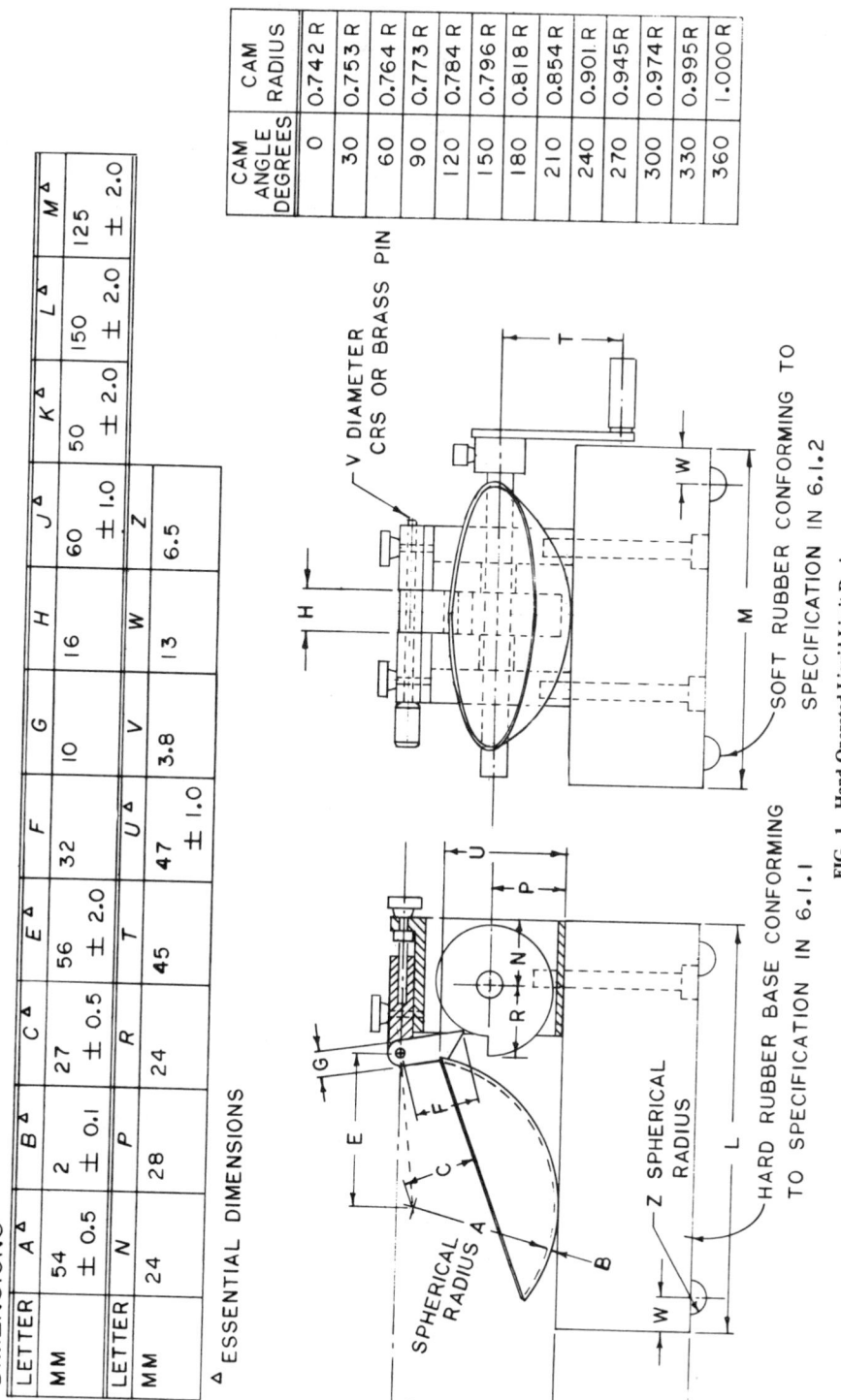

FIG. 1 Hand-Operated Liquid Limit Device

DIMENSIONS

LETTER	A △	B △	C △	D △	E △	F △
MM	2 ± 0.1	11 ± 0.2	40 ± 0.5	8 ± 0.1	50 ± 0.5	2 ± 0.1
LETTER	G	H	J	K △	L △	N
MM	10 MINIMUM	13	60	10 ±0.05	60 DEG ± 1 DEG	20

△ ESSENTIAL DIMENSIONS

□ BACK AT LEAST 15 MM FROM TIP

NOTE : DIMENSION A SHOULD BE 1.9-2.0 AND DIMENSION D
SHOULD BE 8.0-8.1 WHEN NEW TO ALLOW FOR
ADEQUATE SERVICE LIFE

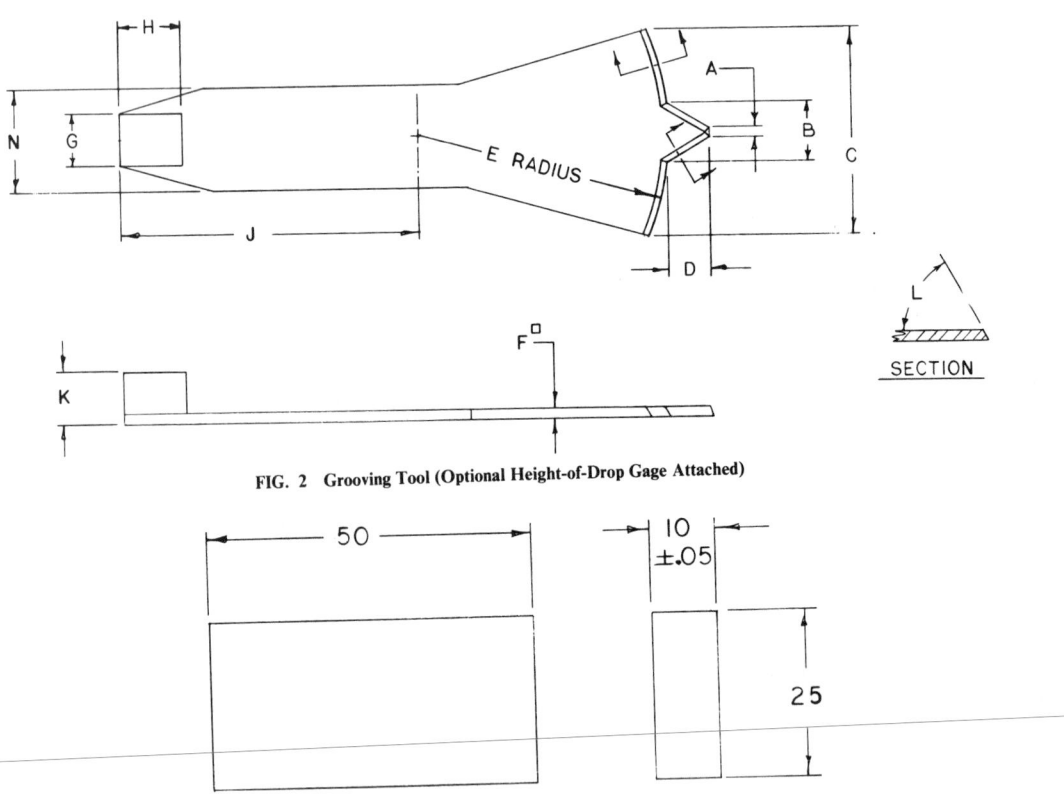

FIG. 2 Grooving Tool (Optional Height-of-Drop Gage Attached)

DIMENSIONS IN MILLIMETRES

FIG. 3 Height of Drop Gage

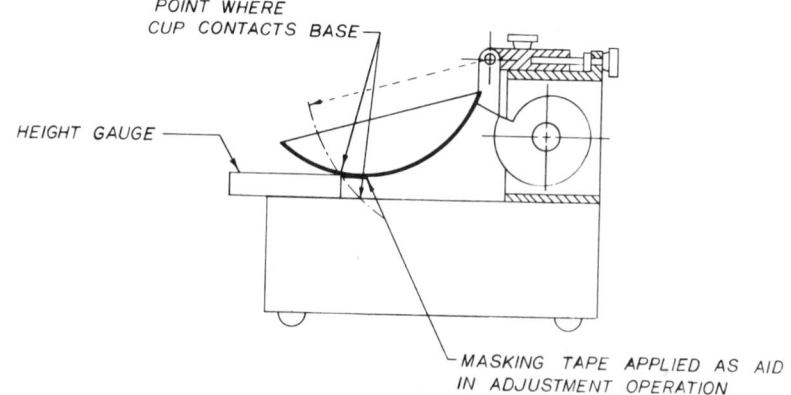

POINT WHERE
CUP CONTACTS BASE

HEIGHT GAUGE

MASKING TAPE APPLIED AS AID
IN ADJUSTMENT OPERATION

FIG. 4 Calibration for Height of Drop

FIG. 5 Grooved Soil Pat in Liquid Limit Device

FIG. 6   Soil Pat After Groove Has Closed

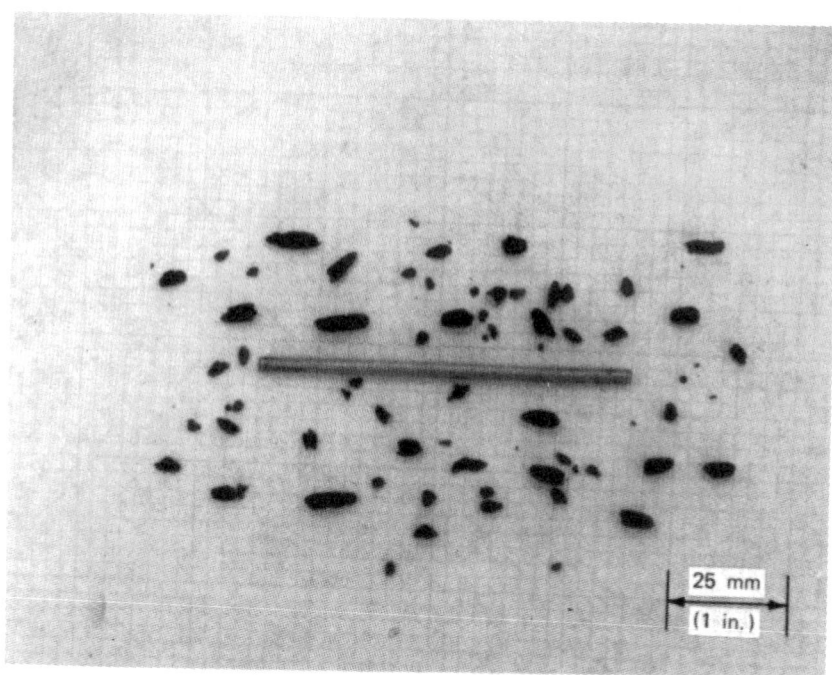

FIG. 7   Lean Clay Soil at the Plastic Limit

The American Society for Testing and Materials takes no position respecting the validity of any patent rights asserted in connection with any item mentioned in this standard. Users of this standard are expressly advised that determination of the validity of any such patent rights, and the risk of infringement of such rights, are entirely their own responsibility.

This standard is subject to revision at any time by the responsible technical committee and must be reviewed every five years and if not revised, either reapproved or withdrawn. Your comments are invited either for revision of this standard or for additional standards and should be addressed to ASTM Headquarters. Your comments will receive careful consideration at a meeting of the responsible technical committee, which you may attend. If you feel that your comments have not received a fair hearing you should make your views known to the ASTM Committee on Standards, 1916 Race St., Philadelphia, Pa. 19103.

Standard Test Method for
CLASSIFICATION OF SOILS FOR ENGINEERING PURPOSES[1]

This standard is issued under the fixed designation D 2487; the number immediately following the designation indicates the year of original adoption or, in the case of revision, the year of last revision. A number in parentheses indicates the year of last reapproval. A superscript epsilon (ε) indicates an editorial change since the last revision or reapproval.

This test method has been approved for use by agencies of the Department of Defense and for listing in the DOD Index of Specifications and Standards.

1. Scope

1.1 This test method describes a system for classifying mineral and organo-mineral soils for engineering purposes based on laboratory determination of particle-size characteristics, liquid limit, and plasticity index and shall be used when precise classification is required.

NOTE 1—Use of this standard will result in a single classification group symbol and group name except when a soil contains 5 to 12 % fines or when the plot of the liquid limit and plasticity index values falls into the crosshatched area of the plasticity chart. In these two cases, a dual symbol is used, for example, GP-GM, CL-ML. When the laboratory test results indicate that the soil is close to another soil classification group, the borderline condition can be indicated with two symbols separated by a slash. The first symbol should be the one based on this standard, for example, CL/CH, GM/SM, SC/CL. Borderline symbols are particularly useful when the liquid limit value of clayey soils is close to 50. These soils can have expansive characteristics and the use of a borderline symbol (CL/CH, CH/CL) will alert the user of the assigned classifications of expansive potential.

1.2 The group symbol portion of this sytem is based on laboratory tests performed on the portion of a soil sample passing the 3-in. (75-mm) sieve (see Specification E 11).

1.3 As a classification system, this test method is limited to naturally occurring soils.

NOTE 2—The group names and symbols used in this test method may be used as a descriptive system applied to such materials as shale, claystone, shells, crushed rock, etc. See Appendix X2.

1.4 This test method is for qualitative application only.

NOTE 3—When quantitative information is required for detailed designs of important structures, this test method must be supplemented by laboratory tests or other quantitative data to determine performance characteristics under expected field conditions.

1.5 The system is based on the widely recognized Unified Soil Classification System which was adopted by several U.S. Government agencies in 1952 as an outgrowth of the Airfield Classification System developed by A. Casagrande.[2]

1.6 *This standard may involve hazardous materials, operations, and equipment. This standard does not purport to address all of the safety problems associated with its use. It is the responsibility of whoever uses this standard to consult and establish appropriate safety and health practices and determine the applicability of regulatory limitations prior to use.*

2. Applicable Documents

2.1 *ASTM Standards:*

C 117 Test Method for Material Finer Than 75-μm (No. 200) Sieve in Mineral Aggregates by Washing[3]

C 136 Method for Sieve Analysis of Fine and Coarse Aggregates[3]

C 702 Methods for Reducing Field Samples of Aggregate to Testing Size[3]

D 420 Recommended Practice for Investigating and Sampling Soil and Rock for Engineering Purposes[4]

D 421 Practice for Dry Preparation of Soil Samples for Particle-Size Analysis and De-

---

[1] This test method is under the jurisdiction of ASTM Committee D-18 on Soil and Rock and is the direct responsibility of Subcommittee D18.07 on Identification and Classification of Soils.

Current edition approved Oct. 25, 1985. Published December 1985. Originally published as D 2487 – 66 T. Last previous edition D 2487 – 83.

[2] Casagrande, A., "Classification and Identification of Soils," *Transactions*, ASCE, 1948, p. 901.

[3] *Annual Book of ASTM Standards*, Vol 04.02.

[4] *Annual Book of ASTM Standards*, Vol 04.08.

termination of Soil Constants[4]

D422 Method of Particle-Size Analysis of Soils[4]

D653 Terms and Symbols Relating to Soil and Rock[4]

D1140 Test Method for Amount of Material in Soils Finer than the No. 200 (75-μm) Sieve[4]

D2216 Method for Laboratory Determination of Water (Moisture) Content of Soil, Rock, and Soil-Aggregate Mixtures[4]

D2217 Practice for Wet Preparation of Soil Samples for Particle-Size Analysis and Determination of Soil Constants[4]

D2488 Practice for Description and Identification of Soils (Visual-Manual Procedure)[4]

D4318 Test Method for Liquid Limit, Plastic Limit, and Plasticity Index of Soils[4]

E11 Specification for Wire-Cloth Sieves for Testing Purposes[3]

3. Summary of Method

3.1 As illustrated in Table 1, this classification system identifies three major soil divisions: coarse-grained soils, fine-grained soils, and highly organic soils. These three divisions are further subdivided into a total of 15 basic soil groups.

3.2 Based on the results of visual observations and prescribed laboratory tests, a soil is catalogued according to the basic soil groups, assigned a group symbol(s) and name, and thereby classified. The flow charts, Fig. 1 for fine-grained soils, and Fig. 2 for coarse-grained soils, can be used to assign the appropriate group symbol(s) and name.

4. Significance and Use

4.1 This test method classifies soils from any geographic location into categories representing the results of prescribed laboratory tests to determine the particle-size characteristics, the liquid limit, and the plasticity index.

4.2 The assigning of a group name and symbol(s) along with the descriptive information required in Practice D2488 can be used to describe a soil to aid in the evaluation of its significant properties for engineering use.

4.3 The various groupings of this classification system have been devised to correlate in a general way with the engineering behavior of soils. This test method provides a useful first step in any field or laboratory investigation for geotechnical engineering purposes.

4.4 This test method may also be used as an aid in training personnel in the use of Practice D2488.

5. Terminology

5.1 *Definitions*—Except as listed below, all definitions are in accordance with Terms and Symbols D653.

NOTE 4—For particles retained on a 3-in. (75-mm) U.S. standard sieve, the following definitions are suggested:

Cobbles—particles of rock that will pass a 12-in. (300-mm) square opening and be retained on a 3-in. (75-mm) U.S. standard sieve, and

Boulders—particles of rock that will not pass a 12-in. (300-mm) square opening

5.1.1 *gravel*—particles of rock that will pass a 3-in. (75-mm) sieve and be retained on a No. 4 (4.75-mm) U.S. standard sieve with the following subdivisions:

Coarse—passes 3-in. (75-mm) sieve and retained on ¾-in. (19-mm) sieve, and

Fine—passes ¾-in. (19-mm) sieve and retained on No. 4 (4.75-mm) sieve.

5.1.2 *sand*—particles of rock that will pass a No. 4 (4.75-mm) sieve and be retained on a No. 200 (75-μm) U.S. standard sieve with the following subdivisions:

Coarse—passes No. 4 (4.75-mm) sieve and retained on No. 10 (2.00-mm) sieve,

Medium—passes No. 10 (2.00-mm) sieve and retained on No. 40 (425-μm) sieve, and

Fine—passes No. 40 (425-μm) sieve and retained on No. 200 (75-μm) sieve.

5.1.3 *clay*—soil passing a No. 200 (75-μm) U.S. standard sieve that can be made to exhibit plasticity (putty-like properties) within a range of water contents and that exhibits considerable strength when air dry. For classification, a clay is a fine-grained soil, or the fine-grained portion of a soil, with a plasticity index equal to or greater than 4, and the plot of plasticity index versus liquid limit falls on or above the "A" line.

5.1.4 *silt*—soil passing a No. 200 (75-μm) U.S. standard sieve that is nonplastic or very slightly plastic and that exhibits little or no strength when air dry. For classification, a silt is a fine-grained soil, or the fine-grained portion of a soil, with a plasticity index less than 4 or if the plot of plasticity index versus liquid limit falls below the "A" line.

5.1.5 *organic clay*—a clay with sufficient organic content to influence the soil properties. For classification, an organic clay is a soil that would be classified as a clay except that its liquid limit value after oven drying is less than 75 % of its liquid limit value before oven drying.

5.1.6 *organic silt*—a silt with sufficient organic content to influence the soil properties. For classification, an organic silt is a soil that would be classified as a silt except that its liquid limit value after oven drying is less than 75 % of its liquid limit value before oven drying.

5.1.7 *peat*—a soil composed of vegetable tissue in various stages of decomposition usually with an organic odor, a dark-brown to black color, a spongy consistency, and a texture ranging from fibrous to amorphous.

5.2 *Descriptions of Terms Specific to This Standard:*

5.2.1 *coefficient of curvature*, Cc—the ratio $(D_{30})^2/(D_{10} \times D_{60})$, where D_{60}, D_{30}, and D_{10} are the particle diameters corresponding to 60, 30, and 10 % finer on the cumulative particle-size distribution curve, respectively.

5.2.2 *coefficient of uniformity*, Cu—the ratio D_{60}/D_{10}, where D_{60} and D_{10} are the particle diameters corresponding to 60 and 10 % finer on the cumulative particle-size distribution curve, respectively.

6. Apparatus

6.1 In addition to the apparatus that may be required for obtaining and preparing the samples and conducting the prescribed laboratory tests, a plasticity chart, similar to Fig. 3, and a cumulative particle-size distribution curve, similar to Fig. 4, are required.

NOTE 5—The "U" line shown on Fig. 3 has been empirically determined to be the approximate "upper limit" for natural soils. It is a good check against erroneous data, and any test results that plot above or to the left of it should be verified.

7. Sampling

7.1 Samples shall be obtained and identified in accordance with a method or methods, recommended in Recommended Practice D 420 or by other accepted procedures.

7.2 For accurate identification, the minimum amount of test sample required for this test method will depend on which of the laboratory tests need to be performed. Where only the particle-size analysis of the sample is required, specimens having the following minimum dry weights are required:

Maximum Particle Size, Sieve Opening	Minimum Specimen Size, Dry Weight
4.75 mm (No. 4)	100 g (0.25 lb)
9.5 mm (⅜ in.)	200 g (0.5 lb)
19.0 mm (¾ in.)	1.0 kg (2.2 lb)
38.1 mm (1½ in.)	8.0 kg (18 lb)
75.0 mm (3 in.)	60.0 kg (132 lb)

Whenever possible, the field samples should have weights two to four times larger than shown.

7.3 When the liquid and plastic limit tests must also be performed, additional material will be required sufficient to provide 150 g to 200 g of soil finer than the No. 40 (425-μm) sieve.

7.4 If the field sample or test specimen is smaller than the minimum recommended amount, the report shall include an appropriate remark.

8. Classification of Peat

8.1 A sample composed primarily of vegetable tissue in various stages of decomposition and has a fibrous to amorphous texture, a dark-brown to black color, and an organic odor should be designated as a highly organic soil and shall be classified as peat, PT, and not subjected to the classification procedures described hereafter.

9. Preparation for Classification

9.1 Before a soil can be classified according to this test method, generally the particle-size distribution of the minus 3-in. (75-mm) material and the plasticity characteristics of the minus No. 40 (425-μm) sieve material must be determined. See 9.8 for the specific required tests.

9.2 The preparation of the soil specimen(s) and the testing for particle-size distribution and liquid limit and plasticity index shall be in accordance with accepted standard procedures. Two procedures for preparation of the soil specimens for testing for soil classification purposes are given in Appendixes X3 and X4. Appendix X3 describes the wet preparation method and is the preferred method for cohesive soils that have never dried out and for organic soils.

9.3 When reporting soil classifications determined by this test method, the preparation and test procedures used shall be reported or referenced.

9.4 Although the test procedure used in determining the particle-size distribution or other considerations may require a hydrometer analysis of the material, a hydrometer analysis is not necessary for soil classification.

9.5 The percentage (by dry weight) of any plus 3-in. (75-mm) material must be determined and reported as auxiliary information.

9.6 The maximum particle size shall be determined (measured or estimated) and reported as auxiliary information.

9.7 When the cumulative particle-size distribution is required, a set of sieves shall be used which include the following sizes (with the largest size commensurate with the maximum particle size) with other sieve sizes as needed or required to define the particle-size distribution:

> 3-in. (75-mm)
> ¾-in.(19.0-mm)
> No. 4 (4.75-mm)
> No. 10 (2.00-mm)
> No. 40 (425-μm)
> No. 200 (75-μm)

9.8 The tests required to be performed in preparation for classification are as follows:

9.8.1 For soils estimated to contain less than 5 % fines, a plot of the cumulative particle-size distribution curve of the fraction coarser than the No. 200 (75-μm) sieve is required. The cumulative particle-size distribution curve may be plotted on a graph similar to that shown in Fig. 4.

9.8.2 For soils estimated to contain 5 to 15 % fines, a cumulative particle-size distribution curve, as described in 9.8.1, is required, and the liquid limit and plasticity index are required.

9.8.2.1 If sufficient material is not available to determine the liquid limit and plasticity index, the fines should be estimated to be either silty or clayey using the procedures described in Practice D 2488 and so noted in the report.

9.8.3 For soils estimated to contain 15 % or more fines, a determination of the percent fines, percent sand, and percent gravel is required, and the liquid limit and plasticity index are required. For soils estimated to contain 90 % fines or more, the percent fines, percent sand, and percent gravel may be estimated using the procedures described in Practice D 2488 and so noted in the report.

10. Preliminary Classification Procedure

10.1 Class the soil as fine-grained if 50 % or more by dry weight of the test specimen passes the No. 200 (75-μm) sieve and follow Section 11.

10.2 Class the soil as coarse-grained if more than 50 % by dry weight of the test specimen is retained on the No. 200 (75-μm) sieve and follow Section 12.

11. Procedure for Classification of Fine-Grained Soils (50 % or more by dry weight passing the No. 200 (75-μm) sieve)

11.1 The soil is an inorganic clay if the position of the plasticity index versus liquid limit plot, Fig. 3, falls on or above the "A" line, the plasticity index is greater than 4, and the presence of organic matter does not influence the liquid limit as determined in 11.3.2.

11.1.1 Classify the soil as a *lean clay*, CL, if the liquid limit is less than 50. See area identified as CL on Fig. 3.

11.1.2 Classify the soil as a *fat clay*, CH, if the liquid limit is 50 or greater. See area identified as CH on Fig. 3.

NOTE 6—In cases where the liquid limit exceeds 110 or the plasticity index exceeds 60, the plasticity chart may be expanded by maintaining the same scale on both axes and extending the "A" line at the indicated slope.

11.1.3 Classify the soil as a *silty clay*, CL-ML, if the position of the plasticity index versus liquid limit plot falls on or above the "A" line and the plasticity index is in the range of 4 to 7. See area identified as CL-ML on Fig. 3.

11.2 The soil is an inorganic silt if the position of the plasticity index versus liquid limit plot, Fig. 3, falls below the "A" line or the plasticity index is less than 4, and presence of organic matter does not influence the liquid limit as determined in 11.3.2.

11.2.1 Classify the soil as a *silt*, ML, if the liquid limit is less than 50. See area identified as ML on Fig. 3.

11.2.2 Classify the soil as an *elastic silt*, MH, if the liquid limit is 50 or greater. See area identified as MH on Fig. 3.

11.3 The soil is an organic silt or clay if organic matter is present in sufficient amounts to influence the liquid limit as determined in 11.3.2.

11.3.1 If the soil has a dark color and an organic odor when moist and warm, a second liquid limit test shall be performed on a test specimen which has been oven dried at 110 ± 5°C to a constant weight, typically over night.

11.3.2 The soil is an organic silt or organic clay if the liquid limit after oven drying is less than 75 % of the liquid limit of the original specimen determined before oven drying (see Procedure B of Practice D 2217).

11.3.3 Classify the soil as an *organic silt* or *organic clay*, OL, if the liquid limit (not oven dried) is less than 50 %. Classify the soil as an *organic silt*, OL, if the plasticity index is less than 4, or the position of the plasticity index versus liquid limit plot falls below the "A" line. Classify the soil as an *organic clay*, OL, if the plasticity index is 4 or greater and the position of the plasticity index versus liquid limit plot falls on or above the "A" line. See area identified as OL (or CL-ML) on Fig. 3.

11.3.4 Classify the soil as an *organic clay* or *organic silt*, OH, if the liquid limit (not oven dried) is 50 or greater. Classify the soil as an *organic silt*, OH, if the position of the plasticity index versus liquid limit plot falls below the "A" line. Classify the soil as an *organic clay*, OH, if the position of the plasticity index versus liquid-limit plot falls on or above the "A" line. See area identified as OH on Fig. 3.

11.4 If less than 30 % but 15 % or more of the test specimen is retained on the No. 200 (75-μm) sieve, the words "with sand" or "with gravel" (whichever is predominant) shall be added to the group name. For example, lean clay with sand, CL; silt with gravel, ML. If the percent of sand is equal to the percent of gravel, use "with sand."

11.5 If 30 % or more of the test specimen is retained on the No. 200 (75-μm) sieve, the words "sandy" or "gravelly" shall be added to the group name. Add the word "sandy" if 30 % or more of the test specimen is retained on the No. 200 (75-μm) sieve and the coarse-grained portion is predominantly sand. Add the word "gravelly" if 30 % or more of the test specimen is retained on the No. 200 (75-μm) sieve and the coarse-grained portion is predominantly gravel. For example, sandy lean clay, CL; gravelly fat clay, CH; sandy silt, ML. If the percent of sand is equal to the percent of gravel, use "sandy."

12. Procedure for Classification of Coarse-Grained Soils (more than 50 % retained on the No. 200 (75-μm) sieve)

12.1 Class the soil as gravel if more than 50 % of the coarse fraction [plus No. 200 (75-μm) sieve] is retained on the No. 4 (4.75-mm) sieve.

12.2 Class the soil as sand if 50 % or more of the coarse fraction [plus No. 200 (75-μm) sieve] passes the No. 4 (4.75-mm) sieve.

12.3 If 12 % or less of the test specimen passes the No. 200 (75-μm) sieve, plot the cumulative particle-size distribution, Fig. 4, and compute the coefficient of uniformity, Cu, and coefficient of curvature, Cc, as given in Eqs 1 and 2.

$$Cu = D_{60}/D_{10} \tag{1}$$

$$Cc = (D_{30})^2/(D_{10} \times D_{60}) \tag{2}$$

where:
D_{10}, D_{30}, and D_{60} = the particle-size diameters corresponding to 10, 30, and 60 %, respectively, passing on the cumulative particle-size distribution curve, Fig. 4.

NOTE 7—It may be necessary to extrapolate the curve to obtain the D_{10} diameter.

12.3.1 If less than 5 % of the test specimen passes the No. 200 (75-μm) sieve, classify the soil as a *well-graded gravel*, GW, or *well-graded sand*, SW, if Cu is greater than 4.0 for gravel or greater than 6.0 for sand, and Cc is at least 1.0 but not more than 3.0.

12.3.2 If less than 5 % of the test specimen passes the No. 200 (75-μm) sieve, classify the soil as *poorly graded gravel*, GP, or *poorly graded sand*, SP, if either the Cu or the Cc criteria for well-graded soils are not satisfied.

12.4 If more than 12 % of the test specimen passes the No. 200 (75-μm) sieve, the soil shall be considered a coarse-grained soil with fines. The fines are determined to be either clayey or silty based on the plasticity index versus liquid limit plot on Fig. 3. (See 9.8.2.1 if insufficient material available for testing).

12.4.1 Classify the soil as a *clayey gravel*, GC, or *clayey sand*, SC, if the fines are clayey, that is, the position of the plasticity index versus liquid limit plot, Fig. 3, falls on or above the "A" line and the plasticity index is greater than 7.

12.4.2 Classify the soil as a *silty gravel*, GM, or *silty sand*, SM, if the fines are silty, that is, the position of the plasticity index versus liquid limit plot, Fig. 3, falls below the "A" line or the plasticity index is less than 4.

12.4.3 If the fines plot as a silty clay, CL-ML, classify the soil as a *silty*, *clayey gravel*, GC-GM, if it is a gravel or a *silty*, *clayey sand*, SC-SM, if it is a sand.

12.5 If 5 to 12 % of the test specimen passes the No. 200 (75-μm) sieve, give the soil a dual classification using two group symbols.

12.5.1 The first group symbol shall correspond to that for a gravel or sand having less than 5 % fines (GW, GP, SW, SP), and the second symbol shall correspond to a gravel or sand having more than 12 % fines (GC, GM, SC, SM).

12.5.2 The group name shall correspond to the first group symbol plus "with clay" or "with silt" to indicate the plasticity characteristics of the fines. For example, well-graded gravel with clay, GW-GC; poorly graded sand with silt, SP-SM (See 9.8.2.1 if insufficient material available for testing).

NOTE 8—If the fines plot as a *silty clay*, CL-ML, the second group symbol should be either GC or SC. For example, a poorly graded sand with 10 % fines, a liquid limit of 20, and a plasticity index of 6 would be classified as a poorly graded sand with silty clay, SP-SC.

12.6 If the specimen is predominantly sand or gravel but contains 15 % or more of the other coarse-grained constituent, the words "with gravel" or "with sand" shall be added to the group name. For example, poorly graded gravel with sand, clayey sand with gravel.

12.7 If the field sample contained any cobbles or boulders or both, the words "with cobbles," or "with cobbles and boulders" shall be added to the group name. For example, silty gravel with cobbles, GM.

13. Report

13.1 The report should include the group name, group symbol, and the results of the laboratory tests. The particle-size distribution shall be given in terms of percent of gravel, sand, and fines. The plot of the cumulative particle-size distribution curve shall be reported if used in classifying the soil. Report appropriate descriptive information according to the procedures in Practice D 2488. A local or commercial name or geologic interpretation for the material may be added at the end of the descriptive information if identified as such. The test procedures used shall be referenced.

NOTE 9—*Example: Clayey Gravel with Sand and Cobbles* (GC)—46 % fine to coarse, hard, subrounded gravel; 30 % fine to coarse, hard, subrounded sand; 24 % clayey fines, LL = 38, PI = 19; weak reaction with HCl; original field sample had 4 % hard, subrounded cobbles; maximum dimension 150 mm.

In-Place Conditions—firm, homogeneous, dry, brown,
Geologic Interpretation—alluvial fan.

NOTE 10—Other examples of soil descriptions are given in Appendix X1.

14. Precision and Bias

14.1 This test method provides qualitative data only; therefore, a precision and bias statement is nonapplicable.

TABLE 1 Soil Classification Chart

			Criteria for Assigning Group Symbols and Group Names Using Laboratory Tests[A]	Soil Classification	
				Group Symbol	Group Name[B]
Coarse-Grained Soils More than 50 % retained on No. 200 sieve	Gravels More than 50 % of coarse fraction retained on No. 4 sieve	Clean Gravels Less than 5 % fines[C]	Cu ≥ 4 and 1 ≤ Cc ≤ 3[E]	GW	Well-graded gravel[F]
			Cu < 4 and/or 1 > Cc > 3[E]	GP	Poorly graded gravel[F]
		Gravels with Fines More than 12 % fines[C]	Fines classify as ML or MH	GM	Silty gravel[F,G,H]
			Fines classify as CL or CH	GC	Clayey gravel[F,G,H]
	Sands 50 % or more of coarse fraction passes No. 4 sieve	Clean Sands Less than 5 % fines[D]	Cu ≥ 6 and 1 ≤ Cc ≤ 3[E]	SW	Well-graded sand[I]
			Cu < 6 and/or 1 > Cc > 3[E]	SP	Poorly graded sand[I]
		Sands with Fines More than 12 % fines[D]	Fines classify as ML or MH	SM	Silty sand[G,H,I]
			Fines classify as CL or CH	SC	Clayey sand[G,H,I]
Fine-Grained Soils 50 % or more passes the No. 200 sieve	Silts and Clays Liquid limit less than 50	inorganic	PI > 7 and plots on or above "A" line[J]	CL	Lean clay[K,L,M]
			PI < 4 or plots below "A" line[J]	ML	Silt[K,L,M]
		organic	Liquid limit − oven dried / Liquid limit − not dried < 0.75	OL	Organic clay[K,L,M,N] / Organic silt[K,L,M,O]
	Silts and Clays Liquid limit 50 or more	inorganic	PI plots on or above "A" line	CH	Fat clay[K,L,M]
			PI plots below "A" line	MH	Elastic silt[K,L,M]
		organic	Liquid limit − oven dried / Liquid limit − not dried < 0.75	OH	Organic clay[K,L,M,P] / Organic silt[K,L,M,Q]
Highly organic soils			Primarily organic matter, dark in color, and organic odor	PT	Peat

[A] Based on the material passing the 3-in. (75-mm) sieve.

[B] If field sample contained cobbles or boulders, or both, add "with cobbles or boulders, or both" to group name.

[C] Gravels with 5 to 12 % fines require dual symbols:
GW-GM well-graded gravel with silt
GW-GC well-graded gravel with clay
GP-GM poorly graded gravel with silt
GP-GC poorly graded gravel with clay

[D] Sands with 5 to 12 % fines require dual symbols:
SW-SM well-graded sand with silt
SW-SC well-graded sand with clay
SP-SM poorly graded sand with silt
SP-SC poorly graded sand with clay

[E] $Cu = D_{60}/D_{10}$ $Cc = \dfrac{(D_{30})^2}{D_{10} \times D_{60}}$

[F] If soil contains ≥ 15 % sand, add "with sand" to group name.

[G] If fines classify as CL-ML, use dual symbol GC-GM, or SC-SM.

[H] If fines are organic, add "with organic fines" to group name.

[I] If soil contains ≥ 15 % gravel, add "with gravel" to group name.

[J] If Atterberg limits plot in hatched area, soil is a CL-ML, silty clay.

[K] If soil contains 15 to 29 % plus No. 200, add "with sand" or "with gravel," whichever is predominant.

[L] If soil contains ≥ 30 % plus No. 200, predominantly sand, add "sandy" to group name.

[M] If soil contains ≥ 30 % plus No. 200, predominantly gravel, add "gravelly" to group name.

[N] PI ≥ 4 and plots on or above "A" line.

[O] PI < 4 or plots below "A" line.

[P] PI plots on or above "A" line.

[Q] PI plots below "A" line.

318

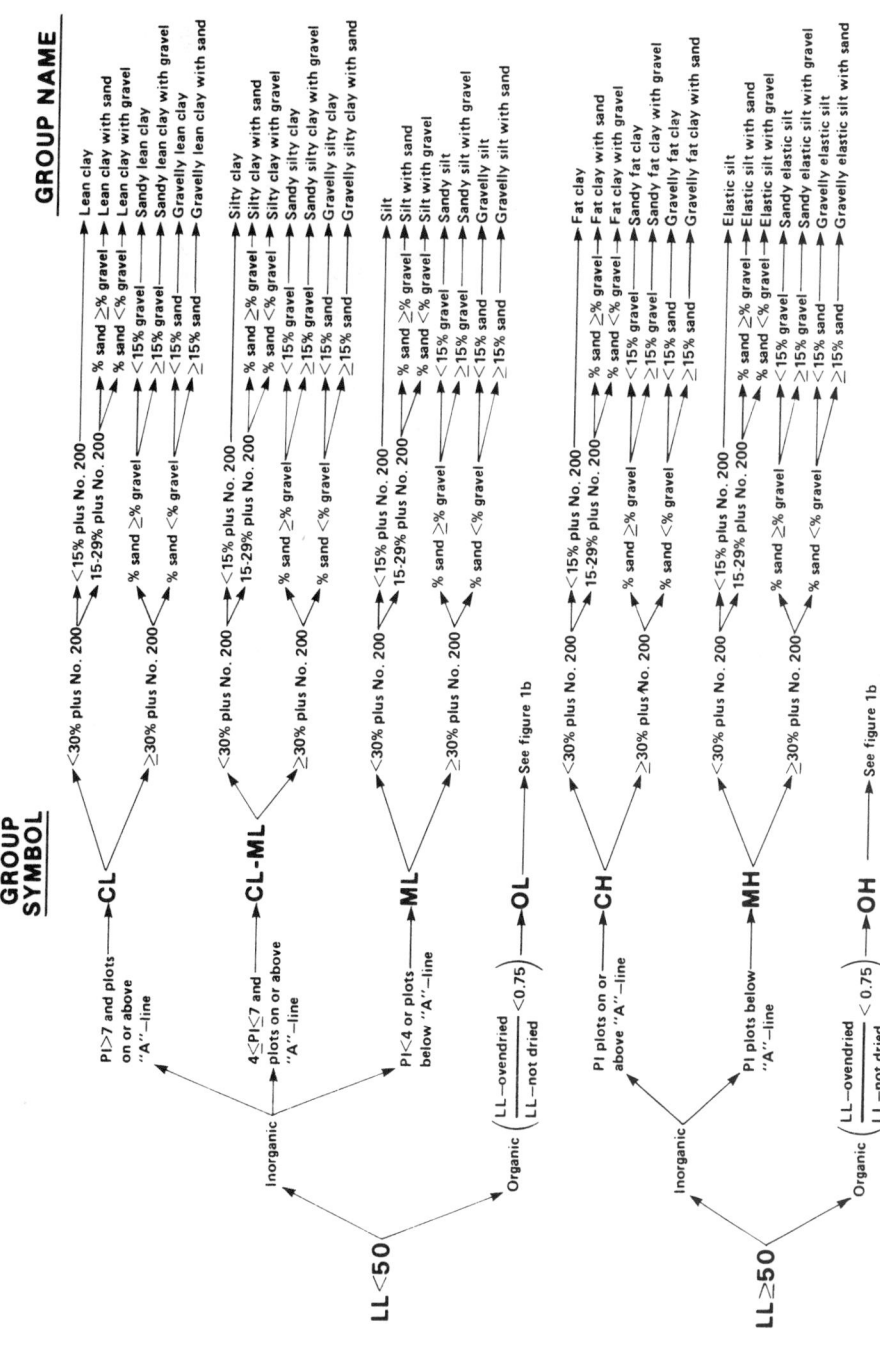

FIG. 1a Flow Chart for Classifying Fine-Grained Soil (50 % or More Passes No. 200 Sieve)

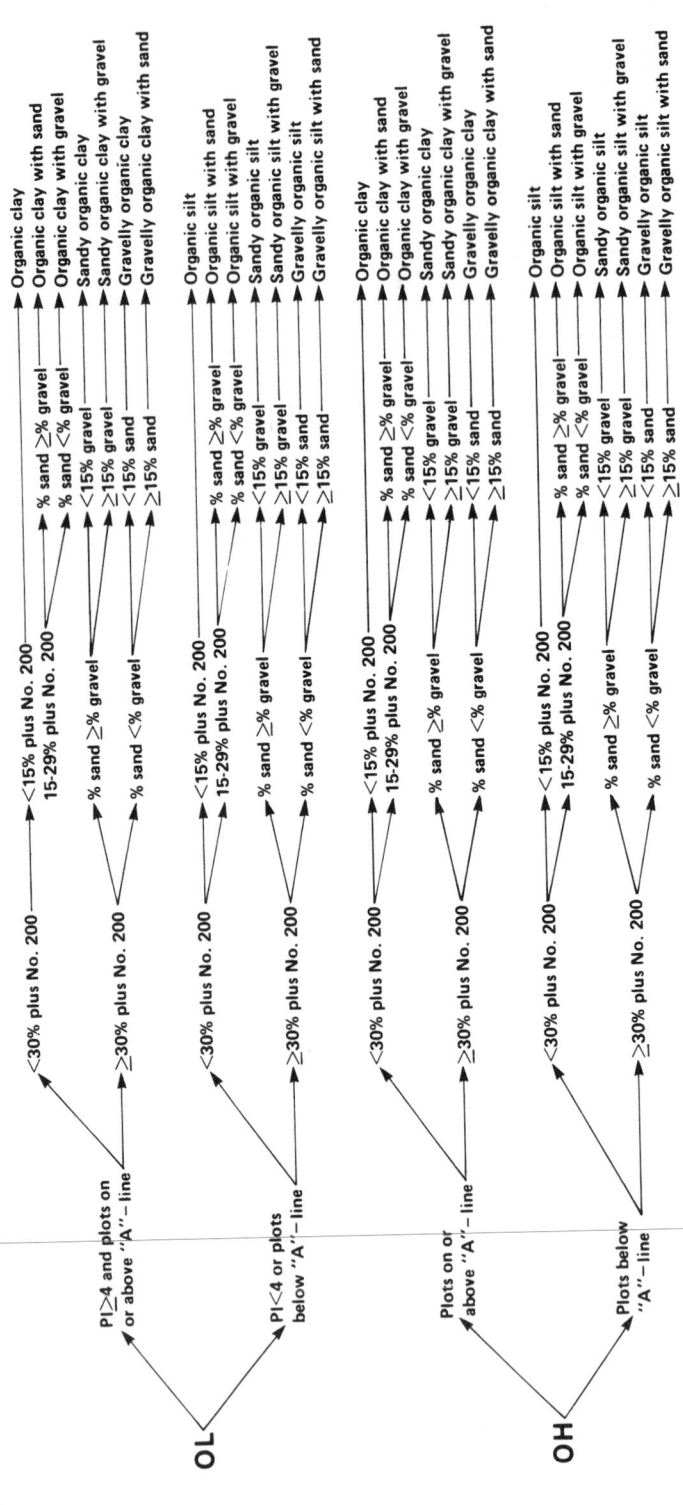

FIG. 1b Flow Chart for Classifying Organic Fine-Grained Soil (50 % or More Passes No. 200 Sieve)

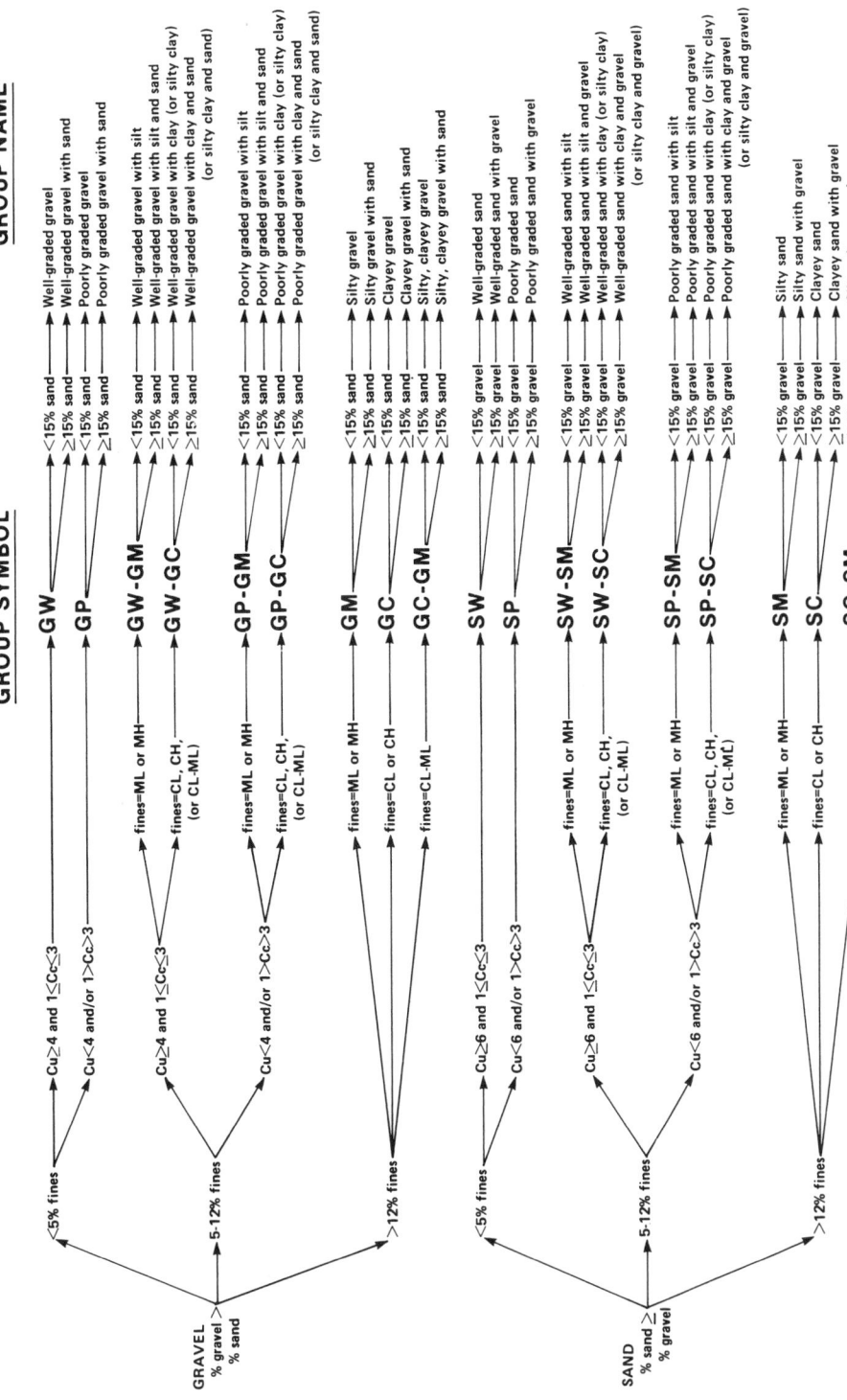

FIG. 2 Flow Chart for Classifying Coarse-Grained Soils (More Than 50 % Retained on No. 200 Sieve)

321

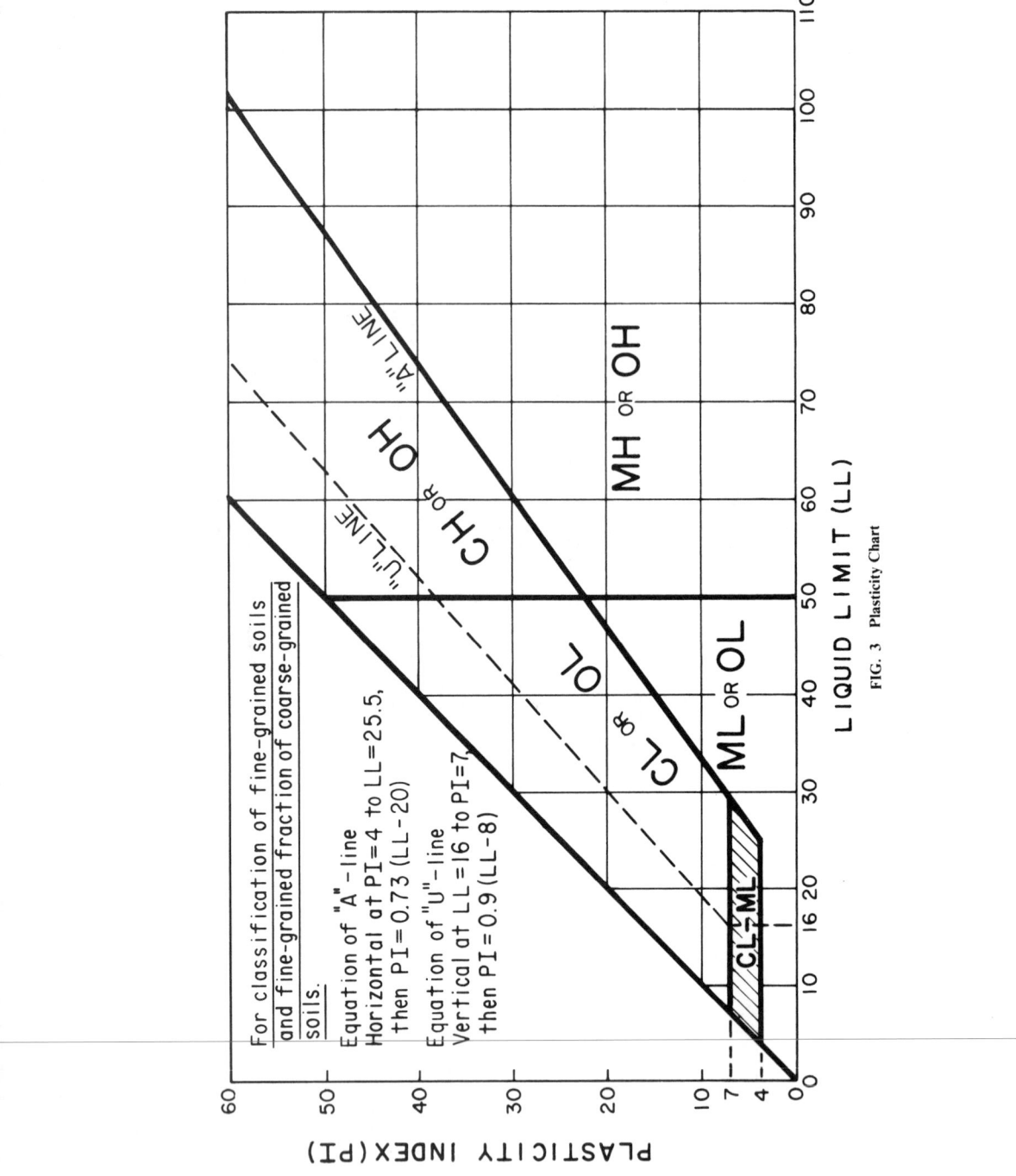

FIG. 3 Plasticity Chart

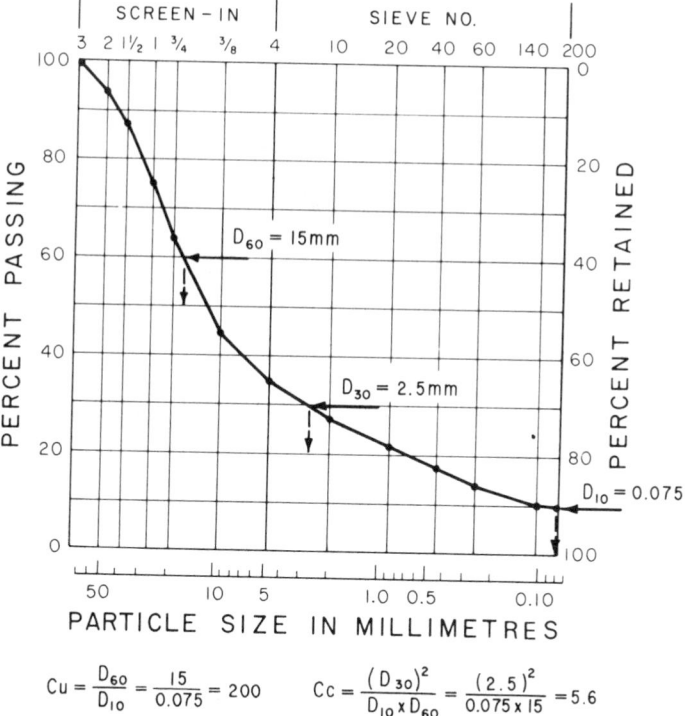

$$Cu = \frac{D_{60}}{D_{10}} = \frac{15}{0.075} = 200 \qquad Cc = \frac{(D_{30})^2}{D_{10} \times D_{60}} = \frac{(2.5)^2}{0.075 \times 15} = 5.6$$

FIG. 4 Cumulative Particle-Size Plot

APPENDIXES

(Nonmandatory Information)

X1. EXAMPLES OF DESCRIPTIONS USING SOIL CLASSIFICATION

X1.1 The following examples show how the information required in 13.1 can be reported. The appropriate descriptive information from Practice D 2488 is included for illustrative purposes. The additional descriptive terms that would accompany the soil classification should be based on the intended use of the classification and the individual circumstances.

X1.1.1 *Well-Graded Gravel with Sand (GW)*—73 % fine to coarse, hard, subangular gravel; 23 % fine to coarse, hard, subangular sand; 4 % fines; Cc = 2.7, Cu = 12.4.

X1.1.2 *Silty Sand with Gravel (SM)*—61 % predominantly fine sand; 23 % silty fines, LL = 33, PI = 6; 16 % fine, hard, subrounded gravel; no reaction with HCl; (field sample smaller than recommended). *In-Place Conditions*—Firm, stratified and contains lenses of silt 1 to 2 in. thick, moist, brown to gray; in-place

density = 106 lb/ft³ and in-place moisture = 9 %.

X1.1.3 *Organic Clay (OL)*—100 % fines, LL (not dried) = 32, LL (oven dried) = 21, PI (not dried) = 10; wet, dark brown, organic odor, weak reaction with HCl.

X1.1.4 *Silty Sand with Organic Fines (SM)*—74 % fine to coarse, hard, subangular reddish sand; 26 % organic and silty dark-brown fines, LL (not dried) = 37, LL (oven dried) = 26, PI (not dried) = 6, wet, weak reaction with HCl.

X1.1.5 *Poorly Graded Gravel with Silt, Sand, Cobbles and Boulders (GP-GM)*—78 % fine to coarse, hard, subrounded to subangular gravel; 16 % fine to coarse, hard, subrounded to subangular sand; 6 % silty (estimated) fines; moist, brown; no reaction with HCl; original field sample had 7 % hard, subrounded cobbles and 2 % hard, subrounded boulders with a maximum dimension of 18 in.

323

X2. USING SOIL CLASSIFICATION AS A DESCRIPTIVE SYSTEM FOR SHALE, CLAYSTONE, SHELLS, SLAG, CRUSHED ROCK, ETC.

X2.1 The group names and symbols used in this test method may be used as a descriptive system applied to materials that exist in situ as shale, claystone, sandstone, siltstone, mudstone, etc., but convert to soils after field or laboratory processing (crushing, slaking, etc.).

X2.2 Materials such as shells, crushed rock, slag, etc., should be identified as such. However, the procedures used in this method for describing the particle size and plasticity characteristics may be used in the description of the material. If desired, a classification in accordance with this test method may be assigned to aid in describing the material.

X2.3 If a classification is used, the group symbol(s) and group names should be placed in quotation marks or noted with some type of distinguishing symbol. See examples.

X2.4 Examples of how soil classifications could be incorporated into a description system for materials that are not naturally occurring soils are as follows:

X2.4.1 *Shale Chunks*—Retrieved as 2 to 4-in. pieces of shale from power auger hole, dry, brown, no reaction with HCl. After laboratory processing by slaking in water for 24 h, material classified as "Sandy Lean Clay (CL)"—61 % clayey fines, LL = 37, PI = 16; 33 % fine to medium sand; 6 % gravel-size pieces of shale.

X2.4.2 *Crushed Sandstone*—Product of commercial crushing operation; "Poorly Graded Sand with Silt (SP-SM)"—91 % fine to medium sand; 9 % silty (estimated) fines; dry, reddish-brown, strong reaction with HCl.

X2.4.3 *Broken Shells*—62 % gravel-size broken shells; 31 % sand and sand-size shell pieces; 7 % fines; would be classified as "Poorly Graded Gravel with Sand (GP)".

X2.4.4 *Crushed Rock*—Processed gravel and cobbles from Pit No. 7; "Poorly Graded Gravel (GP)"—89 % fine, hard, angular gravel-size particles; 11 % coarse, hard, angular sand-size particles, dry, tan; no reaction with HCl; Cc = 2.4, Cu = 0.9.

X3. PREPARATION AND TESTING FOR CLASSIFICATION PURPOSES BY THE WET METHOD

X3.1 This appendix describes the steps in preparing a soil sample for testing for purposes of soil classification using a wet-preparation procedure.

X3.2 Samples prepared in accordance with this procedure should contain as much of their natural water content as possible and every effort should be made during obtaining, preparing, and transporting the samples to maintain the natural moisture.

X3.3 The procedures to be followed in this test method assume that the field sample contains fines, sand, gravel, and plus 3-in. (75-mm) particles and the cumulative particle-size distribution plus the liquid limit and plasticity index values are required (see 9.8). Some of the following steps may be omitted when they are not applicable to the soil being tested.

X3.4 If the soil contains plus No. 200 (75-µm) particles that would degrade during dry sieving, use a test procedure for determining the particle-size characteristics that prevents this degradation.

X3.5 Since this classification system is limited to the portion of a sample passing the 3-in. (75-mm) sieve, the plus 3-in. (75-mm) material shall be removed prior to the determination of the particle-size characteristics and the liquid limit and plasticity index.

X3.6 The portion of the field sample finer than the 3-in. (75-mm) sieve shall be obtained as follows:

X3.6.1 Separate the field sample into two fractions on a 3-in. (75-mm) sieve, being careful to maintain the natural water content in the minus 3-in. (75-mm) fraction. Any particles adhering to the plus 3-in. (75-mm) particles shall be brushed or wiped off and placed in the fraction passing the 3-in. (75-mm) sieve.

X3.6.2 Determine the air-dry or oven-dry weight of the fraction retained on the 3-in. (75-mm) sieve. Determine the total (wet) weight of the fraction passing the 3-in. (75-mm) sieve.

X3.6.3 Thoroughly mix the fraction passing the 3-in. (75-mm) sieve. Determine the water content, in accordance with Method D 2216, of a representative specimen with a minimum dry weight as required in 7.2. Save the water-content specimen for determination of the particle-size analysis in accordance with X3.8.

X3.6.4 Compute the dry weight of the fraction passing the 3-in. (75-mm) sieve based on the water content and total (wet) weight. Compute the total dry weight of the sample and calculate the percentage of material retained on the 3-in. (75-mm) sieve.

X3.7 Determine the liquid limit and plasticity index as follows:

X3.7.1 If the soil disaggregates readily, mix on a clean, hard surface and select a representative sample by quartering in accordance with Methods C 702.

X3.7.1.1 If the soil contains coarse-grained particles coated with and bound together by tough clayey material, take extreme care in obtaining a representative portion of the No. 40 (425-µm) fraction. Typically, a larger portion than normal has to be selected, such as the minimum weights required in 7.2.

X3.7.1.2 To obtain a representative specimen of a basically cohesive soil, it may be advantageous to pass the soil through a ¾-in. (19-mm) sieve or other convenient size so the material can be more easily mixed and then quartered or split to obtain the representative specimen.

X3.7.2 Process the representative specimen in accordance with Procedure B of Practice D 2217.

X3.7.3 Perform the liquid-limit test in accordance with Test Method D 4318, except the soil shall not be air dried prior to the test.

X3.7.4 Perform the plastic-limit test in accordance with Test Method D 4318, except the soil shall not be air dried prior to the test, and calculate the plasticity index.

X3.8 Determine the particle-size distribution as follows:

X3.8.1 If the water content of the fraction passing the 3-in. (75-mm) sieve was required (X3.6.3), use the water-content specimen for determining the particle-size distribution. Otherwise, select a representative specimen in accordance with Methods C 702 with a minimum dry weight as required in 7.2.

X3.8.2 If the cumulative particle-size distribution including a hydrometer analysis is required, determine the particle-size distribution in accordance with Method D 422. See 9.7 for the set of required sieves.

X3.8.3 If the cumulative particle-size distribution without a hydrometer analysis is required, determine the particle-size distribution in accordance with Method C 136. See 9.7 for the set of required sieves. The specimen should be soaked until all clayey aggregations have softened and then washed in accordance with Test Method C 117 prior to performing the particle-size distribution.

X3.8.4 If the cumulative particle-size distribution is not required, determine the percent fines, percent sand, and percent gravel in the specimen in accordance with Test Method C 117, being sure to soak the specimen long enough to soften all clayey aggregations, followed by Method C 136 using a nest of sieves which shall include a No. 4 (4.75-mm) sieve and a No. 200 (75-µm) sieve.

X3.8.5 Calculate the percent fines, percent sand, and percent gravel in the minus 3-in. (75-mm) fraction for classification purposes.

X4. AIR-DRIED METHOD OF PREPARATION OF SOILS FOR TESTING FOR CLASSIFICATION PURPOSES

X4.1 This appendix describes the steps in preparing a soil sample for testing for purposes of soil classification when air-drying the soil before testing is specified or desired or when the natural moisture content is near that of an air-dried state.

X4.2 If the soil contains organic matter or mineral colloids that are irreversibly affected by air drying, the wet-preparation method as described in Appendix X3 should be used.

X4.3 Since this classification system is limited to the portion of a sample passing the 3-in. (75-mm) sieve, the plus 3-in. (75-mm) material shall be removed prior to the determination of the particle-size characteristics and the liquid limit and plasticity index.

X4.4 The portion of the field sample finer than the 3-in. (75-mm) sieve shall be obtained as follows:

X4.4.1 Air dry and weigh the field sample.

X4.4.2 Separate the field sample into two fractions on a 3-in. (75-mm) sieve.

X4.4.3 Weigh the two fractions and compute the percentage of the plus 3-in. (75-mm) material in the field sample.

X4.5 Determine the particle-size distribution and liquid limit and plasticity index as follows (see 9.8 for when these tests are required):

X4.5.1 Thoroughly mix the fraction passing the 3-in. (75-mm) sieve.

X4.5.2 If the cumulative particle-size distribution including a hydrometer analysis is required, determine the particle-size distribution in accordance with Method D 422. See 9.7 for the set of sieves that is required.

X4.5.3 If the cumulative particle-size distribution without a hydrometer analysis is required, determine the particle-size distribution in accordance with Test Method D 1140 followed by Method C 136. See 9.7 for the set of sieves that is required.

X4.5.4 If the cumulative particle-size distribution is not required, determine the percent fines, percent sand, and percent gravel in the specimen in accordance with Test Method D 1140 followed by Method C 136 using a nest of sieves which shall include a No. 4 (4.75-mm) sieve and a No. 200 (75-µm) sieve.

X4.5.5 If required, determine the liquid limit and the plasticity index of the test specimen in accordance with Test Method D 4318.

Standard Practice for
CLASSIFICATION OF SOILS AND SOIL-AGGREGATE MIXTURES FOR HIGHWAY CONSTRUCTION PURPOSES[1]

This standard is issued under the fixed designation D 3282; the number immediately following the designation indicates the year of original adoption or, in the case of revision, the year of last revision. A number in parentheses indicates the year of last reapproval. A superscript epsilon (ε) indicates an editorial change since the last revision or reapproval.

1. Scope

1.1 This practice describes a procedure for classifying mineral and organomineral soils into seven groups based on laboratory determination of particle-size distribution, liquid limit, and plasticity index. It may be used when a precise engineering classification is required, especially for highway construction purposes. Evaluation of soils within each group is made by means of a *group index*, which is a value calculated from an empirical formula.

NOTE 1—The group classification, including the group index, should be useful in determining the relative quality of the soil material for use in earthwork structures, particularly embankments, subgrades, sub-bases, and bases. However, for the detailed design of important structures, additional data concerning strength or performance characteristics of the soil under field conditions will usually be required.

1.2 *This standard may involve hazardous materials, operations, and equipment. This standard does not purport to address all of the safety problems associated with its use. It is the responsibility of whoever uses this standard to consult and establish appropriate safety and health practices and determine the applicability of regulatory limitations prior to use.*

2. Applicable Documents

2.1 *ASTM Standards:*

D 420 Recommended Practice for Investigating and Sampling Soil and Rock for Engineering Purposes[2]

D 421 Practice for Dry Preparation of Soil Samples for Particle-Size Analysis and Determination of Soil Constants[2]

)422 Method for Particle-Size Analysis of Soils[2]

D 423 Test Method for Liquid Limit of Soils[3]

D 424 Test Method for Plastic Limit and Plasticity Index of Soils[3]

D 653 Terms and Symbols Relating to Soil and Rock Mechanics[2]

D 1140 Test Method for Amount of Material in Soils Finer than the No. 200 (75-μm) Sieve[2]

D 1452 Practice for Soil Investigation and Sampling by Auger Borings[2]

D 1586 Method for Penetration Test and Split-Barrel Sampling of Soils[2]

D 1587 Practice for Thin-Walled Tube Sampling of Soils[2]

D 2217 Practice for Wet Preparation of Soil Samples for Particle Size Analysis and Determination of Soil Constants[2]

3. Significance and Use

3.1 The practice described classifies soils from any geographic location into groups (including group indexes) based on the results of prescribed laboratory tests to determine the particle-size characteristics, liquid limit, and plasticity index.

3.2 The assigning of a group symbol and group index can be used to aid in the evaluation of the significant properties of the soil for high-

[1] This practice is under the jurisdiction of ASTM Committee D-18 on Soil and Rock and is the direct responsibility of Subcommittee D18.07 on Identification and Classification of Soils.

Current edition approved Nov. 28, 1983. Published January 1984. Originally published as D 3282 – 73. Last previous edition D 3282 – 73 (1978).

[2] *Annual Book of ASTM Standards*, Vol 04.08.

[3] Discontinued, see *1983 Annual Book of ASTM Standards*, Vol 04.08.

way and airfield purposes.

3.3 The various groupings of this classification system correlate in a general way with the engineering behavior of soils. Also, in a general way, the engineering behavior of a soil varies inversely with its group index. Therefore, this practice provides a useful first step in any field or laboratory investigation for geotechnical engineering purposes.

4. Descriptions of Terms Specific to This Standard

4.1 The following terms are frequently used in this standard. These terms differ slightly from those given in Definitions D 653, but are used here to maintain consistency with common highway usage.

4.1.1 *boulders*—rock fragments, usually rounded by weathering or abrasion, that will be retained on a 3-in. (75-mm) sieve.

4.1.2 *gravel*—particles of rock that will pass a sieve with 3-in. (75-mm) square openings and be retained on a No. 10 (2-mm) sieve.

4.1.3 *coarse sand*—particles of rock or soil that will pass a No. 10 (2-mm) sieve and be retained on a No. 40 (425-μm) sieve.

4.1.4 *fine sand*—particles of rock or soil that will pass a No. 40 (425-μm) sieve and be retained on a No. 200 (75-μm) sieve.

4.1.5 *silt-clay (combined silt and clay)*—fine soil and rock particles that will pass a No. 200 (75-μm) sieve.

4.1.5.1 *silty*—fine-grained material that has a plasticity index of 10 or less.

4.1.5.2 *clayey*—fine-grained material that has a plasticity index of 11 or more.

5. Apparatus

5.1 *Apparatus for Preparation of Samples*—See Practice D 421 or Practice D 2217.

5.2 *Apparatus for Particle-Size Analysis*—See Test Method D 1140 and Method D 422.

5.3 *Apparatus for Liquid Limit Test*—See Test Method D 423.

5.4 *Apparatus for Plastic Limit Test*—See Test Method D 424.

6. Sampling

6.1 Conduct field investigations and sampling in accordance with one or more of the following procedures:

6.1.1 Recommended Practice D 420,

6.1.2 Practice D 1452,

6.1.3 Method D 1586,

6.1.4 Practice D 1587.

7. Test Sample

7.1 Test samples shall represent that portion of the field sample finer than the 3-in. (75-mm) sieve and shall be obtained as follows:

7.1.1 Air-dry the field sample,

7.1.2 Weigh the field sample,

7.1.3 Separate the field sample into two fractions on a 3-in. (75-mm) sieve,

7.1.4 Weigh the fraction retained on the 3-in. (75-mm) sieve. Compute the percentage of plus 3-in. material in the field sample, and note this percentage as auxiliary information, and

7.1.5 Thoroughly mix the fraction passing the 3-in. (75-mm) sieve and select the test samples.

NOTE 2—If visual examination indicates that no boulder size material is present, omit 7.1.

7.2 Prepare the test sample in accordance with Practice D 421 or Practice D 2217. Determine the percentage of the sample finer than a No. 10 (2-mm) sieve.

NOTE 3—It is recommended that the method for wet preparation be used for soils containing organic matter or irreversible mineral colloids.

8. Testing Procedure

8.1 Determine the percentage of the test sample finer than a No. 200 (75-μm) sieve in accordance with Test Method D 1140 or Method D 422.

NOTE 4—For granular materials the percentage of the sample finer than a No. 40 (425-μm) sieve must also be determined.

8.2 Determine the liquid limit and the plasticity index of a portion of the test sample passing a No. 40 (425-μm) sieve in accordance with Test Method D 423 and Test Method D 424.

9. Classification Procedure

9.1 Using the test limits determined in Section 8, classify the soil into the appropriate group or subgroup or both in accordance with Tables 1 or 2. Use Fig. 1 to classify silt-clay materials on the basis of liquid limit and plasticity index values.

NOTE 5—All limiting values are shown as whole numbers. If fractional numbers appear on test reports, convert to the nearest whole numbers for the purpose of classification.

9.1.1 With the required test data available, proceed from left to right in Tables 1 or 2 and the correct classification will be found by the process of elimination. The first group from the left into which the test data will fit is the correct classification.

NOTE 6—Classification of materials in the various groups applies only to the fraction passing the 3-in. (75-mm) sieve. Therefore, any specification regarding the use of A-1, A-2, or A-3 materials in construction should state whether boulders (retained on 3-in. sieve) are permitted.

10. Description of Classification Groups

10.1 *Granular Materials,* containing 35 % or less passing the No. 200 (75-μm) sieve:

10.1.1 *Group A-1*—The typical material of this group is a well-graded mixture of stone fragments or gravel, coarse sand, fine sand, and a nonplastic or feebly-plastic soil binder. However, this group also includes stone fragments, gravel, coarse sand, volcanic cinders, etc., without a soil binder.

10.1.1.1 Subgroup A-1-a includes those materials consisting predominantly of stone fragments or gravel, either with or without a well-graded binder of fine material.

10.1.1.2 Subgroup A-1-b includes those materials consisting predominantly of coarse sand, either with or without a well-graded soil binder.

10.1.2 *Group A-3*—The typical material of this group is fine beach sand or fine desert-blow sand without silty or clay fines, or with a very small amount of nonplastic silt. This group also includes stream-deposited mixtures of poorly-graded fine sand and limited amounts of coarse sand and gravel.

10.1.3 *Group A-2*—This group includes a wide variety of "granular" materials which are borderline between the materials falling in Groups A-1 and A-3, and the silt-clay materials of Groups A-4, A-5, A-6, and A-7. It includes all materials containing 35 % or less passing a No. 200 (75-μm) sieve which cannot be classified in Groups A-1 or A-3, due to the fines content or the plasticity indexes, or both, in excess of the limitations for those groups.

10.1.3.1 Subgroups A-2-4 and A-2-5 include various granular materials containing 35 % or less passing a No. 200 (75-μm) sieve and with a minus No. 40 (425-μm) portion having the characteristics of Groups A-4 and A-5, respectively. These groups include such materials as gravel and coarse sand with silt contents or plasticity indexes in excess of the limitations of Group A-1 and fine sand with nonplastic-silt content in excess of the limitations of Group A-3.

10.1.3.2 Subgroups A-2-6 and A-2-7 include materials similar to those described under Subgroups A-2-4 and A-2-5, except that the fine portion contains plastic clay having the characteristics of the A-6 or A-7 group, respectively.

10.2 *Silt-Clay Materials,* containing more than 35 % passing a No. 200 (75-μm) sieve:

10.2.1 *Group A-4*—The typical material of this group is a nonplastic or moderately plastic silty soil usually having 75 % or more passing a No. 200 (75-μm) sieve. This group also includes mixtures of fine silty soil and up to 64 % of sand and gravel retained on a No. 200 sieve.

10.2.2 *Group A-5*—The typical material of this group is similar to that described under Group A-4, except that it is usually of diatomaceous or micaceous character and may be highly elastic as indicated by the high liquid limit.

10.2.3 *Group A-6*—The typical material of this group is a plastic clay soil usually having 75 % or more passing a No. 200 (75-μm) sieve. This group also includes mixtures of fine clayey soil and up to 64 % of sand and gravel retained on a No. 200 sieve. Materials of this group usually have a high volume change between wet and dry states.

10.2.4 *Group A-7*—The typical material of this group is similar to that described under Group A-6, except that it has the high liquid limits characteristic of Group A-5 and may be elastic as well as subject to high-volume change.

10.2.4.1 Subgroup A-7-5 includes those materials with moderate plasticity indexes in relation to the liquid limit and which may be highly elastic as well as subject to considerable volume change.

10.2.4.2 Subgroup A-7-6 includes those materials with high plasticity indexes in relation to liquid limit and which are subject to extremely high volume change.

NOTE 7—Highly organic soils (peat or muck) may be classified in Group A-8. Classification of these materials is based on visual inspection and is not dependent on the percentage passing the No. 200 (75-μm) sieve, liquid limit, or plasticity index. The material is composed primarily of partially-decayed organic matter, generally has a fibrous texture, a dark brown or black color, and an odor of decay. These organic materials are unsuitable for use in embankments and subgrades. They are highly compressible and have low strength.

11. Group Index Computation

11.1 The classifications obtained from Tables 1 or 2 may be modified by the addition of a group-index value. Group-index values should always be shown in parentheses after the group symbol as A-2-6(3), A-4(5), A-6(12), A-7-5(17), etc.

11.1.1 Calculate the group index from the following empirical formula:

$$\text{Group index} = (F - 35)[0.2 + 0.005(LL - 40)] \\ + 0.01(F - 15)(PI - 10)$$

where:

F = percentage passing No. 200 (75-μm) sieve, expressed as a whole number (this percentage is based only on the material passing the 3-in. (75-mm) sieve),

LL = liquid limit, and

PI = plasticity index.

11.1.2 If the calculated group index is negative, report the group index as zero (0).

11.1.3 If the soil is nonplastic and when the liquid limit cannot be determined, report the group index as zero (0).

11.1.4 Report the group index to the nearest whole number.

11.1.5 The group index value may be estimated using Fig. 2 by determining the partial group index due to the liquid limit and that due to the plasticity index, then obtaining the total of the two partial group indexes.

11.1.6 The group index of soils in the A-2-6 and A-2-7 subgroups shall be calculated using only the PI portion of the formula (or Fig. 2).

11.2 The following examples illustrate the calculations for the group index:

11.2.1 Assume that an A-6 material has 55 % passing a No. 200 (75-μm) sieve, a liquid limit of 40, and a plasticity index of 25, then:

$$\text{Group index} = (55 - 35)[0.2 + 0.005(40 - 40)] \\ + [0.01(55 - 15)(25 - 10)] \\ = 4.0 + 6.0 = 10$$

11.2.2 Assume that an A-7 material has 80 % passing a No. 200 (75-μm) sieve, a liquid limit of 90, and a plasticity index of 50, then:

$$\text{Group index} = (80 - 35)[0.2 \pm 0.005(90 - 40)] \\ + [0.01(80 - 15)(50 - 10)] \\ = 20.3 + 26.0 = 46.3 \text{ (report as 46)}$$

11.2.3 Assume that an A-4 material has 60 % passing a No. 200 (75-μm) sieve, a liquid limit of 25, and a plasticity index of 1, then

$$\text{Group index} = (60 - 35)[0.2 + 0.005(25 - 40)] \\ + [0.01(60 - 15)(1 - 10)] \\ = 25 \times (0.2 - 0.075) + 0.01(45)(-9) \\ = 3.1 - 4.1 = -1.0 \text{ (report as 0)}$$

11.2.4 Assume that an A-2-7 material has

30 % passing a No. 200 (75-μm) sieve, a liquid limit of 50, and a plasticity index of 30, then

$$\text{Group index} = 0.01(30 - 15)(30 - 10) \\ = 3.0 \text{ or 3 (note that only the } PI \text{ portion of the formula was used)}$$

12. Discussion of Group Index

12.1 The empirical group index formula devised for approximate within-group evaluation of the "clayey-granular materials" and the "silt-clay materials" is based on the following assumptions:

12.1.1 Materials falling within Groups A-1-a, A-1-b, A-2-4, A-2-5, and A-3 are satisfactory as subgrade when properly drained and compacted under moderate thickness of pavement (base or surface course, or both) of a type suitable for traffic to be carried or can be made satisfactory by additions of small amounts of natural or artificial binders.

12.1.2 Materials falling within the "clayey granular" Groups A-2-6 and A-2-7 and the "silt-clay" Groups A-4, A-5, A-6, and A-7 will range in quality as subgrade from the approximate equivalent of the good A-2-4 and A-2-5 subgrades to fair and poor subgrades requiring a layer of subbase material or an increased thickness of base course over that required in 12.1.1, in order to furnish adequate support for traffic loads.

12.1.3 A minimum of 35 % passing a No. 200 (75-μm) sieve is assumed to be critical if plasticity is neglected, but the critical minimum is only 15 % when affected by plasticity indexes greater than 10.

12.1.4 Liquid limits of 40 and above are assumed to be critical.

12.1.5 Plasticity indexes of 10 and above are assumed to be critical.

12.2 There is no upper limit of group index value obtained by use of the formula: The adopted critical values of percentage passing the No. 200 (75-μm) sieve, liquid limit, and plasticity index, are based on an evaluation of subgrade, subbase, and base-course materials by several highway organizations that use the tests involved in this classification system.

12.3 Under average conditions of good drainage and thorough compaction, the supporting value of a material as subgrade may be assumed as an inverse ratio to its group index; that is, a group index of 0 indicates a "good" subgrade material and a group index of 20 or greater indicates a "very poor" subgrade material.

TABLE 1 Classification of Soils and Soil-Aggregate Mixtures

General Classification	Granular Materials (35 % or less passing No. 200)			Silt-Clay Materials (More than 35 % passing No. 200)			
Group Classification	A-1	A-3[A]	A-2	A-4	A-5	A-6	A-7
Sieve analysis, % passing:							
No. 10 (2.00 mm)	...	...	...	...	...	...	...
No. 40 (425 μm)	50 max	51 min	...	...	...	...	...
No. 200 (75 μm)	25 max	10 max	35 max	36 min	36 min	36 min	36 min
Characteristics of fraction passing No. 40 (425 μm):							
Liquid limit	...	...	B	40 max	41 min	40 max	41 min
Plasticity index	6 max	N.P.	B	10 max	10 max	11 min	11 min
General rating as subgrade	Excellent to Good			Fair to Poor			

[A]The placing of A-3 before A-2 is necessary in the "left to right elimination process" and does not indicate superiority of A-3 over A-2.

[B]See Table 2 for values.

Reprinted with permission of American Association of State Highway and Transportation Officials

TABLE 2 Classification of Soils and Soil-Aggregate Mixtures

General Classification	Granular Materials (35 % or less passing No. 200)							Silt-Clay Materials (More than 35 % passing No. 200)			
	A-1		A-3	A-2				A-4	A-5	A-6	A-7
Group classification	A-1-a	A-1-b		A-2-4	A-2-5	A-2-6	A-2-7				A-7-5, A-7-6
Sieve analysis, % passing:											
No. 10 (2.00 mm)	50 max	...	...	...	...	...	...	...	...	...	...
No. 40 (425 μm)	30 max	50 max	51 min	...	...	...	...	...	...	...	...
No. 200 (75 μm)	15 max	25 max	10 max	35 max	35 max	35 max	35 max	36 min	36 min	36 min	36 min
Characteristics of fraction passing No. 40 (425 μm):											
Liquid limit			...	40 max	41 min	40 max	41 min	40 max	41 min	40 max	41 min
Plasticity index	6 max		N.P.	10 max	10 max	11 min	11 min	10 max	10 max	11 min	11 min[a]
Usual types of significant constituent materials	Stone Fragments, Gravel and Sand		Fine Sand	Silty or Clayey Gravel and Sand				Silty Soils		Clayey Soils	
General rating as subgrade	Excellent to Good							Fair to Poor			

[a]Plasticity index of A-7-5 subgroup is equal to or less than *LL* minus 30. Plasticity index of A-7-6 subgroup is greater than *LL* minus 30 (see Fig. 1).

Reprinted with permission of American Association of State Highway and Transportation Officials.

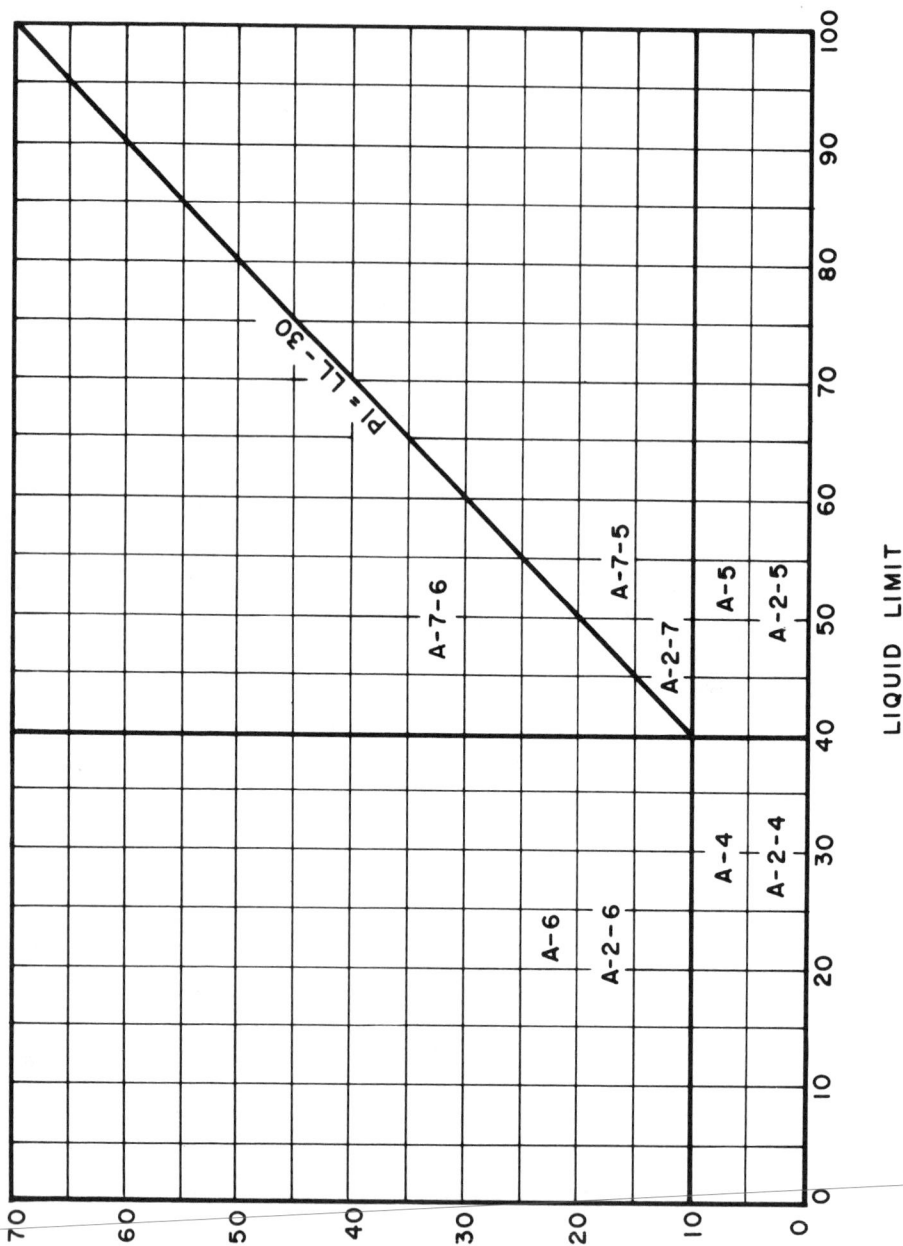

PLASTICITY INDEX

NOTE—A-2 soils contain less than 35 % finer than 200 sieve.

FIG. 1 Liquid Limit and Plasticity Index Ranges for Silt-Clay Materials

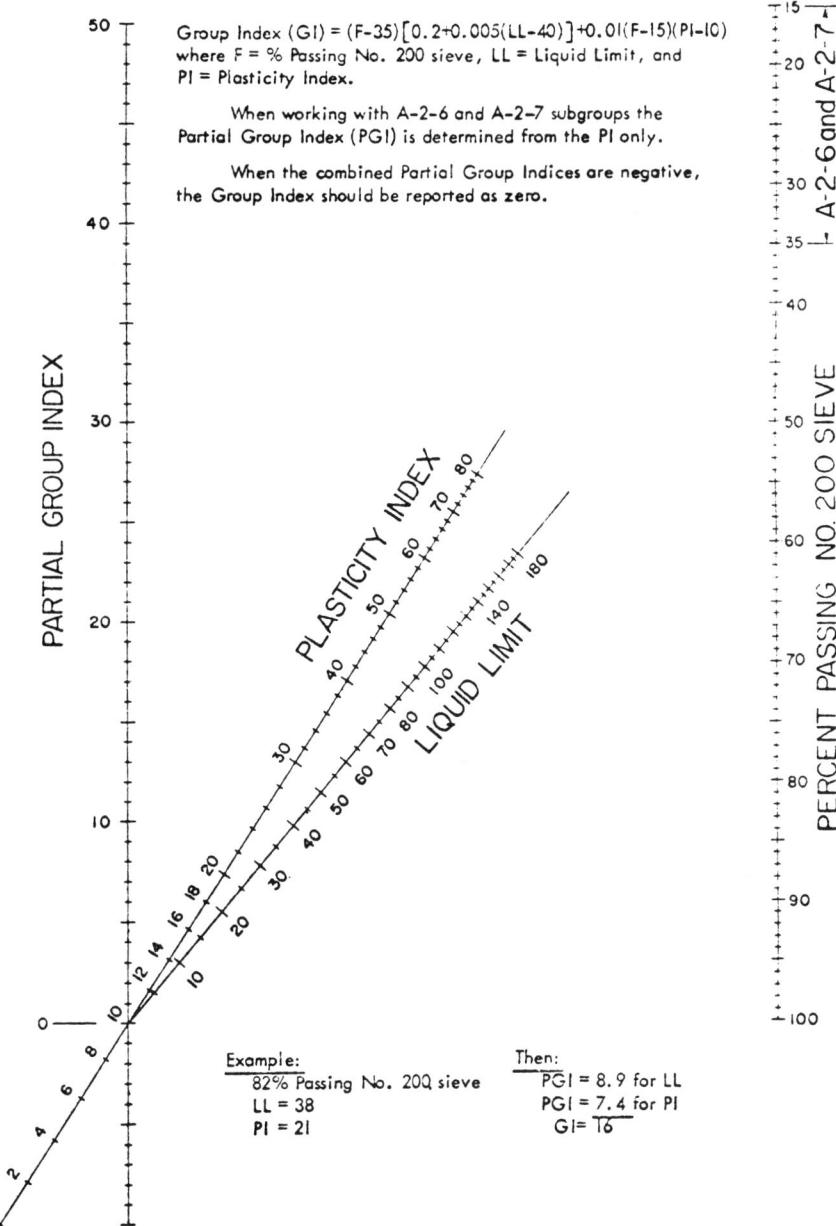

Group Index (GI) = (F-35)[0.2+0.005(LL-40)]+0.01(F-15)(PI-10) where F = % Passing No. 200 sieve, LL = Liquid Limit, and PI = Plasticity Index.

When working with A-2-6 and A-2-7 subgroups the Partial Group Index (PGI) is determined from the PI only.

When the combined Partial Group Indices are negative, the Group Index should be reported as zero.

Example:
82% Passing No. 200 sieve
LL = 38
PI = 21

Then:
PGI = 8.9 for LL
PGI = 7.4 for PI
GI = 16

Reprinted with permission of American Association of State Highway and Transportation Officials

FIG. 2 Group Index Chart

Designation: D 698 – 78

An American National Standard

Standard Test Methods for
MOISTURE-DENSITY RELATIONS OF SOILS AND SOIL-AGGREGATE MIXTURES USING 5.5-lb (2.49-kg) RAMMER AND 12-in. (305-mm) DROP[1]

This standard is issued under the fixed designation D 698; the number immediately following the designation indicates the year of original adoption or, in the case of revision, the year of last revision. A number in parentheses indicates the year of last reapproval. A superscript epsilon (ϵ) indicates an editorial change since the last revision or reapproval.

These methods have been approved for use by agencies of the Department of Defense and for listing in the DoD Index of Specifications and Standards

1. Scope

1.1 These laboratory compaction methods cover the determination of the relationship between the moisture content and density of soils and soil-aggregate mixtures (Note 1) when compacted in a mold of a given size with a 5.5-lb (2.49-kg) rammer dropped from a height of 12 in. (305 mm) (Note 2). Four alternative procedures are provided as follows:

1.1.1 *Method A*—A 4-in. (101.6-mm) mold; material passing a No. 4 (4.75-mm) sieve;

1.1.2 *Method B*—A 6-in. (152.4-mm) mold; material passing a No. 4 (4.75-mm) sieve;

1.1.3 *Method C*—A 6-in. (152.4-mm) mold; material passing a ¾-in. (19.0-mm) sieve; and

1.1.4 *Method D*—A 6-in. (152.4-mm) mold; material passing a ¾-in. (19.0-mm) sieve, corrected by replacement for material retained on a ¾-in. sieve.

NOTE 1—Soils and soil-aggregate mixtures should be regarded as natural occurring fine- or coarse-grained soils or composites or mixtures of natural soils, or mixtures of natural and processed soils or aggregates such as silt, gravel, or crushed rock.

NOTE 2—These laboratory compaction test methods when used on soils and soil-aggregates which are not free-draining will, in most cases, establish a well-defined optimum moisture content and maximum density (see Section 7). However, for free-draining soils and soil-aggregate mixtures, these methods will not, in many cases, produce a well-defined moisture-density relationship and the maximum density obtained will generally be less than that obtained by vibratory methods.

1.2 The method to be used should be indicated in the specifications for the material being tested. If no method is specified, the provisions of Section 5 shall govern.

2. Applicable Documents

2.1 *ASTM Standards:*

C 127 Test Method for Specific Gravity and Absorption of Coarse Aggregate[2]

D 854 Test Method for Specific Gravity of Soils[3]

D 2168 Methods for Calibration of Laboratory Mechanical-Rammer Soil Compactors[3]

D 2216 Method for Laboratory Determination of Water (Moisture) Content of Soil, Rock, and Soil-Aggregate Mixtures[3]

D 2487 Test Method for Classification of Soils for Engineering Purposes[3]

D 2488 Practice for Description and Identification of Soils (Visual-Manual Procedure)[3]

E 11 Specification for Wire-Cloth Sieves for Testing Purposes[4]

3. Apparatus

3.1 *Molds*—The molds shall be cylindrical in shape, made of rigid metal and be within the capacity and dimensions indicated in 3.1.1 or 3.1.2. The molds may be the "split" type, consisting either of two half-round sections, or a

---

[1] These methods are under the jurisdiction of ASTM Committee D-18 on Soil and Rock.

Current edition approved April 27, 1978. Published July 1978. Originally published as D 698 – 42 T. Last previous edition D 698 – 70.

[2] *Annual Book of ASTM Standards*, Vols 04.02 and 04.03.

[3] *Annual Book of ASTM Standards*, Vol 04.08.

[4] *Annual Book of ASTM Standards*, Vols 04.01, 04.02, 04.06, 05.05, and 14.02.

section of pipe split along one element, which can be securely locked together to form a cylinder meeting the requirements of this section. The molds may also be the "taper" type, providing the internal diameter taper is uniform and is not more than 0.200 in./linear ft (16.7 mm/linear m) of mold height. Each mold shall have a base plate assembly and an extension collar assembly, both made of rigid metal and constructed so they can be securely attached to or detached from the mold. The extension collar assembly shall have a height extending above the top of the mold of at least 2 in. (50.8 mm), which may include an upper section that flares out to form a funnel provided there is at least a $^3/_4$-in. (19-mm) straight cylindrical section beneath it.

3.1.1 *Mold*, 4.0 in. (101.6 mm) in diameter, having a capacity of $^1/_{30} \pm 0.0004$ ft^3 (944 $\pm$ 11 cm^3) and conforming to Fig. 1.

3.1.2 *Mold*, 6.0 in. (152.4 mm) in diameter, having a capacity of $^1/_{13.333} \pm 0.0009$ ft^3 (2124 $\pm$ 25 cm^3) and conforming to Fig. 2.

3.1.3 The average internal diameter, height, and volume of each mold shall be determined before initial use and at intervals not exceeding 1000 times the mold is filled. The mold volume shall be calculated from the average of at least six internal diameter and three height measurements made to the nearest 0.001 in. (0.02 mm), or from the amount of water required to completely fill the mold, corrected for temperature variance in accordance with Table 1. If the average internal diameter and volume are not within the tolerances shown in Figs. 1 or 2, the mold shall not be used. The determined volume shall be used in computing the required densities.

3.2 *Rammer* — The rammer may be either manually operated (see 3.2.1) or mechanically operated (see 3.2.2). The rammer shall fall freely through a distance of 12.0 $\pm$ $^1/_{16}$ in. (304.8 $\pm$ 1.6 mm) from the surface of the specimen. The manufactured weight of the rammer shall be 5.5 $\pm$ 0.02 lb (2.49 $\pm$ 0.01 kg). The specimen contact face shall be flat.

3.2.1 *Manual Rammer* — The specimen contact face shall be circular with a diameter of 2.000 $\pm$ 0.005 in. (50.80 $\pm$ 0.13 mm). The rammer shall be equipped with a guide-sleeve which shall provide sufficient clearance so that the free fall of the rammer shaft and

head will not be restricted. The guidesleeve shall have four vent holes at each end (eight holes total) located with centers $^3/_4$ $\pm$ $^1/_{16}$ in. (19.0 $\pm$ 1.6 mm) from each end and spaced 90 deg apart. The minimum diameter of the vent holes shall be $^3/_8$ in. (9.5 mm).

3.2.2 *Mechanical Rammer* — The rammer shall operate mechanically in such a manner as to provide uniform and complete coverage of the specimen surface. There shall be 0.10 $\pm$ 0.03 in. (2.5 $\pm$ 0.8 mm) clearance between the rammer and the inside surface of the mold at its smallest diameter. When used with the 4.0-in. (101.6-mm) mold, the specimen contact face shall be circular with a diameter of 2.000 $\pm$ 0.005 in. (50.80 $\pm$ 0.13 mm). When used with the 6.0-in. (152.4-mm) mold, the specimen contact face shall have the shape of a section of a circle of a radius equal to 2.90 $\pm$ 0.02 in. (73.7 $\pm$ 0.5 mm). The sector face rammer shall operate in such a manner that the vertex of the sector is positioned at the center of the specimen. The mechanical rammer shall be calibrated and adjusted, as necessary, in accordance with 3.2.3.

3.2.3 *Calibration and Adjustment* — The mechanical rammer shall be calibrated, and adjusted as necessary, before initial use; near the end of each period during which the mold was filled 1000 times; before reuse after anything, including repairs, which may affect the test results significantly; and whenever the test results are questionable. Each calibration and adjustment shall be in accordance with Methods D 2168.

3.3 *Sample Extruder* (optional) — A jack, frame or other device adapted for the purpose of extruding compacted specimens from the mold.

3.4 *Balances* — A balance or scale of at least 20-kg capacity sensitive to $\pm$ 1 g and a balance of at least 1000-g capacity sensitive to $\pm$0.01 g.

3.5 *Drying Oven*, thermostatically controlled, preferably of the forced-draft type, capable of maintaining a temperature of 230 $\pm$ 9°F (110 $\pm$ 5°C) for determining the moisture content of the compacted specimen.

3.6 *Straightedge* — A stiff metal straightedge of any convenient length but not less than 10 in. (254 mm). The scraping edge shall have a straightness tolerance of $\pm$0.005 in. ($\pm$0.13 mm) and shall be beveled if it is thicker than $^1/_8$ in.

(3 mm).

3.7 *Sieves*, 3-in. (75-mm), ³/₄-in. (19.0-mm) and No. 4 (4.75-mm), conforming to the requirements of Specification E 11.

3.8 *Mixing Tools*—Miscellaneous tools such as mixing pan, spoon, trowel, spatula, etc., or a suitable mechanical device for thoroughly mixing the sample of soil with increments of water.

4. Procedure

4.1 *Specimen Preparation*—Select a representative portion of quantity adequate to provide, after sieving, an amount of material weighing as follows: Method A—25 lb (11 kg); Methods B, C, and D—50 lb (23 kg). Prepare specimens in accordance with either 4.1.1 through 4.1.3 or 4.1.4.

4.1.1 *Dry Preparation Procedure*—If the sample is too damp to be friable, reduce the moisture content by drying until the material is friable; see 4.1.2. Drying may be in air or by the use of a drying apparatus such that the temperature of the sample does not exceed 140°F (60°C). After drying (if required), thoroughly break up the aggregations in such a manner as to avoid reducing the natural size of the particles. Pass the material through the specified sieve as follows: Methods A and B— No. 4 (4.75-mm); Methods C and D—³/₄-in. (19.0-mm). Correct for oversize material in accordance with Section 5, if Method D is specified.

4.1.2 Whenever practicable, soils classified as ML, CL, OL, GC, SC, MH, CH, OH and PT by Test Method D 2487 shall be prepared in accordance with 4.1.4.

4.1.3 Prepare a series of at least four specimens by adding increasing amounts of water to each sample so that the moisture contents vary by approximately 1½ %. The moisture contents selected shall bracket the optimum moisture content, thus providing specimens which, when compacted, will increase in mass to the maximum density and then decrease in density (see 7.2 and 7.3). Thoroughly mix each specimen to ensure even distribution of moisture throughout and then place in a separate covered container and allow to stand prior to compaction in accordance with Table 2. For the purpose of selecting a standing time, it is not required to perform the actual classification procedures described in Test Method D 2487 (except in the case of referee testing), if previous data exist which provide a basis for classifying the sample.

4.1.4 *Moist Preparation Method*—The following alternate procedure is recommended for soils classified as ML, CL, OL, GC, SC, MH, CH, OH, and PT by Test Method D 2487. Without previously drying the sample, pass it through the ³/₄-in. (19.0-mm) and No. 4 (4.75-mm) sieves. Correct for oversize material in accordance with Section 5, if Method D is specified. Prepare a series of at least four specimens having moisture contents that vary by approximately 1½ %. The moisture contents selected shall bracket the optimum moisture content, thus providing specimens which, when compacted, will increase in mass to the maximum density and then decrease in density (see 7.2 and 7.3). To obtain the appropriate moisture content of each specimen, the addition of a predetermined amount of water (see 4.1.3) or the removal of a predetermined amount of moisture by drying may be necessary. Drying may be in air or by the use of a drying apparatus such that the temperature of the specimen does not exceed 140°F (60°C). The prepared specimens shall then be thoroughly mixed and stand, as specified in 4.1.3 and Table 2, prior to compaction.

NOTE 3—With practice, it is usually possible to visually judge the point of optimum moisture closely enough so that the prepared specimens will bracket the point of optimum moisture content.

4.2 *Specimen Compaction*—Select the proper compaction mold, in accordance with the method being used, and attach the mold extension collar. Compact each specimen in three layers of approximately equal height. Each layer shall receive 25 blows in the case of the 4-in. (101.6-mm) mold; each layer shall receive 56 blows in the case of the 6-in. (152.4-mm) mold. The total amount of material used shall be such that the third compacted layer is slightly above the top of the mold, but not exceeding ¼ in. (6 mm). During compaction the mold shall rest on a uniform rigid foundation, such as provided by a cylinder or cube of concrete weighing not less than 200 lb (91 kg).

4.2.1 In operating the manual rammer, care shall be taken to avoid rebound of the rammer from the top end of the guidesleeve.

The guidesleeve shall be held steady and within 5 deg of the vertical. Apply the blows at a uniform rate not exceeding approximately 1.4 s per blow and in such a manner as to provide complete coverage of the specimen surface.

4.2.2 Following compaction, remove the extension collar; carefully trim the compacted specimen even with the top of the mold by means of the straightedge and determine the mass of the specimen. Divide the mass of the compacted specimen and mold, minus the mass of the mold, by the volume of the mold (see 3.1.3). Record the result as the wet density, γ_m, in pounds per cubic foot (or kilograms per cubic metre) of the compacted specimen.

4.2.3 Remove the material from the mold. Determine moisture content in accordance with Method D 2216, using either the whole specimen or alternatively a representative specimen of the whole specimen. The whole specimen must be used when the permeability of the compacted specimen is high enough so that the moisture content is not distributed uniformly throughout. If the whole specimen is used, break it up to facilitate drying. Obtain the representative specimen by slicing the compacted specimen axially through the center and removing 100 to 500 g of material from one of the cut faces.

4.2.4 Repeat 4.2 through 4.2.3 for each specimen prepared.

5. Oversize Corrections

5.1 If 30 % or more of the sample is retained on a $^3/_4$-in. (19.0-mm) sieve, then none of the methods described under these methods shall be used for the determination of either maximum density or optimum moisture content.

5.2 *Methods A and B*—The material retained on the No. 4 (4.75-mm) sieve is discarded and no oversize correction is made. However, it is recommended that if the amount of material retained is 7 % or greater, Method C be used instead.

5.3 *Method C*—The material retained on the $^3/_4$-in. (19.0-mm) sieve is discarded and no oversize correction is made. However, if the amount of material retained is 10 % or greater, it is recommended that Method D be used instead.

5.4 *Method D:*

5.4.1 This method shall not be used unless the amount of material retained on the $^3/_4$-in. (19.0-mm) sieve is 10 % or greater. When the amount of material retained on the $^3/_4$-in. sieve is less than 10 %, use Method C.

5.4.2 Pass the material retained on the $^3/_4$-in. (19.0-mm) sieve through a 3-in. or 75-mm sieve. Discard the material retained on the 3-in. sieve. The material passing the 3-in. sieve and retained on the $^3/_4$-in. sieve shall be replaced with an equal amount of material passing a $^3/_4$-in. sieve and retained on a No. 4 (4.75-mm) sieve. The material for replacement shall be taken from an unused portion of the sample.

6. Calculations

6.1 Calculate the moisture content and the dry density of each compacted specimen as follows:

$$w = [(A - B)/(B - C)] \times 100$$

and

$$\gamma_d = [\gamma_m/(w + 100)] \times 100$$

where:

w = moisture content in percent of the compacted specimens,
A = mass of container and moist specimen,
B = mass of container and oven-dried specimen,
C = mass of container,
γ_d = dry density, in pounds per cubic foot (or kilograms per cubic metre) of the compacted specimen, and
γ_m = wet density, in pounds per cubic foot (or kilograms per cubic metre) of the compacted specimen.

7. Moisture-Density Relationship

7.1 From the data obtained in 6.1, plot the dry density values as ordinates with corresponding moisture contents as abscissas. Draw a smooth curve connecting the plotted points. Also draw a curve termed the "curve of complete saturation" or "zero air voids curve" on this plot. This curve represents the relationship between dry density and corresponding moisture contents when the voids are completely filled with water. Values of dry density and corresponding moisture contents for plotting the curve of complete saturation can be computed using the following equation:

$$w_{sat} = [(62.4/\gamma_d) - (1/G_s)] \times 100$$

where:

w_{sat} moisture content in percent for complete saturation,

γ_d dry density in pounds per cubic foot (or kilograms per cubic metre),

G_s specific gravity of the material being tested (see Note 4), and

62.4   density of water in pounds per cubic foot (or kilograms per cubic metre).

NOTE 4—The specific gravity of the material can either be assumed or based on the weighted average values of: (a) the specific gravity of the material passing the No. 4 (4.75-mm) sieve in accordance with Test Method D 854; and (b) the apparent specific gravity of the material retained on the No. 4 sieve in accordance with Test Method C 127.

7.2 *Optimum Moisture Content, w_o*—The moisture content corresponding to the peak of the curve drawn as directed in 7.1 shall be termed the "optimum moisture content."

7.3 *Maximum Density, γ_{max}*—The dry density in pounds per cubic foot (or kilograms per cubic metre) of the sample at "optimum moisture content" shall be termed "maximum density."

8. Report

8.1 The report shall include the following.

8.1.1 Method used (Method A, B, C, or D).

8.1.2 Optimum moisture content.

8.1.3 Maximum density.

8.1.4 Description of rammer (whether manual or mechanical).

8.1.5 Description of appearance of material used in test, based on Practice D 2488 (Test Method D 2487 may be used as an alternative).

8.1.6 Origin of material used in test.

8.1.7 Preparation procedure used (moist or dry).

9. Precision

9.1 Criteria for judging the acceptability of the maximum density and optimum moisture content test results are given in Table 3. The standard deviation, s, is calculated from the equation:

$$s^2 = \frac{1}{n-1} \sum_1^n (x - \bar{x})^2$$

where:

n = number of determinations,

x = individual value of each determination, and

$\bar{x}$ = numerical average of the determinations.

9.2 Criteria for assigning standard deviation values for single-operator precision are not available at the present time.

TABLE 1 Volume of Water per Gram based on Temperature[1]

Temperature, °C (°F)	Volume of Water, ml/g
12 (53.6)	1.00048
14 (57.2)	1.00073
16 (60.8)	1.00103
18 (64.4)	1.00138
20 (68.0)	1.00177
22 (71.6)	1.00221
24 (75.2)	1.00268
26 (78.8)	1.00320
28 (82.4)	1.00375
30 (86.0)	1.00435
32 (89.6)	1.00497

[1] Values other than shown may be obtained by referring to the *Handbook of Chemistry and Physics*, Chemical Rubber Publishing Co., Cleveland, Ohio.

TABLE 2 Dry Preparation Method—Standing Times

Classification D 2487	Minimum Standing Time, h
GW, GP, SW, SP	no requirement
GM, SM	3
ML, CL, OL, GC, SC	18
MH, CH, OH, PT	36

TABLE 3 Precision

	Standard Deviation, s	Acceptable Range of Two Results, Expressed as Percent of Mean Value[A]
Single-operator precision:		
Maximum density	. . .	1.9
Optimum moisture content	. . .	9.5
Multilaboratory precision:		
Maximum density	±1.66	4.0
Optimum moisture content	±0.86	15.0

[A] This column indicates a limiting range of values which should not be exceeded by the difference between any two results, expressed as a percentage of the average value. In cooperative test programs it has been determined that 95 % of the tests do not exceed the limiting acceptable ranges shown below. All values shown in this table are based on average test results from a variety of different soils and are subject to future revision.

TABLE 4 Metric Equivalents for Figs. 1 and 2

in.	mm
0.016	0.41
0.026	0.66
1/32	0.80
1/16	1.6
1/8	3.2
1/4	6.4
11/32	8.7
3/8	9.5
1/2	12.7
5/8	15.9
2	50.8
2 1/2	63.5
4	101.6
4 1/4	108.0
4 1/2	114.3
4.584	116.43
6	152.4
6 1/2	165.1
8	203.2

ft³	cm³
1/30	944
0.004	11
1/13.333	2124
0.0009	25

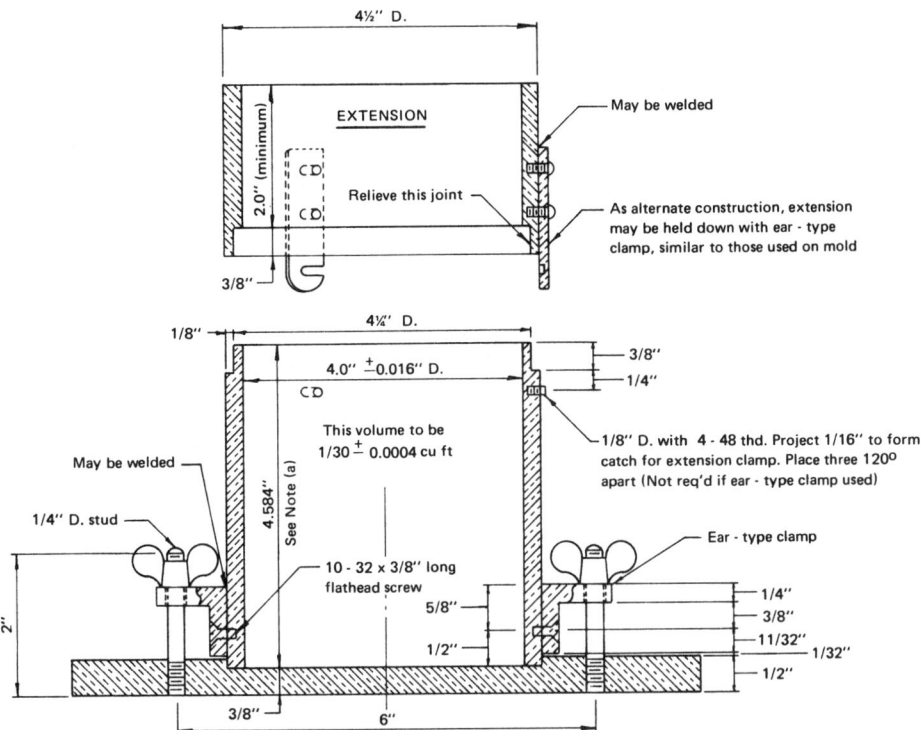

NOTE 1—The tolerance on the height is governed by the allowable volume and diameter tolerances.

NOTE 2—The methods shown for attaching the extension collar to the mold and the mold to the base plate are recommended. However, other methods are acceptable, providing the attachments are equally as rigid as those shown.

FIG. 1 Cylindrical Mold, 4.0-in. for Soil Tests (see Table 4 for metric equivalents).

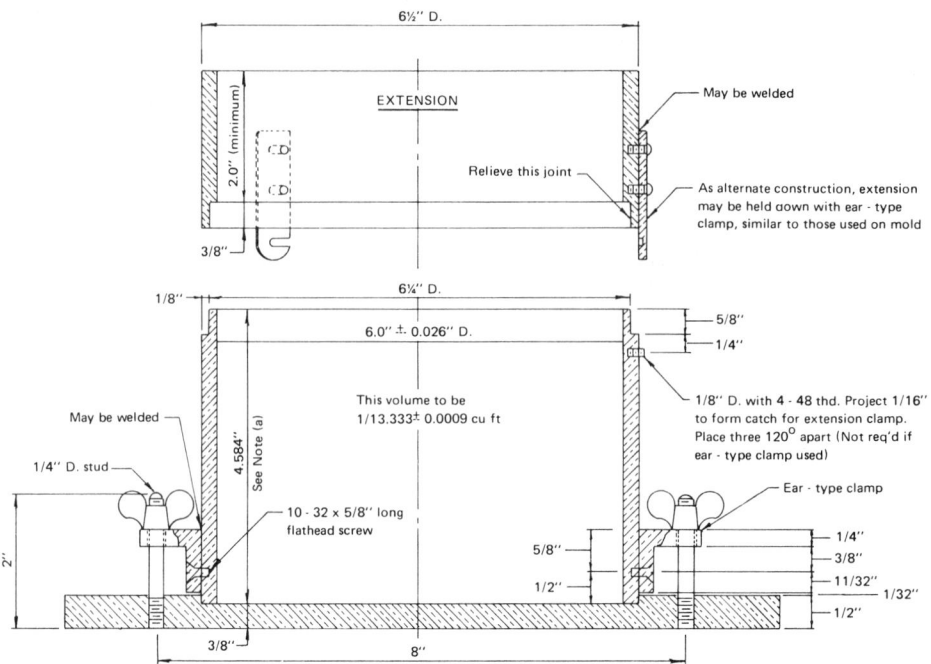

NOTE 1 — The tolerance on the height is governed by the allowable volume and diameter tolerances.

NOTE 2 — The methods shown for attaching the extension collar to the mold and the mold to the base plate are recommended. However, other methods are acceptable, providing the attachments are equally as rigid as those shown.

FIG. 2 Cylindrical Mold, 6.0-in. for Soil Tests (see Table 4 for metric equivalents).

Designation: D 1557 – 78

Standard Test Methods for
MOISTURE-DENSITY RELATIONS OF SOILS AND SOIL-AGGREGATE MIXTURES USING 10-lb (4.54-kg) RAMMER AND 18-in. (457-mm) DROP[1]

This standard is issued under the fixed designation D 1557; the number immediately following the designation indicates the year of original adoption or, in the case of revision, the year of last revision. A number in parentheses indicates the year of last reapproval. A superscript epsilon (ϵ) indicates an editorial change since the last revision or reapproval.

These methods have been approved for use by agencies of the Department of Defense and for listing in the DoD Index of Specifications and Standards.

1. Scope

1.1 These laboratory compaction methods cover the determination of the relationship between the moisture content and density of soils and soil-aggregate mixtures (Note 1) when compacted in a mold of a given size with a 10-lb (4.54-kg) rammer dropped from a height of 18 in. (457 mm) (Note 2). Four alternative procedures are provided as follows:

1.1.1 *Method A*—A 4-in. (101.6-mm) mold; material passing a No. 4 (4.75-mm) sieve;

1.1.2 *Method B*—A 6-in. (152.4-mm) mold; material passing a No. 4 (4.75-mm) sieve;

1.1.3 *Method C*—A 6-in. (152.4-mm) mold; material passing a ¾-in. (19.0-mm) sieve; and

1.1.4 *Method D*—A 6-in. (152.4-mm) mold; material passing a ¾-in. (19.0-mm) sieve, corrected by replacement for material retained on a ¾-in. sieve.

NOTE 1—Soils and soil-aggregate mixtures should be regarded as natural occurring fine- or coarse-grained soils or composites or mixtures of natural soils, or mixtures of natural and processed soils or aggregates such as silt, gravel, or crushed rock.

NOTE 2—These laboratory compaction test methods when used on soils and soil-aggregates which are not free-draining will, in most cases, establish a well-defined optimum moisture content and maximum density (see Section 7). However, for free-draining soils and soil-aggregate mixtures, these methods will not, in many cases, produce a well-defined moisture-density relationship and the maximum density obtained will generally be less than that obtained by vibratory methods.

1.2 The method to be used should be indicated in the specifications for the material being tested. If no method is specified, the provisions of Section 5 shall govern.

2. Applicable Documents

2.1 *ASTM Standards:*

C 127 Test Method for Specific Gravity and Absorption of Coarse Aggregate[2]

D 854 Test Method for Specific Gravity of Soils[3]

D 2168 Methods for Calibration of Laboratory Mechanical-Rammer Soil Compactors[3]

D 2216 Method for Laboratory Determination of Water (Moisture) Content of Soil, Rock, and Soil-Aggregate Mixtures[3]

D 2487 Test Method for Classification of Soils for Engineering Purposes[3]

D 2488 Practice for Description and Identification of Soils (Visual-Manual Procedure)[3]

E 11 Specification for Wire-Cloth Sieves for Testing Purposes[4]

3. Apparatus

3.1 *Molds*—The molds shall be cylindrical in shape, made of rigid metal and be within the capacity and dimensions indicated in 3.1.1 or 3.1.2. The molds may be the "split" type, consisting either of two half-round sections, or a section of pipe split along one element,

---

[1] These methods are under the jurisdiction of ASTM Committee D-18 on Soil and Rock.

Current edition approved April 27, 1978. Published July 1978. Originally published as D 1557 – 58 T. Last previous edition D 1557 – 70.

[2] *Annual Book of ASTM Standards*, Vols 04.02 and 04.03.

[3] *Annual Book of ASTM Standards*, Vol 04.08.

[4] *Annual Book of ASTM Standards*, Vols 04.01, 04.02, 04.06, 05.05, and 14.02.

which can be securely locked together to form a cylinder meeting the requirements of this section. The molds may also be the "taper" type, providing the internal diameter taper is uniform and is not more than 0.200 in./linear ft (16.7 mm/linear m) of mold height. Each mold shall have a base plate assembly and an extension collar assembly, both made of rigid metal and constructed so they can be securely attached to or detached from the mold. The **extension collar assembly shall have a height extending above the top of the mold of at least 2 in. (50 mm) which may include an upper section that flares out to form a funnel providing there is at least a ¾-in. (19-mm) straight cylindrical section beneath it.**

3.1.1 *Mold*, 4.0 in. (101.6 mm) in diameter, having a capacity of $^1/_{30} \pm 0.0004$ ft³ (944 ± 11 cm³) and conforming to Fig. 1.

3.1.2 *Mold*, 6.0 in. (152.4 mm) in diameter, having a capacity of $^1/_{13.333} \pm 0.0009$ ft³ (2124 ± 25 cm³) and conforming to Fig. 2.

3.1.3 The average internal diameter, height, and volume of each mold shall be determined before initial use and at intervals not exceeding 1000 times the mold is filled. The mold volume shall be calculated from the average of at least six internal diameter and three height measurements made to the nearest 0.001 in. (0.02 mm), or from the amount of water required to completely fill the mold, corrected for temperature variance in accordance with Table 1. If the average internal diameter and volume are not within the tolerances shown in Figs. 1 or 2, the mold shall not be used. The determined volume shall be used in computing the required densities.

3.2 *Rammer* — The rammer may be either manually operated (see 3.2.1) or mechanically operated (see 3.2.2). The rammer shall fall freely through a distance of 18.0 ± $^1/_{16}$ in. (457.2 ± 1.6 mm) from the surface of the specimen. The manufactured weight of the rammer shall be 10.00 ± 0.02 lb (4.54 ± 0.01 kg). The specimen contact face shall be flat.

3.2.1 *Manual Rammer* — The specimen contact face shall be circular with a diameter of 2.000 ± 0.005 in. (50.80 ± 0.13 mm). The rammer shall be equipped with a guide-sleeve which shall provide sufficient clearance so that the free fall of the rammer shaft and head will not be restricted. The guidesleeve

shall have four vent holes at each end (eight holes total) located with centers $^3/_4 \pm ^1/_{16}$ in. (19.0 ± 1.6 mm) from each end and spaced 90 deg apart. The minimum diameter of the vent holes shall be $^3/_8$ in. (9.5 mm).

3.2.2 *Mechanical Rammer* — The rammer shall operate mechanically in such a manner as to provide uniform and complete coverage of the specimen surface. There shall be 0.10 ± 0.03 in. (2.5 ± 0.8 mm) clearance between the rammer and the inside surface of the mold at its smallest diameter. When used with the 4.0-in. (101.6-mm) mold, the specimen contact face shall be circular with a diameter of 2.000 ± 0.005 in. (50.80 ± 0.13 mm). When used with the 6.0-in. (152.4-mm) mold, the specimen contact face shall have the shape of a section of a circle of a radius equal to 2.90 ± 0.02 in. (73.7 ± 0.5 mm). The sector face rammer shall operate in such a manner that the vertex of the sector is positioned at the center of the specimen. The mechanical rammer shall be calibrated and adjusted, as necessary, in accordance with 3.2.3.

3.2.3 *Calibration and Adjustment* — The mechanical rammer shall be calibrated, and adjusted as necessary, before initial use; near the end of each period during which the mold was filled 1000 times; before reuse after anything, including repairs, which may affect the test results significantly; and whenever the test results are questionable. Each calibration and adjustment shall be in accordance with Methods D 2168.

3.3 *Sample Extruder* (optional) — A jack, frame, or other device adapted for the purpose of extruding compacted specimens from the mold.

3.4 *Balances* — A balance or scale of at least 20-kg capacity sensitive to ±1 g and a balance of at least 1000-g capacity sensitive to ±0.01 g.

3.5 *Drying Oven*, thermostatically-controlled, preferably of the forced-draft type, capable of maintaining a temperature of 230 ± 9°F (110 ± 5°C) for determining the moisture content of the compacted specimen.

3.6 *Straightedge* — A stiff metal straightedge of any convenient length but not less than 10 in. (254 mm). The scraping edge shall have a straightness tolerance of ±0.005 in. (±0.13 mm) and shall be beveled if it is thicker then $^1/_8$ in. (3 mm).

342

3.7 *Sieves*, 3-in. (75-mm), ¾-in. (19.0-mm), and No. 4 (4.75-mm), conforming to the requirements of Specification E 11.

3.8 *Mixing Tools*—Miscellaneous tools such as mixing pan, spoon, trowel, spatula, etc., or a suitable mechanical device for thoroughly mixing the sample of soil with increments of water.

4. Procedure

4.1 *Specimen Preparation*—Select a representative portion of quantity adequate to provide, after sieving, an amount of material weighing as follows: Methods A—25 lb (11 kg); Methods B, C, and D—50 lb (23 kg). Prepare specimens in accordance with either 4.1.1 through 4.1.3 or 4.1.4.

4.1.1 *Dry Preparation Procedure*—If the sample is too damp to be friable, reduce the moisture content by drying until the material is friable; see 4.1.2. Drying may be in air or by the use of a drying apparatus such that the temperature of the sample does not exceed 140°F (60°C). After drying (if required), thoroughly break up the aggregations in such a manner as to avoid reducing the natural size of the particles. Pass the material through the specified sieve as follows: Methods A and B—No. 4 (4.75-mm); Methods C and D—¾-in. (19.0-mm). Correct for oversize material in accordance with Section 5, if Method D is specified.

4.1.2 Whenever practicable, soils classified as ML, CL, OL, GC, SC, MH, CH, OH and PT by Test Method D 2487 shall be prepared in accordance with 4.1.4.

4.1.3 Prepare a series of at least four specimens by adding increasing amounts of water to each sample so that the moisture contents vary by approximately 1½ %. The moisture contents selected shall bracket the optimum moisture content, thus providing specimens which, when compacted, will increase in mass to the maximum density and then decrease in density (see 7.2 and 7.3). Thoroughly mix each specimen to ensure even distribution of moisture throughout and then place in a separate covered container and allow to stand prior to compaction in accordance with Table 2. For the purpose of selecting a standing time, it is not required to perform the actual classification procedures described in Test Method D 2487 (except in the case of referee

testing), if previous data exist which provide a basis for classifying the sample.

4.1.4 *Moist Preparation Method*—The following alternate procedure is recommended for soils classified as ML, CL, OL, GC, SC, MH, CH, OH and PT by Test Method D 2487. Without previously drying the sample, pass it through the ¾-in. (19.0-mm) and No. 4 (4.75-mm) sieves. Correct for oversize material in accordance with Section 5, if Method D is specified. Prepare a series of at least four specimens having moisture contents that vary by approximately 1½ %. The moisture contents selected shall bracket the optimum moisture content, thus providing specimens which, when compacted, will increase in mass to the maximum density and then decrease in density (see 7.2 and 7.3). To obtain the appropriate moisture content of each specimen, the addition of a predetermined amount of water (see 4.1.3) or the removal of a predetermined amount of moisture by drying may be necessary. Drying may be in air or by the use of a drying apparatus such that the temperature of the specimen does not exceed 140°F (60°C). The prepared specimens shall then be thoroughly mixed and stand, as specified in 4.1.3 and Table 2, prior to compaction.

NOTE 3 – With practice, it is usually possible to visually judge the point of optimum moisture closely enough so that the prepared specimens will bracket the point of optimum moisture content.

4.2 *Specimen Compaction*—Select the proper compaction mold, in accordance with the method being used, and attach the mold extension collar. Compact each specimen in five layers of approximately equal height. Each layer shall receive 25 blows in the case of the 4-in. (101.6-mm) mold; each layer shall receive 56 blows in the case of the 6-in. (152.4-mm) mold. The total amount of material used shall be such that the fifth compacted layer is slightly above the top of the mold, but not exceeding ¼ in. (6 mm). During compaction the mold shall rest on a uniform rigid foundation, such as provided by a cylinder or cube of concrete weighing not less than 200 lb (91 kg).

4.2.1 In operating the manual rammer, care shall be taken to avoid rebound of the rammer from the top end of the guidesleeve.

The guidesleeve shall be held steady and within 5 deg of the vertical. The blows shall be applied at a uniform rate not exceeding approximately 1.4 s per blow and in such a manner as to provide complete and uniform coverage of the specimen surface.

4.2.2 *Mold Sizes*—The mold size used shall be as follows: Method A, 4-in. (101.6-mm); Methods B, C, and D, 6-in. (152.4-mm).

4.2.3 Following compaction, remove the extension collar; carefully trim the compacted specimen even with the top of the mold by means of the straightedge and determine the mass of the specimen. Divide the mass of the compacted specimen and mold, minus the mass of the mold, by the volume of the mold (see 3.1.3). Record the result as the wet density, γ_m, in pounds per cubic foot (or kilograms per cubic metre) of the compacted specimen.

4.2.4 Remove the material from the mold. Determine moisture content in accordance with Method D 2216, using either the whole compacted specimen or alternatively a representative specimen of the whole specimen. The whole specimen must be used when the permeability of the compacted specimen is high enough so that the moisture content is not distributed uniformly throughout. If the whole specimen is used, break it up to facilitate drying. Obtain the representative specimen by slicing the compacted specimen axially through the center and removing 100 to 500 g of material from one of the cut faces.

4.2.5 Repeat 4.2 through 4.2.4 for each specimen prepared.

5. Oversize Corrections

5.1 If 30 % or more of the sample is retained on a ³/₄-in. (19.0-mm) sieve, then none of the methods described under these methods shall be used for the determination of either maximum density or optimum moisture content.

5.2 *Methods A and B*—The material retained on the No. 4 (4.75-mm) sieve is discarded and no oversize correction is made. However, it is recommended that if the amount of material retained is 7 % or greater, Method C be used instead.

5.3 *Method C*—The material retained on the ³/₄-in. (19.0-mm) sieve is discarded and no oversize correction is made. However, if

the amount of material retained is 10 % or greater, it is recommended that Method D be used instead.

5.4 *Method D:*

5.4.1 This method shall not be used unless the amount of material retained on the ³/₄-in. (19.0-mm) sieve is 10 % or greater. When the amount of material retained on the ³/₄-in. sieve is less than 10 %, use Method C.

5.4.2 Pass the material retained on the ³/₄-in. (19.0-mm) sieve through a 3-in. or 75-mm sieve. Discard the material retained on the 3-in. sieve. The material passing the 3-in. sieve and retained on the ³/₄-in. sieve shall be replaced with an equal amount of material passing a ³/₄-in. sieve and retained on a No. 4 (4.75-mm) sieve. The material for replacement shall be taken from an unused portion of the sample.

6. Calculations

6.1 Calculate the moisture content and the dry density of each compacted specimen as follows:

$$w = [(A - B)/(B - C)] \times 100$$

and

$$\gamma_d = [\gamma_m/(w + 100)] \times 100$$

where:

w = moisture content in percent of the compacted specimens,

A = mass of contained and moist specimen,

B = mass of container and oven-dried specimen,

C = mass of container,

γ_d = dry density, in pounds per cubic foot (or kilograms per cubic metre) of the compacted specimen, and

γ_m = wet density, in pounds per cubic foot (or kilograms per cubic metre) of the compacted specimen.

7. Moisture-Density Relationship

7.1 From the data obtained in 6.1, plot the dry density values as ordinates with corresponding moisture contents as abscissas. Draw a smooth curve connecting the plotted points. Also draw a curve termed the "curve of complete saturation" or "zero air voids curve" on this plot. This curve represents the relationship between dry density and corresponding moisture contents when the voids are completely filled with water. Values of dry density and corresponding moisture contents for plotting the curve of complete saturation can be computed using the following equation:

$$w_{sat} = [(62.4/\gamma_d) - (1/G_s)] \times 100$$

where:

w_{sat} = moisture content in percent for complete saturation,

γ_d = dry density in pounds per cubic foot (or kilograms per cubic metre),

G_s = specific gravity of the material being tested (see Note 4), and

62.4 = density of water in pounds per cubic foot (or kilograms per cubic metre).

NOTE 4—The specific gravity of the material can either be assumed or based on the weighted average values of: (a) the specific gravity of the material passing the No. 4 (4.75-mm) sieve in accordance with Test Method D 854; and (b) the apparent specific gravity of the material retained on the No. 4 (4.75-mm) sieve in accordance with Test Method C 127.

7.2 *Optimum Moisture Content, w_o* —The moisture content corresponding to the peak of the curve drawn as directed in 7.1 shall be termed the "optimum moisture content."

7.3 *Maximum Density, γ_{max}* — The dry density in pounds per cubic foot (or kilograms per cubic metre) of the sample at "optimum moisture content" shall be termed "maximum density."

8. Report

8.1 The report shall include the following:

8.1.1 Method used (Method A, B, C, or D).

8.1.2 Optimum moisture content.

8.1.3 Maximum density.

8.1.4 Description of rammer (whether manual or mechanical).

8.1.5 Description of appearance of material used in test, based on Practice D 2488 (Test Method D 2487 may be used as an alternative).

8.1.6 Origin of material used in test.

8.1.7 Preparation procedure used (moist or dry).

9. Precision

9.1 Criteria for judging the acceptability of the maximum density and optimum moisture content test results are given in Table 3. The standard deviation s is calculated from the equation:

$$s^2 = \frac{1}{n-1} \sum_1^n (x - \bar{x})^2$$

where:

n = number of determinations;

x = individual value of each determination; and

$\bar{x}$ = numerical average of the determinations.

9.2 Criteria for assigning standard deviation values for single-operator precision are not available at the present time.

TABLE 1 Volume of Water per Gram based on Temperature[A]

Temperature, °C (°F)	Volume of Water, ml/g
12 (53.6)	1.00048
14 (57.2)	1.00073
16 (60.8)	1.00103
18 (64.4)	1.00138
20 (68.0)	1.00177
22 (71.6)	1.00221
24 (75.2)	1.00268
26 (78.8)	1.00320
28 (82.4)	1.00375
30 (86.0)	1.00435
32 (89.6)	1.00497

[A] Values other than shown may be obtained by referring to the *Handbook of Chemistry and Physics,* Chemical Rubber Publishing Co., Cleveland, OH.

TABLE 2 Dry Preparation Method – Standing Times

Classification D 2487	Minimum Standing Time, h
GW, GP, SW, SP	no requirement
GM, SM	3
ML, CL, OL, GC, SC	18
MH, CH, OH, PT	36

	Standard Devia- tion, s	Acceptable Range of Two Results, Ex- pressed as Percent of Mean Value
TABLE 3 Precision		
Single-operator precision:		
Maximum density	...	1.9
Optimum moisture content	...	9.5
Multilaboratory precision:		
Maximum density	±1.66	4.0
Optimum moisture content	±0.86	15.0

[A] This column indicates a limiting range of values which should not be exceeded by the difference between any two results, expressed as a percentage of the average value. In cooperative test programs it has been determined that 95 % of the tests do not exceed the limiting acceptable ranges shown below. All values shown in this table are based on average test results from a variety of different soils and are subject to future revision.

TABLE 4 Metric Equivalents for Figs. 1 and 2

in.	mm
0.016	0.41
0.026	0.66
1/32	0.8
1/16	1.6
1/8	3.2
1/4	6.4
11/32	8.7
3/8	9.5
1/2	12.7
5/8	15.9
2	50.8
2 1/2	63.5
4	101.6
4 1/4	108.0
4 1/2	114.3
4.584	116.43
6	152.4
6 1/2	165.1
8	203.2

ft³	cm³
1/30	944
0.004	11
1/13.333	2124
0.0009	25

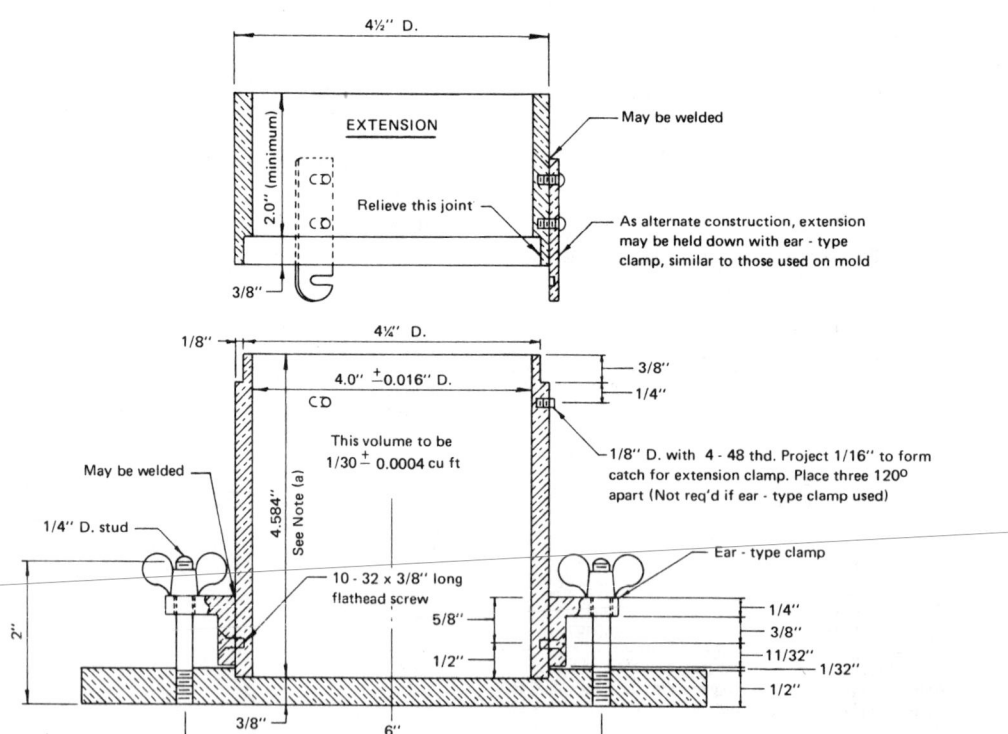

NOTE 1—The tolerance on the height is governed by the allowable volume and diameter tolerances.

NOTE 2—The methods shown for attaching the extension collar to the mold and the mold to the base plate are recommended. However, other methods are acceptable, providing the attachments are equally as rigid as those shown.

FIG. 1 Cylindrical Mold, 4.0-in. for Soil Tests (see Table 4 for metric equivalents).

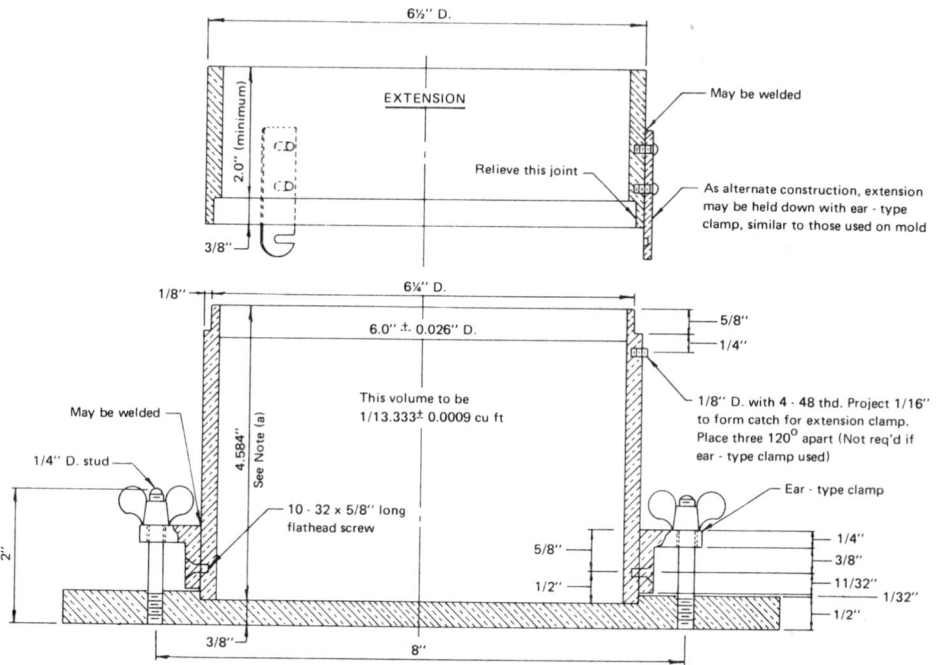

NOTE 1 – The tolerance on the height is governed by the allowable volume and diameter tolerances.

NOTE 2 – The methods shown for attaching the extension collar to the mold and the mold to the base plate are recommended. However, other methods are acceptable, providing the attachments are equally as rigid as those shown.

FIG. 2   Cylindrical Mold, 6.0-in. for Soil Tests (see Table 4 for metric equivalents).

Designation: D 1556 – 82[ϵ1]

Standard Test Method for
DENSITY OF SOIL IN PLACE BY THE SAND-CONE METHOD[1]

This standard is issued under the fixed designation D 1556; the number immediately following the designation indicates the year of original adoption or, in the case of revision, the year of last revision. A number in parentheses indicates the year of last reapproval. A superscript epsilon (ϵ) indicates an editorial change since the last revision or reapproval.

This method has been approved for use by agencies of the Department of Defense and for listing in the DoD Index of Specifications and Standards.

ϵ1 NOTE—Section 7.5.3 was changed editorially in December 1983.

1. Scope

1.1 This method covers the determination of the in-place density of soils.

1.2 Any soil or other material that can be excavated with hand tools can be tested provided the void or pore openings in the mass are small enough to prevent the sand used in the test from entering the natural voids. The soil or other material being tested should have sufficient cohesion or particle attraction to maintain stable sides on a small hole or excavation. It should also be firm enough to withstand the minor pressures exerted in digging the hole and placing the apparatus over it without deforming or sloughing.

1.3 When the moisture content and dry density are to be determined, this method is not to be used in certain soils or materials as indicated in paragraphs 1.3 and 3.5 of Method D 2216.

2. Applicable Documents

2.1 *ASTM Standards:*

C 136 Method for Sieve Analysis of Fine and Coarse Aggregates[2]

D 653 Terms and Symbols Relating to Soil and Rock[3]

D 698 Test Methods for Moisture-Density Relations of Soils and Soil-Aggregate Mixtures Using 5.5-lb (2.49-kg) Rammer and 12-in. (305-mm) Drop[3]

D 1557 Test Methods for Moisture-Density Relations of Soils and Soil-Aggregate Mixtures Using 10-lb (4.54-kg) Rammer and 18-in. (457-mm) Drop[3]

D 2049 Test Method for Relative Density of Cohesionless Soils[4]

D 2216 Method for Laboratory Determination of Water (Moisture) Content of Soil, Rock, and Soil-Aggregate Mixtures[3]

3. Significance and Use

3.1 This method is used widely to determine the density of compacted soils used in the construction of earth embankments, road fill, and structure backfill. It is often used as the basis of acceptance for soils compacted to a specified density or percentage of a maximum density determined by a standard test method.

3.2 This method can be used to determine in-place density of natural soil deposits, aggregates, soil mixtures, or other similar material.

3.3 The use of this method is generally limited to soil in an unsaturated condition. This method is not recommended for soils that are soft or friable (crumble easily) or in a moisture condition such that water seeps into the hand-excavated hole. The accuracy of the test may be affected for soils that deform easily or that may undergo a volume change in the excavated hole from standing or walking near the hole during the test.

4. Apparatus

4.1 *Density Apparatus*, consisting of the following:

---

[1] This method is under the jurisdiction of ASTM Committee D-18 on Soil and Rock and is the direct responsibility of Subcommittee D18.08 on Special and Construction Control Tests.

Current edition approved Nov. 26, 1982. Published January 1983. Originally published as D 1556 – 58 T. Last previous edition D 1556 – 64 (1974).

[2] *Annual Book of ASTM Standards*, Vol 04.02.

[3] *Annual Book of ASTM Standards*, Vol 04.08.

[4] Discontinued, see *1983 Annual Book of ASTM Standards*, Vol 04.08.

4.1.1 A jar or other container having a volume of approximately 1 gal (4000 cm³) or larger.

4.1.2 A detachable appliance consisting of a cylindrical valve with an orifice approximately ½ in. (13 mm) in diameter, having a small metal funnel connecting to a standard gal Mason jar top on one end and a large metal funnel (cone) on the other end. The valve shall have stops to prevent rotating it past the completely open or completely closed positions.

4.1.3 A square or rectangular metal plate with a flanged center hole cast or machined to receive the large funnel (cone) of the appliance described in 4.1.2. The plate shall be flat on the bottom and have sufficient thickness or stiffness to be rigid and shall have sidewalls approximately ⅜ to ½ in. (10 to 13 mm) high.

4.1.4 The details for the apparatus described herein are shown in Fig. 1 and represent the minimum acceptable dimensions suitable for testing soils having maximum particle sizes of approximately 2 in. (50 mm) and test hole volumes of approximately 0.1 ft³ (3000 cm³). When the material being tested contains a small amount of oversize and isolated large particles are encountered, the test can be moved to a new location. Larger apparatus and test hole volumes are needed when particles larger than 2 in. (50 mm) are prevalent.

NOTE 1—The apparatus described here represents a design that has proved satisfactory. Larger apparatus of similar proportions may be used as long as the basic principles of the sand volume determination are observed.

4.2 *Sand*, shall be clean, dry, uniform, uncemented, durable, and free-flowing. Any gradation may be used that has a uniformity coefficient ($C_u = D_{60}/D_{10}$) less than 2.0, a maximum particle size less than 2.00 mm (No. 10 sieve), and less than 3 % by weight passing 250 μm (No. 60 sieve). The particle size distribution (gradation) shall be determined in accordance with Method C 136. Uniform sand is needed to prevent segregation during handling, storage, and use. Sand free of fines and fine sand particles is needed to prevent significant bulk density changes with normal daily changes in atmospheric humidity. Sand comprised of durable, natural subrounded or rounded particles is desirable. Crushed sand or sand having angular particles may not be free flowing, a condition that can cause bridging resulting in inaccurate

density determinations (Note 2). In selecting a sand from a potential source, five separate bulk-density determinations shall be made on each container or bag of sand. To be an acceptable sand, the variation between any determination and the average shall not be greater than 1% of the average. Before using sand in density determinations, it shall be dried, then allowed to reach an air-dried state in the general location where it is to be used. Sand shall not be reused without removing any contaminating soil, checking the gradation, and drying. Bulk-density tests shall be made at intervals not exceeding 14 days, always after any significant changes in atmospheric humidity, before reusing, and before using a new batch from a previously approved supplier (Note 3).

NOTE 2—Some manufactured (crushed) sands such as blasting sand have been successfully used with good reproducibility. The reproducibility of test results using angular sand should be checked under actual testing situations before selecting an angular sand for use.

NOTE 3—Most sands have a tendency to absorb moisture from the atmosphere. A very small amount of absorbed moisture can make a substantial change in bulk density. In areas of high humidity or where the humidity changes often, the bulk density may need to be determined more often than the 14-day maximum interval indicated. The need for more frequent checks can be determined by comparing the results of different bulk-density tests on the same sand made in the area and conditions of use over a period of time.

4.3 *Balances*—A balance or scale of 10-kg capacity readable to 1.0 g and accurate to 2 g from 100 g to 7000 g and 3 g above 7000 g, and a balance of 2000 g capacity readable to 0.1 g and accurate to 0.1 %.

4.4 *Drying Equipment*—Oven, as specified in Method D 2216 (See 6.1.10 and Note 6.)

4.5 *Miscellaneous Equipment*—Knife, small pick, chisel, small trowel, screwdriver, or spoons for digging test hole; buckets with lids, seamless tin or aluminum cans with lids, plastic-lined cloth sacks, or other suitable containers for retaining the density sample, moisture sample, and density sand respectively; thermometer for determining the temperature of water; small, paint-type brush, slide rule or calculator, notebook, etc.

5. Calibration

5.1 Determinations of mass are to be made to the nearest 1 g except for those required for determining moisture content, which shall be made to the nearest 0.1 g.

5.2 Determine the mass of sand required to fill the funnel and base plate as follows:

5.2.1 Put sand in the apparatus and determine the mass of apparatus and sand.

5.2.2 Place the base plate on a clean, level, plane surface. Invert the apparatus and seat the large funnel into the flanged center hole in the base plate, and mark and identify the funnel and plate so that the same funnel and plate can always be matched and reseated in the same position.

5.2.3 Open the valve and keep open until the sand stops running, making sure the apparatus, base plate, or plane surface are not jarred or vibrated before the valve is closed.

5.2.4 Close the valve sharply, determine the mass of the apparatus with remaining sand and calculate the loss of sand. This loss represents the mass of sand required to fill the funnel and base plate.

5.2.5 Repeat the procedures in 5.2.1 to 5.2.4 at least three times. The mass of sand used in the calculations shall be the average of three determinations. The maximum variation between any one determination and the average shall not exceed 1 %.

5.3 Use either Method A or Method B to determine the sand bulk density.

5.4 *Method A*—Determine the bulk density of the sand to be used in the field test as follows:

5.4.1 Select a known-volume container that is approximately the same size and allows the sand to fall approximately the same distance as the hole excavated in making a field test. The $\frac{1}{30}$-ft^3 (944-cm^3) and $\frac{1}{13.333}$-ft^3 (2124-cm^3) molds specified in Test Methods D 698 or the 0.1-ft^3 (2830-cm^3) mold specified in Test Method D 2049 are recommended.

5.4.2 Make measurements of sufficient accuracy to determine the volume of the container to ±1.0 %. The measurement tolerances for the above recommended molds are given in Test Methods D 698 and D 2049.

5.4.3 Flow characteristics through different valve assemblies have been known to cause different bulk-density values. The funnel and valve apparatus used to determine the bulk density of the sand shall be the same as used for making field tests unless other assemblies are determined to provide the same results.

5.4.4 Fill the assembled apparatus with sand.

5.4.5 Determine the mass of the known-volume container when empty.

5.4.6 Support the apparatus over the known-volume container in an inverted position so that the sand falls approximately the same distance as in a field test, and fully open the valve.

5.4.7 Fill the container until it just overflows and close the valve. Carefully strike off excess sand to a smooth level surface at the top of the container. Care must be taken so that the container is not jarred or vibrated before striking off is completed.

5.4.8 Clean any sand from the outside of the container and determine the mass of the known-volume container when full. Determine the net mass of sand by subtracting the mass of the empty container.

NOTE 4—When the known-volume container has the same diameter as the flanged center hole in the metal plate, the procedures in 5.4.6 to 5.4.8 can be simplified by using the plate on top of the known-volume container. Striking off the excess sand is not required when the apparatus with sand is weighed before and after filling the container and the mass required to fill the cone and plate (5.2.5) is subtracted from the difference.

5.4.9 Repeat the procedure in 5.4.4 to 5.4.8 at least three times. The mass used in the calculations shall be the average of three determinations. The maximum variation between any one determination and the average shall not exceed 1 %.

5.5 *Method B*—The bulk density of sand to be used in the field test is determined by first determining the volume of the jar and attachment up to and including the volume of the valve orifice then by using the jar to measure a volume and mass of sand as follows:

5.5.1 In making bulk density determinations, use a glass or other rigid jar that is well-rounded where it tapers toward the opening. Other jars may be used for making in-place density tests.

5.5.2 Use the same funnel and valve assembly for determining the bulk density of sand as will be used in field tests or determine that other assemblies provide the same results.

5.5.3 Use heavy grease or other waterproof substances in those stopcocks and thread assemblies that are not watertight.

5.5.4 Determine the mass of the assembled apparatus and record.

5.5.5 Place the apparatus upright and open the valve.

5.5.6 Fill the apparatus with water until it appears over the valve.

5.5.7 Close valve and remove excess water.

5.5.8 Determine the mass of the apparatus and water, and determine the temperature of the water to the nearest 1°C.

5.5.9 To determine the mass of the water to fill the apparatus, subtract the mass of the apparatus from the mass of the apparatus and the water.

5.5.10 Repeat the procedure described in 5.5.4 to 5.5.9 at least three times. Convert the weight of water, in grams, to millilitres by correcting for the temperature as given in 7.3.1. The volume used shall be the average of three determinations with a maximum variation of 3 mL.

5.5.11 The volume determined in this procedure is constant as long as the jar and attachment are in the same relative position as the previous volume determination. If the two are to be separated, match marks should be made to permit reassembly to this position.

5.5.12 Completely dry the jar and other apparatus and remove any grease or waterproofing substances before proceeding with the following bulk density determination.

5.5.13 Place the empty apparatus upright on a firm level surface, close the valve, and fill the funnel with sand.

5.5.14 Open the valve and, keeping the funnel at least half full of sand, fill the apparatus, making sure the apparatus is not jarred or vibrated before the valve is closed. When the sand stops flowing, close the valve sharply and empty excess sand.

5.5.15 Determine the mass of the apparatus with sand and determine the net mass of sand by subtracting the mass of the apparatus.

5.5.16 Repeat the procedures in 5.5.13 to 5.5.15 at least three times. The mass used in the calculations shall be the average of three determinations. The maximum variation between any one determination and the average shall not exceed 1 %.

6. Procedure

6.1 Determine the density of the soil in place as follows:

6.1.1 Fill the apparatus with sand previously calibrated for bulk density and determine the mass of the apparatus and sand.

6.1.2 Prepare the surface of the location to be tested so that it is a level plane. The base plate makes an excellent tool for striking off the surface to a neat level plane.

6.1.3 Seat the base plate on the plane surface, making sure there is good contact with the ground surface around the edge of the flanged center hole. Mark the outline of the base plate to check for movement during the test, and use nails pushed into the soil adjacent to the edge of the plate or otherwise secure the plate against movement without disturbing the soil to be tested.

6.1.4 In soils where leveling is not successful, a preliminary test shall be run at this point measuring the volume bounded by the funnel plate and ground surface. Fill the space with sand from the apparatus, determine the mass of sand used to fill the space, refill the apparatus, and determine a new initial mass of apparatus and sand before proceeding with the test. After this measurement is completed, carefully brush the sand from the prepared surface.

NOTE 5—A second calibrated apparatus may be taken to the field when this condition is anticipated (instead of refilling and making a second mass determination). The procedure in 6.1.4 may be used for each test when the best possible accuracy is desired, however, it is usually not needed for most production testing where a relatively smooth surface is obtainable.

6.1.5 Dig the test hole inside the center hole in the base plate, being very careful to avoid disturbing the soil that will bound the hole. Test-hole volumes shall be as large as practical to minimize the effects of errors and shall in no case be smaller than the volume indicated in Table 1. The sides of the hole should slope inward slightly toward the bottom that should be reasonably flat or concave. The hole should be kept as free as possible of pockets, overhangs, and sharp obtrusions since these affect the accuracy of the test. Soils that are essentially granular require extreme care and may require digging a conical-shaped test hole. Place all excavated soil and soil loosened during excavation in a container that is marked to identify the test number. Take care to avoid losing any material. Protect this material from any loss of moisture until the mass has been determined and a specimen has been obtained for moisture content determination.

6.1.6 Clean the flange of the center hole in the metal plate, invert the apparatus, and seat the large metal funnel into the flanged hole at the same location as marked during calibration. Open the valve and allow the sand to fill the hole, funnel, and base plate. Take care to avoid jarring or vibrating the apparatus or the ground during this step. When the sand stops flowing, close the valve.

6.1.7 Determine the mass of the apparatus with remaining sand, and calculate the mass of sand used in the test.

6.1.8 Determine the mass of the material that was removed from the test hole.

6.1.9 Mix the material thoroughly and obtain a representative specimen for moisture-content determination or use the entire sample.

6.1.10 Determine the moisture content in accordance with Method D 2216.

NOTE 6—Rapid methods of moisture determination may be used to obtain an approximate value that is later verified or corrected according to the values obtained in accordance with Method D 2216.

6.2 Moisture-content specimens are to be large enough and selected in such a way so as to represent all the material obtained from the test hole. Suggested minimum mass of moisture specimens in relation to maximum particle size are shown in Table 1.

7. Calculations

7.1 Calculations as shown are for using units in grams and cubic centimetres or millilitres. Other units are permissible provided the appropriate conversion factors are used to maintain consistency of units throughout the calculations.

7.2 *Sand Calibration—Method A*:

7.2.1 Calculate the bulk density of the sand as follows:

$$\rho_1 = M_1/V_1$$

where:

ρ_1 = bulk density of the sand, g/cm^3, or multiply by 62.43 for lb/ft^3,

M_1 = mass of sand to fill the known volume container, 5.4.9, g, and

V_1 = volume of the known volume container, 5.4.2, cm^3.

7.3 *Sand Calibration—Method B*:

7.3.1 Calculate the volume of the density apparatus as follows:

$$V_2 = GT$$

where:

V_2 = volume of the density apparatus, mL,

G = mass of water required to fill the apparatus, 5.5.9, and

T = water temperature-volume correction shown in column 3 of Table 2.

7.3.2 Calculate the bulk density of the sand as follows:

$$\rho_1 = M_2/V_2$$

where:

ρ_1 = bulk density of the sand, g/cm^3, or multiply by 62.43 for lb/ft^3,

M_2 = mass of sand required to fill the apparatus 5.5.16, g, and

V_2 = volume of apparatus, 5.5.10.

7.5 *Field Test*:

7.5.1 Calculate the volume of the test hole as follows:

$$V = (M_3 - M_4)/\rho_1$$

where:

V = volume of the test hole, cm^3,

M_3 = mass of sand to fill the test hole, funnel, and base plate, 6.1.7, g,

M_4 = mass of sand to fill the funnel and base plate, 5.2.5, g, and

ρ_1 = bulk density of the sand, 7.2.1 or 7.3.2, g/cm^3.

7.5.2 Calculate the dry mass of material removed from the test hole as follows:

$$M_6 = 100 \, M_5/(w + 100)$$

where:

w = percentage of moisture, in material from test hole, 6.1.10,

M_5 = moist mass of the material from the test hole, 6.1.8,

M_6 = dry mass of material from test hole, g, or multiply by 0.002205 for lb.

7.5.3 Calculate the in-place wet and dry density of the material tested as follows:

$$\rho_m = M_5/V$$
$$\rho_d = M_6/V$$

where:

V = volume of test hole, 7.5.1, cm^3,

M_5 = moist mass of the material from the test hole, 6.1.8, g,

M_6 = dry mass of the material from the test hole, 7.5.2, g,

ρ_m = wet density of the tested material g/cm^3 or multiply by 62.43 for lb/ft^3, and

ρ_d = dry density of the tested material.

NOTE 7—It may be desired to express the in-place density as a percentage of some other density, for example, the laboratory maximum density determined in accordance with Method D 698. This relation can be determined by dividing the in-place density by the maximum density and multiplying by 100.

8. Precision and Accuracy

8.1 The precision and accuracy of this method has not been determined. No available methods provide absolute values of the density of soil in place against which this method can be compared. The variability of soil and the destructive nature of the test do not provide for repetitive duplication of test results to obtain a meaningful statistical evaluation. Accuracy is a function of the care exercised in performing the steps of the test giving particular attention to careful control to systematic repetition of procedures used.

TABLE 1 Minimum Test Hole Volumes and Minimum Moisture Content Samples Based on Maximum Size of Particle

Maximum Particle Size	Minimum Test Hole Volume, cm³	Minimum Test Hole Volume, ft³	Minimum Moisture Content Sample, g
No. 4 Sieve (4.75 mm)	710	0.025	100
½ in. (12.5 mm)	1420	0.050	300
1 in. (25 mm)	2120	0.075	500
2 in. (50 mm)	2830	0.100	1000

TABLE 2 Volume of Water per Gram Based on Temperature

Temperature °C	Temperature °F	Volume of Water, mL/g
12	53.6	1.00048
14	57.2	1.00073
16	60.8	1.00103
18	64.4	1.00138
20	68.0	1.00177
22	71.6	1.00221
24	75.2	1.00268
26	78.8	1.00320
28	82.4	1.00375
30	86.0	1.00435
32	89.6	1.00497

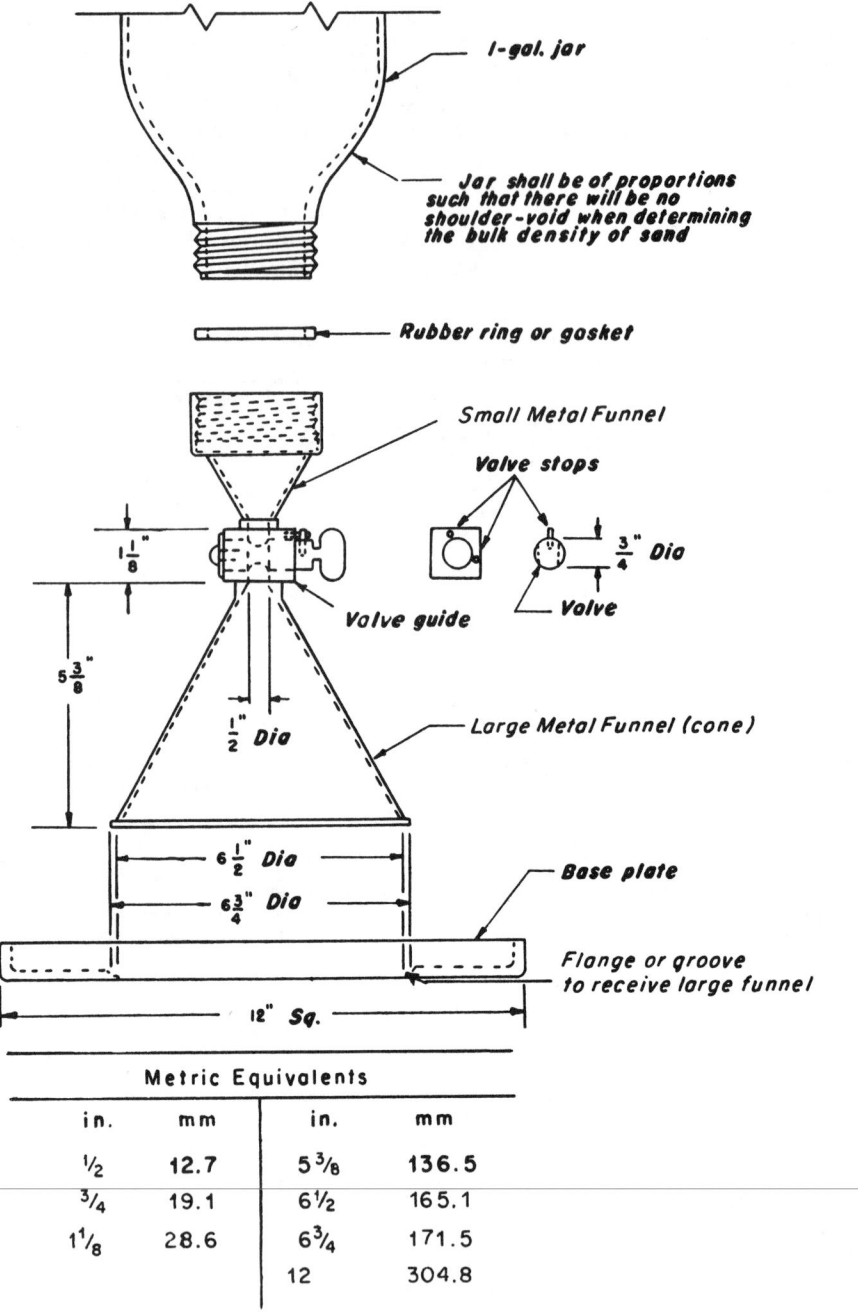

FIG. 1 Density Apparatus

Standard Test Method for
UNIT WEIGHT AND VOIDS IN AGGREGATE[1]

This standard is issued under the fixed designation C 29; the number immediately following the designation indicates the year of original adoption or, in the case of revision, the year of last revision. A number in parentheses indicates the year of last reapproval. A superscript epsilon (ϵ) indicates an editorial change since the last revision or reapproval.

This method has been approved for use by agencies of the Department of Defense and for listing in the DoD Index of Specifications and Standards.

[e1] NOTE—Section 2 was editorially added in September 1978.
[e2] NOTE—Section 11.1 was editorially changed in May 1982.

1. Scope

1.1 This method covers the determination of unit weight of and voids in fine, coarse, or mixed aggregates.

1.2 The values stated in inch-pound units are to be regarded as the standard. The metric equivalents of inch-pound units may be approximate.

2. Applicable Documents

2.1 *ASTM Standards:*

C 127 Test Method for Specific Gravity and Absorption of Coarse Aggregate[2]

C 128 Test Method for Specific Gravity and Absorption of Fine Aggregate[2]

C 702 Methods for Reducing Field Samples of Aggregate to Testing Size[2]

D 75 Practice for Sampling Aggregates[2]

3. Significance and Use

3.1 This test method is often used to determine necessary unit weight values for use in designing portland cement concrete mixtures.

3.2 Voids in aggregate for concrete may also be calculated by this method.

4. Apparatus

4.1 *Balance*—A balance or scale accurate within 0.1 % of the test load at any point within the range of use. The range of use shall be considered to extend from the weight of the measure empty to the weight of the measure plus its contents at 100 lb/ft³ (1600 kg/m³).

4.2 *Tamping Rod*—A round, straight steel rod, ⅝ in. (16 mm) in diameter and approximately 24 in. (600 mm) in length, having one end rounded to a hemispherical tip of the same diameter as the rod.

4.3 *Measure*—A cylindrical metal measure, preferably provided with handles. It shall be watertight, with the top and bottom true and even, preferably machined to accurate dimensions on the inside, and sufficiently rigid to retain its form under rough usage. The top rim shall be smooth and plane within 0.01 in (0.25 mm) and shall be parallel to the bottom within 0.5° (Note 1). Measures of the two larger sizes listed in Table 1 shall be reinforced around the top with a metal band, to provide an over-all wall thickness of not less than 0.20 in. (5 mm) in the upper 1½ in. (38 mm). The capacity and dimensions of the measure shall conform to the limits in Table 1 or 2 (Note 2).

NOTE 1—The top rim is satisfactorily plane if a 0.01-in. (0.25-mm) feeler gage cannot be inserted between the rim and a piece of ¼-in. (6-mm) or thicker plate glass laid over the measure. The top and bottom are satisfactorily parallel if the slope between pieces of plate glass in contact with the top and bottom does not exceed 1 % in any direction.

NOTE 2—Dimensional tolerances and thicknesses of metal prescribed here are intended to be applied to measures acquired after January 1, 1968. Measures acquired before that date may conform either to this standard or to Section 2(*c*) of Test Method C 29 – 60.[3]

---

[1] This method is under the jurisdiction of ASTM Committee C-9 on Concrete and Concrete Aggregates and is the direct responsibility of Subcommittee C09.03.05 on Methods of Testing and Specifications for Physical Characteristics of Concrete Aggregates.

Current edition approved May 26, 1978. Published July 1978. Originally published as C 29 – 20 T. Last previous edition C 29 – 76.

[2] *Annual Book of ASTM Standards,* Vol 04.02.

[3] Discontinued, see *1966 Book of ASTM Standards,* Part 10.

5. Sampling

5.1 Sampling should generally be accomplished in accordance with Practice D 75 and sample reduction in accordance with Methods C 702.

6. Sample

6.1 Dry the sample of aggregate to essentially constant weight, preferably in an oven at 110 ± 5°C (230 ± 9°F).

7. Calibration of Measure

7.1 Fill the measure with water at room temperature and cover with a piece of plate glass in such a way as to eliminate bubbles and excess water.

7.2 Determine the net weight of water in the measure to an accuracy of ±0.1 %.

7.3 Measure the temperature of the water and determine its unit weight, from Table 3, interpolating if necessary.

7.4 Calculate the factor for the measure by dividing the unit weight of the water by the weight required to fill the measure.

8. Rodding Procedure

8.1 The rodding procedure is applicable to aggregates having a maximum size of 1½ in. (40 mm) or less.

8.1.1 Fill the measure one-third full and level the surface with the fingers. Rod the layer of aggregate with 25 strokes of the tamping rod evenly distributed over the surface. Fill the measure two-thirds full and again level and rod as above. Finally, fill the measure to overflowing and again rod as above. Level the surface of the aggregate with the fingers or a straightedge in such a way that any slight projections of the larger pieces of the coarse aggregate approximately balance the larger voids in the surface below the top of the measure.

8.1.2 In rodding the first layer, do not allow the rod to strike the bottom of the measure forcibly. In rodding the second and third layers, use only enough force to cause the tamping rod to penetrate the previous layer of aggregate.

8.1.3 Weigh the measure and its contents and record the net weight of the aggregate to the nearest 0.1 %. Multiply this weight by the factor calculated as described in 7.4. The product is the compact unit weight of the aggregate by rodding.

9. Jigging Procedure

9.1 The jigging procedure is applicable to aggregates having a maximum size greater than 1½ in. (40 mm) and not to exceed 4 in. (100 mm).

9.1.1 Fill the measure in three approximately equal layers as described in 8.1.1, compacting each layer by placing the measure on a firm base, such as a cement-concrete floor, raising the opposite sides alternately about 2 in. (50 mm), and allowing the measure to drop in such a manner as to hit with a sharp, slapping blow. The aggregate particles, by this procedure, will arrange themselves in a densely compacted condition. Compact each layer by dropping the measure 50 times in the manner described, 25 times on each side. Level the surface of the aggregate with the fingers or a straightedge in such a way that any slight projections of the larger pieces of the coarse aggregate approximately balance the larger voids in the surface below the top of the measure.

9.1.2 Weigh the measure and its contents and record the net weight of the aggregate to the nearest 0.1 %. Multiply this weight by the factor calculated as described in 7.4. The product is the compact unit weight of the aggregate by jigging.

10. Shoveling Procedure

10.1 The shoveling procedure is applicable to aggregates having a maximum size of 4 in. (100 mm) or less.

10.1.1 Fill the measure to overflowing by means of a shovel or scoop, discharging the aggregate from a height not to exceed 2 in. (50 mm) above the top of the measure. Exercise care to prevent, so far as possible, segregation of the particle sizes of which the sample is composed. Level the surface of the aggregate with the fingers or a straightedge in such a way that any slight projections of the larger pieces of the coarse aggregate approximately balance the larger voids in the surface below the top of the measure.

10.1.2 Weigh the measure and its contents and record the net weight of the aggregate to the nearest 0.1 %. Multiply this weight by the factor calculated as described in 7.4. The product is the loose unit weight of the aggregate.

11. Void Content in Aggregate

11.1 Void content in aggregate can be calculated as follows using the unit weight measured by either rodding, jigging, or the shoveling procedures:

$$\text{Voids, \%} = \frac{(A \times W) - B}{A \times W} \times 100$$

where:

A = bulk specific gravity (dry basis) as determined in accordance with Methods C 127 or C 128.

B = unit weight of aggregate as determined in Section 8, Rodding Procedure, Section 9, Jigging Procedure, or Section 10, Shoveling Procedure in lb/ft³ (kg/m³).

W = unit weight of water, 62.4 lb/ft³ (999 kg/m³).

11.2 Report the results computed in 11.1 as:

11.2.1 Voids in aggregate compacted by rodding, %, or

11.2.2 Voids in aggregate compacted by jigging, %, or

11.2.3 Voids in loose aggregate, %.

12. Precision

12.1 The multilaboratory standard deviation has been found to be 1.5 lb/ft³ (24 kg/m³)[4] for nominal ³/₄-in. (19.0-mm) maximum size, normal weight, coarse aggregate using ¹/₂ ft³ (15 litres) measures. Therefore, results of two properly conducted tests from two different laboratories on samples of the same coarse aggregate should not differ by more than 4.2 lb/ft³ (67 kg/m³).[4] The corresponding single-operator standard deviation has been found to be 0.7 lb/ft³ (11 kg/m³).[4] Therefore, results of two properly conducted tests by the same operator on the same coarse aggregate should not differ by more than 2.0 lb/ft³ (32 kg/m³).[4]

[4] These numbers represent, respectively the (1S) and (D2S) limits as described in ASTM Practice C 670, for Preparing Precision Statements for Test Methods for Construction Materials, *Annual Book of ASTM Standards*, Vol 04.02.

TABLE 1   Dimensions of Measures, Inch-Pound System[A]

Capacity, ft³	Inside Diameter, in.	Inside Height, in.	Thicknesses of metal, min, in.		Size of Aggregate, max, in.[B]
			Bottom	Wall	
¹/₁₀	6.0 ± 0.1	6.1 ± 0.1	0.20	0.10	¹/₂
¹/₃	8.0 ± 0.1	11.5 ± 0.1	0.20	0.10	1
¹/₂	10.0 ± 0.1	11.0 ± 0.1	0.20	0.12	1¹/₂
1	14.0 ± 0.1	11.2 ± 0.1	0.20	0.12	4

[A] The indicated size of container may be used to test aggregates cf a maximum nominal size equal to or smaller than that listed.

[B] Based on sieves with square openings.

TABLE 2   Dimensions of Measures, Metric System[A]

Capacity, litres	Inside Diameter, mm	Inside Height, mm	Thicknesses of metal, min, mm		Size of Aggregate, max, mm[B]
			Bottom	Wall	
3	155 ± 2	160 ± 2	5.0	2.5	12.5
10	205 ± 2	305 ± 2	5.0	2.5	25
15	255 ± 2	295 ± 2	5.0	3.0	40
30	355 ± 2	305 ± 2	5.0	3.0	100

[A] The indicated size of container may be used to test aggregates of a maximum nominal size equal to or smaller than that listed.

[B] Based on sieves with square openings.

TABLE 3 Unit Weight of Water

Temperature		lb/ft³	kg/m³
°F	°C		
60	15.6	62.366	999.01
65	18.3	62.336	998.54
70	21.1	62.301	997.97
(73.4)	(23.0)	(62.274)	(997.54)
75	23.9	62.261	997.32
80	26.7	62.216	996.59
85	29.4	62.166	995.83

Standard Method of
MAKING AND CURING CONCRETE TEST SPECIMENS IN THE FIELD[1]

This standard is issued under the fixed designation C 31; the number immediately following the designation indicates the year of original adoption or, in the case of revision, the year of last revision. A number in parentheses indicates the year of last reapproval. A superscript epsilon (ϵ) indicates an editorial change since the last revision or reapproval.

This method has been approved for use by agencies of the Department of Defense and for listing in the DoD Index of Specifications and Standards.

1. Scope

1.1 This method covers procedures for making and curing cylindrical and prismatic specimens using job concrete that can be consolidated by rodding or vibration as described herein.

1.2 The molded specimens shall have the same levels of slump, air content, and percentage of coarse aggregate as the concrete being placed in the work.

1.3 The values stated in inch-pound units are to be regarded as the standard. The metric equivalents given in the standard may be approximate.

1.4 *This standard may involve hazardous materials, operations, and equipment. This standard does not purport to address all of the safety problems associated with its use. It is the responsibility of whoever uses this standard to consult and establish appropriate safety and health practices and determine the applicability of regulatory limitations prior to use.*

2. Applicable Documents

2.1 *ASTM Standards:*
C 143 Test Method for Slump of Portland Cement Concrete[2]
C 172 Method of Sampling Freshly Mixed Concrete[2]
C 173 Test Method for Air Content of Freshly Mixed Concrete by the Volumetric Method[2]
C 192 Method of Making and Curing Concrete Test Specimens in the Laboratory[2]
C 231 Test Method for Air Content of Freshly Mixed Concrete by the Pressure Method[2]
C 470 Specification for Molds for Forming Concrete Test Cylinders Vertically[2]
C 511 Specification for Moist Cabinets, Moist Rooms, and Water Storage Tanks Used in the Testing of Hydraulic Cements and Concretes[2]
C 617 Practice for Capping Cylindrical Concrete Specimens[2]

3. Significance and Use

3.1 This method provides standardized requirements for making, curing, protecting, and transporting concrete test specimens under field conditions.

3.2 If specimen preparation is controlled as stipulated herein, the specimens may be used to develop information for the following purposes:

3.2.1 Checking the adequacy of laboratory mixture proportions for strength,

3.2.2 Serve as the basis for comparison with laboratory, field or in-place tests as the basis for safety and in-structure performance evaluation, and as the basis for form and shoring removal time requirements,

3.2.3 Determination of compliance with strength specifications, and

3.2.4 Determination of time when a structure may be put in service.

4. Apparatus

4.1 *Molds, General*—Molds for specimens or fastenings thereto in contact with the concrete

---

[1] This method is under the jurisdiction of ASTM Committee C-9 on Concrete and Concrete Aggregates and is the direct responsibility of Subcommittee C09.03.01 on Methods of Testing Concrete for Strength.
Current edition approved March 1, 1984. Published May 1984. Originally published as C 31 – 20. Last previous edition C 31 – 83.
[2] *Annual Book of ASTM Standards*, Vol 04.02.

shall be made of steel, cast iron, or other non-absorbent material, nonreactive with concrete containing portland or other hydraulic cements. Molds shall hold their dimensions and shape under conditions of severe use. Molds shall be watertight during use as judged by their ability to hold water poured into them. Provisions for tests of watertightness are given in Section 6 of Specification C 470. A suitable sealant, such as heavy grease, modeling clay, or microcrystalline wax shall be used where necessary to prevent leakage through the joints. Positive means shall be provided to hold base plates firmly to the molds. Molds shall be lightly coated with mineral oil or a suitable nonreactive form release material before use.

4.2 *Cylinder Molds:*

4.2.1 *Molds for Casting Specimens Vertically*—Molds for casting concrete test specimens shall conform to the requirements of Specification C 470.

4.3 *Beam Molds*—Beam molds shall be rectangular in shape and of the dimensions required to produce the specimens stipulated in 5.2. The inside surfaces of the molds shall be smooth. The sides, bottom, and ends shall be at right angles to each other and shall be straight and true and free of warpage. Maximum variation from the nominal cross section shall not exceed ⅛ in. (3.2 mm) for molds with depth or breadth of 6 in. (152 mm) or more. Molds shall produce specimens not more than 1/16 in. (1.6 mm) shorter than the required length in accordance with 5.2, but may exceed it by more than that amount.

4.4 *Tamping Rod*—The rod shall be a round, straight steel rod ⅝ in. (16 mm) in diameter and approximately 24 in. (610 mm) long, with the tamping end rounded to a hemispherical tip of the same diameter. Both ends may be rounded, if preferred.

4.5 *Vibrators*—Internal vibrators may have rigid or flexible shafts, preferably powered by electric motors. The frequency or vibration shall be 7000 vibrations per minute or greater while in use. The outside diameter or side dimension of the vibrating element shall be at least 0.75 in. (19 mm) and not greater than 1.50 in. (38 mm). The combined length of the shaft and vibrating element shall exceed the maximum depth of the section being vibrated by at least 3 in. (76 mm). When external vibrators are used, they should be the table or plank type. The frequency of external vibrators shall be at least 3600 vibrations per minute. For both table and plank vibrators, provision shall be made for clamping the mold securely to the apparatus. A vibrating-reed tachometer should be used to check the frequency of vibration.

4.6 *Mallet*—A mallet with a rubber or rawhide head weighing 1.25 ± 0.50 lb (0.57 ± 0.23 kg) shall be used.

4.7 *Small Tools*—Tools and items which may be required are shovels, pails, trowels, wood float, metal float, blunted trowels, straightedge, feeler gage, scoops, and rules.

4.8 *Slump Apparatus*—The apparatus for measurement of slump shall conform to the requirements of Test Method C 143.

4.9 *Sampling and Mixing Receptacle*—The receptacle shall be a suitable heavy gage metal pan, wheelbarrow, or flat, clean nonabsorbent mixing board of sufficient capacity to allow easy remixing of the entire sample with a shovel or trowel.

4.10 *Air Content Apparatus*—The apparatus for measuring air content shall conform to the requirements of Test Methods C 173 or C 231.

5. Test Specimens

5.1 *Compressive Strength Specimens*—Compressive strength specimens shall be cylinders of concrete cast and hardened in an upright position, with a length equal to twice the diameter. The standard specimen shall be the 6 by 12-in. (152 by 305-mm) cylinder when the maximum size of the coarse aggregate does not exceed 2 in. (50 mm). When the maximum size of the coarse aggregate does exceed 2 in. (50 mm), either the concrete sample shall be treated by wet sieving as described in Method C 172 or the diameter of the cylinder shall be at least three times the nominal maximum size of coarse aggregate in the concrete. Unless required by the project specifications, cylinders smaller than 6 of 12 in. shall not be made in the field.

NOTE 1—The maximum size is the smallest sieve opening through which the entire amount of aggregate is required to pass.

5.2 *Flexural Strength Specimens*—Flexural strength specimens shall be rectangular beams of concrete cast and hardened with long axes horizontal. The length shall be at least 2 in. (50 mm) greater than three times the depth as tested. The ratio of width to depth as molded shall not exceed

1.5. The standard beam shall be 6 by 6 in. (152 by 152 mm) in cross section, and shall be used for concrete with maximum size coarse aggregate up to 2 in. (50 mm). When the nominal maximum size of the coarse aggregate exceeds 2 in. (50 mm), the smaller cross sectional dimension of the beam shall be at least three times the nominal maximum size of the coarse aggregate. Unless required by project specifications, beams made in the field shall not have a width or depth of less than 6 in.

6. Sampling Concrete

6.1 The samples used to fabricate test specimens under this standard shall be obtained in accordance with Method C 172 unless an alternative procedure has been approved.

6.2 Record the identity of the sample with respect to the location of the concrete represented and the time of casting.

7. Slump and Air Content

7.1 *Slump*—Measure the slump of each batch of concrete, from which specimens are made, immediately after remixing in the receptacle, as required in Test Method C 143.

7.2 *Air Content*—Determine the air content in accordance with either Test Method C 173 or Test Method C 231. The concrete used in performing the air content test shall not be used in fabricating test specimens.

8. Molding Specimens

8.1 *Place of Molding*—Mold specimens promptly on a level, rigid, horizontal surface, free from vibration and other disturbances, at a place as near as practicable to the location where they are to be stored. Immediately after being struck off, the specimens shall be moved to the storage place where they will remain undisturbed for the initial curing period. If specimens made in single use mold are moved, lift and support the specimens from the bottom of the molds with a trowel or other device.

8.2 *Placing the Concrete*—Place the concrete in the molds using a scoop, blunted trowel, or shovel. Select each scoopful, trowelful, or shovelful of concrete from the mixing pan to ensure that it is representative of the batch. Remix the concrete in the mixing pan with a shovel or trowel to prevent segregation during the molding of specimens. Move the scoop, trowel, or shovel around the perimeter of the mold opening when adding concrete to ensure an even distribution of the concrete and minimize segregation. Further distribute the concrete by use of a tamping rod prior to the start of consolidation. In placing the final layer the operator shall attempt to add an amount of concrete that will exactly fill the mold after compaction. Do not add nonrepresentative concrete to an underfilled mold.

8.2.1 *Number of Layers*—Make specimens in layers as indicated in Table 1.

8.3 *Consolidation:*

8.3.1 *Methods of Consolidation*—Preparation of satisfactory specimens requires different methods of consolidation. The methods of consolidation are rodding, and internal or external vibration. Base the selection of the method of consolidation on the slump, unless the method is stated in the specifications under which the work is being performed. Rod concretes with a slump greater than 3 in. (75 mm). Rod or vibrate concretes with slump of 1 to 3 in. (25 to 75 mm). Vibrate concretes with slump of less than 1 in. (25 mm). Concretes of such low water content that they cannot be properly consolidated by the methods described herein, or requiring other sizes and shapes of specimens to represent the product or structure, are not covered by this method. Specimens for such concretes shall be made in accordance with the requirements of Method C 192 with regard to specimen size and shape and method of consolidation.

8.3.2 *Rodding*—Place the concrete in the mold, in the required number of layers of approximately equal volume. For cylinders, rod each layer with the rounded end of the rod using the number of strokes specified in Table 2. The number of roddings per layer required for beams is one for each 2-in.^{2} (13-cm^2) top surface area of the specimen. Rod the bottom layer throughout its depth. Distribute the strokes uniformly over the cross section of the mold and for each upper layer allow the rod to penetrate about 1/2 in. (12 mm) into the underlying layer when the depth of the layer is less than 4 in. (100 mm), and about 1 in. (25 mm) when the depth is 4 in. or more. If voids are left by the tamping rod, tap the sides of the mold lightly with the mallet or open hand when using light-gage single-use molds to close the voids. After each layer is rodded, spade the concrete along the sides and ends of beam molds with a trowel or other suitable tool.

8.3.3 *Vibration*—Maintain a uniform time period for duration of vibration for the particular kind of concrete, vibrator, and specimen mold involved. The duration of vibration required will depend upon the workability of the concrete and the effectiveness of the vibrator. Usually sufficient vibration has been applied as soon as the surface of the concrete has become relatively smooth. Continue vibration only long enough to achieve proper consolidation of the concrete. Overvibration may cause segregation. Fill the molds and vibrate in the required number of approximately equal layers. Place all the concrete for each layer in the mold before starting vibration of that layer. When placing the final layer, avoid overfilling by more than 1/4 in. (6 mm). Finish the surface either during or after vibration where external vibration is used. Finish the surface after vibration when internal vibration is used. When the finish is applied after vibration, add only enough concrete with a trowel to overfill the mold about 1/8 in. (3 mm). Work it into the surface and then strike it off.

8.3.3.1 *Internal Vibration*—The diameter of the vibrating element, or thickness of a square vibrating element, shall be in accordance with the requirements of 4.5. For beams, the vibrating element shall not exceed 1/3 of the width of the mold. For cylinders, the ratio of the diameter of the cylinder to the diameter of the vibrating element shall be 4.0 or higher. In compacting the specimen the vibrator shall not be allowed to rest on the bottom or sides of the mold. Carefully withdraw the vibrator in such a manner that no air pockets are left in the specimen. After vibration of each layer tap the sides of the mold with the mallet to ensure removal of large entrapped air bubbles.

8.3.3.2 *Cylinders*—Use three insertions of the vibrator at different points for each layer. Allow the vibrator to penetrate through the layer being vibrated, and into the layer below, approximately 1 in. (25 mm).

8.3.3.3 *Beam*—Insert the vibrator at intervals not exceeding 6 in. (150 mm) along the center line of the long dimension of the specimen. For specimens wider than 6 in., use alternating insertions along two lines. Allow the shaft of the vibrator to penetrate into the bottom layer approximately 1 in. (25 mm).

8.3.4 *External Vibration*—When external vibration is used, take care to ensure that the mold is rigidly attached to or securely held against the vibrating element or vibrating surface.

8.4 *Finishing*—After consolidation, unless the finishing has been performed during the vibration (8.3.3), strike off the surface of the concrete and float or trowel it as required. Perform all finishing with the minimum manipulation necessary to produce a flat even surface that is level with the rim or edge of the mold and that has no depressions or projections larger than 1/8 in. (3.2 mm).

8.4.1 *Cylinders*—After consolidation, finish the top surfaces by striking them off with the tamping rod where the consistency of the concrete permits or with a wood float or trowel. If desired, cap the top surface of freshly made cylinders with a thin layer of stiff portland cement paste which is permitted to harden and cure with the specimen. See section on Capping Materials of Practice C 617.

8.4.2 *Beams*—Beams shall be finished with a wood or metal float.

9. Curing

9.1 *Covering After Finishing*—To prevent evaporation of water from the unhardened concrete, cover the specimens immediately after finishing, preferably with a nonabsorptive, nonreactive plate or a sheet of tough, durable, impervious plastic. Wet burlap may be used for covering, but care must be exercised to keep the burlap wet until the specimens are removed from the molds. Placing a sheet of plastic over the burlap will facilitate keeping it wet. Protect the outside surfaces of cardboard molds from all contact with wet burlap or other sources of water for the first 24 h after cylinders have been molded in them. Water may cause the molds to expand and damage specimens at this early age.

9.2 *Initial Curing*—During the first 24 h after molding, store all test specimens under conditions that maintain the temperature immediately adjacent to the specimens in the range of 60 to 80°F (16 to 27°C) and prevent loss of moisture from the specimens. Storage temperatures may be regulated by means of ventilation or by evaporation of water from sand or burlap (Note 2), or by using heating devices such as stoves, electric light bulbs, or thermostatically controlled heating cables. A temperature record of the specimens may be established by means of maximum-minimum thermometers. Store specimens in tightly constructed, firmly braced wooden boxes, damp sand pits, temporary buildings at construction sites, under wet burlap in favorable weather, or in heavyweight closed plastic bags, or use other suitable methods, provided the foregoing requirements limiting specimen temperature and moisture loss are met. Specimens formed in cardboard molds (4.2.1) shall not be stored for the first 24 h in contact with wet sand or wet burlap or under any other condition that will allow the outside surfaces of the mold to absorb water.

NOTE 2—The temperature within damp sand and under wet burlap or similar materials will always be lower than the temperature in the surrounding atmosphere if evaporation takes place.

9.3 *Curing Cylinders for Checking the Adequacy of Laboratory Mixture Proportions for Strength or as the Basis for Acceptance or for Quality Control*—Remove test specimens made for checking the adequacy of the laboratory mixture proportions for strength, or as the basis for acceptance, from the molds at the end of 20 ± 4 h and stored in a moist condition at 73.4 ± 3°F (23 ± 1.7°C) until the moment of test (Note 2). As applied to the treatment of demolded specimens, moist curing means that the test specimens shall have free water maintained on the entire surface area at all times. This condition is met by immersion in saturated lime water and may be met by storage in a moist room or cabinet meeting the requirements of Specification C 511. Specimens shall not be exposed to dripping or running water.

9.4 *Curing Cylinders for Determining Form Removal Time or When a Structure May Be Put into Service*—Store test specimens made for determining when forms may be removed or when a structure may be put in service in or on the structure as near to the point of use as possible, and shall receive, insofar as practicable, the same protection from the elements on all surfaces as is given to the portions of the structure which they represent. Test specimens in the moisture condition resulting from the specified curing treatment. To meet these conditions, specimens made for the purpose of determining when a structure may be put in service shall be removed from the molds at the time of removal of form work. Follow the provisions of 9.6, where applicable, for removal of specimens from molds.

9.5 *Curing Beams for Checking the Adequacy of Laboratory Mixture Proportions for Strength or as the Basis for Acceptance or for Quality Control*—Remove test specimens made for checking the adequacy of the laboratory mixture proportions for flexural strength, or as the basis for acceptance, or for quality control, from the mold between 20 and 48 h after molding and cure according to the provisions of 9.3 except that storage for a minimum period of 20 h immediately prior to testing shall be in saturated lime water at 73.4 ± 3°F (23 ± 1.7°C). At the end of the curing period, between the time the specimen is removed from curing until testing is completed, prevent drying of the surfaces of the specimen.

NOTE 3—Relatively small amounts of drying of the surface of flexural specimens induce tensile stresses in the extreme fibers that will markedly reduce the indicated flexural strength.

9.6 *Curing Beams for Determining when a Structure May Be Put into Service*—Cure test specimens for determining when a structure may be put into service, as nearly as practicable, in the same manner as the concrete in the structure. At the end of 48 ± 4 h after molding, take the specimens in the molds to a location preferably near a field laboratory and remove from the molds. Store specimens representing pavements or slabs on grade by placing them on the ground as molded, with their top surfaces up. Bank the sides and ends of the specimens with earth or sand that shall be kept damp, leaving the top surfaces exposed to the specified curing treatment. Store specimens representing structure concrete as near the point in the structure they represent as possible and afford them the same temperature protection and moisture environment as the structure. At the end of the curing period leave the specimens in place exposed to the weather in the same manner as the structure. Remove all beam specimens from field storage and stored in lime water at 73.4 ± 3°F (23 ± 1.7°C) for 24 ± 4 h immediately before time of testing to ensure uniform moisture condition from specimen to specimen. Observe the precautions given in 9.5 to guard against drying between time of removal from curing to testing.

10. Transportation of Specimens to Laboratory

10.1 Specimens shall not be transported from the field to the laboratory before completion of the initial curing. Specimens to be transported prior to an age of 48 h shall not be demolded prior to completion of transportation. Prior to transporting, specimens shall be cured and protected as required in Section 9. During transportation, the specimens must be protected with suitable cushioning material to prevent damage from jarring and from damage by freezing temperatures, or moisture loss. Moisture loss may be prevented by wrapping the specimens in plastic or surrounding them with wet sand or wet saw dust. When specimens are received by the laboratory, they shall be removed from molds if not done before shipment and placed in the required standard curing at 73.4 ± 3°F (23 ± 1.7°C).

TABLE 1 Number of Layers Required for Specimens

Specimen Type and Size, as Depth, in. (mm)	Mode of Compaction	Number of Layers	Approximate Depth of Layer, in. (mm)
Cylinders:			
12 (305)	rodding	3 equal	4 (100)
Over 12 (305)	rodding	as required	4 (100)
12 (305) to 18 (460)	vibration	2 equal	half depth of specimens
Over 18 (460)	vibration	3 or more	8 (200) as near as practicable
Beams:			
6 (152) to 8 (200)	rodding	2 equal	half depth of specimen
Over 8 (200)	rodding	3 or more	4 (100)
6 (152) to 8 (200)	vibration	1	depth of specimen
Over 8 (200)	vibration	2 or more	8 (200) as near as practicable

TABLE 2 Number of Roddings to be Used in Molding Cylinder Specimens

Diameter of Cylinder, in. (mm)	Number of Strokes per Layer
6 (152)	25
8 (200)	50
10 (250)	75

Designation: C 127 – 84

Standard Test Method for
SPECIFIC GRAVITY AND ABSORPTION OF COARSE AGGREGATE[1]

This standard is issued under the fixed designation C 127; the number immediately following the designation indicates the year of original adoption or, in the case of revision, the year of last revision. A number in parentheses indicates the year of last reapproval. A superscript epsilon (ϵ) indicates an editorial change since the last revision or reapproval.

This test method has been approved for use by agencies of the Department of Defense and for listing in the DoD Index of Specifications and Standards.

1. Scope

1.1 This test method covers the determination of specific gravity and absorption of coarse aggregate. The specific gravity may be expressed as bulk specific gravity, bulk specific gravity (SSD) (saturated-surface-dry), or apparent specific gravity. The bulk specific gravity (SSD) and absorption are based on aggregate after 24 h soaking in water. This test method is not intended to be used with lightweight aggregates.

1.2 The values stated in acceptable metric unit (SI units and units specifically approved in ASTM E 380 for use with SI units) are to be regarded as the standard.

1.3 *This standard may involve hazardous materials, operations, and equipment. This standard does not purport to address all of the safety problems associated with its use. It is the responsibility of whoever uses this standard to consult and establish appropriate safety and health practices and determine the applicability of regulatory limitations prior to use.*

2. Applicable Documents

2.1 *ASTM Standards:*
C 29 Test Method for Unit Weight and Voids in Aggregate[2]
C 125 Definitions of Terms Relating to Concrete and Concrete Aggregates[2]
C 128 Test Method for Specific Gravity and Absorption of Fine Aggregate[2]
C 136 Method for Sieve Analysis of Fine and Coarse Aggregates[2]
C 566 Test Method for Total Moisture Content of Aggregate by Drying[2]
C 670 Practice for Preparing Precision Statements for Test Methods for Construction Materials[2]
C 702 Methods for Reducing Field Samples of Aggregate to Testing Size[2]
D 75 Practice for Sampling Aggregates[2]
D 448 Specification for Standard Sizes of Coarse Aggregate for Highway Construction[2]
E 11 Specification for Wire-Cloth Sieves for Testing Purposes[3]
E 12 Definitions of Terms Relating to Density and Specific Gravity of Solids, Liquids, and Gases[2]
E 380 Metric Practice[3]
2.2 *American Association of State Highway and Transportation Officials Standard:*[4]
AASHTO No. T 85 Specific Gravity and Absorption of Coarse Aggregate

3. Summary of Method

3.1 A sample of aggregate is immersed in water for approximately 24 h to essentially fill the pores. It is then removed from the water, the water dried from the surface of the particles, and weighed. Subsequently the sample is weighed while submerged in water. Finally the sample is oven-dried and weighed a third time. Using the weights thus obtained and formulas in this test

---

[1] This test method is under the jurisdiction of ASTM Committee C-9 on Concrete and Concrete Aggregates and is the direct responsibility of Subcommittee C09.03.05 on Methods of Testing and Specifications for Physical Characteristics of Concrete Aggregates.
Current edition approved April 27, 1984. Published June 1984. Originally published as C 127 – 36 T. Last previous edition C 127 – 81.
[2] *Annual Book of ASTM Standards*, Vol 04.02.
[3] *Annual Book of ASTM Standards*, Vol 14.02. Excerpts in all volumes.
[4] Available from American Association of State Highway and Transportation Officials, 444 North Capitol St. N.W., Suite 225, Washington, D.C. 20001.

method, it is possible to calculate three types of specific gravity and absorption.

4. Significance and Use

4.1 Bulk specific gravity is the characteristic generally used for calculation of the volume occupied by the aggregate in various mixtures containing aggregate, including portland cement concrete, bituminous concrete, and other mixtures that are proportioned or analyzed on an absolute volume basis. Bulk specific gravity is also used in the computation of voids in aggregate in Test Method C 29. Bulk specific gravity (SSD) is used if the aggregate is wet, that is, if its absorption has been satisfied. Conversely, the bulk specific gravity (oven-dry) is used for computations when the aggregate is dry or assumed to be dry.

4.2 Apparent specific gravity pertains to the relative density of the solid material making up the constituent particles not including the pore space within the particles which is accessible to water.

4.3 Absorption values are used to calculate the change in the weight of an aggregate due to water absorbed in the pore spaces within the constituent particles, compared to the dry condition, when it is deemed that the aggregate has been in contact with water long enough to satisfy most of the absorption potential. The laboratory standard for absorption is that obtained after submerging dry aggregate for approximately 24 h in water. Aggregates mined from below the water table may have a higher absorption, when used, if not allowed to dry. Conversely, some aggregates when used may contain an amount of absorbed moisture less than the 24-h soaked condition. For an aggregate that has been in contact with water and that has free moisture on the particle surfaces, the percentage of free moisture can be determined by deducting the absorption from the total moisture content determined by Test Method C 566.

4.4 The general procedures described in this test method are suitable for determining the absorption of aggregates that have had conditioning other than the 24-h soak, such as boiling water or vacuum saturation. The values obtained for absorption by other methods will be different than the values obtained by the prescribed 24-h soak, as will the bulk specific gravity (SSD).

4.5 The pores in lightweight aggregates may or may not become essentially filled with water after immersion for 24 h. In fact, many such aggregates can remain immersed in water for several days without satisfying most of the aggregates' absorption potential. Therefore, this test method is not intended for use with lightweight aggregate.

5. Definitions

5.1 *specific gravity*—the ratio of the mass (or weight in air) of a unit volume of a material to the mass of the same volume of water at stated temperatures. Values are dimensionless.

5.1.1 *bulk specific gravity*—the ratio of the weight in air of a unit volume of aggregate (including the permeable and impermeable voids in the particles, but not including the voids between particles) at a stated temperature to the weight in air of an equal volume of gas-free distilled water at a stated temperature.

5.1.2 *bulk specific gravity (SSD)*—the ratio of the weight in air of a unit volume of aggregate, including the weight of water within the voids filled to the extent achieved by submerging in water for approximately 24 h (but not including the voids between particles) at a stated temperature, compared to the weight in air of an equal volume of gas-free distilled water at a stated temperature.

5.1.3 *apparent specific gravity*—the ratio of the weight in air of a unit volume of the impermeable portion of aggregate at a stated temperature to the weight in air of an equal volume of gas-free distilled water at a stated temperature.

5.2 *absorption*—the increase in the weight of aggregate due to water in the pores of the material, but not including water adhering to the outside surface of the particles, expressed as a percentage of the dry weight. The aggregate is considered "dry" when it has been maintained at a temperature of 110 ± 5°C for sufficient time to remove all uncombined water.

NOTE 1—The terminology for specific gravity is based on terms in Definitions E 12, and that for absorption is based on that term in Definitions C 125.

6. Apparatus

6.1 *Balance*—A weighing device that is sensitive, readable, and accurate to 0.05 % of the sample weight at any point within the range used for this test, or 0.5 g, whichever is greater. The balance shall be equipped with suitable apparatus for suspending the sample container in water from the center of the weighing platform or pan of the weighing device.

6.2 *Sample Container*—A wire basket of 3.35 mm (No. 6) or finer mesh, or a bucket of approximately equal breadth and height, with a capacity of 4 to 7 L for 37.5-mm (1½-in.) nominal maximum size aggregate or smaller, and a larger container as needed for testing larger maximum size aggregate. The container shall be constructed so as to prevent trapping air when the container is submerged.

6.3 *Water Tank*—A watertight tank into which the sample container may be placed while suspended below the balance.

6.4 *Sieves*—A 4.75-mm (No. 4) sieve or other sizes as needed (see 7.2, 7.3, and 7.4), conforming to Specification E 11.

7. Sampling

7.1 Sample the aggregate in accordance with Practice D 75.

7.2 Thoroughly mix the sample of aggregate and reduce it to the approximate quantity needed using the applicable procedures in Methods C 702. Reject all material passing a 4.75-mm (No. 4) sieve by dry sieving and thoroughly washing to remove dust or other coatings from the surface. If the coarse aggregate contains a substantial quantity of material finer than the 4.75-mm sieve (such as for Size No. 8 and 9 aggregates in Specification D 448), use the 2.36-mm (No. 8) sieve in place of the 4.75-mm sieve. Alternatively, separate the material finer than the 4.75-mm sieve and test the finer material according to Test Method C 128.

7.3 The minimum weight of test sample to be used is given below. In many instances it may be desirable to test a coarse aggregate in several separate size fractions; and if the sample contains more than 15 % retained on the 37.5-mm (1½-in.) sieve, test the material larger than 37.5 mm in one or more size fractions separately from the smaller size fractions. When an aggregate is tested in separate size fractions, the minimum weight of test sample for each fraction shall be the difference between the weights prescribed for the maximum and minimum sizes of the fraction.

Nominal Maximum Size, mm (in.)	Minimum Weight of Test Sample, kg (lb)
12.5 (½) or less	2 (4.4)
19.0 (¾)	3 (6.6)
25.0 (1)	4 (8.8)
37.5 (1½)	5 (11)
50 (2)	8 (18)
63 (2½)	12 (26)
75 (3)	18 (40)
90 (3½)	25 (55)

Nominal Maximum Size, mm (in.)	Minimum Weight of Test Sample, kg (lb)
100 (4)	40 (88)
112 (4½)	50 (110)
125 (5)	75 (165)
150 (6)	125 (276)

7.4 If the sample is tested in two or more size fractions, determine the grading of the sample in accordance with Method C 136, including the sieves used for separating the size fractions for the determinations in this method. In calculating the percentage of material in each size fraction, ignore the quantity of material finer than the 4.75-mm (No. 4) sieve (or 2.36-mm (No. 8) sieve when that sieve is used in accordance with 7.2).

8. Procedure

8.1 Dry the test sample to constant weight at a temperature of 110 ± 5°C (230 ± 9°F), cool in air at room temperature for 1 to 3 h for test samples of 37.5-mm (1½-in.) nominal maximum size, or longer for larger sizes until the aggregate has cooled to a temperature that is comfortable to handle (approximately 50°C). Subsequently immerse the aggregate in water at room temperature for a period of 24 ± 4 h.

NOTE 2—When testing coarse aggregate of large nominal maximum size requiring large test samples, it may be more convenient to perform the test on two or more subsamples, and the values obtained combined for the computations described in Section 9.

8.2 Where the absorption and specific gravity values are to be used in proportioning concrete mixtures in which the aggregates will be in their naturally moist condition, the requirement for initial drying to constant weight may be eliminated, and, if the surfaces of the particles in the sample have been kept continuously wet until test, the 24-h soaking may also be eliminated.

NOTE 3—Values for absorption and bulk specific gravity (SSD) may be significantly higher for aggregate not oven dried before soaking than for the same aggregate treated in accordance with 8.1. This is especially true of particles larger than 75 mm (3 in.) since the water may not be able to penetrate the pores to the center of the particle in the prescribed soaking period.

8.3 Remove the test sample from the water and roll it in a large absorbent cloth until all visible films of water are removed. Wipe the larger particles individually. A moving stream of air may be used to assist in the drying operation. Take care to avoid evaporation of water from aggregate pores during the operation of surface-drying. Weigh the test sample in the saturated surface-dry condition. Record this and all subsequent weights to the nearest 0.5 g or 0.05 % of the sample weight, whichever is greater.

8.4 After weighing, immediately place the saturated-surface-dry test sample in the sample container and determine its weight in water at 23 ± 1.7°C (73.4 ± 3°F), having a density of 997 ± 2 kg/m³. Take care to remove all entrapped air before weighing by shaking the container while immersed.

NOTE 4—The container should be immersed to a depth sufficient to cover it and the test sample during weighing. Wire suspending the container should be of the smallest practical size to minimize any possible effects of a variable immersed length.

8.5 Dry the test sample to constant weight at a temperature of 110 ± 5°C (230 ± 9°F), cool in air at room temperature 1 to 3 h, or until the aggregate has cooled to a temperature that is comfortable to handle (approximately 50°C), and weigh.

9. Calculations

9.1 *Specific Gravity:*

9.1.1 *Bulk Specific Gravity*—Calculate the bulk specific gravity, 23/23°C (73.4/73.4°F), as follows:

$$\text{Bulk sp gr} = A/(B - C)$$

where:

A = weight of oven-dry test sample in air, g,

B = weight of saturated-surface-dry test sample in air, g, and

C = weight of saturated test sample in water, g.

9.1.2 *Bulk Specific Gravity (Saturated-Surface-Dry)*—Calculate the bulk specific gravity, 23/23°C (73.4/73.4°F), on the basis of weight of saturated-surface-dry aggregate as follows:

$$\text{Bulk sp gr (saturated-surface-dry)} = B/(B - C)$$

9.1.3 *Apparent Specific Gravity*—Calculate the apparent specific gravity, 23/23°C (73.4/73.4°F), as follows:

$$\text{Apparent sp gr} = A/(A - C)$$

9.2 *Average Specific Gravity Values*—When the sample is tested in separate size fractions the average value for bulk specific gravity, bulk specific gravity (SSD), or apparent specific gravity can be computed as the weighted average of the values as computed in accordance with 9.1 using the following equation:

$$G = \cfrac{1}{\dfrac{P_1}{100\ G_1} + \dfrac{P_2}{100\ G_2} + \cdots \dfrac{P_n}{100\ G_n}}$$

(see Appendix X1)

where:

G = average specific gravity. All forms of expression of specific gravity can be averaged in this manner.

G_1, $G_2 \ldots G_n$ = appropriate specific gravity values for each size fraction depending on the type of specific gravity being averaged.

P_1, $P_2, \ldots P_n$ = weight percentages of each size fraction present in the original sample.

NOTE 5—Some users of this test method may wish to express the results in terms of density. Density may be determined by multiplying the bulk specific gravity, bulk specific gravity (SSD), or apparent specific gravity by the weight of water (997.5 kg/m³ or 0.9975 Mg/m³ or 62.27 lb/ft³ at 23°C). Some authorities recommend using the density of water at 4°C (1000 kg/m³ or 1.000 Mg/m³ or 62.43 lb/ft³) as being sufficiently accurate. Results should be expressed to three significant figures. The density terminology corresponding to bulk specific gravity, bulk specific gravity (SSD), and apparent specific gravity has not been standardized.

9.3 *Absorption*—Calculate the percentage of absorption, as follows:

$$\text{Absorption, \%} = [(B - A)/A] \times 100$$

9.4 *Average Absorption Value*—When the sample is tested in separate size fractions, the average absorption value is the average of the values as computed in 9.3, weighted in proportion to the weight percentages of the size fractions in the original sample as follows:

$$A = (P_1 A_1/100) + (P_2 A_2/100) + \ldots (P_n A_n/100)$$

where:

A = average absorption, %,

A_1, $A_2 \ldots A_n$ = absorption percentages for each size fraction, and

P_1, $P_2, \ldots P_n$ = weight percentages of each size fraction present in the original sample.

10. Report

10.1 Report specific gravity results to the nearest 0.01, and indicate the type of specific gravity, whether bulk, bulk (saturated-surface-dry), or apparent.

10.2 Report the absorption result to the nearest 0.1 %.

10.3 If the specific gravity and absorption values were determined without first drying the aggregate, as permitted in 8.2, it shall be noted in the report.

11. Precision

11.1 The estimates of precision of this test method listed in Table 1 are based on results from the AASHTO Materials Reference Laboratory Reference Sample Program, with testing conducted by this test method and AASHTO Method T 85. The significant difference between the methods is that Test Method C 127 requires a saturation period of 24 ± 4 h, while Method T 85 requires a saturation period of 15 h minimum. This difference has been found to have an insignificant effect on the precision indices. The data are based on the analyses of more than 100 paired test results from 40 to 100 laboratories.

TABLE 1 Precision

	Standard Deviation (1S)[A]	Acceptable Range of Two Results (D2S)[A]
Single-Operator Precision:		
Bulk specific gravity (dry)	0.009	0.025
Bulk specific gravity (SSD)	0.007	0.020
Apparent specific gravity	0.007	0.020
Absorption[B], %	0.088	0.25
Multilaboratory Precision:		
Bulk specific gravity (dry)	0.013	0.038
Bulk specific gravity (SSD)	0.011	0.032
Apparent specific gravity	0.011	0.032
Absorption[B], %	0.145	0.41

[A] These numbers represent, respectively, the (1S) and (D2S) limits as described in Practice C 670. The precision estimates were obtained from the analysis of combined AASHTO Materials Reference Laboratory reference sample data from laboratories using 15 h minimum saturation times and other laboratories using 24 ± 4 h saturation times. Testing was performed on normal-weight aggregates, and started with aggregates in the oven-dry condition.

[B] Precision estimates are based on aggregates with absorptions of less than 2 %.

APPENDIXES

(Nonmandatory Information)

X1. DEVELOPMENT OF EQUATIONS

X1.1 The derivation of the equation is apparent from the following simplified cases using two solids. Solid 1 has a weight W_1 in grams and a volume V_1 in millilitres; its specific gravity (G_1) is therefore W_1/V_1. Solid 2 has a weight W_2 and volume V_2, and $G_2 = W_2/V_2$. If the two solids are considered together, the specific gravity of the combination is the total weight in grams divided by the total volume in millilitres:

$$G = (W_1 + W_2)/(V_1 + V_2)$$

Manipulation of this equation yields the following:

$$G = \frac{1}{\dfrac{V_1 + V_2}{W_1 + W_2}} = \frac{1}{\dfrac{V_1}{W_1 + W_2} + \dfrac{V_2}{W_1 + W_2}}$$

$$G = \frac{1}{\dfrac{W_1}{W_1 + W_2}\left(\dfrac{V_1}{W_1}\right) + \dfrac{W_2}{W_1 + W_2}\left(\dfrac{V_2}{W_2}\right)}$$

However, the weight fractions of the two solids are:

$$W_1/(W_1 + W_2) = P_1/100 \text{ and } W_2/(W_1 + W_2) = P_2/100$$

and,

$$1/G_1 = V_1/W_1 \text{ and } 1/G_2 = V_2/W_2$$

Therefore,

$$G = 1/[(P_1/100)(1/G_1) + (P_2/100)(1/G_2)]$$

An example of the computation is given in Table X1.1.

TABLE X1.1 Example of Calculation of Average Values of Specific Gravity and Absorption for a Coarse Aggregate Tested in Separate Sizes

Size Fraction, mm (in.)	% in Original Sample	Sample Weight Used in Test, g	Bulk Specific Gravity (SSD)	Absorption, %
4.75 to 12.5 (No. 4 to ½)	44	2213.0	2.72	0.4
12.5 to 37.5 (½ to 1½)	35	5462.5	2.56	2.5
37.5 to 63 (1½ to 2½)	21	12593.0	2.54	3.0

Average Specific Gravity (SSD)

$$G_{SSD} = \frac{1}{\frac{0.44}{2.72} + \frac{0.35}{2.56} + \frac{0.21}{2.54}} = 2.62$$

Average Absorption

$$A = (0.44)(0.4) + (0.35)(2.5) + (0.21)(3.0) = 1.7\%$$

X2. INTERRELATIONSHIPS BETWEEN SPECIFIC GRAVITIES AND ABSORPTION AS DEFINED IN TEST METHODS C 127 AND C 128

X2.1 Let:

S_d = bulk specific gravity (dry basis),
S_s = bulk specific gravity (SSD basis),
S_a = apparent specific gravity, and
A = absorption in %.

X2.2 Then,

$$S_s = (1 + A/100)S_d \tag{1}$$

$$S_a = \frac{1}{\frac{1}{S_d} - \frac{A}{100}} = \frac{S_d}{1 - \frac{AS_d}{100}} \tag{2}$$

$$S_a = \frac{1}{\frac{1 + A/100}{S_s} - \frac{A}{100}} = \frac{S_s}{1 - \left[\frac{A}{100}(S_s - 1)\right]} \tag{2a}$$

$$A = \left(\frac{S_s}{S_d} - 1\right)100 \tag{3}$$

$$A = \left(\frac{S_a - S_s}{S_a(S_s - 1)}\right)100 \tag{4}$$

Standard Test Method for
SPECIFIC GRAVITY AND ABSORPTION OF FINE AGGREGATE[1]

This standard is issued under the fixed designation C 128; the number immediately following the designation indicates the year of original adoption or, in the case of revision, the year of last revision. A number in parentheses indicates the year of last reapproval. A superscript epsilon (ε) indicates an editorial change since the last revision or reapproval.

This method has been approved for use by agencies of the Department of Defense and for listing in the DoD Index of Specifications and Standards.

1. Scope

1.1 This test method covers the determination of bulk and apparent specific gravity, 23/23°C (73.4/73.4°F), and absorption of fine aggregate.

1.2 This test method determines (after 24 h in water) the bulk specific gravity and the apparent specific gravity as defined in Definitions E 12, the bulk specific gravity on the basis of weight of saturated surface-dry aggregate, and the absorption as defined in Definitions C 125.

NOTE 1—The subcommittee is considering revising Test Methods C 127 and C 128 to use the term "density" instead of "specific gravity" for coarse and fine aggregate, respectively.

1.3 The values stated in acceptable metric units (SI units and units specifically approved in ASTM E 380 for use with SI units) are to be regarded as the standard.

1.4 *This standard may involve hazardous materials, operations, and equipment. This standard does not purport to address all of the safety problems associated with its use. It is the responsibility of whoever uses this standard to consult and establish appropriate safety and health practices and determine the applicability of regulatory limitations prior to use.*

2. Applicable Documents

2.1 *ASTM Standards:*
C 29 Test Method for Unit Weight and Voids in Aggregate[2]
C 70 Test Method for Surface Moisture in Fine Aggregate[2]
C 125 Definitions of Terms Relating to Concrete and Concrete Aggregates[2]

C 127 Test Method for Specific Gravity and Absorption of Coarse Aggregate[2]
C 188 Test Method for Density of Hydraulic Cement[3]
C 566 Test Method for Total Moisture Content of Aggregate by Drying[2]
C 670 Practice for Preparing Precision Statements for Test Methods for Construction Materials[2,3]
C 702 Methods for Reducing Field Samples of Aggregate to Testing Size[2]
D 75 Practice for Sampling Aggregates[2]
E 12 Definitions of Terms Relating to Density and Specific Gravity of Solids, Liquids, and Gases[2]
E 380 Metric Practice[4]

2.2 *American Association of State Highway and Transportation Officials Standard:*[5]
AASHTO No. T 84 Specific Gravity and Absorption of Fine Aggregates

3. Significance and Use

3.1 Bulk specific gravity is the characteristic

---

[1] This test method is under the jurisdiction of ASTM Committee C-9 on Concrete and Concrete Aggregates and is the direct responsibility of Subcommittee C09.03.05 on Methods of Testing and Specifications for Physical Characteristics of Concrete Aggregates.
Current edition approved April 27, 1984. Published June 1984. Originally published as C 128 – 36. Last previous edition C 128 – 79ᵉ¹.
[2] *Annual Book of ASTM Standards*, Vol 04.02.
[3] *Annual Book of ASTM Standards*, Vol 04.01.
[4] *Annual Book of ASTM Standards*, Vol 14.02. Excerpts in all volumes.
[5] Available from American Association of State Highway and Transportation Officials, 444 North Capitol St. N.W., Suite 225, Washington, D.C. 20001.

generally used for calculation of the volume occupied by the aggregate in various mixtures containing aggregate including portland cement concrete, bituminous concrete, and other mixtures that are proportioned or analyzed on an absolute volume basis. Bulk specific gravity is also used in the computation of voids in aggregate in Test Method C 29 and the determination of moisture in aggregate by displacement in water in Test Method C 70. Bulk specific gravity determined on the saturated surface-dry basis is used if the aggregate is wet, that is, if its absorption has been satisfied. Conversely, the bulk specific gravity determined on the oven-dry basis is used for computations when the aggregate is dry or assumed to be dry.

3.2 Apparent specific gravity pertains to the relative density of the solid material making up the constituent particles not including the pore space within the particles that is accessible to water. This value is not widely used in construction aggregate technology.

3.3 Absorption values are used to calculate the change in the weight of an aggregate due to water absorbed in the pore spaces within the constituent particles, compared to the dry condition, when it is deemed that the aggregate has been in contact with water long enough to satisfy most of the absorption potential. The laboratory standard for absorption is that obtained after submerging dry aggregate for approximately 24 h in water. Aggregates mined from below the water table may have a higher absorption when used, if not allowed to dry. Conversely, some aggregates when used may contain an amount of absorbed moisture less than the 24 h-soaked condition. For an aggregate that has been in contact with water and that has free moisture on the particle surfaces, the percentage of free moisture can be determined by deducting the absorption from the total moisture content determined by Test Method C 566 by drying.

4. Apparatus

4.1 *Balance*—A balance or scale having a capacity of 1 kg or more, sensitive to 0.1 g or less, and accurate within 0.1 % of the test load at any point within the range of use for this test. Within any 100-g range of test load, a difference between readings shall be accurate within 0.1 g.

4.2 *Pycnometer*—A flask or other suitable container into which the fine aggregate test sample can be readily introduced and in which the volume content can be reproduced within ± 0.1 cm³. The volume of the container filled to mark shall be at least 50 % greater than the space required to accommodate the test sample. A volumetric flask of 500 cm³ capacity or a fruit jar fitted with a pycnometer top is satisfactory for a 500-g test sample of most fine aggregates. A Le Chatelier flask as described in Test Method C 188 is satisfactory for an approximately 55-g test sample.

4.3 *Mold*—A metal mold in the form of a frustum of a cone with dimensions as follows: 40 ± 3 mm inside diameter at the top, 90 ± 3 mm inside diameter at the bottom, and 75 ± 3 mm in height, with the metal having a minimum thickness of 0.8 mm.

4.4 *Tamper*—A metal tamper weighing 340 ± 15 g and having a flat circular tamping face 25 ± 3 mm in diameter.

5. Sampling

5.1 Sampling shall be accomplished in general accordance with Practice D 75.

6. Preparation of Test Specimen

6.1 Obtain approximately 1 kg of the fine aggregate from the sample using the applicable procedures described in Methods C 702.

6.1.1 Dry it in a suitable pan or vessel to constant weight at a temperature of 110 ± 5°C (230 ± 9°F). Allow it to cool to comfortable handling temperature, cover with water, either by immersion or by the addition of at least 6 % moisture to the fine aggregate, and permit to stand for 24 ± 4 h.

6.1.2 As an alternative to 6.1.1, where the absorption and specific gravity values are to be used in proportioning concrete mixtures with aggregates used in their naturally moist condition, the requirement for initial drying to constant weight may be eliminated and, if the surfaces of the particles have been kept wet, the 24-h soaking may also be eliminated.

NOTE 2—Values for absorption and for specific gravity in the saturated surface-dry condition may be significantly higher for aggregate not oven dried before soaking than for the same aggregate treated in accordance with 6.1.1.

6.2 Decant excess water with care to avoid loss of fines, spread the sample on a flat nonabsorbent surface exposed to a gently moving current of warm air, and stir frequently to secure homogeneous drying. If desired, mechanical aids such as tumbling or stirring may be employed to assist in achieving the saturated surface-dry condition. Continue this operation until the test specimen approaches a free-flowing condition. Follow the procedure in 6.2.1 to determine whether or not surface moisture is present on the constituent fine aggregate particles. It is intended that the first trial of the cone test will be made with some surface water in the specimen. Continue drying with constant stirring and test at frequent intervals until the test indicates that the specimen has reached a surface-dry condition. If the first trial of the surface moisture test indicates that moisture is not present on the surface, it has been dried past the saturated surface-dry condition. In this case thoroughly mix a few millilitres of water with the fine aggregate and permit the specimen to stand in a covered container for 30 min. Then resume the process of drying and testing at frequent intervals for the onset of the surface-dry condition.

6.2.1 *Cone Test for Surface Moisture*—Hold the mold firmly on a smooth nonabsorbent surface with the large diameter down. Place a portion of the partially dried fine aggregate loosely in the mold by filling it to overflowing and heaping additional material above the top of the mold by holding it with the cupped fingers of the hand holding the mold. Lightly tamp the fine aggregate into the mold with 25 light drops of the tamper. Each drop should start about 5 mm (0.2 in.) above the top surface of the fine aggregate. Permit the tamper to fall freely under gravitational attraction on each drop. Adjust the starting height to the new surface elevation after each drop and distribute the drops over the surface. Remove loose sand from the base and lift the mold vertically. If surface moisture is still present, the fine aggregate will retain the molded shape. When the fine aggregate slumps slightly it indicates that it has reached a surface-dry condition. Some angular fine aggregate or material with a high proportion of fines may not slump in the cone test upon reaching a surface-dry condition. This may be the case if fines become airborne upon dropping a handful of the sand from the cone test 100 to 150 mm onto a surface. For these materials the saturated surface-dry condition should be considered as the point that one side of the fine aggregate slumps slightly upon removing the mold.

NOTE 3—The following criteria have also been used on materials that do not readily slump:

(1) *Provisional Cone Test*—Fill the cone mold as described in 6.2.1 except only use 10 drops of the tamper. Add more fine aggregate and use 10 drops of the tamper again. Then add material two more times using 3 and 2 drops of the tamper, respectively. Level off the material even with the top of the mold, remove loose material from the base; and lift the mold vertically.

(2) *Provisional Surface Test*—If airborne fines are noted when the fine aggregate is such that it will not slump when it is at a moisture condition, add more moisture to the sand, and at the onset of the surface-dry condition, with the hand lightly pat approximately 100 g of the material on a flat, dry, clean, dark or dull nonabsorbent surface such as a sheet of rubber, a worn oxidized, galvanized, or steel surface, or a black-painted metal surface. After 1 to 3 s remove the fine aggregate. If noticeable moisture shows on the test surface for more than 1 to 2 s then surface moisture is considered to be present on the fine aggregate.

(3) Colorimetric procedures described by Kandhal and Lee, Highway Research Record No. 307, p. 44.

(4) For reaching the saturated surface-dry condition on a single size material that slumps when wet, hard-finish paper towels can be used to surface dry the material until the point is just reached where the paper towel does not appear to be picking up moisture from the surfaces of the fine aggregate particles.

7. Procedure

7.1 Partially fill the pycnometer with water. Immediately introduce into the pycnometer 500.0 g of saturated surface-dry fine aggregate prepared as described in Section 6, and fill with additional water to approximately 90 % of capacity. If a weight other than 500 g is used, insert the actual weight in place of the figure "500" wherever it appears in the appropriate formulas. Roll, invert, and agitate the pycnometer to eliminate all air bubbles. Adjust its temperature to 23 ± 1.7°C (73.4 ± 3°F), if necessary by immersion in circulating water, and bring the water level in the pycnometer to its calibrated capacity. Determine total weight of the pycnometer, specimen, and water. Record this and all other weights to the nearest 0.1 g.

NOTE 4—It normally takes about 15 to 20 min to eliminate air bubbles.

7.1.1 *Alternative to Weighing in 7.1*—The quantity of added water necessary to fill the pycnometer at the required temperature may be determined volumetrically using a buret accurate to 0.15 mL. Compute the total weight of the pycnometer, specimen, and water as follows:

$$C = 0.9975\ V_a + 500 + W$$

where:

C = weight of pycnometer with specimen and water to calibration mark, g,

V_a = volume of water added to pycnometer, mL, and

W = weight of the pycnometer empty, g.

7.1.2 *Alternative to the Procedure in 7.1*—Use a Le Chatelier flask initially filled with water to a point on the stem between the 0 and the 1-mL mark. Record this initial reading with the flask and contents within the temperature range of 23 ± 1.7°C (73.4 ± 3°F). Add 55.0 g of fine aggregate in the saturated surface-dry condition (or other weight as necessary to result in raising the water level to some point on the upper series of gradation). After all fine aggregate has been introduced, place the stopper in the flask and roll the flask in an inclined position, or gently whirl it in a horizontal circle so as to dislodge all entrapped air, continuing until no further bubbles rise to the surface. Take a final reading with the flask and contents within 1°C (1.8°F) of the original temperature.

7.2 Remove the fine aggregate from the pycnometer, dry to constant weight at a temperature of 110 ± 5°C (230 ± 9°F), cool in air at room temperature for 1 ± ½ h, and weigh to the nearest 0.1 g.

7.2.1 If the Le Chatelier flask method is used, a separate sample portion is needed for the determination of absorption. Weigh a separate 500.0-g portion of the saturated surface-dry fine aggregate, dry to constant weight, and reweigh as described in 7.2.

7.3 Determine the weight of the pycnometer filled to its calibration capacity with water at 23 ± 1.7°C (73.4 ± 3°F).

7.3.1 *Alternative to Weighing in 7.3*—The quantity of water necessary to fill the empty pycnometer at the required temperature may be determined volumetrically using a buret accurate to 0.15 mL. Calculate the weight of the pycnometer filled with water as follows:

$$B = 0.9975\ V + W$$

where:

B = weight of flask filled with water, g,

V = volume of flask, mL, and

W = weight of the flask empty, g.

8. Bulk Specific Gravity

8.1 Calculate the bulk specific gravity, 23/ 23°C (73.4/73.4°F), as defined in Definitions E 12, as follows:

$$\text{Bulk sp gr} = A/(B + 500 - C)$$

where:

A = weight of oven-dry specimen in air, g,

B = weight of pycnometer filled with water, g, and

C = weight of pycnometer with specimen and water to calibration mark, g.

8.1.1 If the Le Chatelier flask method was used, calculate the bulk specific gravity, 23/23°C, as follows:

$$\text{Bulk sp gr} = \frac{55[1 - ((500 - A)/A)]}{0.9975\ (R_2 - R_1)}$$

where:

R_1 = initial reading of water level in Le Chatelier flask, and

R_2 = final reading of water level in Le Chatelier flask.

9. Bulk Specific Gravity (Saturated Surface-Dry Basis)

9.1 Calculate the bulk specific gravity, 23/ 23°C (73.4/73.4°F), on the basis of weight of saturated surface-dry aggregate as follows:

Bulk sp gr (saturated surface-dry basis)
$$= 500/(B + 500 - C)$$

9.1.1 If the Le Chatelier flask method was used, calculate the bulk specific gravity, 23/23°C, on the basis of saturated surface-dry aggregate as follows:

Bulk sp gr (saturated surface-dry basis)

$$= \frac{55}{0.9975\ (R_2 - R_1)}$$

10. Apparent Specific Gravity

10.1 Calculate the apparent specific gravity, 23/23°C (73.4/73.4°F), as defined in Definitions E 12, as follows:

$$\text{Apparent sp gr} = A/(B + A - C)$$

11. Absorption

11.1 Calculate the percentage of absorption, as defined in Definitions C 125, as follows:

$$\text{Absorption, \%} = [(500 - A)/A] \times 100$$

12. Report

12.1 Report specific gravity results to the nearest 0.01 and absorption to the nearest 0.1 %.

The Appendix gives mathematical interrelationships among the three types of specific gravities and absorption. These may be useful in checking the consistency of reported data or calculating a value that was not reported by using other reported data.

12.2 If the fine aggregate was tested in a naturally moist condition other than the oven dried and 24 h-soaked condition, report the source of the sample and the procedures used to prevent drying prior to testing.

13. Precision

13.1 The estimates of precision of this test method (listed in Table 1) are based on results from the AASHTO Materials Reference Laboratory Reference Sample Program, with testing conducted by this test method and AASHTO Method T 84. The significant difference between the methods is that Test Method C 128 requires a saturation period of 24 ± 4 h, and Method T 84 requires a saturation period of 15 to 19 h. This difference has been found to have an insignificant effect on the precision indices. The data are based on the analyses of more than 100 paired test results from 40 to 100 laboratories.

TABLE 1 Precision

	Standard Deviation (1S)[A]	Acceptable Range of Two Results (D2S)[A]
Single-Operator Precision:		
Bulk specific gravity (dry)	0.011	0.032
Bulk specific gravity (SSD)	0.0095	0.027
Apparent specific gravity	0.0095	0.027
Absorption[B], %	0.11	0.31
Multilaboratory Precision:		
Bulk specific gravity (dry)	0.023	0.066
Bulk specific gravity (SSD)	0.020	0.056
Apparent specific gravity	0.020	0.056
Absorption[B], %	0.23	0.66

[A] These numbers represent, respectively, the (1S) and (D2S) limits as described in Practice C 670. The precision estimates were obtained from the analysis of combined AASHTO Materials Research Laboratory reference sample data from laboratories using 15 to 19 h saturation times and other laboratories using 24 ± 4 h saturation time. Testing was performed on normal weight aggregates, and started with aggregates in the oven-dry condition.

[B] Precision estimates are based on aggregates with absorptions of less than 1 % and may differ for manufactured fine aggregates and fine aggregates having absorption values greater than 1 %.

APPENDIX

(Nonmandatory Information)

X1. INTERRELATIONSHIPS BETWEEN SPECIFIC GRAVITIES AND ABSORPTION AS DEFINED IN TEST METHODS C 127 AND C 128

X1.1 Let:

S_d = bulk specific gravity (dry-basis),
S_s = bulk specific gravity (SSD-basis),
S_a = apparent specific gravity, and
A = absorption in %.

Then:

(1) $\qquad S_s = (1 + A/100)S_d$

(2) $\qquad S_a = \dfrac{1}{\dfrac{1}{S_d} - \dfrac{A}{100}} = \dfrac{S_d}{1 - \dfrac{AS_d}{100}}$

(2a) $\qquad$ or $S_a = \dfrac{1}{\dfrac{1 + A/100}{S_s} - \dfrac{A}{100}}$

$\qquad\qquad\quad = \dfrac{S_s}{1 - \dfrac{A}{100}(S_s - 1)}$

(3) $\qquad A = \left(\dfrac{S_s}{S_d} - 1\right) 100$

(4) $\qquad A = \left(\dfrac{S_a - S_s}{S_d(S_s - 1)}\right) 100$

Standard Method for
SIEVE ANALYSIS OF FINE AND COARSE AGGREGATES[1]

This standard is issued under the fixed designation C 136; the number immediately following the designation indicates the year of original adoption or, in the case of revision, the year of last revision. A number in parentheses indicates the year of last reapproval. A superscript epsilon (ϵ) indicates an editorial change since the last revision or reapproval.

This method has been approved for use by agencies of the Department of Defense and for listing in the DoD Index of Specifications and Standards.

1. Scope

1.1 This method covers the determination of the particle size distribution of fine and coarse aggregates by sieving.

1.2 Some specifications for aggregates which reference this method contain grading requirements including both coarse and fine fractions. Instructions are included for sieve analysis of such aggregates.

1.3 The values stated in acceptable metric units (SI units and units specifically approved in ASTM E 380 for use with SI units) are to be regarded as the standard. The values in parentheses are provided for information purposes only.

1.4 *This standard may involve hazardous materials, operations, and equipment. This standard does not purport to address all of the safety problems associated with its use. It is the responsibility of whoever uses this standard to consult and establish appropriate safety and health practices and determine the applicability of regulatory limitations prior to use.*

2. Applicable Documents

2.1 *ASTM Standards:*
C 117 Test Method for Materials Finer Than 75-μm (No. 200) Sieve in Mineral Aggregates by Washing[2]
C 670 Practice for Preparing Precision Statements for Test Methods for Construction Materials[2,3]
C 702 Methods for Reducing Field Samples of Aggregate to Testing Size[2]
D 75 Practice for Sampling Aggregates[2]
E 11 Specification for Wire-Cloth Sieves for Testing Purposes[2,3]
E 380 Metric Practice[4]

2.2 *American Association of State Highway and Transportation Officials Standard:*
AASHTO No. T 27 Sieve Analysis of Fine and Coarse Aggregates[5]

3. Summary of Method

3.1 A weighed sample of dry aggregate is separated through a series of sieves of progressively smaller openings for determination of particle size distribution.

4. Significance and Use

4.1 This method is used primarily to determine the grading of materials proposed for use as aggregates or being used as aggregates. The results are used to determine compliance of the particle size distribution with applicable specification requirements and to provide necessary data for control of the production of various aggregate products and mixtures containing aggregates. The data may also be useful in developing relationships concerning porosity and packing.

4.2 Accurate determination of material finer than the 75-μm (No. 200) sieve cannot be achieved by use of this method alone. Test

---

[1] This method is under the jurisdiction of ASTM Committee C-9 on Concrete and Concrete Aggregates and is the direct responsibility of Subcommittee C09.03.05 on Methods of Testing and Specifications for Physical Characteristics of Concrete Aggregates.
Current edition approved Oct. 26, 1984. Published December 1984. Originally published as C 136 – 38 T. Last previous edition C 136 – 83.
[2] *Annual Book of ASTM Standards*, Vol 04.02.
[3] *Annual Book of ASTM Standards*, Vol 04.01.
[4] *Annual Book of ASTM Standards*, Vol 14.02. Excerpts in all volumes.
[5] Available from American Association of State Highway and Transportation Officials, 444 North Capitol St. N.W., Suite 225, Washington, DC 20001.

Method C 117 for material finer than 75-μm sieve by washing should be employed.

5. Apparatus

5.1 *Balances*—Balances or scales used in testing fine and coarse aggregate shall have readability and accuracy as follows:

5.1.1 For fine aggregate, readable to 0.1 g and accurate to 0.1 g or 0.1 % of the test load, whichever is greater, at any point within the range of use.

5.1.2 For coarse aggregate, or mixtures of fine and coarse aggregate, readable and accurate to 0.5 g or 0.1 % of the test load, whichever is greater, at any point within the range of use.

5.2 *Sieves*—The sieves shall be mounted on substantial frames constructed in a manner that will prevent loss of material during sieving. The sieves shall conform to Specification E 11. Sieves with openings larger than 125 mm (5 in.) shall have a permissible variation in average opening of ±2 % and shall have a nominal wire diameter of 8.0 mm (5/16 in.) or larger.

NOTE 1—It is recommended that sieves mounted in frames larger than standard 203-mm (8 in.) diameter frames be used for testing coarse aggregate.

5.3 *Mechanical Sieve Shaker*—A mechanical sieve shaker, if used, shall impart a vertical, or lateral and vertical, motion to the sieve, causing the particles thereon to bounce and turn so as to present different orientations to the sieving surface. The sieving action shall be such that the criterion for adequacy of sieving described in 7.4 is met in a reasonable time period.

NOTE 2—Use of a mechanical sieve shaker is recommended when the size of the sample is 20 kg or greater, and may be used for smaller samples, including fine aggregate. Excessive time (more than approximately 10 min) to achieve adequate sieving may result in degradation of the sample. The same mechanical sieve shaker may not be practical for all sizes of samples, since the large sieving area needed for practical sieving of a large nominal size coarse aggregate very likely could result in loss of a portion of the sample if used for a small sample of coarse aggregate or fine aggregate.

5.4 *Oven*—An oven of appropriate size capable of maintaining a uniform temperature of 110 ± 5°C (230 ± 9°F).

6. Sampling

6.1 Sample the aggregate in accordance with Practice D 75. The weight of the field sample shall be the weight shown in Practice D 75 or four times the weight required in 6.4 and 6.5 (except as modified in 6.6), whichever is greater.

6.2 Thoroughly mix the sample and reduce it to an amount suitable for testing using the applicable procedures described in Methods C 702. The sample for test shall be approximately of the weight desired when dry and shall be the end result of the reduction. Reduction to an exact predetermined weight shall not be permitted.

NOTE 3—Where sieve analysis, including determination of material finer than the 75-μm sieve, is the only testing proposed, the size of the sample may be reduced in the field to avoid shipping excessive quantities of extra material to the laboratory.

6.3 *Fine Aggregate*—The test sample of fine aggregate shall weigh, after drying, approximately the following amount:

Aggregate with at least 95 % passing a 2.36-mm (No. 8) sieve	100 g
Aggregate with at least 85 % passing a 4.75-mm (No. 4) sieve and more than 5 % retained on a 2.36-mm (No. 8) sieve	500 g

6.4 *Coarse Aggregate*—The weight of the test sample of coarse aggregate shall conform with the following:

Nominal Maximum Size, Square Openings, mm (in.)	Minimum Weight of Test Sample, kg (lb)
9.5 (⅜)	1 (2)
12.5 (½)	2 (4)
19.0 (¾)	5 (11)
25.0 (1)	10 (22)
37.5 (1½)	15 (33)
50 (2)	20 (44)
63 (2½)	35 (77)
75 (3)	60 (130)
90 (3½)	100 (220)
100 (4)	150 (330)
112 (4½)	200 (440)
125 (5)	300 (660)
150 (6)	500 (1100)

6.5 *Coarse and Fine Aggregate Mixtures*—The weight of the test sample of coarse and fine aggregate mixtures shall be the same as for coarse aggregate in 6.4.

6.6 The size of sample required for aggregates with large nominal maximum size is such as to preclude testing except with large mechanical sieve shakers. However, the intent of this method will be satisfied for samples of aggregate larger than 50 mm nominal maximum size if a smaller weight of sample is used, provided that the criterion for acceptance or rejection of the material is based on the average of results of several samples, such that the sample size used times the number of samples averaged equals the minimum weight of sample shown in 6.4.

6.7 In the event that the amount of material finer than the 75-µm (No. 200) sieve is to be determined by Test Method C 117, proceed as follows:

6.7.1 For aggregates with a nominal maximum size of 12.5 mm (1/2 in.) or less, use the same test sample for testing by Test Method C 117 and this method. First test the sample in accordance with Test Method C 117 through the final drying operation, then dry sieve the sample as stipulated in 7.2 through 7.7 of this method.

6.7.2 For aggregates with a nominal maximum size greater than 12.5 mm (1/2 in.), a single test sample may be used as described in 6.7.1, or separate test samples may be used for Test Method C 117 and this method.

6.7.3 Where the specifications require determination of the total amount of material finer than the 75-µm sieve by washing and dry sieving, use the procedure described in 6.7.1.

7. Procedure

7.1 Dry the sample to constant weight at a temperature of 110 ± 5°C (230 ± 9°F).

NOTE 4—For control purposes, particularly where rapid results are desired, it is generally not necessary to dry coarse aggregate for the sieve analysis test. The results are little affected by the moisture content unless: (1) the nominal maximum size is smaller than about 12.5 mm (½ in.); (2) the coarse aggregate contains appreciable material finer than 4.75 mm (No. 4); or (3) the coarse aggregate is highly absorptive (a lightweight aggregate, for example). Also, samples may be dried at the higher temperatures associated with the use of hot plates without affecting results, provided steam escapes without generating pressures sufficient to fracture the particles, and temperatures are not so great as to cause chemical breakdown of the aggregate.

7.2 Suitable sieve sizes shall be selected to furnish the information required by the specifications covering the material to be tested. The use of additional sieves may be desirable to provide other information, such as fineness modulus, or to regulate the amount of material on a sieve. Nest the sieves in order of decreasing size of opening from top to bottom and place the sample on the top sieve. Agitate the sieves by hand or by mechanical apparatus for a sufficient period, established by trial or checked by measurement on the actual test sample, to meet the criterion for adequacy or sieving described in 7.4.

7.3 Limit the quantity of material on a given sieve so that all particles have opportunity to reach sieve openings a number of times during the sieving operation. For sieves with openings smaller than 4.75-mm (No. 4), the weight retained on any sieve at the completion of the sieving operation shall not exceed 6 kg/m^2 (4 g/in.^{2}) of sieving surface. For sieves with openings 4.75 mm (No. 4) and larger, the weight in kg/m^2 of sieving surface shall not exceed the product of 2.5 × (sieve opening in mm). In no case shall the weight be so great as to cause permanent deformation of the sieve cloth.

NOTE 5—The 6 kg/m^2 amounts to 194 g for the usual 203-mm (8 in.) diameter sieve. The amount of material retained on a sieve may be regulated by (1) the introduction of a sieve with larger openings immediately above the given sieve or (2) testing the sample in a number of increments.

7.4 Continue sieving for a sufficient period and in such manner that, after completion, not more than 1 weight % of the residue on any individual sieve will pass that sieve during 1 min of continuous hand sieving performed as follows: Hold the individual sieve, provided with a snug-fitting pan and cover, in a slightly inclined position in one hand. Strike the side of the sieve sharply and with an upward motion against the heel of the other hand at the rate of about 150 times per minute, turn the sieve about one sixth of a revolution at intervals of about 25 strokes. In determining sufficiency of sieving for sizes larger than the 4.75-mm (No. 4) sieve, limit material on the sieve to a single layer of particles. If the size of the mounted testing sieves makes the described sieving motion impractical, use 203-mm (8 in.) diameter sieves to verify the sufficiency of sieving.

7.5 In the case of coarse and fine aggregate mixtures, the portion of the sample finer than the 4.75-mm (No. 4) sieve may be distributed among two or more sets of sieves to prevent overloading of individual sieves.

7.5.1 Alternatively, the portion finer than the 4.75-mm (No. 4) sieve may be reduced in size using a mechanical splitter according to Methods C 702. If this procedure is followed, compute the weight of each size increment of the original sample as follows:

$$A = \frac{W_1}{W_2} \times B$$

where:

A = weight of size increment on total sample basis,

W_1 = weight of fraction finer than 4.75-mm (No. 4) sieve in total sample,

W_2 = weight of reduced portion of material finer than 4.75-mm (No. 4) sieve actually sieved, and

B = weight of size increment in reduced portion sieved.

7.6 Unless a mechanical sieve shaker is used, hand sieve particles larger than 75 mm (3 in.) by determining the smallest sieve opening through which each particle will pass. Start the test on the smallest sieve to be used. Rotate the particles, if necessary, in order to determine whether they will pass through a particular opening; however, do not force particles to pass through an opening.

7.7 Determine the weight of each size increment by weighing on a scale or balance conforming to the requirements specified in 5.1 to the nearest 0.1 % of the total original dry sample weight. The total weight of the material after sieving should check closely with original weight of sample placed on the sieves. If the amounts differ by more than 0.3 %, based on the original dry sample weight, the results should not be used for acceptance purposes.

7.8 If the sample has previously been tested by Test Method C 117, add the weight finer than the 75-μm (No. 200) sieve determined by that method to the weight passing the 75-μm (No. 200) sieve by dry sieving of the same sample in this method.

8. Calculation

8.1 Calculate percentages passing, total percentages retained, or percentages in various size fractions to the nearest 0.1 % on the basis of the total weight of the initial dry sample. If the same test sample was first tested by Test Method C 117, include the weight of material finer than the 75-μm (No. 200) size by washing in the sieve analysis calculation; and use the total dry sample weight prior to washing in Test Method C 117 as the basis for calculating all the percentages.

8.2 Calculate the fineness modulus, when required, by adding the total percentages of material in the sample that is coarser than each of the following sieves (cumulative percentages retained), and dividing the sum by 100: 150-μm (No. 100), 300-μm (No. 50), 600-μm (No. 30), 1.18-mm (No. 16), 2.36-mm (No. 8), 4.75-mm (No. 4), 9.5-mm (3/8-in.), 19.0-mm (3/4-in.), 37.5-mm (1½-in.), and larger, increasing in the ratio of 2 to 1.

9. Report

9.1 Depending upon the form of the specifications for use of the material under test, the report shall include the following:

9.1.1 Total percentage of material passing each sieve, or

9.1.2 Total percentage of material retained on each sieve, or

9.1.3 Percentage of material retained between consecutive sieves.

9.2 Report percentages to the nearest whole number, except if the percentage passing the 75-μm (No. 200) sieve is less than 10 %, it shall be reported to the nearest 0.1 %.

9.3 Report the fineness modulus, when required, to the nearest 0.01.

10. Precision

10.1 The estimates of precision of this method listed in Table 1 are based on results from the AASHTO Materials Reference Laboratory Reference Sample Program, with testing conducted by this method and AASHTO Method T 27. While there are differences in the minimum weight of the test sample required for other nominal maximum sizes of aggregate, no differences entered into the testing to affect the determination of these precision indices. The data are based on the analyses of more than 100 paired test results from 40 to 100 laboratories. The values in the table are given for different ranges of percentage of aggregate passing one sieve and retained on the next finer sieve.

TABLE 1 Precision

	% of Size Fraction Between Consecutive Sieves	Coefficient of Variation (1S %), %[B]	Standard Deviation (1S), %[A]	Acceptable Range of Test Results	
				(D2S %)[B] % of Avg.	(D2S),[A] %
Coarse Aggregates:[C]					
Single-Operator	0 to 3	30[D]	. . .	85[D]	. . .
Precision	3 to 10		1.4[D]		4.0[D]
	10 to 20		0.95		2.7
	20 to 50		1.38		3.9
Multilaboratory	0 to 3	35[D]	. . .	99[D]	. . .
Precision	3 to 10		1.06		3.0
	10 to 20		1.66		4.7
	20 to 30		2.01		5.7
	30 to 40		2.44		6.9
	40 to 50		3.18		9.0
Fine Aggregates:					
Single-Operator	0 to 3		0.14		0.4
Precision	3 to 10		0.43		1.2
	10 to 20		0.60		1.7
	20 to 30		0.64		1.8
	30 to 40		0.71		2.0
	40 to 50		. . .		. . .
Multilaboratory	0 to 3		0.21		0.6
Precision	3 to 10		0.57		1.6
	10 to 20		0.95		2.7
	20 to 30		1.24		3.5
	30 to 40		1.41		4.0
	40 to 50		. . .		. . .

[A] These numbers represent, respectively, the (1S) and (D2S) limits as described in Practice C 670.

[B] These numbers represent, respectively, the (1S %) and (D2S %) limits as described in Practice C 670.

[C] The precision estimates are based on coarse aggregates with nominal maximum size of 19.0 mm (¾ in.).

[D] These values are from precision indices first included in Method C 136 – 77. Other indices were developed in 1982 from more recent AASHTO Materials Reference Laboratory sample data, which did not provide sufficient information to revise the values so noted.

Standard Specification for
PORTLAND CEMENT[1]

This standard is issued under the fixed designation C 150; the number immediately following the designation indicates the year of original adoption or, in the case of revision, the year of last revision. A number in parentheses indicates the year of last reapproval. A superscript epsilon (ϵ) indicates an editorial change since the last revision or reapproval.

This specification has been approved for use by agencies of the Department of Defense and for listing in the DoD Index of Specifications and Standards.

1. Scope

1.1 This specification covers eight types of portland cement, as follows (see Note):

1.1.1 *Type I*—For use when the special properties specified for any other type are not required.

1.1.2 *Type IA*—Air-entraining cement for the same uses as Type I, where air-entrainment is desired.

1.1.3 *Type II*—For general use, more especially when moderate sulfate resistance or moderate heat of hydration is desired.

1.1.4 *Type IIA*—Air-entraining cement for the same uses as Type II, where air-entrainment is desired.

1.1.5 *Type III*—For use when high early strength is desired.

1.1.6 *Type IIIA*—Air-entraining cement for the same use as Type III, where air-entrainment is desired.

1.1.7 *Type IV*—For use when a low heat of hydration is desired.

1.1.8 *Type V*—For use when high sulfate resistance is desired.

NOTE—Attention is called to the fact that cements conforming to the requirements for all of these types may not be carried in stock in some areas. In advance of specifying the use of other than Type I cement, it should be determined whether the proposed type of cement is or can be made available.

1.2 The values stated in inch-pound units are to be regarded as the standard.

2. Applicable Documents

2.1 *ASTM Standards:*

C 33 Specification for Concrete Aggregates[2]

C 109 Test Method for Compressive Strength of Hydraulic Cement Mortars (Using 2-in. or 50-mm Cube Specimens)[3]

C 114 Methods for Chemical Analysis of Hydraulic Cement[3]

C 115 Test Method for Fineness of Portland Cement by the Turbidimeter[3]

C 151 Test Method for Autoclave Expansion of Portland Cement[3]

C 183 Methods of Sampling and Acceptance of Hydraulic Cement[3]

C 185 Test Method for Air Content of Hydraulic Cement Mortar[3]

C 186 Test Method for Heat of Hydration of Hydraulic Cement[3]

C 191 Test Method for Time of Setting of Hydraulic Cement by Vicat Needle[3]

C 204 Test Method for Fineness of Portland Cement by Air Permeability Apparatus[3]

C 219 Terminology Relating to Hydraulic Cement[3]

C 226 Specification for Air-Entraining Additions for Use in the Manufacture of Air-Entraining Portland Cement[3]

C 265 Test Method for Calcium Sulfate in Hydrated Portland Cement Mortar[3]

C 266 Test Method for Time of Setting of Hydraulic Cement by Gillmore Needles[3]

C 451 Test Method for Early Stiffening of Portland Cement (Paste Method)[3]

C 452 Test Method for Potential Expansion of Portland Cement Mortars Exposed to Sulfate[3]

C 465 Specification for Processing Additions

---

[1] This specification is under the jurisdiction of ASTM Committee C-1 on Cement and is the direct responsibility of Subcommittee C01.10 on Portland Cement.

Current edition approved May 31, 1985. Published July 1985. Originally published as C 150 – 40 T. Last previous edition C 150 – 84a.

[2] *Annual Book of ASTM Standards*, Vol 04.02.

[3] *Annual Book of ASTM Standards*, Vol 04.01.

for Use in the Manufacture of Hydraulic Cements[3]

C 563 Test Method for Optimum SO_3 in Portland Cement[3]

3. Definitions

3.1 *portland cement*—a hydraulic cement produced by pulverizing clinker consisting essentially of hydraulic calcium silicates, usually containing one or more of the forms of calcium sulfate as an interground addition.

3.2 *air-entraining portland cement*—a hydraulic cement produced by pulverizing clinker consisting essentially of hydraulic calcium silicates, usually containing one or more of the forms of calcium sulfate as an interground addition, and with which there has been interground an air-entraining addition.

4. Ordering Information

4.1 Orders for material under this specification shall include the following:

4.1.1 This specification number and date,

4.1.2 Type or types allowable. If no type is specified, Type I shall be supplied,

4.1.3 Any optional chemical requirements from Table 1A, if desired,

4.1.4 Type of setting-time test required, Vicat or Gilmore. If not specified, the Vicat shall be used,

4.1.5 Any optional physical requirements from Table 2A, if desired.

5. Additions

5.1 The cement covered by this specification shall contain no addition except as follows:

5.1.1 Water or calcium sulfate, or both, may be added in amounts such that the limits shown in Table 1 for sulfur trioxide and loss-on-ignition shall not be exceeded.

5.1.2 At the option of the manufacturer, processing additions may be used in the manufacture of the cement, provided such materials in the amounts used have been shown to meet the requirements of Specification C 465.

5.1.3 Air-entraining portland cement shall contain an interground addition conforming to the requirements of Specification C 226.

6. Chemical Composition

6.1 Portland cement of each of the eight types shown in Section 1 shall conform to the respective standard chemical requirements prescribed

in Table 1. In addition, optional chemical requirements are shown in Table 1A.

7. Physical Properties

7.1 Portland cement of each of the eight types shown in Section 1 shall conform to the respective standard physical requirements prescribed in Table 2. In addition, optional physical requirements are shown in Table 2A.

8. Sampling

8.1 When the purchaser desires that the cement be sampled and tested to verify compliance with this specification, sampling and testing should be performed in accordance with Methods C 183.

8.2 Methods C 183 are not designed for manufacturing quality control and are not required for manufacturer's certification.

9. Test Methods

9.1 Determine the applicable properties enumerated in this specification in accordance with the following methods:

9.1.1 *Air Content of Mortar*—Test Method C 185.

9.1.2 *Chemical Analysis*—Methods C 114.

9.1.3 *Strength*—Test Method C 109.

9.1.4 *False Set*—Test Method C 451.

9.1.5 *Fineness by Air Permeability*—Method C 204.

9.1.6 *Fineness by Turbidimeter*—Test Method C 115.

9.1.7 *Heat of Hydration*—Test Method C 186.

9.1.8 *Autoclave Expansion*—Test Method C 151.

9.1.9 *Time of Setting by Gillmore Needles*—Test Method C 266.

9.1.10 *Time of Setting by Vicat Needles*—Test Method C 191.

9.1.11 *Sulfate Expansion*—Test Method C 452.

9.1.12 *Calcium Sulfate in Mortar*—Test Method C 265.

9.1.13 *Optimum SO_3*—Test Method C 563.

10. Inspection

10.1 Inspection of the material shall be made as agreed upon by the purchaser and the seller as part of the purchase contract.

11. Rejection

11.1 The cement may be rejected if it fails to

meet any of the requirements of this specification.

11.2 Cement remaining in bulk storage at the mill, prior to shipment, for more than 6 months, or cement in bags in local storage in the hands of a vendor for more than 3 months, after completion of tests, may be retested before use and may be rejected if it fails to conform to any of the requirements of this specification.

11.3 Packages varying more than 3 % from the weight marked thereon may be rejected; and if the average weight of packages in any shipment, as shown by weighing 50 packages taken at random, is less than that marked on the packages, the entire shipment may be rejected.

12. Manufacturer's Statement

12.1 At the request of the purchaser, the manufacturer shall state in writing the nature, amount, and identity of the air-entraining agent used, and of any processing addition that may have been used, and also, if requested, shall supply test data showing compliance of such air-entraining addition with the provisions of Specification C 226, and of any such processing addition with Specification C 465.

13. Packaging and Package Marking

13.1 When the cement is delivered in packages, the words "Portland Cement," the type of cement, the name and brand of the manufacturer, and the weight of the cement contained therein shall be plainly marked on each package. When the cement is an air-entraining type, the words "air-entraining" shall be plainly marked on each package. Similar information shall be provided in the shipping documents accompanying the shipment of packaged or bulk cement. All packages shall be in good condition at the time of inspection.

14. Storage

14.1 The cement shall be stored in such a manner as to permit easy access for proper inspection and identification of each shipment, and in a suitable weather-tight building that will protect the cement from dampness and minimize warehouse set.

15. Manufacturer's Certification

15.1 Upon request of the purchaser in the contract or order, a manufacturer's report shall be furnished at the time of shipment stating the results of tests made on samples of the material taken during production or transfer and certifying that the applicable requirements of this specification have been met.

TABLE 1 Standard Chemical Requirements

Cement Type[A]	I and IA	II and IIA	III and IIIA	IV	V
Silicon dioxide (SiO_2), min, %	...	20.0	...	...	...
Aluminum oxide (Al_2O_3), max, %	...	6.0	...	...	...
Ferric oxide (Fe_2O_3), max, %	...	6.0	...	6.5	...
Magnesium oxide (MgO), max, %	6.0	6.0	6.0	6.0	6.0
Sulfur trioxide (SO_3),[B] max, %					
When (C_3A)[C] is 8 % or less	3.0	3.0	3.5	2.3	2.3
When (C_3A)[C] is more than 8 %	3.5	[D]	4.5	[D]	[D]
Loss on ignition, max, %	3.0	3.0	3.0	2.5	3.0
Insoluble residue, max, %	0.75	0.75	0.75	0.75	0.75
Tricalcium silicate (C_3S)[C] max, %	...	...	...	35	...
Dicalcium silicate (C_2S)[C] min, %	...	...	...	40	...
Tricalcium aluminate (C_3A)[C] max, %	...	8	15	7	5[E]
Tetracalcium aluminoferrite plus twice the tricalcium aluminate[C] ($C_4AF + 2(C_3A)$), or solid solution ($C_4AF + C_2F$), as applicable, max, %	...	...	...	...	25[E]

[A] See Note.

[B] There are cases where optimum SO_3 for a particular cement is close to or in excess of the limit in this specification. When it has been demonstrated by Test Method C 563 that the optimum SO_3 exceeds a value 0.5 % less than the specification limit, an additional amount of SO_3, is permissible provided that, when the cement with the additional calcium sulfate is tested by Test Method C 265, the calcium sulfate in the hydrated mortar at 24 ± ¼ h expressed as SO_3 does not exceed 0.50 g/L. When the manufacturer supplies cement under this provision, he will, upon request, supply supporting data to the purchaser.

[C] The expressing of chemical limitations by means of calculated assumed compounds does not necessarily mean that the oxides are actually or entirely present as such compounds.

When expressing compounds, C = CaO, S = SiO_2, A = Al_2O_3, F = Fe_2O_3. For example, $C_3A = 3CaO \cdot Al_2O_3$.

Titanium dioxide and phosphorus pentoxide (TiO_2 and P_2O_5) shall be included with the Al_2O_3 content. The value historically and traditionally used for Al_2O_3 in calculating potential compounds for specification purposes is the ammonium hydroxide group minus ferric oxide ($R_2O_3 - Fe_2O_3$) as obtained by classical wet chemical methods. This procedure includes as Al_2O_3 the TiO_2, P_2O_5 and other trace oxides which precipitate with the ammonium hydroxide group in the classical wet chemical methods. Many modern instrumental methods of cement analysis determine aluminum or aluminum oxide directly without the minor and trace oxides included by the classical method. Consequently, for consistency and to provide comparability with historic data and among various analytical methods, when calculating potential compounds for specification purposes, those using methods which determine Al or Al_2O_3 directly should add to the determined Al_2O_3 weight quantities of P_2O_5, TiO_2 and any other oxide except Fe_2O_3 which would precipitate with the ammonium hydroxide group when analyzed by the classical method and which is present in an amount of 0.05 weight % or greater. The weight percent of minor or trace oxides to be added to Al_2O_3 by those using direct methods may be obtained by actual analysis of those oxides in the sample being tested or estimated from historical data on those oxides on cements from the same source, provided that the estimated values are identified as such.

When the ratio of percentages of aluminum oxide to ferric oxide is 0.64 or more, the percentages of tricalcium silicate, dicalcium silicate, tricalcium aluminate, and tetracalcium aluminoferrite shall be calculated from the chemical analysis as follows:

Tricalcium silicate = (4.071 × % CaO) − (7.600 × % SiO_2) − (6.718 × % Al_2O_3) − (1.430 × % Fe_2O_3) − (2.852 × % SO_3)

Dicalcium silicate = (2.867 × % SiO_2) − (0.7544 × % C_3S)

Tricalcium aluminate = (2.650 × % Al_2O_3) − (1.692 × % Fe_2O_3)

Tetracalcium aluminoferrite = 3.043 × % Fe_2O_3

When the alumina-ferric oxide ratio is less than 0.64, a calcium aluminoferrite solid solution (expressed as ss($C_4AF + C_2F$)) is formed. Contents of this solid solution and of tricalcium silicate shall be calculated by the following formulas:

ss($C_4AF + C_2F$) = (2.100 × % Al_2O_3) + (1.702 × % Fe_2O_3)

Tricalcium silicate = (4.071 × % CaO) − (7.600 × % SiO_2) − (4.479 × % Al_2O_3) − (2.859 × % Fe_2O_3) − (2.852 × % SO_3).

No tricalcium aluminate will be present in cements of this composition. Dicalcium silicate shall be calculated as previously shown.

In the calculation of all compounds the oxides determined to the nearest 0.1 % shall be used.

All values calculated as described in this note shall be reported to the nearest 1 %.

[D] Not applicable.

[E] Does not apply when the sulfate expansion limit in Table 2A is specified.

TABLE 1A Optional Chemical Requirements

NOTE—These optional requirements apply only when specifically requested.

Cement Type[A]	I and IA	II and IIA	III and IIIA	IV	V	Remarks
Tricalcium aluminate (C₃A),[B] max, %	...	...	8	...	...	for moderate sulfate resistance
Tricalcium aluminate (C₃A),[B] max, %	...	...	5	...	...	for high sulfate resistance
Sum of tricalcium silicate and tricalcium aluminate,[B] max, %	...	58[C]	...	...	...	for moderate heat of hydration
Alkalies (Na₂O + 0·658K₂O), max, %	0.60[D]	0.60[D]	0.60[D]	0.60[D]	0.60[D]	low-alkali cement

[A] See Note.

[B] The expressing of chemical limitations by means of calculated assumed compounds does not necessarily mean that the oxides are actually or entirely present as such compounds.

When expressing compounds, $C = CaO$, $S = SiO_2$, $A = Al_2O_3$, $F = Fe_2O_3$. For example, $C_3A = 3CaO \cdot Al_2O_3$.

Titanium dioxide and phosphorus pentoxide (TiO_2 and P_2O_5) shall be included with the Al_2O_3 content. The value historically and traditionally used for Al_2O_3 in calculating potential compounds for specification purposes is the ammonium hydroxide group minus ferric oxide ($R_2O_3 - Fe_2O_3$) as obtained by classical wet chemical methods. This procedure includes as Al_2O_3 the TiO_2, P_2O_5 and other trace oxides which precipitate with the ammonium hydroxide group in the classical wet chemical methods. Many modern instrumental methods of cement analysis determine aluminum or aluminum oxide directly without the minor and trace oxides included by the classical method. Consequently, for consistency and to provide comparability with historic data and among various analytical methods, when calculating potential compounds for specification purposes, those using methods which determine Al or Al_2O_3 directly should add to the determined Al_2O_3 weight quantities of P_2O_5, TiO_2 and any other oxide except Fe_2O_3 which would precipitate with the ammonium hydroxide group when analyzed by the classical method and which is present in an amount of 0.05 weight % or greater. The weight percent of minor or trace oxides to be added to Al_2O_3 by those using direct methods may be obtained by actual analysis of those oxides in the sample being tested or estimated from historical data on those oxides on cements from the same source, provided that the estimated values are identified as such.

When the ratio of percentages of aluminum oxide to ferric oxide is 0.64 or more, the percentages of tricalcium silicate, dicalcium silicate, tricalcium aluminate and tetracalcium aluminoferrite shall be calculated from the chemical analysis as follows:

Tricalcium silicate = $(4.071 \times \% CaO) - (7.600 \times \% SiO_2) - (6.718 \times \% Al_2O_3) - (1.430 \times \% Fe_2O_3) - (2.852 \times \% SO_3)$

Dicalcium silicate = $(2.867 \times \% SiO_2) - (0.7544 \times \% C_3S)$

Tricalcium aluminate = $(2.650 \times \% Al_2O_3) - (1.692 \times \% Fe_2O_3)$

Tetracalcium aluminoferrite = $3.043 \times \% Fe_2O_3$

When the alumina-ferric oxide ratio is less than 0.64, a calcium aluminoferrite solid solution (expressed as ss ($C_4AF + C_2F$)) is formed. Contents of this solid solution and of tricalcium silicate shall be calculated by the following formulas:

ss($C_4AF + C_2F$) = $(2.100 \times \% Al_2O_3) + (1.702 \times \% Fe_2O_3)$

Tricalcium silicate = $(4.071 \times \% CaO) - (7.600 \times \% SiO_2) - (4.479 \times \% Al_2O_3) - (2.859 \times \% Fe_2O_3) - (2.852 \times \% SO_3)$.

No tricalcium aluminate will be present in cements of this composition. Dicalcium silicate shall be calculated as previously shown.

In the calculation of all compounds the oxides determined to the nearest 0.1 % shall be used.

All values calculated as described in this note shall be reported to the nearest 1 %.

[C] This limit applies when moderate heat of hydration is required and the limit for heat of hydration in Table 2A is not specified.

[D] This limit may be specified when the cement is to be used in concrete with aggregates that may be deleteriously reactive. Reference should be made to Specification C 33 for suitable criteria of deleterious reactivity.

TABLE 2 Standard Physical Requirements

Cement Type[A]	I	IA	II	IIA	III	IIIA	IV	V
Air content of mortar,[B] volume %:								
max	12	22	12	22	12	22	12	12
min	...	16	...	16	...	16	...	...
Fineness,[C] specific surface, m^2/kg† (alternative methods):								
Turbidimeter test, min	160†	160†	160†	160†	...	...	160†	160†
Air permeability test, min	280†	280†	280†	280†	...	...	280†	280†
Autoclave expansion, max, %	0.80	0.80	0.80	0.80	0.80	0.80	0.80	0.80
Strength, not less than the values shown for the ages indicated below:[D]								
Compressive strength, psi (MPa):								
1 day	...	...	...	...	1800 (12.4)	1450 (10.0)	...	...
3 days	1800 (12.4)	1450 (10.0)	1500 (10.3) 1000^F (6.9)^F	1200 (8.3) 800^F (5.5)^F	3500 (24.1)	2800 (19.3)	...	1200 (8.3)
7 days	2800 (19.3)	2250 (15.5)	2500 (17.2) 1700^F (11.7)^F	2000 (13.8) 1350^F (9.3)^F	...	...	1000 (6.9)	2200 (15.2)
28 days	...	...	...	...	...	...	2500 (17.2)	3000 (20.7)
Time of setting (alternative methods):[E]								
Gillmore test:								
Initial set, min, not less than	60	60	60	60	60	60	60	60
Final set, min, not more than	600	600	600	600	600	600	600	600
Vicat test:[G]								
Time of setting, min, not less than	45	45	45	45	45	45	45	45
Time of setting, min, not more than	375	375	375	375	375	375	375	375

[A] See Note.

[B] Compliance with the requirements of this specification does not necessarily ensure that the desired air content will be obtained in concrete.

[C] Either of the two alternative fineness methods may be used at the option of the testing laboratory. However, when the sample fails to meet the requirements of the air-permeability test, the turbidimeter test shall be used, and the requirements in this table for the turbidimetric method shall govern.

[D] The strength at any specified test age shall be not less than that attained at any previous specified test age.

[E] The purchaser should specify the type of setting-time test required. In case he does not so specify, the requirements of the Vicat test only shall govern.

[F] When the optional heat of hydration or the chemical limit on the sum of the tricalcium silicate and tricalcium aluminate is specified.

[G] The time of setting is that described as initial setting time in Test Method C 191.

† These values were revised in 1978 because of the change in unit from cm^2/g to m^2/kg.

NOTE—These optional requirements apply only when specifically requested.

Cement Type[A]	I	IA	II	IIA	III	IIIA	IV	V
False set, final penetration, min, %	50	50	50	50	50	50	50	50
Heat of hydration:								
7 days, max, cal/g (kJ/kg)	...	...	70 (290)[B]	70 (290)[B]	...	...	60 (250)	...
28 days, max, cal/g (kJ/kg)	...	...	...	...	...	...	70 (290)	...
Strength, not less than the values shown:								
Compressive strength, psi (MPa)								
28 days	4000 (27.6)	3200 (22.1)	4000 (27.6) 3200[B] (22.1)[B]	3200 (22.1) 2560[B] (17.7)[B]	...	...	...	...
Sulfate expansion,[C] 14 days, max, %	...	...	...	...	...	...	...	0.040

[A] See Note.

[B] When the heat of hydration requirements are specified, the sum of the tricalcium silicate and tricalcium aluminate shall not be specified. These strength requirements apply when either heat of hydration requirements or the sum of tricalcium silicate and tricalcium aluminate are specified.

[C] When the sulfate expansion is specified, it shall be instead of the limits of C₃A and C₄AF + 2 C₃A listed in Table 1.

The American Society for Testing and Materials takes no position respecting the validity of any patent rights asserted in connection with any item mentioned in this standard. Users of this standard are expressly advised that determination of the validity of any such patent rights, and the risk of infringement of such rights, are entirely their own responsibility.

This standard is subject to revision at any time by the responsible technical committee and must be reviewed every five years and if not revised, either reapproved or withdrawn. Your comments are invited either for revision of this standard or for additional standards and should be addressed to ASTM Headquarters. Your comments will receive careful consideration at a meeting of the responsible technical committee, which you may attend. If you feel that your comments have not received a fair hearing you should make your views known to the ASTM Committee on Standards, 1916 Race St., Philadelphia, Pa. 19103.

Designation: D 198 – 84$^{\epsilon 1}$

Standard Methods of
STATIC TESTS OF TIMBERS IN STRUCTURAL SIZES[1]

This standard is issued under the fixed designation D 198; the number immediately following the designation indicates the year of original adoption or, in the case of revision, the year of last revision. A number in parentheses indicates the year of last reapproval. A superscript epsilon (ε) indicates an editorial change since the last revision or reapproval.

$^{\epsilon 1}$ NOTE—Editorial changes were made in the Appendixes in April 1985.

INTRODUCTION

Numerous evaluations of structural members of solid sawn timber have been conducted in accordance with ASTM Methods D 198 – 27. While the importance of continued use of a satisfactory standard should not be underestimated, the original standard (1927) was designed primarily for sawn material such as solid wood bridge stringers and joists. With the advent of laminated timbers, wood-plywood composite members, and even reinforced and prestressed timbers, a procedure adaptable to a wider variety of wood structural members is required.

The present standard expands the original standard to permit its application to wood members of all types. It provides methods of evaluation under loadings other than flexure in recognition of the increasing need for improved knowledge of properties under such loadings as tension to reflect the increasing use of dimension lumber in the lower chords of trusses. The standard establishes practices that will permit correlation of results from different sources through the use of a uniform procedure. Provision is made for varying the procedure to take account of special problems.

1. Scope

1.1 These methods cover the evaluation of timbers in structural size by various testing procedures.

1.2 The methods appear in the following order:

	Sections
Flexure	4 to 11
Compression (Short Column)	12 to 19
Compression (Long Member)	20 to 27
Tension	28 to 35
Torsion	36 to 43
Shear Modulus	44 to 51

1.3 Notations and symbols relating to the various testing procedures are given in Table X1.1.

1.4 *This standard may involve hazardous materials, operations, and equipment. This standard does not purport to address all of the safety problems associated with its use. It is the responsibility of whoever uses this standard to consult and establish appropriate safety and health practices and determine the applicability of regulatory limitations prior to use.*

2. Applicable Documents

2.1 *ASTM Standards:*

D 9 Definitions of Terms Relating to Wood[2]

D 1165 Nomenclature of Domestic Hardwoods and Softwoods[2]

D 2016 Test Methods for Moisture Content of Wood[2]

D 2395 Test Methods for Specific Gravity of Wood and Wood-Base Materials[2]

E 4 Methods of Load Verification of Testing Machines[3]

---

[1] These methods are under the jurisdiction of ASTM Committee D-7 on Wood and are under the jurisdiction of Subcommittee D07.09 on Methods of Testing.

Current edition approved Feb. 24, 1984. Published April 1984. Originally published as D 198 – 24. Last previous edition D 198 – 76.

[2] *Annual Book of ASTM Standards*, Vol 04.09.

[3] *Annual Book of ASTM Standards*, Vol 03.01.

E 6 Definitions of Terms Relating to Methods of Mechanical Testing[3]

E 83 Method of Verification and Classification of Extensometers[3]

3. Definitions

3.1 See Definitions E 6, Definitions D 9, and Nomenclature D 1165. A few related terms not covered in the above standards are as follows:

3.1.1 *span*—the total distance between reactions on which a beam is supported to accommodate a transverse load (Fig. 1).

3.1.2 *shear span*—two times the distance between a reaction and the nearest load point for a symmetrically loaded beam (Fig. 1).

3.1.3 *depth of beam*—that dimension of the beam which is perpendicular to the span and parallel to the direction in which the load is applied (Fig. 1).

3.1.4 *span-depth ratio*—the numerical ratio of total span divided by beam depth.

3.1.5 *shear span-depth ratio*—the numerical ratio of shear span divided by beam depth.

3.1.6 *structural wood beam*—solid wood, laminated wood, or composite structural members for which strength, stiffness, or both are primary criteria for the intended application and which usually are used in full length and in cross-sectional sizes greater than nominal 2 by 2 in. (38 by 38 mm).

3.1.7 *composite wood beam*—a laminar construction comprising a combination of wood and other simple or complex materials assembled and intimately fixed in relation to each other so as to use the properties of each to attain specific structural advantage for the whole assembly.

FLEXURE

4. Scope

4.1 This method covers the determination of the flexural properties of structural beams made of solid or laminated wood, or of composite constructions. The method is intended primarily for beams of rectangular cross section but is also applicable to beams of round and irregular shapes, such as round posts, I-beams, or other special sections.

5. Summary of Method

5.1 The structural member, usually a straight or a slightly cambered beam of rectangular cross section, is subjected to a bending moment by supporting it near its ends, at locations called reactions, and applying transverse loads symmetrically imposed between these reactions. The beam is deflected at a prescribed rate, and coordinate observations of loads and deflections are made until rupture occurs.

6. Significance and Use

6.1 The flexural properties established by this method provide:

6.1.1 Data for use in development of grading rules and specifications.

6.1.2 Data for use in development of working stresses for structural members.

6.1.3 Data on the influence of imperfections on mechanical properties of structural members.

6.1.4 Data on strength properties of different species or grades in various structural sizes.

6.1.5 Data for use in checking existing equations or hypotheses relating to the structural behavior of beams.

6.1.6 Data on the effects of chemical or environmental conditions on mechanical properties.

6.1.7 Data on effects of fabrication variables such as depth, taper, notches, or type of end joint in laminations.

6.1.8 Data on relationships between mechanical and physical properties.

6.2 Procedures are described here in sufficient detail to permit duplication in different laboratories so that comparisons of results from different sources will be valid. Special circumstances may require deviation from some details of these procedures. Any variations shall be carefully described in the report (see Section 11).

7. Apparatus

7.1 *Testing Machine*—A device that provides (*1*) a rigid frame to support the specimen yet permit its deflection without restraint, (*2*) a loading head through which the force is applied without high stress concentrations in the beam, and (*3*) a force-measuring device that is calibrated to ensure accuracy in accordance with Methods E 4.

7.2 *Support Apparatus:*

7.2.1 *Reaction Bearing Plates*—The beam shall be supported by metal bearing plates to prevent damage to the beam at the point of contact between beam and reaction support (Fig. 1). The size of the bearing plates may vary with the size and shape of the beam. For rectangular beams as large as 12 in. (305 mm) deep by 6 in.

(152 mm) wide, the recommended size of bearing plate is ½ in. (13 mm) thick by 6 in. (152 mm) lengthwise and extending entirely across the width of the beam.

7.2.2 *Reaction Bearing Roller*—The bearing plates shall be supported by either rollers and a fixed knife edge reaction or a rocker type-knife edge reaction so that shortening and rotation of the beam about the reaction due to deflection will be unrestricted (Fig. 1).

7.2.3 *Reaction Bearing Alignment*—Provisions shall be made at the reaction to allow for initial twist in the length of the beam. If the bearing surfaces of the beam at its reactions are not parallel, the beam shall be shimmed or the individual bearing plates shall be rotated about an axis parallel to the span to provide full bearing across the width of the specimen (Fig. 2).

7.2.4 *Lateral Support*—Specimens that have a depth-to-width ratio of three or greater are subject to lateral instability during loading, thus requiring lateral support. Support shall be provided at least at points located about half-way between the reaction and the load point. Additional supports may be used as required. Each support shall allow vertical movement without frictional restraint but shall restrict lateral deflection (Fig. 3).

7.3 *Load Apparatus:*

7.3.1 *Load Bearing Blocks*—The load shall be applied through bearing blocks (Fig. 1) across the full beam width which are of sufficient thickness to eliminate high stress concentrations at places of contact between beam and bearing blocks. The loading surface of the blocks shall have a radius of curvature equal to two to four times the beam depth for a chord length at least equal to the depth of the beam. Load shall be applied to the blocks in such a manner that the blocks may rotate about an axis perpendicular to the span (Fig. 4). Provisions such as rotatable bearings or shims shall be made to ensure full contact between the beam and both loading blocks. Metal bearing plates and rollers shall be used in conjunction with one load bearing block to permit beam deflection without restraint (Fig. 4). The size of these plates and rollers may vary with the size and shape of the beam, the same as for the reaction bearing plates. Beams having circular or irregular cross sections shall have bearing blocks which distribute the load uniformly to the bearing surface and permit, unrestrained deflections.

7.3.2 *Load Points*—The total load on the beam shall be applied equally at two points equidistant from the reactions. The two load points will normally be at a distance from their reaction equal to one third of the span, but for special purposes other distances may be specified.

NOTE 1—One of the objectives of two-point loading is to subject the portion of the beam between load points to a uniform bending moment, free of shear, and with comparatively small loads at the load points. For example, loads applied at one-third span length from reactions would be less than if applied at one-fourth span length from reaction to develop a moment of similar magnitude. When loads are applied at the one-third points the moment distribution of the beam simulates that for loads uniformly distributed across the span to develop a moment of similar magnitude. If loads are applied at the outer one-fourth points of the span, the maximum moment and shear are the same as the maximum moment and shear for the same total load uniformly distributed across the span.

7.4 *Deflection Apparatus:*

7.4.1 *General*—For either apparent or true modulus of elasticity calculations, devices shall be provided by which the deflection of the neutral axis of the beam at the center of the span is measured with respect to either the reaction or between cross sections free of shear deflections.

7.4.2 *Wire Deflectometer*—Deflection may be read directly by means of a wire stretched taut between two nails driven into the neutral axis of the beam directly above the reactions and extending across a scale attached at the neutral axis of the beam at midspan. Deflections may be read with a telescope or reading glass to magnify the area where the wire crosses the scale. When a reading glass is used, a reflective surface placed adjacent to the scale will help to avoid parallax.

7.4.3 *Yoke Deflectometer*—A satisfactory device commonly used for short, small beams or to measure deflection of the center of the beam with respect to any point along the neutral axis consists of a lightweight U-shaped yoke suspended between nails driven into the beam at its neutral axis and a dial micrometer attached to the center of the yoke with its stem attached to a nail driven into the beam at midspan at the neutral axis. Further modification of this device may be attained by replacing the dial micrometer with a deflection transducer for automatic recording (Fig. 4).

7.4.4 *Accuracy*—The devices shall be such as to permit measurements to the nearest 0.01 in. (0.25 mm) on spans greater than 3 ft (0.9 m) and

0.001 in. (0.03 mm) on spans less than 3 ft (0.9 m).

8. Test Specimen

8.1 *Material*—The test specimen shall consist basically of a beam which may be solid wood, laminated wood, or a composite construction of wood or of wood combined with plastics or metals in sizes that are usually used in structural applications.

8.2 *Identification*—Material or materials of the test specimen shall be identified as fully as possible by including the origin or source of supply, species, and history of drying and conditioning, chemical treatment, fabrication, and other pertinent physical or mechanical details which may affect the strength. Details of this information shall depend on the material or materials in the beam. For example, the solid wooden beams would be identified by the character of the wood, that is, species, source, etc., whereas composite wooden beams would be identified by the characteristics of the dissimilar materials and their size and location in the beam.

8.3 *Specimen Measurements*—The weight and dimensions as well as moisture content of the specimen shall be accurately determined before test. Weights and dimensions (length and cross section) shall be measured to three significant figures. Sufficient measurements of the cross section shall be made along the length of the beam to describe the width and depth of rectangular specimen and to accurately describe the critical section or sections of nonuniform beams. The physical characteristics of the specimen as described by its density and moisture content may be determined in accordance with Test Methods D 2395 and Method A of Test Methods D 2016.

8.4 *Specimen Description.*—The inherent imperfections or intentional modifications of the composition of the beam shall be fully described by recording the size and location of such factors as knots, checks, and reinforcements. Size and location of intentional modifications such as placement of laminations, glued joints, and reinforcing steel shall be recorded during the fabrication process. The size and location of imperfections in the interior of any beam must be deduced from those on the surface, especially in the case of large sawn members. A sketch or photographic record shall be made of each face

and the ends showing the size, location, and type of growth characteristics, including slope of grain, knots, distribution of sapwood and heartwood, location of pitch pockets, direction of annual rings, and such abstract factors as crook, bow, cup, or twist which might affect the strength of the beam.

8.5 *Rules for Determination of Specimen Length*—The cross-sectional dimensions of solid wood structural beams and composite wooden beams usually have established sizes, depending upon the manufacturing process and intended use, so that no modification of these dimensions is involved. The length, however, will be established by the type of data desired. The span length is determined from knowledge of beam depth, the distance between load points, as well as the type and orientation of material in the beam. The total beam length shall also include an overhang or extension beyond each reaction support so that the beam can accommodate the bearing plates and rollers and will not slip off the reactions during test.

NOTE 2—Some evaluations will require simulation of a specific design condition where nonnormal overhang is involved. In such instances the report shall include a complete description of test conditions, including overhang at each support.

8.5.1 The span length of beams intended primarily for evaluation of shear properties shall be such that the shear span is relatively short. Beams of wood of uniform rectangular cross section having the ratio of a/h less than five are in this category and provide a high percentage of shear failures.

NOTE 3—If approximate values of modulus of rupture S_R and shear strength τ_m are known, a/h values should be less than $S_R/4\tau_m$, assuming that when $a/h = S_R/4\tau_m$ the beam will fail at the same load in either shear or in extreme outer fibers.

8.5.2 The span length of beams intended primarily for evaluation of flexural properties shall be such that the shear span is relatively long. Beams of wood of uniform rectangular cross section having a/h ratios of from 5:1 to 12:1 are in this category.

NOTE 4—The a/h values should be somewhat greater than $S_R/4\tau_m$ so that the beams do not fail in shear but should not be so large that beam deflections cause sizable thrust of reactions and thrust values need to be taken into account. A suggested range of a/h values is between approximately $0.5\ S_R/\tau_m$ and $1.2\ S_R/\tau_m$. In this category, shear distortions affect the total

392

deflection, so that flexural properties may be corrected by formulae provided in the Appendix.

8.5.3 The span length of beams intended primarily for evaluation of only the deflection of specimen due to bending moment shall be such that the shear span is long. Wood beams of uniform rectangular cross section in this category have a/h ratios greater than 12:1.

NOTE 5—The shear stresses and distortions are assumed to be small so that they can be neglected; hence the a/h ratio is suggested to be greater than S_R/τ_m.

9. Procedure

9.1 *Conditioning*—Unless otherwise indicated in the research program or material specification, condition the test specimen to constant weight so it is in moisture equilibrium under the desired environmental conditions. Approximate moisture contents with moisture meters or measure more accurately by weights of samples according to Method A of Test Methods D 2016.

9.2 *Test Setup*—Determine the size of the specimen, the span, and the shear span in accordance with 7.3.2 and 8.5. Locate the beam symmetrically on its supports with load bearing and reaction bearing blocks as described in 7.2 to 7.4. The beams shall be adequately supported laterally according to 7.2.4. Set apparatus for measuring deflections in place (see 7.4). Full contact shall be attained between support bearings, loading blocks, and the beam surface.

9.3 *Speed of Testing*—Conduct the test at a constant rate to achieve maximum load in about 10 min, but maximum load should be reached in not less than 6 min nor more than 20 min. A constant rate of outer strain, z, of 0.0010 in./in.·min (0.001 mm/mm·min) will usually permit the tests of wood members to be completed in the prescribed time. The rate of motion of the movable head of the test machine corresponding to this suggested rate of strain when two symmetrical concentrated loads are employed may be computed from the following equation:

$$N = Za(3L - 4a)/3h$$

9.4 *Load-Deflection Curves:*

9.4.1 Obtain load-deflection data with apparatus described in 7.4.1. Note the load and deflection at first failure, at the maximum load, and at points of sudden change. Continue loading until complete failure or an arbitrary terminal load has been reached.

9.4.2 If additional deflection apparatus is provided to measure deflection over a second distance, l, according to 7.4.1, such load-deflection data shall be obtained only up to the proportional limit.

9.5 *Record of Failures*—Describe failures in detail as to type, manner and order of occurrence, and position in beam. Record descriptions of the failures and relate them to drawings or photographs of the beam referred to in 8.4. Also record notations as the order of their occurrence on such references. Hold the section of the beam containing the failure for examination and reference until analysis of the data has been completed.

10. Calculations

10.1 Compute physical and mechanical properties and their appropriate adjustments for the beam in accordance with the relationships in the Appendix X2.

11. Report

11.1 The report shall include the following:

11.1.1 Complete identification of the solid wood or composite construction, including species, origin, shape and form, fabrication procedure, type and location of imperfections or reinforcements, and pertinent physical or chemical characteristics relating to the quality of the material,

11.1.2 History of seasoning and conditioning,

11.1.3 Loading conditions to portray the load, support mechanics, lateral supports, if used, and type of equipment,

11.1.4 Deflection apparatus,

11.1.5 Depth and width of the specimen or pertinent cross-sectional dimensions,

11.1.6 Span length and shear span distance,

11.1.7 Rate of load application,

11.1.8 Computed physical and mechanical properties, including specific gravity and moisture content, flexural strength, stress at proportional limit, modulus of elasticity, and a statistical measure of variability of these values,

11.1.9 Data for composite beams include shear and bending moment values and deflections,

11.1.10 Description of failure, and

11.1.11 Details of any deviations from the prescribed or recommended methods as outlined in the standard.

COMPRESSION PARALLEL TO GRAIN (SHORT COLUMN, NO LATERAL SUPPORT, $l/r < 17$)

12. Scope

12.1 This method covers the determination of the compressive properties of elements taken from structural members made of solid or laminated wood, or of composite constructions when such an element has a slenderness ratio (length to least radius of gyration) of less than 17. The method is intended primarily for members of rectangular cross section but is also applicable to irregularly shaped studs, braces, chords, round posts, or special sections.

13. Summary of Method

13.1 The structural member is subjected to a force uniformly distributed on the contact surface of the specimen in a direction generally parallel to the longitudinal axis of the wood fibers, and the force generally is uniformly distributed throughout the specimen during loading to failure without flexure along its length.

14. Significance and Use

14.1 The compressive properties obtained by axial compression will provide information similar to that stipulated for flexural properties under Section 6.

14.2 The compressive properties parallel to grain include modulus of elasticity, stress at proportional limit, compressive strength, and strain data beyond proportional limit.

15. Apparatus

15.1 *Testing Machine*—Any device having the following is suitable:

15.1.1 *Drive Mechanism*—A drive mechanism for imparting to a movable loading head a uniform, controlled velocity with respect to the stationary base.

15.1.2 *Load Indicator*—A load-indicating mechanism capable of showing the total compressive force on the specimen. This force-measuring system shall be calibrated to ensure accuracy in accordance with Methods E 4.

15.2 *Bearing Blocks*—Bearing blocks shall be used to apply the load uniformly over the two contact surfaces and to prevent eccentric loading on the specimen. At least one spherical bearing block shall be used to ensure uniform bearing. Spherical bearing blocks may be used on either or both ends of the specimen, depending on the degree of parallelism of bearing surfaces (Fig. 5). The radius of the sphere shall be as small as practicable, in order to facilitate adjustment of the bearing plate to the specimen, and yet large enough to provide adequate spherical bearing area. This radius is usually one to two times the greatest cross section dimension. The center of the sphere shall be on the plane of the specimen contact surface. The size of the compression plate shall be larger than the contact surface. It has been found convenient to provide an adjustment for moving the specimen on its bearing plate with respect to the center of spherical rotation to ensure axial loading.

15.3 *Compressometer:*

15.3.1 *Gage Length*—For modulus of elasticity calculations, a device shall be provided by which the deformation of the specimen is measured with respect to specific paired gage points defining the gage length. To obtain test data representative of the test material as a whole, such paired gage points shall be located symmetrically on the lengthwise surface of the specimen as far apart as feasible, yet at least one times the larger cross-sectional dimension from each of the contact surfaces. At least two pairs of such gage points on diametrically opposite sides of the specimen shall be used to measure the average deformation.

15.3.2 *Accuracy*—The device shall be able to measure changes in deformation to three significant figures. Since gage lengths vary over a wide range, the measuring instruments should conform to their appropriate class in Methods E 83.

16. Test Specimen

16.1 *Material*—The test specimen shall consist basically of a structural timber which may be solid wood, laminated wood, or a composite construction of wood or of wood combined with plastics or metals in sizes that are commercially used in structural applications, that is in sizes greater than nominal 2 by 2-in. (386 by 38-mm) cross section (see 3.1.6).

16.2 *Identification*—Material or materials of the test specimen shall be as fully described as that for beams in 8.2.

16.3 *Specimen Dimensions*—The weight and dimensions, as well as moisture content of the specimen, shall be accurately measured before test. Weights and dimensions (length and cross section) shall be measured to three significant figures. Sufficient measurements of the cross section shall be made along the length of the specimen to describe shape characteristics and to determine the smallest section. The physical characteristics of the specimen, as described by its density and moisture content, may be determined in accordance with Test Methods D 2395 and Method A of Test Methods D 2016, respectively.

16.4 *Specimen Description*—The inherent imperfections and intentional modifications shall be described as for beams in 8.4.

16.5 *Specimen Length*—The length of the specimen shall be such that the compressive force continues to be uniformly distributed throughout the specimen during loading—hence no flexure occurs. To meet this requirement, the specimen shall be a short column having a maximum length, l, less than 17 times the least radius of gyration, r, of the cross section of the specimen (see compressive notations). The minimum length of the specimen for stress and strain measurements shall be greater than three times the larger cross section dimension or about ten times the radius of gyration.

17. Procedure

17.1 *Conditioning*—Unless otherwise indicated in the research program or material specification, condition the test specimen to constant weight so it is at moisture equilibrium, under the desired environment. Approximate moisture contents with moisture meters or measure more accurately by weights of samples in accordance with Method A of Test Methods D 2016.

17.2 *Test Setup:*

17.2.1 *Bearing Surfaces*—After the specimen length has been calculated in accordance with 17.5, cut the specimen to the proper length so that the contact surfaces are plane, parallel to each other, and normal to the long axis of the specimen. Furthermore, the axis of the specimen shall be generally parallel to the fibers of the wood.

NOTE 6—A sharp fine-toothed saw of either the crosscut or "novelty" crosscut type has been used satisfactorily for obtaining the proper end surfaces. Power equipment with accurate table guides is especially recommended for this work

NOTE 7—It is desirable to have failures occur in the body of the specimen and not adjacent to the contact surface. Therefore, the cross-sectional areas adjacent to the loaded surface may be reinforced.

17.2.2 *Centering*—First geometrically center the specimens on the bearing plates and then adjust the spherical seats so that the specimen is loaded uniformly and axially.

17.3 *Speed of Testing*—For measuring load-deformation data, apply the load at a constant rate of head motion so that the fiber strain is 0.001 in./in.·min ± 25 % (0.001 mm/mm·min). For measuring only compressive strength, the test may be conducted at a constant rate to achieve maximum load in about 10 min, but not less than 5 nor more than 20 min.

17.4 *Load-Deformation Curves*—If load-deformation data have been obtained, note the load and deflection at first failure, at changes in slope of curve, and at maximum load.

17.5 *Records*—Record the maximum load, as well as a description and sketch of the failure relating the latter to the location of imperfections in the specimen. Re-examine the section of the specimen containing the failure during analysis of the data.

18. Calculations

18.1 Physical and mechanical properties shall be computed in accordance with Definitions E 6, and as follows (see compressive notations):

18.1.1 Stress at proportional limit = P'/A in pounds per square inch (MPa).

18.1.2 Compressive strength = P/A in pounds per square inch (MPa).

18.1.3 Modulus of elasticity = $P'/A\epsilon$ in pounds per square inch (MPa).

19. Report

19.1 The report shall include the following:

19.1.1 Complete identification,

19.1.2 History of seasoning and conditioning,

19.1.3 Load apparatus,

19.1.4 Deflection apparatus,

19.1.5 Length and cross-section dimensions,

19.1.6 Gage length,

19.1.7 Rate of load application,

19.1.8 Computed physical and mechanical properties, including specific gravity and moisture content, compressive strength, stress at proportional limit, modulus of elasticity, and a statistical measure of variability of these values,

19.1.9 Description of failure, and

19.1.10 Details of any deviations from the prescribed or recommended methods as outlined in the standard.

COMPRESSION PARALLEL TO GRAIN (CRUSHING STRENGTH OF LATERALLY SUPPORTED LONG MEMBER, EFFECTIVE $l'/r < 17$)

20. Scope

20.1 This method covers the determination of the compressive properties of structural members made of solid or laminated wood, or of composite constructions when such a member has a slenderness ratio (length to least radius of gyration) of more than 17, and when such a member is to be evaluated in full size but with lateral supports which are spaced to produce an effective slenderness ratio, l'/r, of less than 17. The method is intended primarily for members of rectangular cross section but is also applicable to irregularly shaped studs, braces, chords, round posts, or special sections.

21. Summary of Method

21.1 The structural member is subjected to a force uniformly distributed on the contact surface of the specimen in a direction generally parallel to the longitudinal axis of the wood fibers, and the force generally is uniformly distributed throughout the specimen during loading to failure without flexure along its length.

22. Significance and Use

22.1 The compressive properties obtained by axial compression will provide information similar to that stipulated for flexural properties under Section 6.

22.2 The compressive properties parallel to grain include modulus of elasticity, stress at proportional limit, compressive strength, and strain data beyond proportional limit.

23. Apparatus

23.1 *Testing Machine*—Any device having the following is suitable:

23.1.1 *Drive Mechanism*—A drive mechanism for imparting to a movable loading head a uniform, controlled velocity with respect to the stationary base.

23.1.2 *Load Indicator*—A load-indicating mechanism capable of showing the total compressive force on the specimen. This force-measuring system shall be calibrated to ensure accuracy in accordance with Methods E 4.

23.2 *Bearing Blocks*—Bearing blocks shall be used to apply the load uniformly over the two contact surfaces and to prevent eccentric loading on the specimen. One spherical bearing block shall be used to ensure uniform bearing, or a rocker-type bearing block shall be used on each end of the specimen with their axes of rotation at 0° to each other (Fig. 6). The radius of the sphere shall be as small as practicable, in order to facilitate adjustment of the bearing plate to the specimen, and yet large enough to provide adequate spherical bearing area. This radius is usually one to two times the greatest cross section dimension. The center of the sphere shall be on the plane of the specimen contact surface. The size of the compression plate shall be larger than the contact surface.

23.3 *Lateral Support:*

23.3.1 *General*—Evaluation of the crushing strength of long structural members requires that they be supported laterally to prevent buckling during the test without undue pressure against the sides of the specimen. Furthermore, the support shall not restrain either the longitudinal compressive deformation or load during test. The support shall be either continuous or intermittent. Intermittent supports shall be spaced so that the distance, l', between supports is less than 17 times the least radius of gyration of the cross section.

23.3.2 *Rectangular Members*—The general rules for structural members apply to rectangular structural members. However, the effective column length as controlled by intermittent support spacing on flatwise face need not equal that on edgewise face. The minimum spacing of the supports on the flatwise face shall be 17 times the least radius of gyration of the cross section which is about the centroidal axis parallel to flat face. And the minimum spacing of the supports on the edgewise face shall be 17 times the other radius of gyration (Fig. 6). A satisfactory method of providing lateral support for 2-in. (38-mm) dimension stock is shown in Fig. 7. A 27-in. (686-mm) I-beam provides the frame for the test machine. Small I-beams provide reactions for longitudinal pressure. A pivoted top I-beam provides lateral support on one flatwise face, while the web of the large I-beam provides the other. In between these steel members, metal guides on 3-in. (7.6-cm) spacing (hidden from view) attached to plywood fillers provide the flatwise support and contact surface. In between the flanges of the 27-in. (686-mm) I-beam, fingers and wedges provide edgewise lateral support.

23.4 *Compressometer:*

23.4.1 *Gage Length*—For modulus of elasticity calculations, a device shall be provided by which the deformation of the specimen is measured with respect to specific paired gage points defining the gage length. To obtain data representative of the test material as a whole, such paired gage points shall be located symmetrically on the lengthwise surface of the specimen as far apart as feasible, yet at least one times the larger cross-sectional dimension from each of the contact surfaces. At least two pairs of such gage points on diametrically opposite sides of the specimen shall be used to measure the average deformation.

23.4.2 *Accuracy*—The device shall be able to measure changes in deformation to three significant figures. Since gage lengths vary over a wide range, the measuring instruments should conform to their appropriate class in Method E 83.

24. Test Specimen

24.1 *Material*—The test specimen shall consist basically of a structural timber which may be solid wood, laminated wood, or it may be a composite construction of wood or of wood combined with plastics or metals in sizes that are commercially used in structural applications, that is, in sizes greater than nominal 2 by 2-in. (38 by 38-mm) cross section (see 3.1.6).

24.2 *Identification*—Material or materials of the test specimen shall be as fully described as that for beams in 8.2.

24.3 *Specimen Dimensions*—The weight and dimensions, as well as moisture content of the specimen, shall be accurately measured before test. Weights and dimensions (length and cross section) shall be measured to three significant figures. Sufficient measurements of the cross section shall be made along the length of the specimen to describe shape characteristics and to determine the smallest section. The physical characteristics of the specimen, as described by its density and moisture content, may be determined in accordance with Test Methods D 2395 and Method A of Test Methods D 2016, respectively.

24.4 *Specimen Description*—The inherent imperfections and intentional modifications shall be described as for beams in 8.4.

24.5 *Specimen Length*—The cross-sectional and length dimensions of structural members usually have established sizes, depending on the manufacturing process and intended use, so that no modification of these dimensions is involved. Since the length has been approximately established, the full length of the member shall be tested, except for trimming or squaring the bearing surface (see 25.2.1).

25. Procedure

25.1 *Preliminary*—Unless otherwise indicated in the research program or material specification, the test specimen shall be conditioned to constant weight so it is at moisture equilibrium, under the desired environment. Moisture contents may be approximated with moisture meters or more accurately measured by weights of samples according to Method A of Test Methods D 2016.

25.2 *Test Setup:*

25.2.1 *Bearing Surfaces*—The bearing surfaces of the specimen shall be cut so that the contact surfaces are plane, parallel to each other, and normal to the long axis of the specimen.

25.2.2 *Setup Method*—After physical measurements have been taken and recorded, the specimen shall be placed in the testing machine between the bearing blocks at each end and between the lateral supports on the four sides. The contact surfaces shall be geometrically centered on the bearing plates and then the spherical seats shall be adjusted for full contact. A slight longitudinal pressure shall be applied to hold the specimen while the lateral supports are adjusted and fastened to conform to the warp, twist, or bend of the specimen.

25.3 *Speed of Testing*—For measuring load-deformation data, the load shall be applied at a constant rate of head motion so that the fiber strain is 0.001 in./in.·min ± 25 % (0.001 mm/mm·min). For measuring only compressive strength, the test may be conducted at a constant rate to achieve maximum load in about 10 min, but not less than 5 nor more than 20 min.

25.4 *Load-Deformation Curves*—If load-deformation data have been obtained, load and deflection at first failure, at changes in slope of curve, and at maximum load should be noted.

25.5 *Records*—The maximum load shall be recorded, as well as a description and sketch of the failure relating the latter to the location of imperfections in the specimen. The section of the specimen containing the failure should be reexamined during analysis of the data.

26. Calculations

26.1 Physical and mechanical properties shall be computed in accordance with Definitions E 6 and as follows (see compressive notations):

26.1.1 Stress at proportional limit = P'/A in pounds per square inch (MPa).

26.1.2 Compressive strength = P/A in pounds per square inch (MPa).

26.1.3 Modulus of elasticity = $P'/A\epsilon$ in pounds per square inch (MPa).

27. Report

27.1 The report shall include the following:

27.1.1 Complete identification,

27.1.2 History of seasoning conditioning,

27.1.3 Load apparatus,

27.1.4 Deflection apparatus,

27.1.5 Length and cross section dimensions,

27.1.6 Gage length,

27.1.7 Rate of load application,

27.1.8 Computed physical and mechanical properties, including specific gravity of moisture content, compressive strength, stress at proportional limit, modulus of elasticity, and a statistical measure of variability of these values,

27.1.9 Description of failure, and

27.1.10 Details of any deviations from the prescribed or recommended methods as outlined in the standard.

TENSION PARALLEL TO GRAIN

28. Scope

28.1 This method covers the determination of the tensile properties of structural elements made primarily of lumber equal to and greater than nominal 1 in. (19 mm) thick.

29. Summary of Method

29.1 The structural member is clamped at the extremities of its length and subjected to a tensile load so that in sections between clamps the tensile forces shall be axial and generally uniformly distributed throughout the cross sections without flexure along its length.

30. Significance and Use

30.1 The tensile properties obtained by axial tension will provide information similar to that stipulated for flexural properties in Section 6.

30.2 The tensile properties obtained include modulus of elasticity, stress at proportional limit, tensile strength, and strain data beyond proportional limit.

31. Apparatus

31.1 *Testing Machine*—Any device having the following is suitable:

31.1.1 *Drive Mechanism*—A drive mechanism for imparting to a movable clamp a uniform, controlled velocity with respect to a stationary clamp.

31.1.2 *Load Indicator*—A load-indicating mechanism capable of showing the total tensile force on the test section of the tension specimen. This force-measuring system shall be calibrated to ensure accuracy in accordance with Methods E 4.

31.1.3 *Grips*—Suitable grips or fastening devices shall be provided which transmit the tensile load from the movable head of the drive mechanism to one end of the test section of the tension specimen, and similar devices shall be provided to transmit the load from the stationary mechanism to the other end of the test section of the specimen. Such devices shall not apply a bending moment to the test section, allow slippage under load, inflict damage, or inflict stress concentrations to the test section. Such devices may be either plates bonded to the specimen or unbonded plates clamped to the specimen by various pressure modes.

31.1.3.1 *Grip Alignment*—The fastening device shall apply the tensile loads to the test section of the specimen without applying a bending moment. For ideal test conditions, the grips should be self-aligning, that is, they should be attached to the force mechanism of the machine in such a manner that they will move freely into axial alignment as soon as the load is applied, and thus apply uniformly distributed forces along the test section and across the test cross section (Fig. 8(*a*)). For less ideal test conditions, each grip should be gimbaled about one axis which should be perpendicular to the wider surface of the rectangular cross section of the test specimen, and the axis of rotation should be through the fastened area (Fig. 8(*b*)). When neither self-aligning grips nor single gimbaled grips are available, the specimen may be clamped in the heads of a universal-type testing machine with wedge-type jaws (Fig. 8(*c*)). A method of providing approximately full spherical alignment has three axes of rotation, not necessarily concurrent but, however, having a common axis longitudinal and through the centroid of the specimen (Figs. 8(*d*) and 9).

31.1.3.2 *Contact Surface*—The contact surface between grips and test specimen shall be such that slippage does not occur. A smooth texture on the grip surface should be avoided, as well as very rough and large projections which damage the contact surface of the wood. Grips that are surfaced with a coarse emery paper (60X aluminum oxide emery belt) have been found satisfactory for softwoods. However, for hardwoods, grips may have to be glued to the specimen to prevent slippage.

31.1.3.3 *Contact Pressure*—For unbonded grip devices, lateral pressure should be applied to the jaws of the grip so that slippage does not occur between grip and specimen. Such pressure may be applied by means of bolts or wedge-shaped jaws, or both. Wedge-shaped jaws, such as those shown on Fig. 10, which slip on the inclined plane to produce contact pressure have been found satisfactory. To eliminate stress concentration or compressive damage at the tip end of the jaw, the contact pressure should be reduced to zero. The variable thickness jaws (Fig. 10), which cause a variable contact surface and which produce a lateral pressure gradient, have been found satisfactory.

31.1.4 *Extensometer:*

31.1.4.1 *Gage Length*—For modulus of elasticity determinations, a device shall be provided by which the elongation of the test section of the specimen is measured with respect to specific paired gage points defining the gage length. To obtain data representative of the test material as a whole, such gage points shall be symmetrically located on the lengthwise surface of the specimen as far apart as feasible, yet at least two times the larger cross-sectional dimension from each jaw edge. At least two pairs of such gage points on diametrically opposite sides of the specimen shall be used to measure the average deformation.

31.1.4.2 *Accuracy*—The device shall be able to measure changes in elongation to three significant figures. Since gage lengths vary over a wide range, the measuring instruments should conform to their appropriate class in Method E 83.

32. Test Specimen

32.1 *Material*—The test specimen shall consist basically of a structural timber which may be solid wood, laminated wood, or it may be a composite construction of wood or wood combined with plastics or metals in sizes that are commercially used in structural "tensile" applications, that is, in sizes equal to and greater than nominal 1-in. (32-mm) thick lumber.

32.2 *Identification*—Material or materials of the test specimen shall be fully described as beams in 8.2.

32.3 *Specimen Description*—The specimen shall be described in a manner similar to that outlined in 8.3 and 8.4.

32.4 *Specimen Length*—The tension specimen, which has its long axis parallel to grain in the wood, shall have a length between grips equal to at least eight times the larger cross-sectional dimension when tested in self-aligning grips (see 31.1.3.1). However, when tested without self-aligning grips, it is recommended that the length between grips be at least 20 times the greater cross-sectional dimension.

33. Procedure

33.1 *Conditioning*—Unless otherwise indicated, the specimen shall be conditioned as outlined in 9.1.

33.2 *Test Setup*—After physical measurements have been taken and recorded, place the specimen in the grips of the load mechanism, taking care to have the long axis of the specimen and the grips coincide. The grips should securely clamp the specimen with either bolts or wedge-shaped jaws. If the latter are employed, apply a small preload to ensure that all jaws move an equal amount and maintain axial-alignment of specimen and grips. If either bolts or wedges are employed tighten the grips evenly and firmly to the degree necessary to prevent slippage. Under load, continue the tightening if necessary, even crushing the wood perpendicular to grain, so that no slipping occurs and a tensile failure occurs outside the jaw contact area.

33.3 *Speed of Testing*—For measuring load-elongation data, apply the load at a constant rate of head motion so that the fiber strain in the test section between jaws is 0.0006 in./in.·min ± 25 % (.0006 mm/mm·min). For measuring only tensile strength, the load may be applied at a constant rate of grip motion so that maximum load is achieved in about 10 min but not less than 5 nor more than 20 min.

33.4 *Load-Elongation Curves*—If load-elongation data have been obtained throughout the test, correlate changes in specimen behavior, such as appearance of cracks or splinters, with elongation data.

33.5 *Records*—Record the maximum load, as well as a description and sketch of the failure relating the latter to the location of imperfections in the test section. Re-examine the section containing the failure during analysis of data.

34. Calculations

34.1 Compute physical and mechanical properties in accordance with Definitions E 6, and as follows (see tensile notations):

34.1.1 Stress at proportional limit = P'/A in pounds per square inch (MPa).

34.1.2 Tensile strength = P/A in pounds per square inch (MPa).

34.1.3 Modulus of elasticity = $P'/A\epsilon$ in pounds per square inch (MPa).

35. Report

35.1 The report shall include the following:

35.1.1 Complete identification,

35.1.2 History of seasoning,

35.1.3 Load apparatus, including type of end condition,

35.1.4 Deflection apparatus,

35.1.5 Length and cross-sectional dimensions,

35.1.6 Gage length,

35.1.7 Rate of load application,

35.1.8 Computed properties,

35.1.9 Description of failures, and

35.1.10 Details of any deviations from the prescribed or recommended methods as outlined in the standard.

TORSION

36. Scope

36.1 This method covers the determination of the torsional properties of structural elements made of solid or laminated wood, or of composite constructions. The method is intended primarily for structural element or rectangular cross section but is also applicable to beams of round or irregular shapes.

37. Summary of Method

37.1 The structural element is subjected to a torsional moment by clamping it near its ends and applying opposing couples to each clamping device. The element is deformed at a prescribed rate and coordinate observations of torque and twist are made for the duration of the test.

38. Significance and Use

38.1 The torsional properties obtained by twisting the structural element will provide information similar to that stipulated for flexural properties under Section 6.

38.2 The torsional properties of the element include an apparent modulus of rigidity of the element as a whole, stress at proportional limit, torsional strength and twist beyond proportional limit.

39. Apparatus

39.1 *Testing Machine*—Any device having the following is suitable:

39.1.1 *Drive Mechanism*—A drive mechanism for imparting an angular displacement at a uniform rate between a movable clamp on one end of the element and another clamp at the other end.

39.1.2 *Torque Indicator*—A torque-indicating mechanism capable of showing the total couple on the element. This measuring system shall be calibrated to ensure accuracy in accordance with Methods E 4.

39.2 *Support Apparatus:*

39.2.1 *Clamps*—Each end of the element shall be securely held by metal plates of sufficient bearing area and strength to grip the element with a vise-like action without slippage, damage, or stress concentrations in the test section when the torque is applied to the assembly. The plates of the clamps shall be symmetrical about the longitudinal axis of the cross section of the element.

39.2.2 *Clamp Supports*—Each of the clamps shall be supported by roller bearings or bearing blocks that allow the structural element to rotate about its natural longitudinal axis. Such supports may be ball bearings in a rigid frame of a torque-testing machine (Figs. 11 and 12) or they may be bearing blocks (Figs. 13 and 14) on the stationary and movable frames of a universal-type test machine. Either type of support shall allow the transmission of the couple without friction to the torque measuring device, and shall allow freedom for longitudinal movement of the element during the twisting. Apparatus of Fig. 13 is not suitable for large amounts of twist unless the angles are measured at each end to enable proper torque calculation.

39.2.3 *Frame*—The frame of the torque-testing machine shall be capable of providing the reaction for the drive mechanism, the torque indicator, and the bearings. The framework necessary to provide these reactions in a universal-type test machine shall be two rigid steel beams attached to the movable and stationary heads forming an X. The extremities of the X shall bear on the lever arms attached to the test element (Fig. 13).

39.3 *Troptometer:*

39.3.1 *Gage Length*—For modulus of rigidity calculations, a device shall be provided by which the angle of twist of the element is measured with respect to specific paired gage points defining the gage length. To obtain test data representative of the element as a whole, such paired gage points shall be located symmetrically on the lengthwise surface of the element as far apart as feasible, yet at least two times the larger cross-sectional dimension from each of the clamps. A yoke (Fig. 16) or other suitable device (Fig. 12) shall be firmly attached at each gage point to permit measurement of the angle of twist. The angle of twist is measured by observing the relative rotation of the two yokes or other devices at the gage points with the aid of any suitable apparatus including a light beam (Fig. 12), dials (Fig. 14), or string and scale (Figs. 15 and 16).

39.3.2 *Accuracy*—The device shall be able to measure changes in twist to three significant figures. Since gage lengths may vary over a wide range, the measuring instruments should conform to their appropriate class in Method E 83.

40. Test Element

40.1 *Material*—The test element shall consist basically of a structural timber, which may be solid wood, laminated wood, or a composite construction of wood or wood combined with plastics or metals in sizes that are commercially used in structural applications.

40.2 *Identification*—Material or materials of the test element shall be as fully described as for beams in 8.2.

40.3 *Element Measurements*—The weight and dimensions as well as the moisture content shall be accurately determined before test. Weights and dimensions (length and cross section) shall be measured to three significant figures. Sufficient measurements of the cross section shall be made along the length of the specimen to describe characteristics and to determine the smallest cross section. The physical characteristics of the element, as described by its density

and moisture content, may be determined in accordance with Test Methods D 2395 and Test Methods D 2016, Method A, respectively.

40.4 *Element Description*—The inherent imperfections and intentional modifications shall be described as for beams in 8.4.

40.5 *Element Length*—The cross-sectional dimensions of solid wood structural elements and composite elements usually are established, depending upon the manufacturing process and intended use so that normally no modification of these dimensions is involved. However, the length of the specimen shall be at least eight times the larger cross-sectional dimension.

41. Procedure

41.1 *Conditioning*—Unless otherwise indicated in the research program or material specification, condition the test element to constant weight so it is at moisture equilibrium under the desired environment. Approximate moisture contents with moisture meters, or measure more accurately by weights of samples in accordance with Test Methods D 2016, Method A.

41.2 *Test Setups*—After physical measurements have been taken and recorded, place the element in the clamps of the load mechanism, taking care to have the axis of rotation of the clamps coincide with the longitudinal centroidal axis of the element. The clamps should be tightened to securely hold the element in either type of testing machine. If the tests are made in a universal-type test machine, the bearing blocks shall be equal distances from the axis of rotation of the element.

41.3 *Speed of Testing*—For measuring torque-twist data, apply the load at a constant rate of head motion so that the angular detrusion of the outer fibers in the test section between gage points is about 0.004 radian per inch of length (0.16 radian per metre of length) per minute ±50 %. For measuring only shear strength, the torque may be applied at a constant rate of twist so that maximum torque is achieved in about 10 min but not less than 5 nor more than 20 min.

41.4 *Torque-Twist Curves*—If torque-twist data have been obtained, torque and twist at first failure, at changes in slope of curve, and at maximum torque should be noted.

41.5 *Record of Failures*—Describe failures in detail as to type, manner and order of occurrence, angle with the grain, and position in the test element. Record descriptions relating to imperfections in the element. The section of the element containing the failure should be reexamined during analysis of the data.

42. Calculations

42.1 Compute physical and mechanical properties in accordance with definitions in E 6 and relationships in Table X3.1 and X3.2.

43. Report

43.1 The report shall include the following:

43.1.1 Complete identification,

43.1.2 History of seasoning and conditioning,

43.1.3 Apparatus for applying and measuring torque,

43.1.4 Apparatus for measuring angle of twist,

43.1.5 Length and cross-section dimensions,

43.1.6 Gage length,

43.1.7 Rate of twist applications,

43.1.8 Computed properties, and

43.1.9 Description of failures.

SHEAR MODULUS

44. Scope

44.1 This method covers the determination of the modulus of rigidity (G) or shear modulus of structural beams made of solid or laminated wood. Application to composite constructions can only give a measure of the apparent or effective shear modulus. The method is intended primarily for beams of rectangular cross section but is also applicable to other sections with appropriate modification of equation coefficients.

45. Summary of Method

45.1 The structural member, usually a straight or a slightly cambered beam of rectangular cross section, is subjected to a bending moment by supporting it at two locations called reactions, and applying a single transverse load midway between these reactions. The beam is deflected at a prescribed rate and a single observation of coordinate load and deflection is taken. This procedure is repeated on at least four different spans.

46. Significance and Use

46.1 The shear modulus established by this method will provide information similar to that stipulated for flexural properties under Section 6.

47. Apparatus

47.1 The test machine and specimen configuration, supports, and loading are identical to Section 7 with the following exception:

47.1.1 The load shall be applied as a single, concentrated load midway between the reactions.

48. Test Specimen

48.1 See Section 8.

49. Procedure

49.1 *Conditions*—See Section 9.1.

49.2 *Test Setup*—The specimen shall be positioned in the test machine as described in Section 9.2 and loaded in center point bending over at least four different spans with the same cross section at the center of each. The spans shall be chosen so as to give approximately equal increments of $(h/L)^2$ between them, within the range of 0.035 to 0.0025. The applied load must be sufficient to provide a reliable estimate of the initial bending stiffness of the specimen, but in no instance shall exceed the proportional limit or shear capacity of the specimen.

NOTE 8—Span to depth ratios of 5.5, 6.5, 8.5, and 20 meet the $(h/L)^2$ requirements of this section.

49.3 *Load-Deflection Measurements*—Obtain load-deflection data with the apparatus described in 7.4.1. One data point is required on each span tested.

49.4 *Records*—Span to depth ratios chosen and load levels achieved on each span shall be recorded.

49.5 *Speed of Testing*—See Section 9.3.

50. Calculations

50.1 Shear modulus is determined by plotting $1/E_f$ (where E_f is the apparent modulus of elasticity calculated under center point loading) versus $(h/L)^2$ for each span tested. As indicated in Fig. 17 and in Appendix X4, shear modulus is proportional to the slope of the best-fit line between these points.

51. Report

51.1 See Section 11.

PRECISION AND BIAS

52. Precision and Bias

52.1 The precision and bias of these methods are being established.

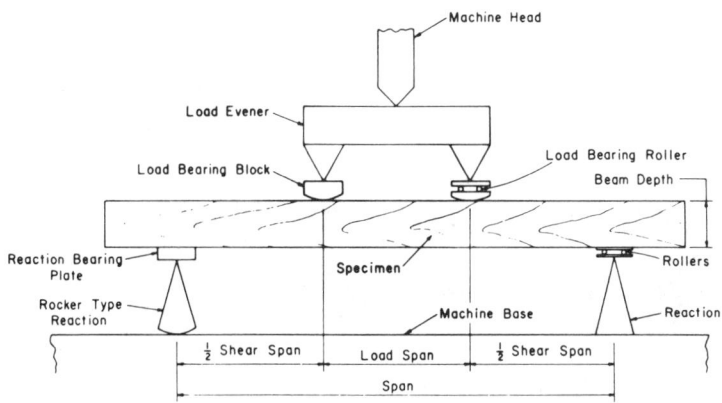

FIG. 1 Flexure Method

FIG. 2 Example of Bearing Plate, *A*, Rollers, *B*, and Reaction-Alignment-Rocker, *C*, for Small Beams

FIG. 3 Example of Lateral Support for Long, Deep Beams

FIG. 4 Example of Curved Loading Block, *A*, Load-Alignment Rocker, *B*, Roller-Curved Loading Block, *C*, Load Evener, *D*, and Deflection-Measuring Apparatus, *E*

FIG. 5 Compression of a Wood Structural Element

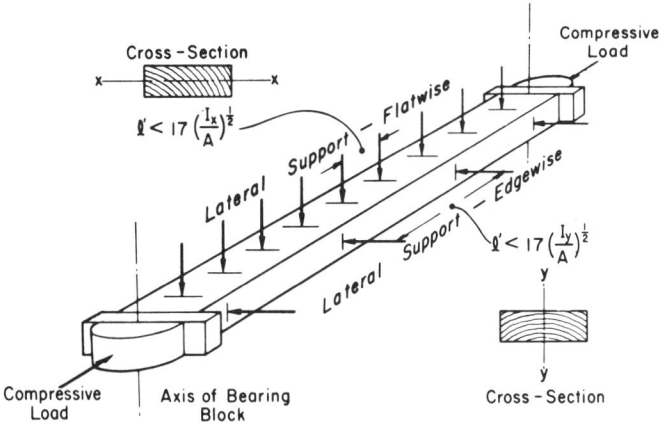

FIG. 6 Minimum Spacing of Lateral Supports of Long Columns

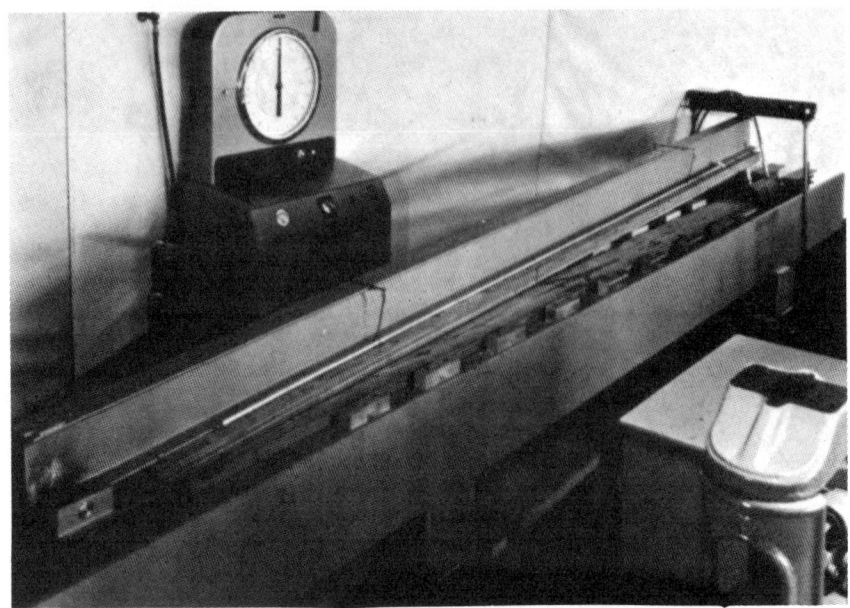

FIG. 7 Compression of Long Slender Structural Member

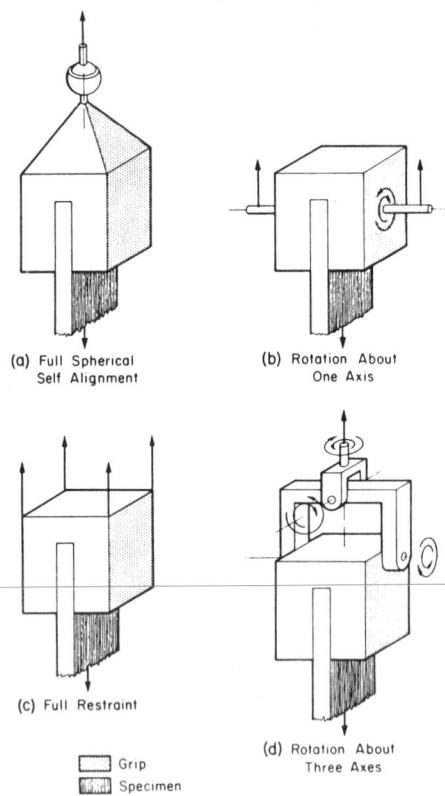

(a) Full Spherical
    Self Alignment

(b) Rotation About
    One Axis

(c) Full Restraint

☐ Grip
▨ Specimen

(d) Rotation About
    Three Axes

FIG. 8 Types of Tension Grips for Structural Members

FIG. 9   Horizontal Tensile Grips for 2 by 10-in. Structural Members

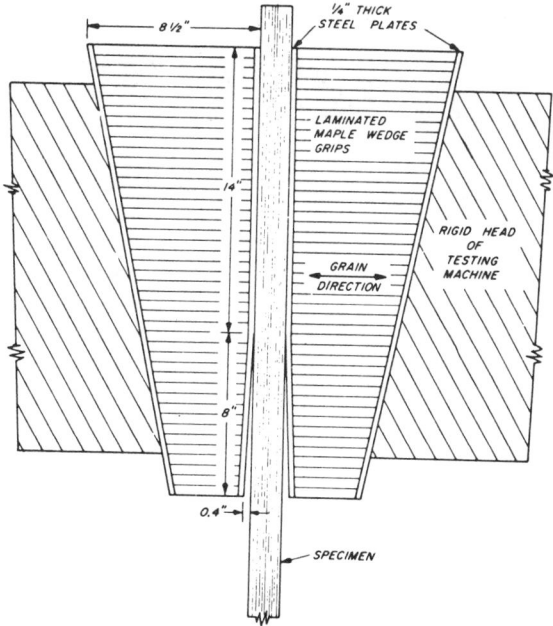

FIG. 10   Side View of Wedge Grips Used to Anchor Full-Size Structurally, Graded Tension Specimens

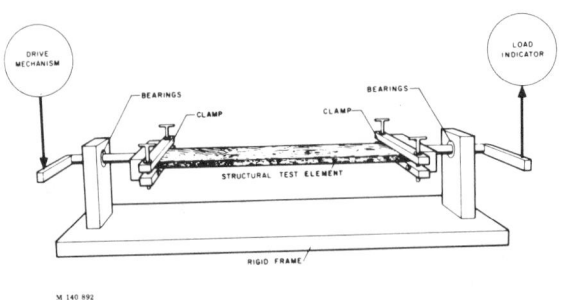

FIG. 11 Fundamentals of a Torsional Test Machine

FIG. 12 Example of Torque-Testing Machine (Torsion test in apparatus meeting specification requirements)

408

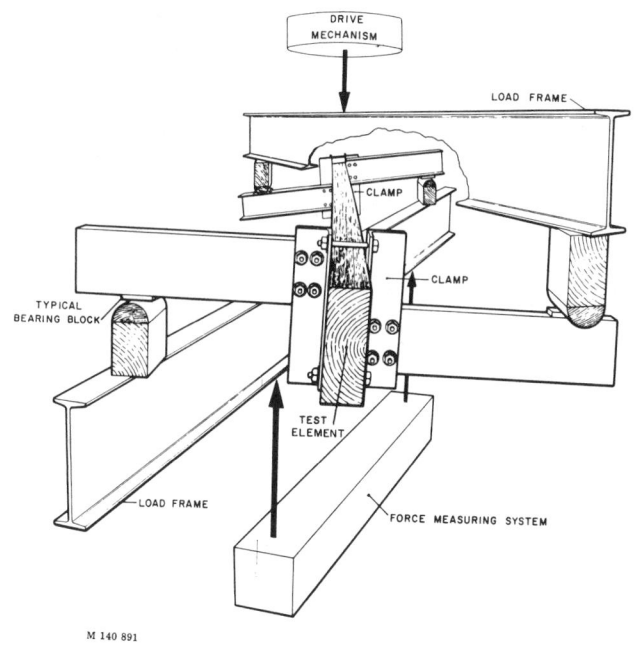

M 140 891

FIG. 13   Schematic Diagram of a Torsion Test Made in a Universal-Type Test Machine

FIG. 14   Example of Torsion Test of Structural Beam in a Universal-Type Test Machine

FIG. 15 Torsion Test with Yoke-Type Troptometer

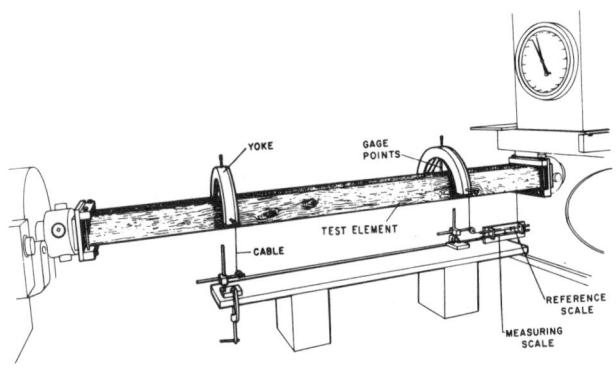

M 140 890

FIG. 16 Troptometer Measuring System

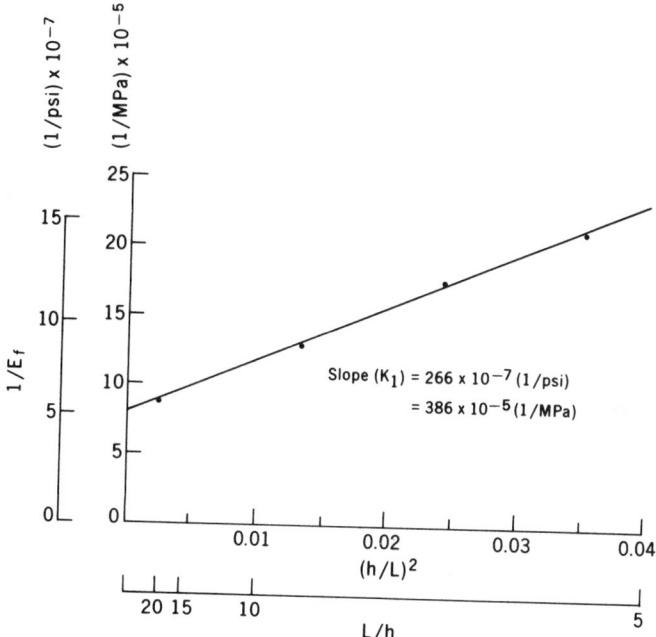

FIG. 17 Determination of Shear Modulus

APPENDIXES

(Nonmandatory Information)

X1. PHYSICAL PROPERTIES

TABLE X1.1 Physical Properties

Specific gravity (at test), $G_g = CW_g/V$	Test Methods D 2395
Specific gravity (ovendry), $G_d = G_g/(100 + MC)$	
Moisture content (% of dry weight), $MC = 100 \, (W_g/W_d - 1)$	Method A of Test Methods D 2016

GENERAL NOTATIONS

A — Cross-sectional area, in.^{2} (mm^2).

C — 0.061, a constant for use when W_g is measured in grams in equation for specific gravity. 22.7, a constant for use when W_g is measured in pounds in equation for specific gravity.

ϵ — Strain at proportional limit, in./in. (mm/mm).

G_d — Specific gravity (ovendry).

G_g — Specific gravity (at test).

I — Moment of inertia of the cross section about a designated axis, in.^{4} (mm^4).

N — Rate of motion of movable head, in./min (mm/min)

n — Number of specimens in sample.

S — Estimated standard deviation = $[(\sum X^2 - n\bar{X}^2)/(n - 1)]^{1/2}$.

V — Volume, in.^{3} (mm^3).

W_g — Weight of moisture specimen (at test), lb (g).

W_d — Weight of moisture specimen (ovendry), lb (g).

X — Individual values.

$\bar{X}$ — Average of n individual values.

FLEXURAL NOTATIONS

a — Distance from reaction to nearest load point, in. (mm) ($\frac{1}{2}$ shear span).

A_m — Area of graph paper under the load-deflection curve from zero load to maximum load in.^{2} (mm^2) when deflection is measured between reaction and center of span.

A_t — Area of graph paper under load-deflection curve from zero load to failing load or arbitrary terminal load, in.^{2} (mm^2), when deflection is measured between reaction and center of span.

b — Width of beam, in. (mm).

c — Distance from neutral axis of beam to extreme outer fiber, in. (mm).

G — Modulus of rigidity in shear in psi (MPa).

h — Depth of beam, in. (mm).

k — Graph paper scale constant for converting unit area of graph paper to load-deflection units.

l — Span of the beam that is used to measure deflections caused only by the bending moment, that is, no shear distortions, in. (mm).

L — Span of beam, in. (mm).

M — Maximum bending moment at maximum load, lbf·in. (N·m).

M' — Maximum bending moment at proportional limit load, lbf·in. (N·m).

P — Maximum transverse load on beam, lbf (N).

P' — Load on beam at proportional limit, lbf (N).

S_f — Fiber stress at proportional limit, psi (MPa).

S_R — Modulus of rupture.

z — Rate of fiber strain, in./in. (mm/mm), of outer fiber length per min.

Δ — Deflection of beam, in. (mm), at neutral axis between reaction and center of beam at the proportional limit, in. (mm).

Δ_l — Deflection of the beam measured at midspan over distance l, in. (mm).

τ_m — Maximum shear strength, psi (MPa).

COMPRESSIVE NOTATIONS

l — Length of compression column, in. (mm).

l' — Effective length of column between supports for lateral stability, in. (mm).

P — Maximum compressive load, lbf (N).

P' — Compressive load at proportional limit, lbf (N).

r — Radius of gyration = $[(I)/(A)]^{1/2}$, in. (mm).

TENSILE NOTATIONS

P — Maximum tensile load, lbf (N).

P' — Tensile load at proportional limit, lbf (N).

SHEAR NOTATIONS

E — Modulus of elasticity.

E_f — Apparent E, center point loading.

G — Modulus of rigidity (shear modulus)

I — Moment of inertia

P' — Load on beam at deflection, Δ', lbf (N) (Below proportional limit).

Δ' — Deflection of beam, in. (mm).

K — Shear coefficient. Defined in Table X4.1.

K_1 — Slope of line through multiple test data plotted on $(h/L)^2$ versus $(1/E_f)$.

X2. FLEXURE

TABLE X2.1 Flexure Formulas[A]

Mechanical Properties	General	Two-Point Loading Rectangular Beam	Third-Point Loading Rectangular Beam
Fiber stress at proportional limit, S_f	$\dfrac{M'c}{I}$	$\dfrac{3P'a}{bh^2}$	$\dfrac{P'L}{bh^2}$
Modulus of rupture, S_R	$\dfrac{Mc}{I}$	$\dfrac{3Pa}{bh^2}$	$\dfrac{PL}{bh^2}$
Modulus of elasticity, E_f (apparent E)	$\dfrac{P'a}{48I\Delta}(3L^2 - 4a^2)$	$\dfrac{P'a}{4bh^3\Delta}(3L^2 - 4a^2)$	$\dfrac{P'L^3}{4.7bh^3\Delta}$
Modulus of elasticity, E_G (true E, shear corrected)	$\dfrac{M'l^2}{8I\Delta_l}$	$\dfrac{P'a(3L^2 - 4a^2)}{4bh^3\Delta\left(1 - \dfrac{3P'a}{5bhG\Delta}\right)}$ $\dfrac{3P'al^2}{4bh^3\Delta_l}$	$\dfrac{P'L^3}{4.7bh^3\Delta\left(1 - \dfrac{P'L}{5bhG\Delta}\right)}$ $\dfrac{P'Ll^2}{4bh^3\Delta_l}$
Work to proportional limit per unit of volume, W_k		$\dfrac{P'\Delta}{2Lbh}\left[\dfrac{4a(3L - 4a) + \dfrac{24h^2E_G}{10G}}{3L^2 - 4a^2 + \dfrac{24h^2E_G}{10G}}\right]$	$\dfrac{P'\Delta}{2Lbh}\left[\dfrac{\dfrac{20}{9}L^2 + \dfrac{24h^2E_G}{10G}}{\dfrac{23}{9}L^2 + \dfrac{24h^2E_G}{10G}}\right]$
Approximate work to maximum load per unit of volume, W_m		$\dfrac{KA_m}{Lbh}\left[\dfrac{4a(3L - 4a) + \dfrac{24h^2E_G}{10G}}{3L^2 - 4a^2 + \dfrac{24h^2E_G}{10G}}\right]$	$\dfrac{KA_m}{Lbh}\left[\dfrac{\dfrac{20}{9}L^2 + \dfrac{24h^2E_G}{10G}}{\dfrac{23}{9}L^2 + \dfrac{24h^2E_G}{10G}}\right]$
Approximate total work per unit of volume, W_t		$\dfrac{KA_t}{Lbh}\left[\dfrac{4a(3L - 4a) + \dfrac{24h^2E_G}{10G}}{3L^2 - 4a^2 + \dfrac{24h^2E_G}{10G}}\right]$	$\dfrac{KA_t}{Lbh}\left[\dfrac{\dfrac{20}{9}L^2 + \dfrac{24h^2E_G}{10G}}{\dfrac{23}{9}L^2 + \dfrac{24h^2E_G}{10G}}\right]$
Shear stress, τ_m		$\dfrac{3}{4}\dfrac{P}{bh}$	$\dfrac{0.75G}{bh}$

[A] For wooden beams having uniform cross section throughout their length.

X3. TORSION

TABLE X3.1 Torsion Formulas[A]

Mechanical Properties	Cross Section			
	Circle	Square	Rectangle	General[B]
Fiber shear stress of greatest intensity at middle of long side; at proportional limit, S_s'	$2T'/\pi r^3$ (1A)	$4.808\, T'/w^3$ (1B)	$8\gamma T'/\mu wt^2$ (1C)	T'/Q (1D)
Fiber shear strength of greatest intensity at middle of long side, S_s	$2T/\pi r^3$ (2A)	$4.808\, T/w^3$ (2B)	$8\gamma T/\mu wt^2$ (2C)	T/Q (2D)
Fiber shear strength at middle of short side, S_s''			$8\gamma_1 T/\mu t^3$ (3C)	
Apparent modulus of rigidity, G	$2L_gT'/\pi r^4\theta$ (4A)	$7.11\, L_gT'/w^4\theta$ (4B)	$16\, L_gT'/wt^3[(16/3) - \lambda(t/w)]\,\theta$ (4C)	$L_gT'/\theta K$ (4D)

[A] From NACA rep. 334.

[B] Values of "Q" and "K" may be found in Roark, R. J., *Formulas for Stress and Strain*, McGraw-Hill, 1965, p. 194.

TABLE X3.2 Factors for Calculating Torsional Rigidity and Stress of Rectangular Prisms[A]

Ratio of Sides Column 1	λ Column 2	μ Column 3	γ Column 4	γ₁ Column 5
1.00	3.08410	2.24923	1.35063	1.35063
1.05	3.12256	2.35908	1.39651	
1.10	3.15653	2.46374	1.43956	
1.15	3.18554	2.56330	1.47990	
1.20	3.21040	2.65788	1.51753	
1.25	3.23196	2.74772	1.55268	1.13782
1.30	3.25035	2.83306	1.58544	
1.35	3.26632	2.91379	1.61594	
1.40	3.28002	2.99046	1.64430	
1.45	3.29171	3.06319	1.67265	
1.50	3.30174	3.13217	1.69512	0.97075
1.60	3.31770	3.25977	1.73889	0.91489
1.70	3.32941	3.37486	1.77649	
1.75	3.33402	3.42843	1.79325	0.84098
1.80	3.33798	3.47890	1.80877	
1.90	3.34426	3.57320	1.83643	
2.00	3.34885	3.65891	1.86012	0.73945
2.25	3.35564	3.84194	1.90543	
2.50	3.35873	3.98984	1.93614	0.59347
2.75	3.36023	4.11143	1.95687	
3.00	3.36079	4.21307	1.97087	
3.33	. . .	. . .	. . .	0.44545
3.50	3.36121	4.37299	1.98672	
4.00	3.36132	4.49300	1.99395	0.37121
4.50	3.36133	4.58639	1.99724	
5.00	3.36133	4.66162	1.99874	0.29700
6.00	3.36133	4.77311	1.99974	
6.67	3.36133	. . .	. . .	0.22275
7.00	3.36133	4.85314	1.99995	
8.00	3.36133	4.91317	1.99999	0.18564
9.00	3.36133	4.95985	2.00000	
10.00	3.36133	4.99720	2.00000	0.14858
20.00	3.36133	5.16527	2.00000	0.07341
50.00	3.36133	5.26611	2.00000	
100.00	3.36133	5.29972	2.00000	
∞	3.36133	5.33333	2.00000	0.00000

[A] Table I, "Factors for calculating torsional rigidity and stress of rectangular prisms," from National Advisory Committee for Aeronautics Report No. 334, "The Torsion of Members Having Sections Common in Aircraft Construction," by G. W. Trayer and H. W. March about 1929.

Torsion Notations

G	Apparent modulus of rigidity, psi (MPa).
K	Stiffness-shape factor.
L_g	Gage length of torsional element, in. (mm).
Q	Stress-shape factor.
r	Radius, in. (mm).
S_s	Fiber shear stress of greatest intensity at middle of long side at proportional limit, psi (MPa).
S_s	Fiber shear strength of greatest intensity at middle of long side at maximum torque, psi (MPa).
S_s''	Fiber shear strength at middle of short side at maximum torque, psi (MPa).
T	Twisting moment or torque, lbf·in. (N·m).
T'	Torque at proportional limit, lbf·in. (N·m).
t	Thickness, in. (mm).
w	Width of element, in. (mm).
γ	St. Venant constant, column 4, Table X4.
γ_1	St. Venant constant, column 5, Table X4.
θ	Total angle of twist, radians (in./in. or mm/mm).
λ	St. Venant constant, column 2, Table X4.
μ	St. Venant constant, column 3, Table X4.

X4. SHEAR MODULUS

X4.1 The elastic deflection of a prismatic beam under a single center point load is:

$$\Delta = \frac{PL^3}{48EI} + \frac{PL}{4GA'} \qquad (1)$$

where:

Δ = deflection at midspan,
P = applied load,
L = span,
E = modulus of elasticity,
I = moment of inertia,
G = modulus of rigidity (shear modulus), and
A' = modified shear area.

X4.2 All parameters are self-explanatory with the exception of the modified shear area. The modified

414

shear area is the product of the cross-sectional area, A, and a shear coefficient, K.[4] The shear coefficient relates the effective transverse shear strain to the average shear stress on the section. "K" is defined as the ratio of average shear strain on a section to shear strain at the centroid. Shear coefficients have been calculated and tabulated for a variety of beam configurations.

X4.2.1 Introducing K into Eq 1:

$$\Delta = \frac{PL^3}{48EI} + \frac{PL}{4GKA} \qquad (2)$$

X4.3 Often the relationship between deflection and elastic constants is simplified by ignoring the shear contribution, or the second term in Eq 2. The remaining elastic constant is called the "apparent" modulus of elasticity, E_f:

$$\Delta = \frac{PL^3}{48E_fI} \qquad (3)$$

X4.4 At the same deflection the apparent modulus of elasticity can be expressed in terms of the true elastic constants:

$$\frac{PL^3}{48E_fI} = \frac{PL^3}{48EI} + \frac{PL}{4GKA} \qquad (4)$$

X4.5 For a rectangular section of width, b, and depth, h, Eq 4 reduces to:

$$\frac{L^2}{E_fh^2} = \frac{L^2}{Eh^2} + \frac{1}{KG} \qquad (5)$$

X4.6 Multiplying both sides of Eq 5 by $(h/L)^2$ yields:

$$\frac{1}{E_f} = \frac{1}{E} + \frac{1}{KG}(h/L)^2 \qquad (6)$$

X4.6.1 Equation 6 can be graphed by substituting $y = 1/E_f$ and $x = (h/L)^2$. In the resulting $y = mx + b$ graph, the slope of a line connecting multiple data points is equal to $1/KG$.

X4.7 For a circular section of diameter, h, Eq 4 reduces to:

$$\frac{1}{E_f} = \frac{1}{E} + \frac{3}{4KG}(h/L)^2 \qquad (7)$$

X4.8 Using values for $K = (10(1 + v))/(12 + 11v)$ (rectangular) and $K = (6(1 + v))/(7 + 6v)$ (circular) and Poisson's ratios ranging from 0.05 to 0.5 yield:[4]

Rectangular: $K = 0.84$ to 0.86, and

Circular: $K = 0.86$ to 0.90.

X4.9 On plots of $1/E_f$ versus $(h/L)^2$, shear modulus, G, can be expressed in terms of the slope of the line connecting multiple observations. If the slope is called K_1, then:

$G = 1.17/K1$ to $1.20/K_1$ (rectangular), and

$G = 1.48/K1$ to $1.55/K_1$ (circular).

X4.10 As CIB/RILEM has already proposed $1.2/K_1$ for rectangular beams, the corresponding value for circular beams, $1.55/K_1$, should be used.

X4.11 Determination of shear modulus for other beam cross sections must start at Eq 4, substituting appropriate values for I, A, and K.

---

[4] Cowper, G. R., "The Shear Coefficient in Timoshenko's Beam Theory," *Journal of Applied Mechanics*, ASME, 1966, pp. 335–340.

TABLE X4.1 Shear Modulus Formulas

Mechanical Property	Formula
Modulus of elasticity, E_f (apparent E, center point loading)	$\dfrac{P'L^3}{48I\Delta'}$
Shear modulus, G^A	
Rectangular section	$1.2/K_1{}^B$
Circular section	$1.55/K_1$

[A] Based on solution of the equation $\Delta = (PL^3/48EI) + (PL/4KGA)$. K is tabulated for other cross sections by Cowper, G. R., "The Shear Coefficient in Timoshenko's Beam Theory," *of Applied Mechanics*, ASME, 1966, pp. 335–340.
[B] K_1 = Slope of the line plotted through the test values as shown in Fig. 17.

Designation: D 1559 – 82

Standard Test Method for
RESISTANCE TO PLASTIC FLOW OF BITUMINOUS
MIXTURES USING MARSHALL APPARATUS[1]

This standard is issued under the fixed designation D 1559; the number immediately following the designation indicates the year of original adoption or, in the case of revision, the year of last revision. A number in parentheses indicates the year of last reapproval. A superscript epsilon (ε) indicates an editorial change since the last revision or reapproval.

1. Scope

1.1 This method covers the measurement of the resistance to plastic flow of cylindrical specimens of bituminous paving mixture loaded on the lateral surface by means of the Marshall apparatus. This method is for use with mixtures containing asphalt cement, asphalt cut-back or tar, and aggregate up to 1-in. (25.4-mm) maximum size.

2. Significance and Use

2.1 This method is used in the laboratory mix design of bituminous mixtures. Specimens are prepared in accordance with the method and tested for maximum load and flow. Density and voids properties may also be determined on specimens prepared in accordance with the method. The testing section of this method can also be used to obtain maximum load and flow for bituminous paving specimens cored from pavements or prepared by other methods. These results may differ from values obtained on specimens prepared by this method.

3. Apparatus

3.1 *Specimen Mold Assembly*—Mold cylinders 4 in. (101.6 mm) in diameter by 3 in. (76.2 mm) in height, base plates, and extension collars shall conform to the details shown in Fig. 1. Three mold cylinders are recommended.

3.2 *Specimen Extractor*, steel, in the form of a disk with a diameter not less than 3.95 in. (100 mm) and ½ in. (13 mm) thick for extracting the compacted specimen from the specimen mold with the use of the mold collar. A suitable bar is required to transfer the load from the ring dynamometer adapter to the extension collar while extracting the specimen.

3.3 *Compaction Hammer*—The compaction hammer (Fig. 2) shall have a flat, circular tamping face and a 10-lb (4536-g) sliding weight with a free fall of 18 in.(457.2 mm). Two compaction hammers are recommended.

NOTE 1—The compaction hammer may be equipped with a finger safety guard as shown in Fig. 2.

3.4 *Compaction Pedestal*—The compaction pedestal shall consist of an 8 by 8 by 18-in. (203.2 by 203.2 by 457.2-mm) wooden post capped with a 12 by 12 by 1-in. (304.8 by 304.8 by 25.4-mm) steel plate. The wooden post shall be oak, pine, or other wood having an average dry weight of 42 to 48 lb/ft^3 (0.67 to 0.77 g/cm^3). The wooden post shall be secured by four angle brackets to a solid concrete slab. The steel cap shall be firmly fastened to the post. The pedestal assembly shall be installed so that the post is plumb and the cap is level.

3.5 *Specimen Mold Holder*, mounted on the compaction pedestal so as to center the compaction mold over the center of the post. It shall hold the compaction mold, collar, and base plate securely in position during compaction of the specimen.

3.6 *Breaking Head*—The breaking head (Fig. 3) shall consist of upper and lower cylindrical segments or test heads having an inside radius of curvature of 2 in. (50.8 mm) accurately machined. The lower segment shall be

[1] This method is under the jurisdiction of ASTM Committee D-4 on Road and Paving Materials and is the direct responsibility of Subcommittee D04.20 on Mechanical Tests of Bituminous Mixes.
Current edition approved April 30, 1982. Published August 1982. Originally published as D 1559 – 58. Last previous edition D 1559 – 76.

mounted on a base having two perpendicular guide rods or posts extending upward. Guide sleeves in the upper segment shall be in such a position as to direct the two segments together without appreciable binding or loose motion on the guide rods.

3.7 *Loading Jack*—The loading jack (Fig. 4) shall consist of a screw jack mounted in a testing frame and shall produce a uniform vertical movement of 2 in. (50.8 mm)/min. An electric motor may be attached to the jacking mechanism.

NOTE 2—Instead of the loading jack, a mechanical or hydraulic testing machine may be used provided the rate of movement can be maintained at 2 in. (50.8 mm)/min while the load is applied.

3.8 *Ring Dynamometer Assembly*—One ring dynamometer (Fig. 4) of 5000-lb (2267-kg) capacity and sensitivity of 10 lb (4.536 kg) up to 1000 lb (453.6 kg) and 25 lb (11.340 kg) between 1000 and 5000 lb (453.6 and 2267 kg) shall be equipped with a micrometer dial. The micrometer dial shall be graduated in 0.0001 in. (0.0025 mm). Upper and lower ring dynamometer attachments are required for fastening the ring dynamometer to the testing frame and transmitting the load to the breaking head.

NOTE 3—Instead of the ring dynamometer assembly, any suitable load-measuring device may be used provided the capacity and sensitivity meet the above requirements.

3.9 *Flowmeter*—The flowmeter shall consist of a guide sleeve and a gage. The activating pin of the gage shall slide inside the guide sleeve with a slight amount of frictional resistance. The guide sleeve shall slide freely over the guide rod of the breaking head. The flowmeter gage shall be adjusted to zero when placed in position on the breaking head when each individual test specimen is inserted between the breaking head segments. Graduations of the flowmeter gage shall be in 0.01-in. (0.25-mm) divisions.

NOTE 4—Instead of the flowmeter, a micrometer dial or stress-strain recorder graduated in 0.001 in. (0.025 mm) may be used to measure flow.

3.10 *Ovens or Hot Plates*—Ovens or hot plates shall be provided for heating aggregates, bituminous material, specimen molds, compaction hammers, and other equipment to the required mixing and molding temperatures. It is recommended that the heating units be thermostatically controlled so as to maintain the

required temperature within 5°F (2.8°C). Suitable shields, baffle plates or sand baths shall be used on the surfaces of the hot plates to minimize localized overheating.

3.11 *Mixing Apparatus*—Mechanical mixing is recommended. Any type of mechanical mixer may be used provided it can be maintained at the required mixing temperature and will provide a well-coated, homogeneous mixture of the required amount in the allowable time, and further provided that essentially all of the batch can be recovered. A metal pan or bowl of sufficient capacity and hand mixing may also be used.

3.12 *Water Bath*—The water bath shall be at least 6 in. (152.4 mm) deep and shall be thermostatically controlled so as to maintain the bath at 140 ± 1.8°F (60 ± 1.0°C) or 100 ± 1.8°F (37.8 ± 1°C). The tank shall have a perforated false bottom or be equipped with a shelf for supporting specimens 2 in. (50.8 mm) above the bottom of the bath.

3.13 *Air Bath*—The air bath for asphalt cutback mixtures shall be thermostatically controlled and shall maintain the air temperature at 77°F ± 1.8°F (25 ± 1.0°C).

3.14 *Miscellaneous Equipment:*

3.14.1 *Containers* for heating aggregates, flat-bottom metal pans or other suitable containers.

3.14.2 *Containers* for heating bituminous material, either gill-type tins, beakers, pouring pots, or saucepans may be used.

3.14.3 *Mixing Tool*, either a steel trowel (garden type) or spatula, for spading and hand mixing.

3.14.4 *Thermometers* for determining temperatures of aggregates, bitumen, and bituminous mixtures. Armored-glass or dial-type thermometers with metal stems are recommended. A range from 50 to 400°F (9.9 to 204°C), with sensitivity of 5°F (2.8°C) is required.

3.14.5 *Thermometers* for water and air baths with a range from 68 to 158°F (20 to 70°C) sensitive to 0.4°F (0.2°C).

3.14.6 *Balance*, 2-kg capacity, sensitive to 0.1 g, for weighing molded specimens.

3.14.7 *Balance*, 5-kg capacity, sensitive to 1.0 g, for batching mixtures.

3.14.8 *Gloves* for handling hot equipment.

3.14.9 *Rubber Gloves* for removing specimens from water bath.

3.14.10 *Marking Crayons* for identifying specimens.

3.14.11 *Scoop*, flat bottom, for batching aggregates.

3.14.12 *Spoon*, large, for placing the mixture in the specimen molds.

4. Test Specimens

4.1 *Number of Specimens*—Prepare at least three specimens for each combination of aggregates and bitumen content.

4.2 *Preparation of Aggregates*—Dry aggregates to constant weight at 221 to 230°F (105 to 110°C) and separate the aggregates to dry-sieving into the desired size fractions.[2] The following size fractions are recommended:

1 to ¾ in. (25.0 to 19.0 mm)
¾ to ⅜ in. (19.0 to 9.5 mm)
⅜ in. to No. 4 (9.5 mm to 4.75 mm)
No. 4 to No. 8 (4.75 mm to 2.36 mm)
Passing No. 8 (2.36 mm)

4.3 *Determination of Mixing and Compacting Temperatures:*

4.3.1 The temperatures to which the asphalt cement and asphalt cut-back must be heated to produce a viscosity of 170 ± 20 cSt shall be the mixing temperature.

4.3.2 The temperature to which asphalt cement must be heated to produce a viscosity of 280 ± 30 cSt shall be the compacting temperature.

4.3.3 From a composition chart for the asphalt cut-back used, determine from its viscosity at 140°F (60°C) the percentage of solvent by weight. Also determine from the chart the viscosity at 140°F (60°C) of the asphalt cut-back after it has lost 50 % of its solvent. The temperature determined from the viscosity temperature chart to which the asphalt cut-back must be heated to produce a viscosity of 280 ± 30 cSt after a loss of 50 % of the original solvent content shall be the compacting temperature.

4.3.4 The temperature to which tar must be heated to produce Engler specific viscosities of 25 ± 3 and 40 ± 5 shall be respectively the mixing and compacting temperature.

4.4 *Preparation of Mixtures:*

4.4.1 Weigh into separate pans for each test specimen the amount of each size fraction required to produce a batch that will result in a compacted specimen 2.5 ± 0.05 in. (63.5 ± 1.27 mm) in height (about 1200 g). Place the pans on the hot plate or in the oven and heat to a temperature not exceeding the mixing temper-

ature established in 4.3 by more than approximately 50°F (28°C) for asphalt cement and tar mixes and 25°F (14°C) for cut-back asphalt mixes. Charge the mixing bowl with the heated aggregate and dry mix thoroughly. Form a crater in the dry blended aggregate and weigh the preheated required amount of bituminous material into the mixture. For mixes prepared with cutback asphalt introduce the mixing blade in the mixing bowl and determine the total weight of the mix components plus bowl and blade before proceeding with mixing. Care must be exercised to prevent loss of the mix during mixing and subsequent handling. At this point, the temperature of the aggregate and bituminous material shall be within the limits of the mixing temperature established in 4.3. Mix the aggregate and bituminous material rapidly until thoroughly coated.

4.4.2 Following mixing, cure asphalt cutback mixtures in a ventilated oven maintained at approximately 20°F (11.1°C) above the compaction temperature. Curing is to be continued in the mixing bowl until the precalculated weight of 50 % solvent loss or more has been obtained. The mix may be stirred in a mixing bowl during curing to accelerate the solvent loss. However, care should be exercised to prevent loss of the mix. Weigh the mix during curing in successive intervals of 15 min initially and less than 10 min intervals as the weight of the mix at 50 % solvent loss is approached.

4.5 *Compaction of Specimens:*

4.5.1 Thoroughly clean the specimen mold assembly and the face of the compaction hammer and heat them either in boiling water or on the hot plate to a temperature between 200 and 300°F (93.3 and 148.9°C). Place a piece of filter paper or paper toweling cut to size in the bottom of the mold before the mixture is introduced. Place the entire batch in the mold, spade the mixture vigorously with a heated spatula or trowel 15 times around the perimeter and 10 times over the interior. Remove the collar and smooth the surface of the mix with a trowel to a slightly rounded shape. Temperatures of the mixtures immediately prior to compaction shall be within the limits of the compacting temper-

---

[2] Detailed requirements for these sieves are given in ASTM Specification E 11, for Wire-Cloth Sieves for Testing Purposes see *Annual Book of ASTM Standards*, Vol 14.02.

ature established in 4.3.

4.5.2 Replace the collar, place the mold assembly on the compaction pedestal in the mold holder, and unless otherwise specified, apply 50 blows with the compaction hammer with a free fall in 18 in. (457.2 mm). During compaction, the operator shall hold the axis of the compaction hammer by hand as nearly perpendicular to the base of the mold assembly as possible. Remove the base plate and collar, and reverse and reassemble the mold. Apply the same number of compaction blows to the face of the reversed specimen. After compaction, remove the base plate and place the sample extractor on that end of the specimen. Place the assembly with the extension collar up in the testing machine, apply pressure to the collar by means of the load transfer bar, and force the specimen into the extension collar. Lift the collar from the specimen. Carefully transfer the specimen to a smooth, flat surface and allow it to stand overnight at room temperature. Weigh, measure, and test the specimen.

NOTE 5—In general, specimens shall be cooled as specified in 4.5.2. When more rapid cooling is desired, table fans may be used. Mixtures that lack sufficient cohesion to result in the required cylindrical shape on removal from the mold immediately after compaction may be cooled in the mold in air until sufficient cohesion has developed to result in the proper cylindrical shape.

5. Procedure

5.1 Bring the specimens prepared with asphalt cement or tar to the specified temperature by immersing in the water bath 30 to 40 min or placing in the oven for 2 h. Maintain the bath or oven temperature at 140 ± 1.8°F (60 ± 1.0°C) for the asphalt cement specimens and 100 ± 1.8°F (37.8 ± 1.0°C) for tar specimens. Bring the specimens prepared with asphalt cutback to the specified temperature by placing them in the air bath for a minimum of 2 h. Maintain the air bath temperature at 77 ± 1.8°F (25 ± 1.0°C). Thoroughly clean the guide rods and the inside surfaces of the test heads prior to making the test, and lubricate the guide rods so that the upper test head slides freely over them. The testing-head temperature shall be maintained between 70 to 100°F (21.1 to 37.8°C) using a water bath when required. Remove the specimen from the water bath, oven, or air bath, and place in the lower segment of the breaking head. Place the upper segment of the breaking head on the specimen, and place the complete assembly in position on the testing machine. Place the flowmeter, where used, in position over one of the guide rods and adjust the flowmeter to zero while holding the sleeve firmly against the upper segment of the breaking head. Hold the flowmeter sleeve firmly against the upper segment of the breaking head while the test load is being applied.

5.2 Apply the load to the specimen by means of the constant rate of movement of the load jack or testing-machine head of 2 in. (50.8 mm)/min until the maximum load is reached and the load decreases as indicated by the dial. Record the maximum load noted on the testing machine or converted from the maximum micrometer dial reading. Release the flowmeter sleeve or note the micrometer dial reading, where used, the instant the maximum load begins to decrease. Note and record the indicated flow value or equivalent units in hundredths of an inch (twenty-five hundredths of a millimetre) if a micrometer dial is used to measure the flow. The elapsed time for the test from removal of the test specimen from the water bath to the maximum load determination shall not exceed 30 s.

NOTE 6—For core specimens, correct the load when thickness is other than 2½ in. (63.5 mm) by using the proper multiplying factor from Table 1.

6. Report

6.1 The report shall include the following information:

6.1.1 Type of sample tested (laboratory sample or pavement core specimen).

NOTE 6—For core specimens, the height of each test specimen in inches (or millimetres) shall be reported.

6.1.2 Average maximum load in pounds-force (or newtons) of at least three specimens, corrected when required.

6.1.3 Average flow value, in hundredths of an inch, twenty-five hundredths of a millimetre, of three specimens, and

6.1.4 Test temperature.

TABLE 1   Stability Correlation Ratios[A]

Volume of Specimen, cm³	Approximate Thickness of Specimen, in.[B]	mm	Correlation Ratio
200 to 213	1	25.4	5.56
214 to 225	1$\frac{1}{16}$	27.0	5.00
226 to 237	1$\frac{1}{8}$	28.6	4.55
238 to 250	1$\frac{3}{16}$	30.2	4.17
251 to 264	1$\frac{1}{4}$	31.8	3.85
265 to 276	1$\frac{5}{16}$	33.3	3.57
277 to 289	1$\frac{3}{8}$	34.9	3.33
290 to 301	1$\frac{7}{16}$	36.5	3.03
302 to 316	1$\frac{1}{2}$	38.1	2.78
317 to 328	1$\frac{9}{16}$	39.7	2.50
329 to 340	1$\frac{5}{8}$	41.3	2.27
341 to 353	1$\frac{11}{16}$	42.9	2.08
354 to 367	1$\frac{3}{4}$	44.4	1.92
368 to 379	1$\frac{13}{16}$	46.0	1.79
380 to 392	1$\frac{7}{8}$	47.6	1.67
393 to 405	1$\frac{15}{16}$	49.2	1.56
406 to 420	2	50.8	1.47
421 to 431	2$\frac{1}{16}$	52.4	1.39
432 to 443	2$\frac{1}{8}$	54.0	1.32
444 to 456	2$\frac{3}{16}$	55.6	1.25
457 to 470	2$\frac{1}{4}$	57.2	1.19
471 to 482	2$\frac{5}{16}$	58.7	1.14
483 to 495	2$\frac{3}{8}$	60.3	1.09
496 to 508	2$\frac{7}{16}$	61.9	1.04
509 to 522	2$\frac{1}{2}$	63.5	1.00
523 to 535	2$\frac{9}{16}$	64.0	0.96
536 to 546	2$\frac{5}{8}$	65.1	0.93
547 to 559	2$\frac{11}{16}$	66.7	0.89
560 to 573	2$\frac{3}{4}$	68.3	0.86
574 to 585	2$\frac{13}{16}$	71.4	0.83
586 to 598	2$\frac{7}{8}$	73.0	0.81
599 to 610	2$\frac{15}{16}$	74.6	0.78
611 to 625	3	76.2	0.76

[A] The measured stability of a specimen multiplied by the ratio for the thickness of the specimen equals the corrected stability for a 2½-in. (63.5-mm) specimen.

[B] Volume-thickness relationship is based on a specimen diameter of 4 in. (101.6 mm).

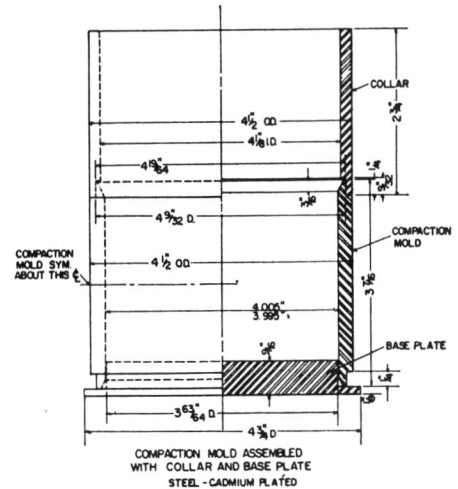

COMPACTION MOLD ASSEMBLED
WITH COLLAR AND BASE PLATE
STEEL - CADMIUM PLATED

Table of Equivalents for Figs. 1 and 3

Inch-Pound Units, in.	Metric Equivalents, mm	Inch-Pound Units, in.	Metric Equivalents, mm	Inch-Pound Units, in.	Metric Equivalents, mm	Inch-Pound Units, in.	Metric Equivalents, mm
0.005	0.11	$^{11}/_{16}$	17.5	$2^{5}/_{16}$	58.7	$4^{1}/_{8}$	104.8
$^{1}/_{32}$	0.8	$^{3}/_{8}$	19.0	$2^{1}/_{2}$	63.5	$4^{9}/_{32}$	108.7
$^{1}/_{16}$	1.6	$^{7}/_{8}$	22.2	$2^{3}/_{4}$	69.8	$4^{19}/_{64}$	109.1
$^{1}/_{8}$	3.2	$^{15}/_{16}$	23.8	$2^{7}/_{8}$	73.0	$4^{1}/_{2}$	114.3
$^{3}/_{16}$	4.8	1	25.4	3	76.2	$4^{5}/_{8}$	117.5
$^{1}/_{4}$	6.4	$1^{7}/_{8}$	28.6	$3^{1}/_{4}$	82.6	$4^{3}/_{4}$	120.6
$^{9}/_{32}$	7.1	$1^{1}/_{4}$	31.8	$3^{7}/_{16}$	87.3	$5^{1}/_{16}$	128.6
$^{3}/_{8}$	9.5	$1^{3}/_{8}$	34.9	$3^{7}/_{8}$	98.4	$5^{1}/_{8}$	130.2
0.496	12.6	$1^{1}/_{2}$	38.1	$3^{63}/_{64}$	101.2	$5^{3}/_{4}$	146.0
0.499	12.67	$1^{5}/_{8}$	41.3	3.990	101.35	6	152.4
$^{1}/_{2}$	12.7	$1^{3}/_{4}$	44.4	3.995	101.47	$6^{1}/_{4}$	158.8
$^{9}/_{16}$	14.3	2	50.8	4	101.6	$7^{5}/_{8}$	193.7
$^{5}/_{8}$	15.9	$2^{1}/_{4}$	57.2	4.005	101.73	27	685.8

FIG. 1 Compaction Mold

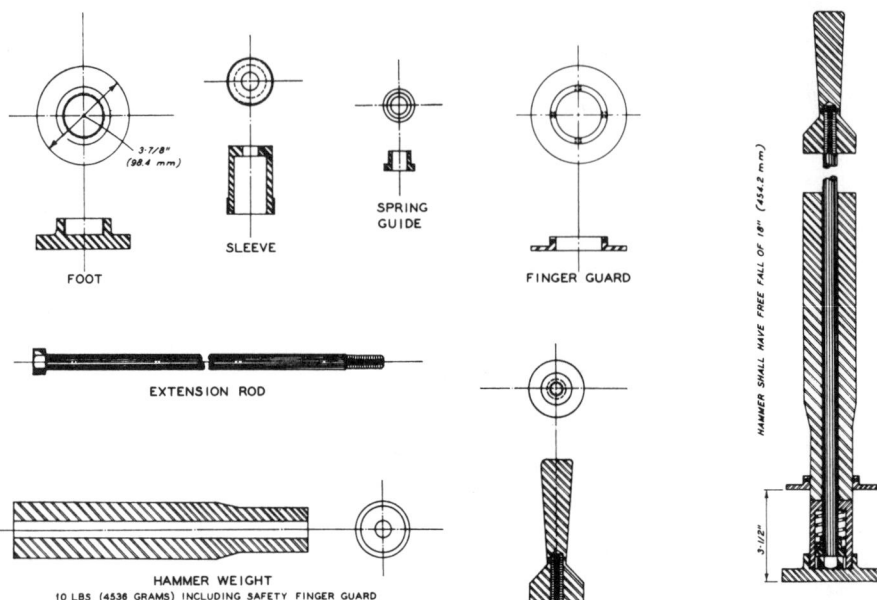

FIG. 2   Compaction Hammer

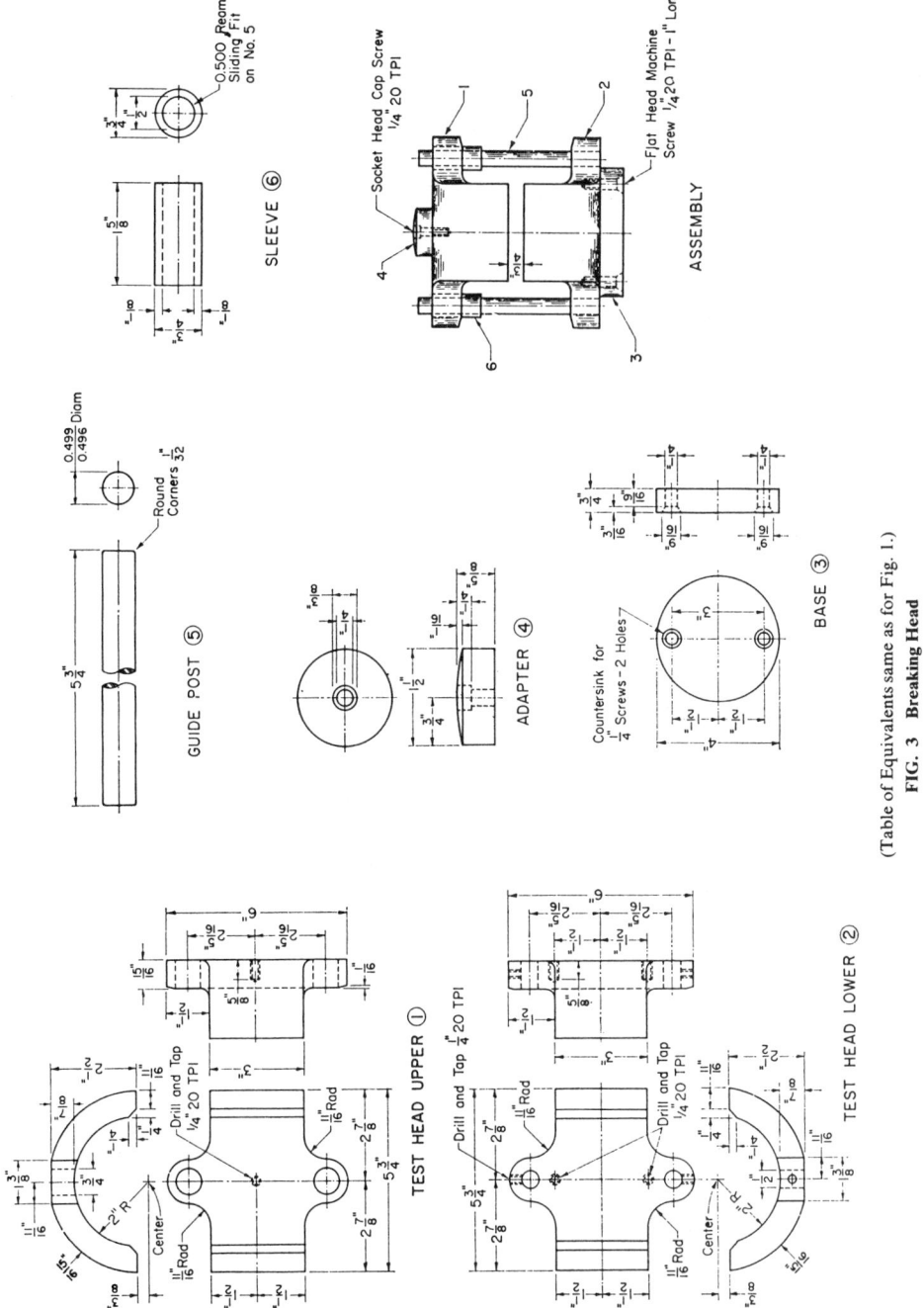

(Table of Equivalents same as for Fig. 1.)

FIG. 3   Breaking Head

SLEEVE ⑥

ASSEMBLY

GUIDE POST ⑤

ADAPTER ④

BASE ③

TEST HEAD UPPER ①

TEST HEAD LOWER ②

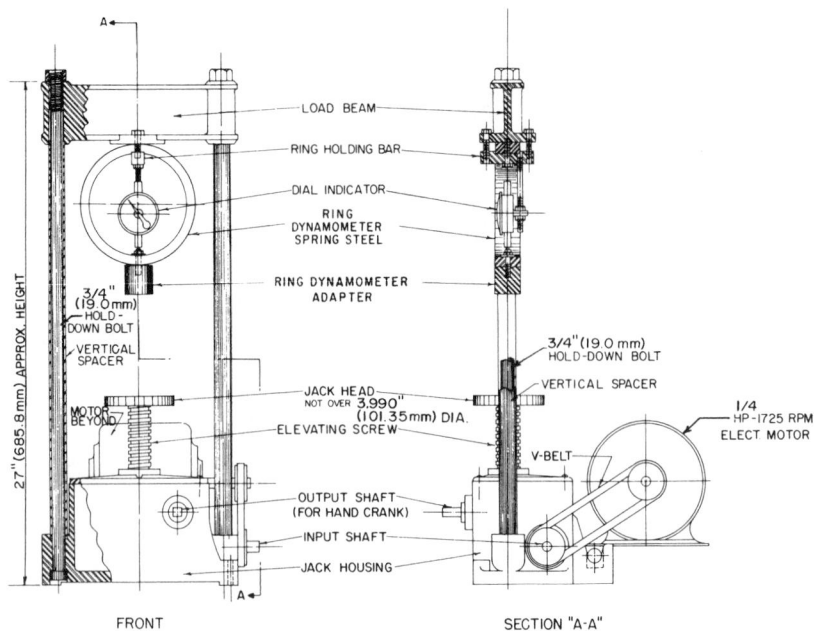

FIG. 4   Compression Testing Machine

Designation: E 8 – 85a

American Association State
Highway and Transportation Officials Standard
AASHTO No.: T 68

Standard Methods of
TENSION TESTING OF METALLIC MATERIALS[1]

This standard is issued under the fixed designation E 8; the number immediately following the designation indicates the year of original adoption or, in the case of revision, the year of last revision. A number in parentheses indicates the year of last reapproval. A superscript epsilon (ϵ) indicates an editorial change since the last revision or reapproval.

These methods have been approved for use by agencies of the Department of Defense to replace method 211.1 of Federal Test Method Standard No. 151b and for listing in the DoD Index of Specifications and Standards.

1. Scope

1.1 These methods cover the tension testing of metallic materials in any form at room temperature, specifically, the methods of determination of yield strength, yield point, tensile strength, elongation, and reduction of area.

NOTE 1—A complete metric companion to Methods E 8 has been developed—E 8M; therefore, no metric equivalents are shown in these methods.

NOTE 2—Exceptions to the provisions of these methods may need to be made in individual specifications or test methods for a particular material. For examples, see Methods and Definitions A 370 and Methods B 557.

1.2 *This standard may involve hazardous materials, operations, and equipment. This standard does not purport to address all of the safety problems associated with its use. It is the responsibility of whoever uses this standard to consult and establish appropriate safety and health practices and determine the applicability of regulatory limitations prior to use.*

2. Applicable Documents

2.1 *ASTM Standards:*

A 370 Methods and Definitions for Mechanical Testing of Steel Products[2]

B 557 Methods of Tension Testing Wrought and Cast Aluminum- and Magnesium-Alloy Products[3]

E 4 Practices of Load Verification of Testing Machines[4]

E 6 Definitions of Terms Relating to Methods of Mechanical Testing[4]

E 29 Recommended Practice for Indicating Which Places of Figures Are to Be Considered Significant in Specified Limiting

Values[5]

E 83 Method of Verification and Classification of Extensometers[4]

E 345 Methods of Tension Testing of Metallic Foil[4]

3. Definitions

3.1 The definitions of terms relating to tension testing appearing in Definitions E 6 shall be considered as applying to the terms used in these methods of tension testing.

4. Significance and Use

4.1 Tension tests provide information on the strength and ductility of materials under uniaxial tensile stresses. This information may be useful in comparisons of materials, alloy development, quality control, and design under certain circumstances.

4.2 The results of tension tests of specimens machined to standardized dimensions from selected portions of a part or material may not totally represent the strength and ductility properties of the entire end product or its in-service behavior in different environments.

4.3 These methods are considered satisfactory for acceptance testing of commercial shipments. The methods have been used extensively in the trade for this purpose.

---

[1] These methods are under the jurisdiction of ASTM Committee E-28 on Mechanical Testing and are the direct responsibility of Subcommittee E 28.04 on Tension Testing.

Current edition approved March 29, and April 26, 1985. Published June 1985. Originally published as E 8 – 24 T. Last previous edition E 8 – 84.

[2] *Annual Book of ASTM Standards*, Vol 01.04.

[3] *Annual Book of ASTM Standards*, Vol 02.03.

[4] *Annual Book of ASTM Standards*, Vol 03.01.

[5] *Annual Book of ASTM Standards*, Vols 03.01 and 14.02.

5. Apparatus

5.1 *Testing Machines*—Machines used for tension testing shall conform to the requirements of Practices E 4. The loads used in determining tensile strength and yield strength or yield point shall be within the loading range of the testing machine as defined in Practices E 4.

5.2 *Gripping Devices:*

5.2.1 *General*—Various types of gripping devices may be used to transmit the measured load applied by the testing machine to the test specimens. To ensure axial tensile stress within the gage length, the axis of the test specimen should coincide with the center line of the heads of the testing machine. Any departure from this requirement may introduce bending stresses that are not included in the usual stress computation (load divided by cross-sectional area).

NOTE 3—The effect of this eccentric loading may be illustrated by calculating the bending moment and stress thus added. For a standard ½-in. diameter specimen, the stress increase is 1.5 percentage points for each 0.001 in. of eccentricity. This error increases to 2.24 percentage points/0.001 in. for a 0.350-in. diameter specimen and to 3.17 percentage points/0.001 in. for a 0.250-in. diameter specimen.

5.2.2 *Wedge Grips*—Testing machines usually are equipped with wedge grips. These wedge grips generally furnish a satisfactory means of gripping long specimens of ductile metal and flat plate test specimens such as those shown in Fig. 6. If, however, for any reason, one grip of a pair advances farther than the other as the grips tighten, an undesirable bending stress may be introduced. When liners are used behind the wedges, they must be of the same thickness and their faces must be flat and parallel. For best results, the wedges should be supported over their entire lengths by the heads of the testing machine. This requires that liners of several thicknesses be available to cover the range of specimen thickness. For proper gripping, it is desirable that the entire length of the serrated face of each wedge be in contact with the specimen. Proper alignment of wedge grips and liners is illustrated in Fig. 1. For short specimens and for specimens of many materials it is generally necessary to use machined test specimens and to use a special means of gripping to ensure that the specimens, when under load, shall be as nearly as possible in uniformly distributed pure axial tension (see 5.2.3, 5.2.4, and 5.2.5).

5.2.3 *Grips for Threaded and Shouldered Specimens and Brittle Materials*—A schematic diagram of a gripping device for threaded-end specimens is shown in Fig. 2, while Fig. 3 shows a device for gripping specimens with shouldered ends. Both of these gripping devices should be attached to the heads of the testing machine through properly lubricated spherical-seated bearings. The distance between spherical bearings should be as great as feasible.

5.2.4 *Grips for Sheet Materials*—The self-adjusting grips shown in Fig. 4 have proved satisfactory for testing sheet materials that cannot be tested satisfactorily in the usual type of wedge grips.

5.2.5 *Grips for Wire*—Grips of either the wedge or snubbing types as shown in Figs. 4 and 5 or flat wedge grips may be used.

5.3 *Dimension-Measuring Devices*—Micrometers and other devices used for measuring linear dimensions shall be accurate to at least one half the smallest unit to which the individual dimension is required to be measured.

6. Test Specimens

6.1 *General:*

6.1.1 Test specimens shall be either substantially full size or machined, as prescribed in the product specifications for the material being tested.

NOTE 4—Dimensions of specimens are given in inch-pound units and metric units. The metric units are not always exact arithmetic equivalents but have been adjusted to provide practical equivalents for critical dimensions while retaining geometric proportionality.

6.1.2 Improperly prepared test specimens often are the reason for unsatisfactory and incorrect test results. It is important, therefore, that care be exercised in the preparation of specimens, particularly in the machining, to assure the desired precision and accuracy in test results.

6.1.3 It is desirable to have the cross-sectional area of the specimen smallest at the center of the reduced section to ensure fracture within the gage length. For this reason, a small taper is permitted in the reduced section of each of the specimens described in the following sections.

6.1.4 For brittle materials it is desirable to have fillets of large radius at the ends of the gage length.

6.2 *Plate-Type Specimens*—The standard plate-type test specimen is shown in Fig. 6. This specimen is used for testing metallic materials in the form of plate, shapes, and flat material having a nominal thickness of ³/₁₆ in. or over. When product specifications so permit, other types of specimens may be used, as provided in 6.3, 6.4, and 6.5.

6.3 *Sheet-Type Specimens:*

6.3.1 The standard sheet-type test specimen is shown in Fig. 6. This specimen is used for testing metallic materials in the form of sheet, plate, flat wire, strip, band, hoop, rectangles, and shapes ranging in nominal thickness from 0.005 to ⁵/₈ in. When product specifications so permit, other types of specimens may be used, as provided in 6.2, 6.4, and 6.5.

NOTE 5—Methods E 345 may be used for tension testing of materials in thicknesses up to 0.0059 in. (0.150 mm).

6.3.2 Pin ends as shown in Fig. 7 may be used. In order to avoid buckling in tests of thin and high-strength materials, it may be neccessary to use stiffening plates at the grip ends.

6.4 *Round Specimens:*

6.4.1 The standard 0.500-in. diameter round test specimen shown in Fig. 8 is used quite generally for testing metallic materials, both cast and wrought.

6.4.2 Figure 8 also shows small-size specimens proportional to the standard specimen. These may be used when it is necessary to test material from which the standard specimen or specimens shown in Fig. 6 cannot be prepared. Other sizes of small round specimens may be used. In any such small-size specimen it is important that the gage length for measurement of elongation be four times the diameter of the specimen.

6.4.3 The shape of the ends of the specimen outside of the gage length shall be suitable to the material and of a shape to fit the holders or grips of the testing machine so that the loads may be applied axially. Figure 9 shows specimens with various types of ends that have given satisfactory results.

6.5 *Specimens for Sheet, Strip, Flat Wire, and Plate*—In testing sheet, plate, flat wire, and strip one of the following types of specimens shall be used:

6.5.1 For material ranging in nominal thickness from 0.005 to ⁵/₈ in., the sheet-type specimen described in 6.3.

NOTE 6—Attention is called to the fact that either of the flat specimens described in 6.2 and 6.3 may be used for material from ³/₁₆ to ⁵/₈ in. in thickness, and

one of the round specimens described in 6.4 may also be used for material ½ in. or more in thickness.

6.5.2 For material having a nominal thickness of ³/₁₆ in. or over (Note 6), the plate-type specimen described in 6.2.

6.5.3 For material having a nominal thickness of ½ in. or over (Note 6), the largest practical size of specimen described in 6.4.

6.6 *Specimens for Wire, Rod, and Bar:*

6.6.1 For round wire and rod, test specimens having the full cross-sectional area of the wire or rod shall be used wherever practicable. The gage length for the measurement of elongation of wire less than ⅛ in. in diameter shall be as prescribed in product specifications. In testing wire of ⅛-in. or larger diameter, unless otherwise specified, a gage length equal to four times the diameter shall be used. The total length of the specimens shall be at least equal to the gage length plus the length of wire required for the full use of the grips employed.

6.6.2 For wire of octagonal, hexagonal, or square cross section, for rod or bar of round cross section where the specimen required in 6.6.1 is not practicable, and for rod or bar of octagonal, hexagonal, or square cross section, one of the following types of specimens shall be used:

6.6.2.1 *Full Cross Section* (Note 7)—It is permissible to reduce the test section slightly with abrasive cloth or paper, or machine it sufficiently to ensure fracture within the gage marks. For material not exceeding 0.188 in. in diameter or distance between flats, the cross-sectional area may be reduced to not less than 90 % of the original area without changing the shape of the cross section. For material over 0.188 in. in diameter or distance between flats, the diameter or distance between flats may be reduced by not more than 0.010 in. without changing the shape of the cross section. Square, hexagonal, or octagonal wire or rod not exceeding 0.188 in. between flats may be turned to a round having a cross-sectional area not smaller than 90 % of the area of the maximum inscribed circle. Fillets, preferably with a radius of ⅜ in., but not less than ⅛ in., shall be used at the ends of the reduced sections. Square, hexagonal, or octagonal rod over 0.188 in. between flats may be turned to a round having a diameter no smaller than 0.010 in. less than the original distance between flats.

NOTE 7—The ends of copper or copper alloy specimens may be flattened 10 to 50 % from the original dimension in a jig similar to that shown in Fig. 10, to facilitate fracture within the gage marks. In flattening the opposite ends of the test specimen, care shall be taken to ensure that the four flattened surfaces are parallel and that the two parallel surfaces on the same side of the axis of the test specimen lie in the same plane.

6.6.2.2 For rod and bar, the largest practical size of round specimen as described in 6.4 may be used in place of a test specimen of full cross section. Unless otherwise specified in the product specification, specimens shall be parallel to the direction of rolling or extrusion.

6.7 *Specimens for Rectangular Bar*—In testing rectangular bar one of the following types of specimens shall be used:

6.7.1 *Full Cross Section*—It is permissible to reduce the width of the specimen throughout the test section with abrasive cloth or paper, or by machining sufficiently to facilitate fracture within the gage marks, but in no case shall the reduced width be less than 90 % of the original. The edges of the midlength of the reduced section not less than ¾ in in length shall be parallel to each other and to the longitudinal axis of the specimen within 0.002 in. Fillets, preferably with a radius of ⅜ in. but not less than ⅛ in. shall be used at the ends of the reduced sections.

6.7.2 Rectangular bar of thickness small enough to fit the grips of the testing machine but of too great width may be reduced in width by cutting to fit the grips, after which the cut surfaces shall be machined or cut and smoothed to ensure failure within the desired section. The reduced width shall be not less than the original bar thickness. Also, one of the types of specimens described in 6.2, 6.3, and 6.4 may be used.

6.8 *Shapes, Structural and Other*—In testing shapes other than those covered by the preceding sections, one of the types of specimens described in 6.2, 6.3, and 6.4 shall be used.

6.9 *Specimens for Pipe and Tube* (Note 8):

6.9.1 For all small tube (Note 8), particularly sizes 1 in. and under in nominal outside diameter, and frequently for larger sizes, except as limited by the testing equipment, it is standard practice to use tension test specimens of full-size tubular sections. Snug-fitting metal plugs shall be inserted far enough into the ends of such tubular specimens to permit the testing machine jaws to grip the specimens properly. The plugs shall not extend into that part of the specimen on which the elongation is measured. Figure 11 shows a suitable form of plug, the location of the plugs in

the specimen, and the location of the specimen in the grips of the testing machine.

NOTE 8—The term "tube" is used to indicate tubular products in general, and includes pipe, tube, and tubing.

6.9.2 For large-diameter tube that cannot be tested in full section, longitudinal tension test specimens shall be cut as indicated in Fig. 12. Specimens from welded tube shall be located approximately 90° from the weld. If the tube-wall thickness is under ¾ in., either a specimen of the form and dimensions shown in Fig. 13 or one of the small-size specimens proportional to the standard ½-in. specimen, as mentioned in 6.4.2 and shown in Fig. 8, shall be used. Specimens of the type shown in Fig. 13 may be tested with grips having a surface contour corresponding to the curvature of the tube. When grips with curved faces are not available, the ends of the specimens may be flattened without heating. If the tube-wall thickness is ¾ in. or over, the standard specimen shown in Fig. 8 shall be used.

6.9.3 Transverse tension test specimens for tube may be taken from rings cut from the ends of the tube as shown in Fig. 14. Flattening of the specimen may be either after separating as in *A*, or before separating as in *B*. Transverse tension test specimens for large tube under ¾ in. in wall thickness shall be either of the small-size specimens shown in Fig. 8 or of the form and dimensions shown for Specimen 2 in Fig. 13. When using the latter specimen, either or both surfaces of the specimen may be machined to secure a uniform thickness, provided not more than 15 % of the normal wall thickness is removed from each surface. For large tube ¾ in. and over in wall thickness, the standard specimen shown in Fig. 8 shall be used for transverse tension tests. Specimens for transverse tension tests on large welded tube to determine the strength of welds shall be located perpendicular to the welded seams, with the welds at about the middle of their lengths.

6.10 *Specimens for Forgings*—For testing forgings, the largest round specimen described in 6.4 shall be used. If round specimens are not feasible, then the largest specimen described in 6.5 shall be used.

6.11 *Specimens for Castings*—In testing castings either the standard specimen shown in Fig. 8 or the specimen shown in Fig. 15 shall be used unless otherwise provided in the product specifications.

428

NOTE 9—Test coupons for castings shall be made as shown in Fig. 3 of Methods and Definitions A 370.

6.12 *Specimen for Malleable Iron*—For testing malleable iron the test specimen shown in Fig. 16 shall be used, unless otherwise provided in the product specifications.

6.13 *Specimen for Die Castings*—For testing die castings the test specimen shown in Fig. 17 shall be used unless otherwise provided in the product specifications.

6.14 *Specimens for Powdered Metals*—For testing powdered metals the test specimens shown in Figs. 18 and 19 shall be used, unless otherwise provided in the product specifications.

6.15 *Gage Length of Test Specimens:*

6.15.1 The gage length for the determination of elongation shall be in accordance with the product specifications for the material being tested. Gage marks shall be stamped lightly with a punch, scribed lightly with dividers or drawn with ink as preferred. For material that is sensitive to the effect of slight notches and for small specimens, the use of layout ink will aid in locating the original gage marks after fracture.

6.15.2 Extensometers with gage lengths equal to or shorter than the nominal gage length (dimension shown as "*G*-Gage Length" in the accompanying figures) of the specimen may be used to determine the yield phenomenon.

6.16 *Location of Test Specimens:*

6.16.1 Unless otherwise specified, the axis of the test specimen shall be located as follows:

6.16.1.1 At the center for products 1½ in. or less in thickness, diameter, or distance between flats.

6.16.1.2 Midway from the center to the surface for products over 1½ in. in thickness, diameter, or distance between flats.

6.16.2 For forgings, specimens shall be taken as provided in the applicable product specifications, either from the predominant or thickest part of the forging from which a coupon can be obtained, or from a prolongation of the forging, or from separately forged coupons representative of the forging. When not otherwise specified, the axis of the specimen shall be parallel to the direction of grain flow.

6.17 *Surface Finish of Specimens*—When materials are tested with surface conditions other than as manufactured, the surface finish of the test specimens shall be as provided in the applicable product specifications.

NOTE 10—Particular attention should be given to the uniformity and quality of surface finishes of specimens for high strength and very low ductility materials since this has been shown to be a factor in the variability of test results.

7. Procedures

7.1 *Measurement of Dimensions of Test Specimens:*

7.1.1 To determine the cross-sectional area of a tension test specimen, measure the dimensions of the cross section at the center of the reduced section except that for referee testing of specimens under ³⁄₁₆ in. in their least dimension, measure the dimensions where the least cross-sectional area is found. Measure and record the cross-sectional dimensions of tension test specimens 0.200 in. and over to the nearest 0.001 in.; the cross-sectional dimensions from 0.100 in. but less than 0.200 in., to the nearest 0.0005 in.; the cross-sectional dimensions from 0.020 in. but less than 0.100 in., to the nearest 0.0001 in.; and when practical, the cross-sectional dimensions less than 0.020 in., to at least the nearest 1 % but in all cases to at least the nearest 0.0001 in.

NOTE 11—Specimens generally have smooth surface finishes. Rough surfaces due to the manufacturing process such as hot rolling, metallic coating, etc., may lead to inaccuracy of the computed areas greater than the measured dimensions would indicate. Therefore, cross-sectional dimensions of test specimens with rough surfaces due to processing may be measured and recorded to the nearest 0.001 in.

7.1.2 Determine cross-sectional areas of full-size test specimens of nonsymmetrical cross sections by weighing a length not less than 20 times the largest cross-sectional dimension and using the value of density of the material. Determine the weight to the nearest 0.5 % or less.

7.1.3 When using specimens of the type shown in Fig. 13 taken from tubes, the cross-sectional area shall be determined as follows:

If $D/W \leq 6$:

$$A = [(W/4) \times (D^2 - W^2)^{1/2}] + [(D^2/4)$$
$$\times \arcsin(W/D)] - [(W/4) \times ((D - 2T)\,2 - W^2)^{1/2}]$$
$$- [((D - 2T)/2)^2 \times \arcsin(W/(D - 2T))]$$

where:

A = exact cross-sectional area, in.²,

W = width of the specimen in the reduced section, in.,

D = measured outside diameter of the tube, in., and

T = measured wall thickness of the specimen,

in. arcsin values to be in radians.

If D/W > 6, the exact equation or the following equation may be used:

$$A = W \times T$$

where:

A = approximate cross-sectional area, in.^{2},

W = width of the specimen in the reduced section, in., and

T = measured wall thickness of the specimen, in.

7.2 *Speed of Testing:*

7.2.1 Speed of testing may be defined (*a*) in terms of free-running crosshead speed (rate of movement of the crosshead of the testing machine when not under load), (*b*) in terms of rate of separation of the two heads of the testing machine during a test, (*c*) in terms of the elapsed time for completing part or all of the test (*d*) in terms of rate of stressing the specimen, or (*e*) in terms of rate of straining the specimen.

NOTE 12—For some materials, the free-running crosshead speed, which is the least accurate, may be adequate, while for other materials one of the remaining methods, listed in increasing order of precision, may be necessary in order to obtain test values within acceptable limits.

7.2.1.1 Specifying suitable numerical limits for speed and selection of the method are the responsibilities of the product committees. Suitable limits for speed of testing should be specified for materials for which the differences resulting from the use of different speeds are of such magnitude that the test results are unsatisfactory for determining the acceptability of the material. In such instances, depending upon the material and the use for which the test results are intended, one or more of the methods described in the following paragraphs is recommended for specifying speed of testing.

NOTE 13—Speed of testing can affect test values because of the rate sensitivity of materials and the temperature-time effects.

7.2.2 *Free-Running Crosshead Speed*—The

allowable limits for the rate of movement of the crosshead of the testing machine, when not under load, shall be specified in inches per inch of length of reduced section (or distance between grips for specimens not having reduced sections) per minute. The limits for the crosshead speed may be further qualified by specifying different limits for various types and sizes of specimens. The average crosshead speed can be experimen-

tally determined by using suitable length measuring and timing devices.

7.2.3 *Rate of Separation of Heads During Tests*—The allowable limits for rate of separation of the heads of the testing machine during a test shall be specified in inches per inch of length of reduced section (or distance between grips for specimens not having reduced sections) per minute. The limits for the rate of separation may be further qualified by specifying different limits for various types and sizes of specimens. Many testing machines are equipped with pacing or indicating devices for the measurement and control of the rate of separation of the heads of the machine during a test, but in the absence of such a device the average rate of separation of the heads can be experimentally determined by using suitable length-measuring and timing devices.

7.2.4 *Elapsed Time*—The allowable limits for the elapsed time from the beginning of loading (or from some specified stress) to the instant of fracture, to the maximum load, or to some other stated stress, shall be specified in minutes or seconds. The elapsed time can be determined with a timing device.

7.2.5 *Rate of Stressing*—The allowable limits for rate of stressing shall be specified in pounds per square inch per minute. Many testing machines are equipped with pacing or indicating devices for the measurement and control of the rate of stressing, but in the absence of such a device the average rate of stressing can be determined with a timing device by observing the time required to apply a known increment of stress.

7.2.6 *Rate of Straining*—The allowable limits for rate of straining shall be specified in inches per inch per minute. Some testing machines are equipped with pacing or indicating devices for the measurement and control of rate of straining, but in the absence of such a device the average rate of straining can be determined with a timing device by observing the time required to effect a known increment of strain.

7.2.7 Unless otherwise specified, any convenient speed of testing may be used up to one half the specified yield strength or yield point, or up to one quarter the specified tensile strength, whichever is smaller. The speed above this point shall be within the limits specified. If different speed limitations are required for use in determining yield strength, yield point, tensile strength, elongation, and reduction of area, they should be stated in the product specifications. In the absence of any specified limitations on speed of testing the following general rules shall apply:

7.2.7.1 The speed of testing shall be such that the loads and strains used in obtaining the test results are accurately indicated.

7.2.7.2 During the conduct of the test to determine yield strength or yield point the rate of stress application shall not exceed 100 000 psi/min.

7.2.7.3 After the yield strength or yield point has been determined, the speed may be increased to a maximum of 0.5 in./in. of gage length per minute. The extensometer and strain rate indicator may be used to set the strain rate prior to its removal. If the extensometer and strain rate indicator are not used to set this strain rate, the speed should be set not to exceed 0.5 in./in. of the length of the reduced section (or distance between the grips for specimens not having reduced sections) per minute.

7.3 *Determination of Yield Strength*—Determine yield strength by either of the methods described in 7.3.1 or 7.3.2.

7.3.1 *Offset Method*—To determine the yield strength by the "offset method," it is necessary to secure data (autographic or numerical) from which a stress-strain diagram may be drawn. Then on the stress-strain diagram (Fig. 20) lay off *Om* equal to the specified value of the "offset", draw *mn* parallel to *OA*, and thus locate *r*, the intersection of *mn* with the stress-strain diagram (Note 14). In reporting values of yield strength obtained by this method, the specified value of offset used should be stated in parentheses after the term yield strength. Thus:

Yield strength (offset = 0.2 %)   = 52 000 psi

In using this method a Class B2 extensometer (see Method E 83) would be sufficiently sensitive for most materials.

7.3.2 *Extension-Under-Load Method*—To determine the yield strength by the extension-under-load (EUL) method, it is necessary to secure data (*1*) by autographic or numerical devices so that a stress-strain diagram may be drawn from which the value of the stress occurring at the specified value of extension may be ascertained, or (*2*) by a device attached to or part of an extensometer that indicates when the specific extension occurs so that the load then occurring may be ascertained; the device attached to the extensometer or the load-indicating device, or both, may be automatic. This method is illus-

trated in Fig. 22. The extension can be satisfactorily determined by the use of a Class B2 extensometer. The stress that occurs at the specified extension shall be reported thus:

Yield strength (EUL = XX) = YY psi

where XX is the specified value of extension.

NOTE 14—If the load drops before the specified offset or extension-under-load is reached, technically the material does not have a yield strength (for that offset or extension-under-load), but the stress at the maximum load attained before the specified offset or extension-under-load is reached may be reported instead of the yield strength.

NOTE 15—When there is disagreement over the results of this test, the offset method for determining yield strength is recommended as the referee method.

7.4 *Determination of Yield Point:*

7.4.1 For material having a "sharp-kneed" stress-strain diagram, determine the yield point by one of the methods described in 7.4.1.1 or 7.4.1.2.

7.4.1.1 *Halt-of-the-Load Method*—Apply an increasing load to the specimen at a uniform deformation rate. When the yield point of the material is reached, the increase of the load stops. At that time, there is a halt or hesitation of the load-indicating mechanism. When the increase in load stops or hesitates, record the corresponding stress as the yield point.

NOTE 16—This method was formerly known as the halt-of-the-pointer method and also as the drop-of-the-beam method.

7.4.1.2 *Autographic Diagram Method*—Obtain a stress-strain diagram by an autographic recording device, and record the stress corresponding to the top of the knee or the point at which the curve drops as the yield point (see Fig. 21).

7.4.2 When test specimens do not exhibit a well-defined disproportionate deformation that characterizes a yield point as measured by the drop-of-the-beam, halt-of-the-pointer, or autographic diagram methods described above, a value equivalent to the yield point in its practical significance may be determined by the following methods and may be substituted for the yield point:

7.4.2.1 *Strain Rate Method*—Attach a Class B2 extensometer to the specimen at the gage marks. When the specimen is in place and the extensometer attached, increase the load at a reasonably uniform rate. Watch the elongation of the specimen as shown by the extensometer and note for this determination the load at which the rate of elongation shows a sudden increase.

7.4.2.2 *Extension - Under - Load Method*—Attach a Class C extensometer to the specimen. When the load producing a specified extension is reached, record the stress corresponding to this load as the yield point, and remove the extensometer. This same value may be obtained from an autographic stress-strain diagram (Fig. 22).

NOTE 17—The appropriate value of the total extension should be specified. For steel with yield point specified not over 80 000 psi, an appropriate value is 0.005 in./in. of gage length. For higher strength steels, yield strength should be specified in preference to yield point.

NOTE 18—A suitable device that automatically determines the load at the specified extension without plotting a stress-strain curve may be used if its accuracy has been demonstrated to be satisfactory.

NOTE 19—When no other means of measuring elongation are available, a pair of dividers or similar device can be used to determine a point of detectable elongation between two gage marks on the specimen. The gage length shall be 2 in. The stress corresponding to the load at the instant of detectable elongation can be recorded as "approximate" yield point.

7.5 *Tensile Strength*—Calculate the tensile strength by dividing the maximum load carried by the specimen during a tension test by the original cross-sectional area of the specimen.

7.6 *Elongation:*

7.6.1 In reporting values of elongation, give both the original gage length and the percentage increase.

7.6.2 When the specified elongation is greater than 3 %, fit ends of the fractured specimen together carefully and measure the distance between the gage marks to the nearest 0.01 in. for gage lengths of 2 in. and under, and to at least the nearest 0.5 % of the gage length for gage lengths over 2 in. A percentage scale reading to 0.5 % of the gage length may be used.

7.6.3 When the *specified* elongation is 3 % or less, determine the elongation of standard round specimens (see Fig. 8) using the following procedure, except that the procedure given in 7.6.2 may be used instead when the *measured* elongation is greater than 3 %.

7.6.3.1 Measure the original gage length of the specimen to the nearest 0.002 in.

7.6.3.2 Remove partly torn fragments that will interfere with fitting together the ends of the fractured specimen or with making the final measurement.

7.6.3.3 Fit the fractured ends together with matched surfaces and apply an end load along the axis of the specimen sufficient to close the fractured ends together. If desired, this load may then be removed carefully, provided the specimen remains intact

NOTE 20—The use of an end load of approximately 2000 psi has been found to give satisfactory results on test specimens of aluminum alloy.

7.6.3.4 Measure the final gage length to the nearest 0.002 in. and report the elongation to the nearest 0.2 %.

7.6.4 Specimens other than the standard specimen described in Fig. 8 are exempt from the requirement of 7.6.3 except as required by the applicable product specification.

7.6.5 If any part of the fracture takes place outside of the middle half of the gage length or in a punched or scribed mark within the reduced section, the elongation value obtained may not be representative of the material. In acceptance testing, if the elongation so measured meets the minimum requirements specified, no further testing is required, but if the elongation is less than the minimum requirements, discard the test and retest.

7.6.6 In determining extension at fracture (elastic plus plastic extension), autographic or automated methods using extensometers may be employed.

7.6.6.1 In determining percent elongation from extension at fracture data, only the plastic extension shall be used. The elastic portion can be estimated graphically or by calculation and then subtracted from the total extension at fracture.

7.7 *Reduction of Area:*

7.7.1 *Specimens with originally circular cross sections*—Fit the ends of the fractured specimen together and measure the reduced diameter to the same accuracy as the original measurement.

NOTE 21—Because of anisotropy, circular cross sections often do not remain circular during straining in tension. The shape is usually elliptical, thus, the area may be calculated by $\pi \cdot d_1 \cdot d_2/4$, where d_1 and d_2 are the major and minor diameters, respectively.

7.7.2 *Specimens with original rectangular cross sections*—Fit the ends of the fractured specimen together and measure the thickness and width at the minimum cross section to the same accuracy as the original measurements.

NOTE 22—Because of the constraint to deformation that occurs at the corners of rectangular specimens, the dimensions at the center of the original flat surfaces are less than those at the corners. The shape of these surfaces are often assumed to be parabolic. When this assumption is made, an effective thickness, t_e, may be calculated as follows: $(t_1 + 4t_2 + t_3)/6$, where t_1 and t_3 are the thicknesses at the corners, and t_2 is the thickness at mid-width. An effective width may be similarly calculated.

7.7.3 Calculate the reduced area based upon the dimensions determined in 7.7.1 or 7.7.2. The difference between the area thus found and the area of the original cross section expressed as a percentage of the original area, is the reduction of area.

7.7.4 If any part of the fracture takes place outside the middle half of the reduced section or in a punched or scribed gage mark within the reduced section, the reduction of area value obtained may not be representative of the material. In acceptance testing, if the reduction of area so calculated meets the minimum requirements specified, no further testing is required, but if the reduction of area is less than the minimum requirements, discard the test results and retest.

7.7.5 Results of measurements of reduction of area shall be rounded using the procedures of Recommended Practice E 29 and any specific procedures in the product specifications. In the absence of a specified procedure, it is recommended that reduction of area test values in the range from 0 to 10 % be rounded to the nearest 0.5 % and test values of 10 % and greater to the nearest 1 %.

7.8 *Rounding reported tests data for yield strength, yield point, and tensile strength*—Test data should be rounded using the procedures of Recommened Practice E 29 and the specific procedures in the product specifications. In the absence of a specified procedure for rounding the test data, one of the procedures described in the following paragraphs is recommended.

7.8.1 For test values up to 50 000 psi, round to the nearest 100 psi; for test values of 50 000 psi and up to 100 000 psi, round to the nearest 500 psi; for test values of 100 000 psi and greater, round to the nearest 1000 psi.

NOTE 23—For steel products, see Methods and Definitions A 370.

7.8.1.1 For test values up to 500 MPa, round to the nearest 1 MPa; for test values of 500 MPa and up to 1000 MPa, round to the nearest 5 MPa; for test values of 1000 MPa and greater, round to the nearest 10 MPa (see Note 21).

7.8.2 For all test values, round to the nearest 100 psi.

NOTE 24—For aluminum- and magnesium-alloy products, see Methods B 557.

7.8.2.1 For all test values, round to the nearest 0.5 MPa (see Note 22).

7.8.3 For all test values, round to the nearest 500 psi.

7.8.4 For all test values, round to the nearest 5 MPa.

7.9 *Replacement of Specimens*—A test specimen may be discarded and a replacement specimen taken from the same sample remnant, if possible, in the following cases:

7.9.1 The original specimen had a poorly machined surface,

7.9.2 The original specimen had the wrong dimensions,

7.9.3 The specimen's properties were changed because of poor machining practice,

7.9.4 The test procedure was incorrect,

7.9.5 The fracture was outside the gage length,

7.9.6 For elongation determinations, the fracture was outside the middle half of the gage length, or

7.9.7 There was a malfunction of the testing equipment.

NOTE 25—The tension specimen is inappropriate for assessing some types of imperfections in the material. Other methods and specimens employing ultrasonics, dye penetrants, radiography, etc., may be considered when flaws such as cracks, flakes, porosity, etc., are revealed during a test and soundness is a condition of acceptance.

8. Report

8.1 Test information on materials not covered by a product specification should be reported in accordance with 9.2 or 9.3, or both.

8.2 Test information to be reported shall include the following when applicable:

8.2.1 Material and specimen identification.

8.2.2 Yield strength and the method used to determine yield strength (see 7.3).

8.2.3 Yield point and the method used to determine yield point (see 7.4).

8.2.4 Tensile strength (see 7.5).

8.2.5 Elongation (report both the original gage length and the percentage increase) (see 7.6.)

8.2.6 Reduction of area (see 7.7).

8.3 Test information to be available on request shall include:

8.3.1 Specimen type (Section 6).

8.3.2 Specimen test section dimension(s).

8.3.3 Equation used to calculate cross-sectional area of rectangular specimens taken from large-diameter tubular products.

8.3.4 Speed and method used to determine speed of testing (see 7.2).

8.3.5 Method used for rounding of test results (see 7.8).

8.3.6 Reasons for replacement specimens (see 7.9).

9. Precision and Bias[6]

9.1 *Precision*—The precision of these methods is being established.[7]

9.2 *Bias*—The bias of these methods includes quantitative estimates of the uncertainties of the dimensional measuring devices, the calibrations of test equipment, and the skill of the operators. At this time, statements on bias should be limited to the documented performance of particular laboratories.

---

[6] Supporting data available on loan from ASTM Headquarters. Request RR:E-28-1004.

[7] Test results that might allow statistical evaluation for this statement are herewith solicited.

Wedge Grips
Upper Head of Testing Machine
Liners-Thickness Varied According to Specimen Thickness to Keep Wedge Grips from Protruding Above or Below Head of Testing Machine
Flat Specimen

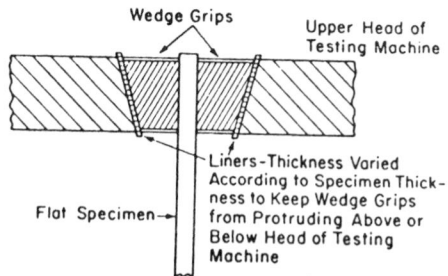

FIG. 1 Wedge Grips with Liners for Flat Specimens

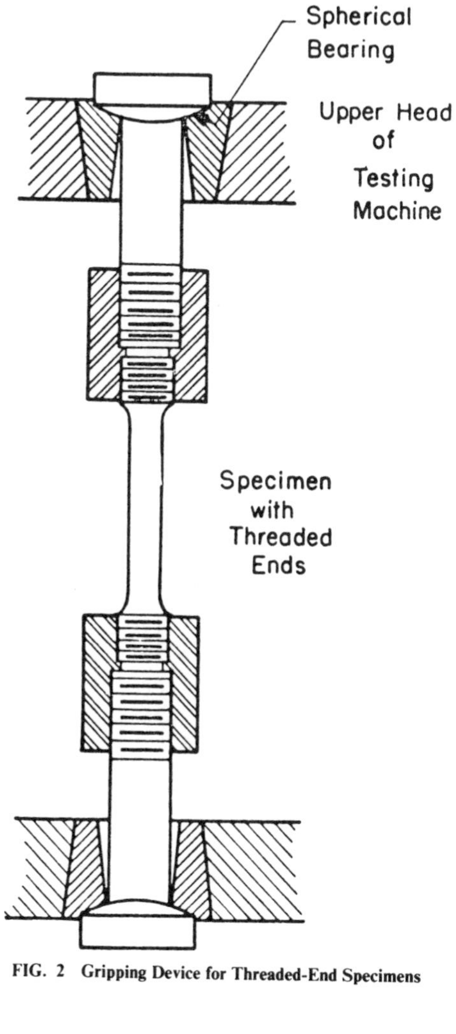

Spherical Bearing
Upper Head of Testing Machine
Specimen with Threaded Ends

FIG. 2 Gripping Device for Threaded-End Specimens

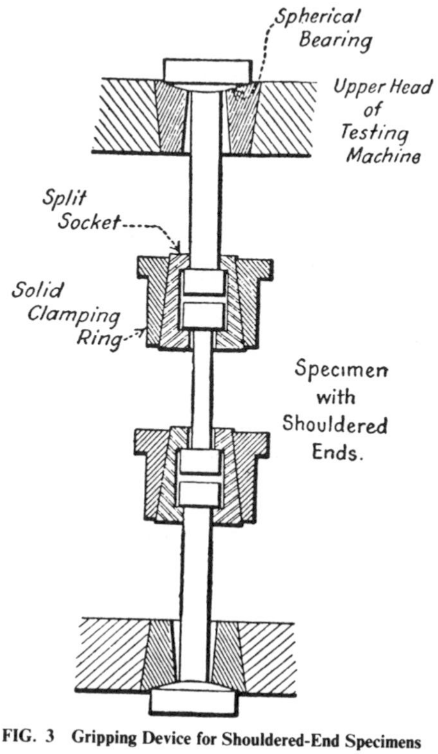

Spherical Bearing
Upper Head of Testing Machine
Split Socket
Solid Clamping Ring
Specimen with Shouldered Ends.

FIG. 3 Gripping Device for Shouldered-End Specimens

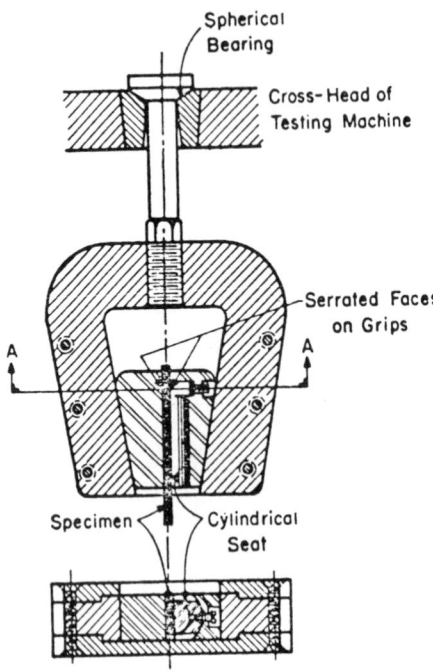

Spherical
Bearing

Cross-Head of
Testing Machine

Serrated Faces
on Grips

A ———— A

Specimen   Cylindrical
Seat

Section A-A—for Sheet and Strip

Specimen

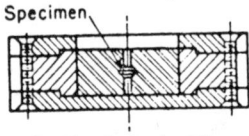

Section A-A— for Wire

FIG. 4   Gripping Devices for Sheet and Wire Specimens

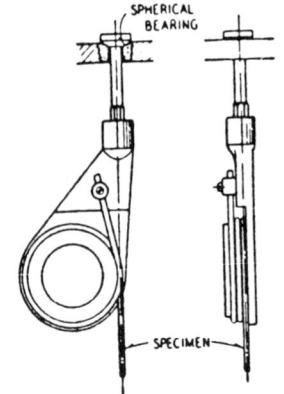

SPHERICAL
BEARING

SPECIMEN

FIG. 5   Snubbing Device for Testing Wire

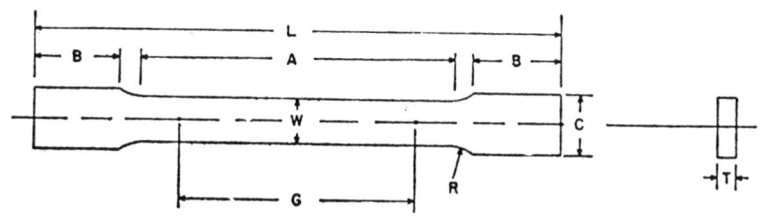

Dimensions

| | Standard Specimens | | Subsize Specimen |
	Plate-Type, 1½-in. Wide	Sheet-Type, ½-in. Wide	¼-in. Wide
	in.	in.	in.
G—Gage length (Notes 1 and 2)	8.00 ± 0.01	2.000 ± 0.005	1.000 ± 0.003
W—Width (Notes 3 and 4)	1½ + ⅛ −¼	0.500 ± 0.010	0.250 ± 0.005
T—Thickness (Note 5)		thickness of material	
R—Radius of fillet, min (Note 6)	1	½	¼
L—Over-all length, min (Notes 2 and 7)	18	8	4
A—Length of reduced section, min	9	2¼	1¼
B—Length of grip section, min (Note 8)	3	2	1¼
C—Width of grip section, approximate (Notes 4 and 9)	2	¾	⅜

NOTE 1—For the 1½-in. wide specimen, punch marks for measuring elongation after fracture shall be made on the flat or on the edge of the specimen and within the reduced section. Either a set of nine or more punch marks 1 in. apart, or one or more pairs of punch marks 8 in. apart may be used.

NOTE 2—When elongation measurements of 1½-in. wide specimens are not required, a minimum length of reduced section (A) of 2¼ in. may be used with all other dimensions similar to the plate-type specimen.

NOTE 3—For the three sizes of specimens, the ends of the reduced section shall not differ in width by more than 0.004, 0.002 or 0.001 in., respectively. Also, there may be a gradual decrease in width from the ends to the center, but the width at each end shall not be more than 0.015, 0.005, or 0.003 in., respectively, larger than the width at the center.

NOTE 4—For each of the three sizes of specimens, narrower widths (W and C) may be used when necessary. In such cases the width of the reduced section should be as large as the width of the material being tested permits; however, unless stated specifically, the requirements for elongation in a product specification shall not apply when these narrower specimens are used.

NOTE 5—The dimension T is the thickness of the test specimen as provided for in the applicable material specifications. Minimum thickness of 1½-in. wide specimens shall be ³⁄₁₆ in. Maximum thickness of ½-in. and ¼-in. wide specimens shall be ⅝ in. and ¼ in., respectively.

NOTE 6—For the 1½-in. wide specimen, a ½-in. minimum radius at the ends of the reduced section is permitted for steel specimens under 100 000 psi in tensile strength when a profile cutter is used to machine the reduced section.

NOTE 7—To aid in obtaining axial loading during testing of ¼-in. wide specimens, the over-all length should be as large as the material will permit, up to 8.00 in.

NOTE 8—It is desirable, if possible, to make the length of the grip section large enough to allow the specimen to extend into the grips a distance equal to two thirds or more of the length of the grips. If the thickness of ½-in. wide specimens is over ⅜ in., longer grips and correspondingly longer grip sections of the specimen may be necessary to prevent failure in the grip section.

NOTE 9—For the three sizes of specimens, the ends of the specimen shall be symmetrical in width with the center line of the reduced section within 0.10, 0.05 and 0.005 in., respectively. However, for referee testing and when required by product specifications, the ends of the ½-in. wide specimen shall be symmetrical within 0.01 in.

NOTE 10—Specimens with sides parallel throughout their length are permitted, except for referee testing, provided: (a) the above tolerances are used; (b) an adequate number of marks are provided for determination of elongation; and (c) when yield strength is determined, a suitable extensometer is used. If the fracture occurs at a distance of less than 2W from the edge of the gripping device, the tensile properties determined may not be representative of the material. In acceptance testing, if the properties meet the minimum requirements specified, no further testing is required, but if they are less than the minimum requirements, discard the test and retest.

FIG. 6 Rectangular Tension Test Specimens

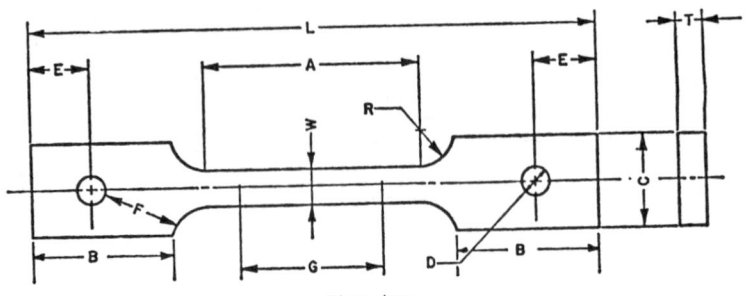

Dimensions

	in.
G—Gage length	2.000 ± 0.005
W—Width (Note 1)	0.500 ± 0.010
T—Thickness, max (Note 2)	⅝
R—Radius of fillet, min (Note 3)	½
L—Over-all length, min	8
A—Length of reduced section, min	2¼
B—Length of grip section, min	2
C—Width of grip section, approximate	2
D—Diameter of hole for pin, min (Note 4)	½
E—Edge distance from pin, approximate	1½
F—Distance from hole to fillet, min	½

NOTE 1—The ends of the reduced section shall differ in width by not more than 0.002 in. There may be a gradual taper in width from the ends to the center, but the width at each end shall be not more than 0.005 in. greater than the width at the center.

NOTE 2—The dimension T is the thickness of the test specimen as stated in the applicable product specifications.

NOTE 3—For some materials, a fillet radius R larger than ½ in. may be needed.

NOTE 4—Holes must be on center line of reduced section, within ±0.002 in.

NOTE 5—Variations of dimensions C, D, E, F, and L may be used that will permit failure within the gage length.

FIG. 7 Pin-Loaded Tension Test Specimen with 2-in. Gage Length

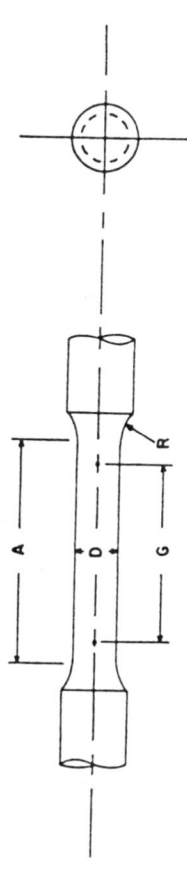

Dimensions

	Standard Specimen	Small-Size Specimens Proportional to Standard			
	in.	in.	in.	in.	in.
Nominal Diameter	0.500	0.350	0.250	0.160	0.113
G—Gage length	2.000 ± 0.005	1.400 ± 0.005	1.000 ± 0.005	0.640 ± 0.005	0.450 ± 0.005
D—Diameter (Note 1)	0.500 ± 0.010	0.350 ± 0.007	0.250 ± 0.005	0.160 ± 0.003	0.113 ± 0.002
R—Radius of fillet, min	³⁄₈	¼	³⁄₁₆	⁵⁄₃₂	³⁄₃₂
A—Length of reduced section, min (Note 2)	2¼	1¾	1¼	¾	⅝

Note 1—The reduced section may have a gradual taper from the ends toward the center, with the ends not more than 1 % larger in diameter than the center (controlling dimension).

Note 2—If desired, the length of the reduced section may be increased to accommodate an extensometer of any convenient gage length. Reference marks for the measurement of elongation should, nevertheless, be spaced at the indicated gage length.

Note 3—The gage length and fillets may be as shown, but the ends may be of any form to fit the holders of the testing machine in such a way that the load shall be axial (see Fig. 9). If the ends are to be held in wedge grips it is desirable, if possible, to make the length of the grip section great enough to allow the specimen to extend into the grips a distance equal to two thirds or more of the length of the grips.

Note 4—On the round specimens in Figs. 8 and 9, the gage lengths are equal to four times the nominal diameter. In some product specifications other specimens may be provided for, but unless the 4-to-1 ratio is maintained within dimensional tolerances, the elongation values may not be comparable with those obtained from the standard test specimen.

Note 5—The use of specimens smaller than 0.250-in. diameter shall be restricted to cases when the material to be tested is of insufficient size to obtain larger specimens or when all parties agree to their use for acceptance testing. Similar specimens require suitable equipment and greater skill in both machining and testing.

Note 6—Five sizes of specimens often used have diameters of approximately 0.505, 0.357, 0.252, 0.160, and 0.113 in., the reason being to permit easy calculations of stress from loads, since the corresponding cross-sectional areas are equal or close to 0.200, 0.100, 0.0500, 0.0200, and 0.0100 in.², respectively. Thus, when the actual diameters agree with these values, the stresses (or strengths) may be computed using the simple multiplying factors 5, 10, 20, 50, and 100, respectively. (The metric equivalents of these five diameters do not result in correspondingly convenient cross-sectional areas and multiplying factors.)

FIG. 8 Standard 0.500-in. Round Tension Test Specimen with 2-in. Gage Length and Examples of Small-Size Specimens Proportional to the Standard Specimen

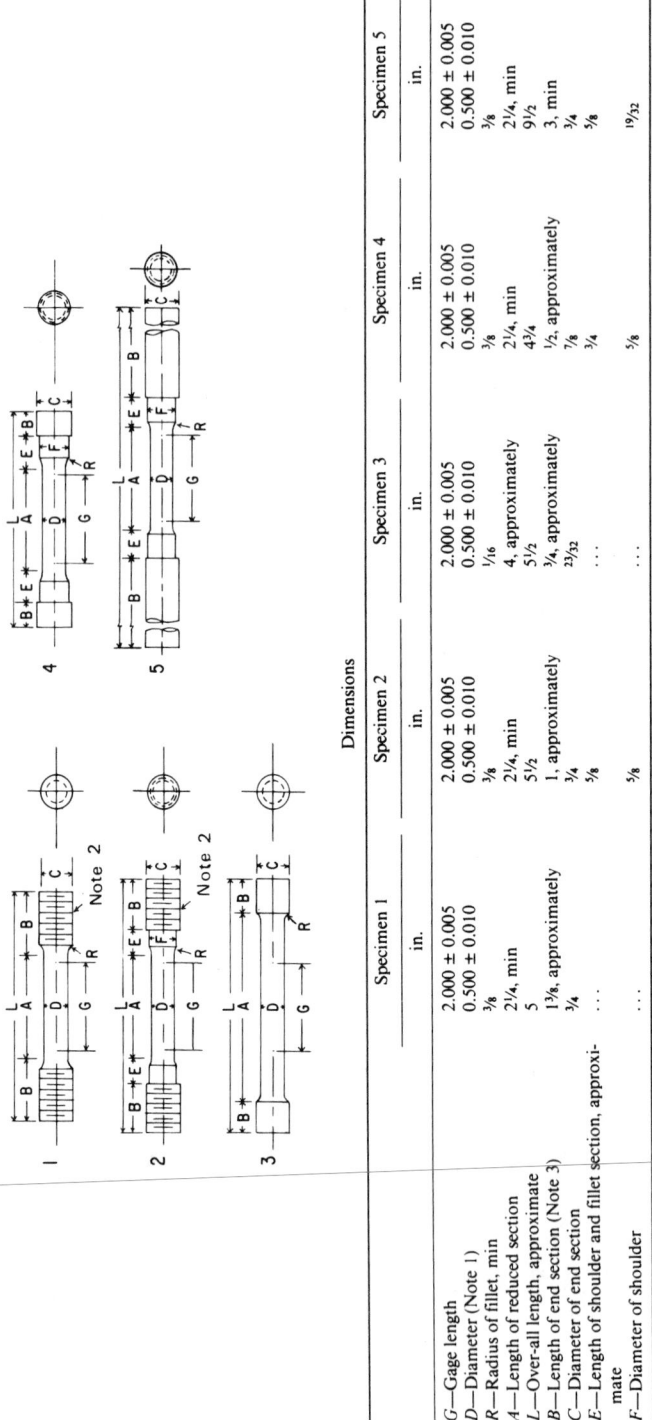

Dimensions

	Specimen 1	Specimen 2	Specimen 3	Specimen 4	Specimen 5
	in.	in.	in.	in.	in.
G—Gage length	2.000 ± 0.005	2.000 ± 0.005	2.000 ± 0.005	2.000 ± 0.005	2.000 ± 0.005
D—Diameter (Note 1)	0.500 ± 0.010	0.500 ± 0.010	0.500 ± 0.010	0.500 ± 0.010	0.500 ± 0.010
R—Radius of fillet, min	⅜	⅜	1/16	⅜	⅜
A—Length of reduced section	2¼, min	2¼, min	4, approximately	2¼, min	2¼, min
L—Over-all length, approximate	5	5½	5½	4¾	9½
B—Length of end section (Note 3)	1⅜, approximately	1, approximately	¾, approximately	½, approximately	3, min
C—Diameter of end section	¾	¾	23/32	⅞	¾
E—Length of shoulder and fillet section, approximate	...	⅝	...	¾	⅝
F—Diameter of shoulder	...	⅝	...	⅝	19/32

FIG. 9   Various Types of Ends for Standard Round Tension Test Specimens

NOTE 1—The reduced section may have a gradual taper from the ends toward the center with the ends not more than 0.005 in. larger in diameter than the center.

NOTE 2—On Specimens 1 and 2, any standard thread is permissible that provides for proper alignment and aids in assuring that the specimen will break within the reduced section.

NOTE 3—On Specimen 5 it is desirable, if possible, to make the length of the grip section great enough to allow the specimen to extend into the grips a distance equal to two thirds or more of the length of the grips.

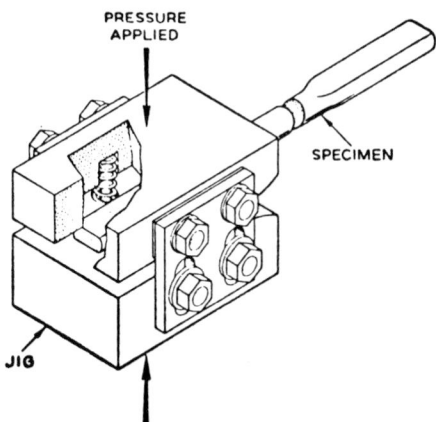

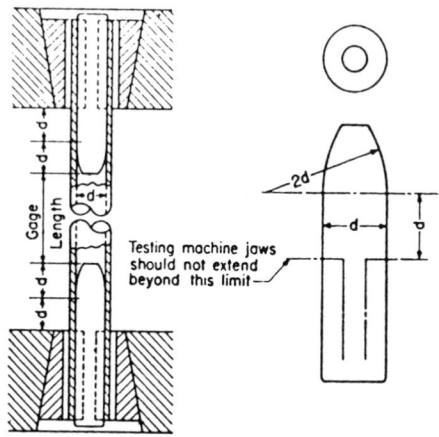

FIG. 10   Squeezing Jig for Flattening Ends of Full-Size
Tension Test Specimens

NOTE—The diameter of the plug shall have a slight taper
from the line limiting the test machine jaws to the curved
section.

FIG. 11   Metal Plugs for Testing Tubular Specimens, Proper
Location of Plugs in Specimen and of Specimen in Heads of
Testing Machine

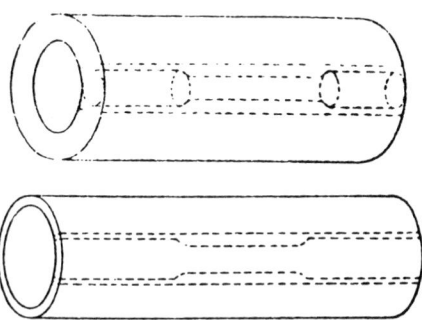

NOTE—The edges of the blank for the specimen shall be cut
parallel to each other.

FIG. 12   Location from Which Longitudinal Tension Test
Specimens Are to be Cut from Large-Diameter Tube

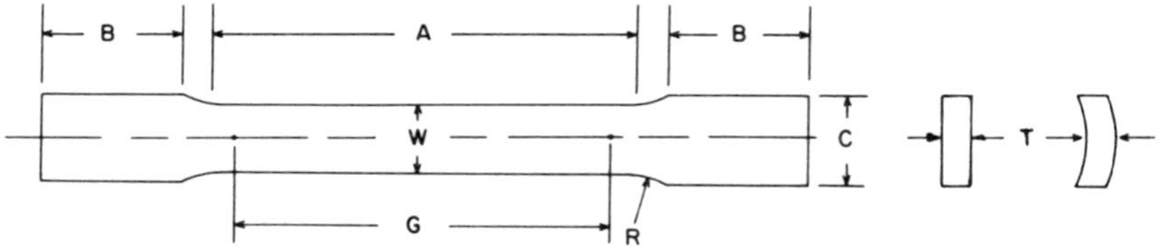

Dimensions

	Specimen 1	Specimen 2	Specimen 3	Specimen 4	Specimen 5	Specimen 6	Specimen 7
	in.	in.	in.	in.	in.	in.	in.
G—Gage length	2.000±0.005	2.000±0.005	8.00±0.01	2.000±0.005	4.000±0.005	2.000±0.005	4.000±0.005
W—Width (Note 1)	0.500±0.010	1½ +⅛ −¼	1½ +⅛ −¼	0.750±0.031	0.750±0.031	1.000±0.062	1.000±0.062
			Measured Thickness of Specimen				
T—Thickness (Note 2)							
R—Radius of fillet, min	½	1	1	1	1	1	1
A—Length of reduced section, min	2¼	2¼	9	2¼	4½	2¼	4½
B—Length of grip section, min (Note 3)	3	3	3	3	3	3	3
C—Width of grip section, approximate (Note 4)	11/16	2	2	1	1	1½	1½

FIG. 13 Tension Test Specimens for Large-Diameter Tubular Products

NOTE 1—The ends of the reduced section shall differ in width by not more than 0.002 in. for Specimen 1, and not more than 0.005 in. for Specimens 2 and 3. There may be a gradual taper in width from the ends to the center, but the width at each end shall be not more than 0.005 in. greater than the width at the center for 2-in. gage length specimens and not more than 0.015 in. greater for 8-in. gage length specimens.

NOTE 2—It is desirable, if possible, to make the length of the grip section great enough to allow the specimen to extend into the grips a distance equal to two thirds or more of the length of the grips.

NOTE 3—The ends of the specimen shall be symmetrical with the center line of the reduced section within 0.05 in. for Specimen No. 1 and 0.10 in. for Specimen Nos. 2 and 3.

NOTE 4—Specimens with sides parallel throughout their length are permitted, except for referee testing, provided: (a) the above tolerances are used; (b) an adequate number of marks are provided for determination of elongation; and (c) when yield strength is determined, a suitable extensometer is used. If the fracture occurs at a distance of less than 2W from the edge of the gripping device, the tensile properties determined may not be representative of the material. If the properties meet the minimum requirements specified, no further testing is required, but if they are less than the minimum requirements, discard the test and retest.

NOTE 5—Specimens with G/W less than 4 should not be used for determination of elongation.

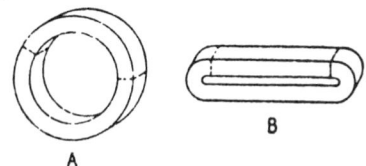

FIG. 14 Location of Transverse Tension Test Specimen in Ring Cut from Tubular Products

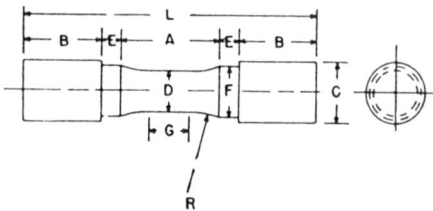

Dimensions

	Specimen 1	Specimen 2	Specimen 3
	in.	in.	in.
G—Length of parallel section			
D—Diameter	0.500 ± 0.010	0.750 ± 0.015	1.25 ± 0.02
R—Radius of fillet, min	1	1	2
A—Length of reduced section, min	1¼	1½	2¼
L—Over-all length, min	3¾	4	6⅜
B—Length of end section, approximate	1	1	1¾
C—Diameter of end section, approximate	¾	1⅛	1⅞
E—Length of shoulder, min	¼	¼	5⁄16
F—Diameter of shoulder	⅝ ± 1⁄64	15⁄16 ± 1⁄64	1⁷⁄16 ± 1⁄64

NOTE—The reduced section and shoulders (dimensions *A*, *D*, *E*, *F*, *G*, and *R*) shall be as shown, but the ends may be of any form of fit the holders of the testing machine in such a way that the load can be axial. Commonly the ends are threaded and have the dimensions *B* and *C* given above.

FIG. 15 Standard Tension Test Specimen for Cast Iron

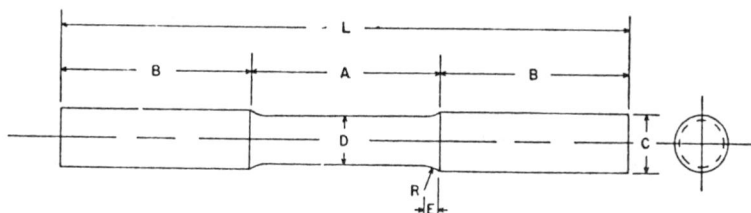

Dimensions

	in.
D—Diameter	⅝
R—Radius of fillet	5⁄16
A—Length of reduced section	2½
L—Over-all length	7½
B—Length of end section	2½
C—Diameter of end section	¾
E—Length of fillet	3⁄16

FIG. 16 Standard Tension Test Specimen for Malleable Iron

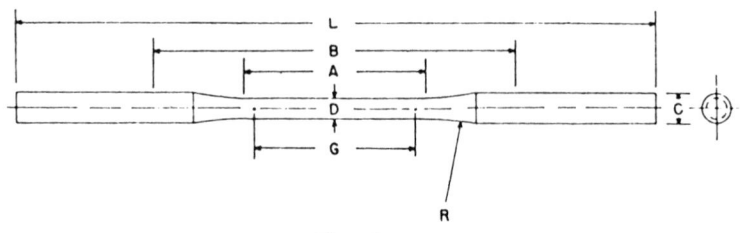

Dimensions

	in.
G—Gage length	2.000 ± 0.005
D—Diameter (see Note)	0.250 ± 0.005
R—Radius of fillet, min	3
A—Length of reduced section, min	2¼
L—Over-all length, min	9
B—Distance between grips, min	4½
C—Diameter of end section, approximate	⅜

NOTE—The reduced section may have a gradual taper from the ends toward the center, with the ends not more than 0.005 in. larger in diameter than the center.

FIG. 17   Standard Tension Test Specimens for Die Castings

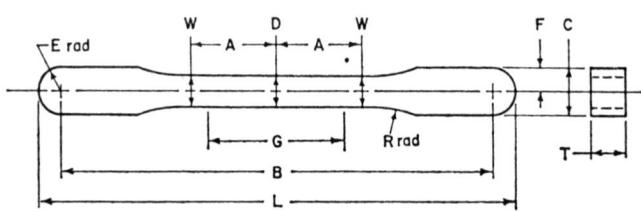

Pressing Area = 1.00 in.²
Dimensions Specified, are Those of the Die

Dimensions

	in.
G—Gage length	1
D—Width at center	0.225 ± 0.001
W—Width at end of reduced section	0.235 ± 0.001
T—Compact to this thickness	0.200 ± 0.250
R—Radius of fillet	1
A—Half-length of reduced section	⅝
B—Grip length	3.187 ± 0.001
L—Over-all length	3.529 ± 0.001
C—Width of grip section	0.343 ± 0.001
F—Half width of grip section	0.1715 ± 0.0010
E—End radius	0.171 ± 0.001

FIG. 18   Standard Flat Unmachined Tension Test Specimens for Powdered Metal Products

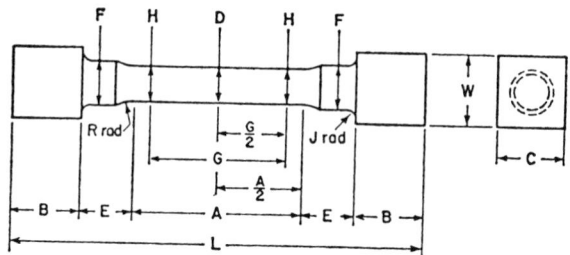

Pressing Area of Unmachined Compact = 1.5 in.²
Machining Recommendations

1. Rough Machine to ⁵⁄₁₆ in. dia
2. Finish Turn 0.250 in. dia with Radii and Taper
3. Polish with 00 Emery Cloth
4. Lap with Crocus Cloth

Dimensions

	in.
G—Gage length	1
D—Diameter at center of reduced section	0.250 ± 0.001
H—Diameter at ends of gage length	0.250 + (0.001/0.002)
R—Radius of fillet	¼
A—Length of reduced section	1¼
L—Over-all length (die cavity length)	3
B—Length of end section	½
C—Compact to this end thickness	0.500 ± 0.050
W—Die cavity width	½
E—Length of shoulder and fillet	⅜
F—Diameter of shoulder	⁵⁄₁₆
J—End fillet radius, max	¹⁄₁₆

NOTE—The gage length and fillets of the specimen shall be as shown. The ends as shown are designed to provide a total pressing area of 1.00 in.² Other end designs are acceptable, and in some cases are required for high-strength sintered materials. Some suggested alternative end designs include:

1. Longer ends, of the same general shape and configuration as the standard, provide more surface area for gripping.

2. Shallow transverse grooves, or ridges, may be pressed in the ends to be gripped by jaws machined to fit the contour of the specimen.

FIG. 19 Standard Round Machined Tension Test Specimen for Powdered Metal Products

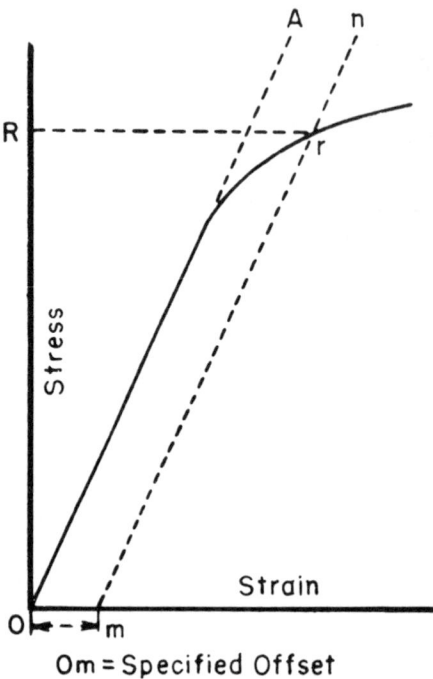

Om = Specified Offset

FIG. 20   Stress-Strain Diagram for Determination of Yield Strength by the Offset Method

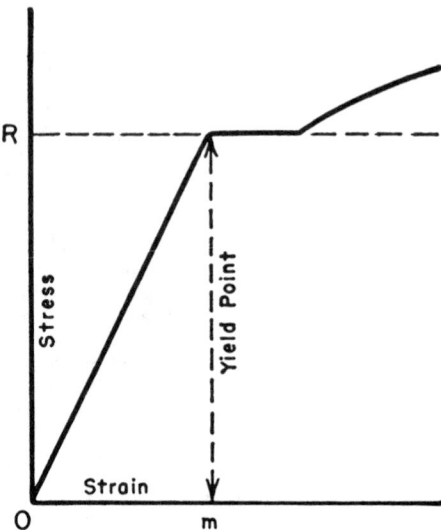

FIG. 21   Stress-Strain Diagram Showing Yield Point Corresponding with Top of Knee

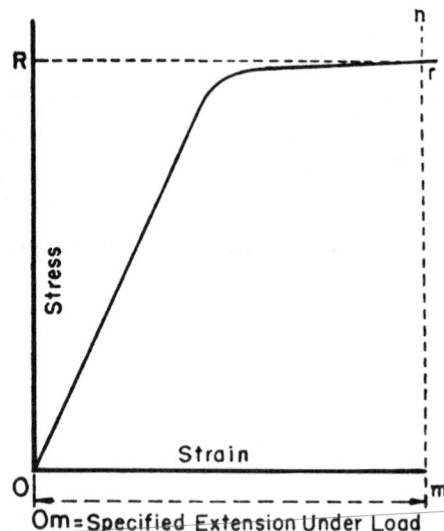

Om = Specified Extension Under Load

FIG. 22   Stress-Strain Diagram for Determination of Yield Strength or Yield Point by the Extension-Under-Load Method

Standard Test Method for
BRINELL HARDNESS OF METALLIC MATERIALS[1]

This standard is issued under the fixed designation E 10; the number immediately following the designation indicates the year of original adoption or, in the case of revision, the year of last revision. A number in parentheses indicates the year of last reapproval. A superscript epsilon (ε) indicates an editorial change since the last revision or reapproval.

This method has been approved for use by agencies of the Department of Defense to replace method 242.1 of Federal Test Method Standard No. 151b and for listing in DoD Index of Specifications and Standards.

1. Scope

1.1 This test method covers the determination of the Brinell hardness of metallic materials, including methods for the verification of Brinell hardness testing machines (Part B) and the calibration of standardized hardness test blocks (Part C). Two general classes of standard tests are recognized:

1.1.1 *Verification, Laboratory, or Referee Tests*, where a high degree of accuracy is required, and

1.1.2 *Routine Tests*, where a somewhat lower degree of accuracy is permissible.

1.2 The values stated in inch-pound units are to be regarded as the standard.

1.3 *This standard may involve hazardous materials, operations, and equipment. This standard does not purport to address all of the safety problems associated with its use. It is the responsibility of whoever uses this standard to consult and establish appropriate safety and health practices and determine the applicability of regulatory limitations prior to use.*

2. Applicable Documents

2.1 *ASTM Standards:*
E 4 Practices for Load Verification of Testing Machines[2]
E 140 Hardness Conversion Tables for Metals (Relationship Between Brinell Hardness, Vickers Hardness, Rockwell Hardness, Rockwell Superficial Hardness, and Knoop Hardness)[2]

3. Definitions

3.1 *Brinell hardness test*—an indentation hardness test using calibrated machines to force a hard ball, under specified conditions, into the surface of the material under test and to measure the diameter of the resulting impression after removal of the load.

3.2 *Brinell hardness number, HB*—a number related to the applied load and to the surface area of the permanent impression made by a ball indenter computed from the equation:

$$HB = 2P/[\pi D(D - \sqrt{D^2 - d^2})]$$

where:
P = applied load, kgf,
D = diameter of the ball, mm, and
d = mean diameter of the impression, mm.

NOTE 1—It is a recognized experimental fact that the Brinell hardness number of nearly all materials is influenced by the magnitude of the indenting load, the diameter of the ball indenter, and the elastic characteristics of the ball. In general, a ball 10 mm in diameter of suitable composition should be used with applied loads 3000, 1500, or 500 kgf, depending upon the hardness of the material to be tested. Although Brinell hardness numbers may vary with the test load used with the 10-mm ball, when smaller balls are used for tests on thin specimens, test results will generally be in agreement with the 10-mm ball test when the ratio of the test load to the square of the ball diameter is held constant (see 3.2 and 6.1).

In Table 1 are given the Brinell hardness numbers corresponding to various diameters of impression for 3000, 1500, and 500-kgf loads, making it unnecessary to calculate for each test the value of the Brinell hardness number by the above equation when these loads are used with a ball 10 mm in diameter.

---

[1] This test method is under the jurisdiction of ASTM Committee E-28 on Mechanical Testing and is the direct responsibility of Subcommittee E 28.06 on Indentation Hardness Testing.

Current edition approved April 27, 1984. Published June 1984. Originally published as E 10 – 24 T. Last previous edition E 10 – 78.

[2] *Annual Book of ASTM Standards*, Vol 03.01.

NOTE 2—The Brinell hardness number followed by the symbol HB without any suffix numbers denotes the following test conditions:

Ball diameter	10 mm
Load	3000 kgf
Duration of loading	10 to 15 s

For other conditions, the hardness number and symbol HB is supplemented by numbers indicating the test conditions in the following order: diameter of ball, load, and duration of loading.

Example—63 HB 10/500/30 indicates a Brinell hardness of 63 measured with a ball of 10-mm diameter and a load of 500 kgf applied for 30 s.

3.3 *verification*—checking or testing to assure conformance with the specification.

3.4 *calibration*—determination of the values of the significant parameters by comparison with values indicated by a reference instrument or by a set of reference standards.

A. GENERAL DESCRIPTION AND TEST PROCEDURE FOR BRINELL HARDNESS TESTS

4. Apparatus

4.1 *Testing Machine*—Equipment for Brinell hardness testing usually consists of a testing machine which supports the test specimen and applies a predetermined indenting load to a ball in contact with the specimen. The magnitude of the indenting load is limited to certain discrete values. The design of the testing machines shall be such that no rocking or lateral movement of the indenter or specimen occurs while the load is being applied. In the use of machines employing a dead-weight system, precautions shall be taken to prevent a momentary overload caused by the inertia of the dead-weight system. The loading system shall be operated with the utmost care as the maximum value of the indenting load is approached to prevent a large acceleration of the dead-weight system.

4.2 *Brinell Balls:*

4.2.1 The standard ball for Brinell hardness testing shall be 10.000 mm in diameter with a deviation from this value of not more than 0.005 mm in any diameter. Smaller balls having the diameters and tolerances indicated in Table 2 may be also used provided the precautions set forth in 7.1 are observed.

4.2.2 A steel ball having a Vickers hardness (HV) of at least 850 (see 4.2.2.1) using a 10-kgf load may be used on material having a HB not over 450, or a carbide ball on material not over 630. The Brinell test is not recommended for material having a HB over 630. The ball shall be polished and free of defects. For laboratory or referee tests the type of ball used shall be specifically reported where Brinell hardness values exceed 200.

4.2.2.1 The maximum value of the mean of the two diagonals of the impression on the ball made by the Vickers indenter shall not exceed the values listed in Table 3.

4.2.3 The ball shall not show a permanent change in diameter greater than 0.005 mm when pressed with the test load against the test specimens. If a ball is used in a test of a specimen which shows a Brinell hardness number greater than the limit for the ball as detailed in 4.2.2, the ball shall be remeasured after the test. Should the ball show a permanent change in diameter greater than that specified above, the results of the test shall be considered unreliable and the ball shall be unsuitable for further use.

4.3 *Measuring Microscope*—The divisions of the micrometer scale of the microscope or other measuring devices used for the measurement of the diameter of the impression shall be such as to permit the direct measuring of the diameter to 0.1 mm and the estimation of the diameter to 0.02 mm.

NOTE 3—This requirement applies to the construction of the microscope only and is not a requirement for measurement of the impression.

5. Test Specimens

5.1 Specimens used in Brinell hardness testing vary greatly in form since it is frequently desirable to make the impression upon a part to be used in the finished product rather than upon a sample test specimen.

5.1.1 *Thickness*—The thickness of the piece tested shall be such that no bulge or other marking showing the effect of the load appears on the side of the piece opposite the impression. In any event the thickness of the specimen shall be at least ten times the depth of the indentation (Table 4). The minimum width shall conform with the requirements of 7.3.

5.1.2 *Finish*—When necessary, the surface on which the impression is to be made shall be filed, ground, machined, or polished with abrasive material so that the edge of the impression shall be clearly enough defined to permit the measurement of the diameter to the specified accuracy (see 8.1). Care should be taken to avoid overheating or cold working the surface.

6. Verification of Apparatus

6.1 *Verification Methods*—The hardness testing machine shall be verified as specified in Part B.

6.1.1 Two acceptable methods of verifying Brinell hardness testing machines are given in Part B.

6.2 *Loading Range*—A Brinell hardness testing machine for laboratory or referee tests is acceptable for use over a loading range within which the machine error does not exceed ±1 %. A Brinell hardness testing machine used for routine testing is acceptable for use over a loading range within which the machine error does not exceed ±2 %.

7. Procedure

7.1 *Magnitude of Test Load*—The load in the standard Brinell test shall be 3000, 1500, or 500 kgf. It is desirable that the test load be of such magnitude that the diameter of the impression be in the range 2.50 to 6.00 mm (25.0 to 60.0 % of ball diameter). A lower limit in impression diameter is necessary because of the reduction in sensitivity of the test with reduction in impression size. The upper limit is influenced by limitations of travel of the indenter in certain types of testers. The thickness and spacing requirements of 5.1.1 and 7.3 may determine the maximum permissible diameter of impression for a specific test. Table 5 gives standard test loads and approximate Brinell hardness numbers for the above range of impression diameters. It is not mandatory that the Brinell test conform to these HB ranges, but it should be realized that different Brinell hardness numbers may be obtained for a given material by using different loads on a 10-mm ball. For the purpose of obtaining a continuous scale of values if may be desirable, however, to use a single load to cover the complete range of hardness for a given class of materials. For softer metals, loads of 250, 125, or 100 kgf are sometimes used. The load used shall be specifically stated in the test report (Note 3). For testing thin or small specimens a ball less than 10 mm in diameter is sometimes used. Such tests (which are not to be regarded as standard tests) will approximate the standard tests more closely if the relation between the applied load, P, measured in kilograms-force, and the diameter of the ball, D, measured in millimetres, is the same as in the standard tests, where:

$P/D^2 = 30$ for 3000-kgf load and 10-mm ball
$P/D^2 = 15$ for 1500-kgf load and 10-mm ball
$P/D^2 = \ \ 5$ for 500-kgf load and 10-mm ball

Example—A 125-kgf test load on a 5-mm diameter ball would approximate a standard 500-kgf test load on a 10-mm diameter ball.

7.1.1 Special tests for soft metals which do not stimulate the standard tests are made with the following load-diameter ratios:

$$P/D^2 = 2.5$$
$$P/D^2 = 1.25$$
$$P/D^2 = 1.0$$

When balls smaller than 10 mm in diameter are used, both the test load and ball size shall be specifically stated in the test report (see 3.2, Note 2, and 10.1). Balls differing in size from the standard 10-mm ball shall conform to the requirements for the material and the permissible variations in diameter specified for the standard ball.

NOTE 4—It is the responsibility of the ASTM committees having jurisdiction over product specifications to specify the load to be used. If no load is specified, it shall be assumed that a load of 3000 kgf is intended.

7.2 *Radius of Curvature*—When indentations are made on a curved surface, the minimum radius of curvature of the surface shall be not less than 1 in. (25 mm) for a 10-mm ball. The diameter of the indentation shall be taken as the mean of the two principal diameters.

7.3 *Spacing of Indentations*—The distance of the center of the indentation from the edge of the specimen or edge of another indentation shall be at least two and one half times the diameter of the indentation.

7.4 *Application of Test Load*—Apply the load to the specimen steadily, without a jerk. In machines employing a dead-weight system take precautions to prevent a momentary overload caused by the inertia of the weights. Also operate the loading system with the utmost care as the maximum value of the indenting load is approached to prevent a large acceleration of the dead-weight system. Apply the full test load for 10 to 15 s except for certain soft metals.

NOTE 5—If for some materials a duration of loading other than 10 to 15 s is necessary, the product specification should specify a definite time of load application. The results of the test should be reported using the nomenclature outlined in Note 2 of this method.[3]

---

[3] See discussion of the Report of Committee E-1, *Proceedings*, Am. Soc. Testing Mats., Vol 24, Part I, 1924, pp. 729 and 730.

7.5 *Alignment*—The angle between the load line and the normal to the specimen shall not exceed 2°.

8. Measurement of Impression

8.1 *Diameter*—In the Brinell hardness test, two diameters of the impression at right angles to each other shall be measured and their mean value used as a basis for calculation of the Brinell hardness number. For routine tests and for tests to determine compliance with a material or product specification, the diameter of the impression shall be read to 0.05 mm.

NOTE 6—These measurements are usually made with a low-magnification portable microscope (approximately 20 ×) having a fixed scale in the eyepiece. If a more accurate determination is needed, as in referee or standardization tests, a laboratory comparator such as a micrometer microscope is required.

9. Conversion to Other Hardness Scales or Tensile Strength Values

9.1 There is no general method for converting accurately Brinell hardness numbers to other hardness scales or tensile strength values. Such conversions are, at best, approximations and therefore should be avoided, except for special cases where a reliable basis for the approximate conversion has been obtained by comparison tests.

NOTE 7—Hardness Conversion Tables E 140, for Metals, give approximate conversion values for specific materials such as steel, austenitic stainless steel, nickel and high-nickel alloys, and cartridge brass.

10. Report

10.1 The report shall include the following information:

10.1.1 The Brinell hardness number,

10.1.2 The test conditions when the Brinell hardness number is determined from loads other than 3000 kgf, ball diameters other than 10 mm, and loading times other than 10 to 15 s (see 3.2, Note 2), and

10.1.3 The type ball used if the Brinell hardness number exceeds 200 (see 4.2.2).

11. Precision and Bias

11.1 An interlaboratory comparison program is now in progress which when completed will be the basis of a statement on precision.

11.2 *Bias*—There is no basis for defining the bias for this test method.

B. VERIFICATION OF BRINELL HARDNESS TESTING MACHINES

12. Scope

12.1 Part B covers two procedures for the verification of Brinell hardness testing machines. These are:

12.1.1 Separate verification of load application, indenter, and the measuring microscope for measuring the diameter of the impression.

12.1.2 Verification of standardized test block method.

12.2 New or rebuilt machines, and machines used for laboratory or referee tests shall be checked by the separate verification method (see 12.1.1).

12.3 Machines used for routine testing may be checked by either verification method.

13. General Requirements

13.1 Before a Brinell hardness testing machine is verified, the machine shall be examined to ensure that:

13.1.1 The machine is properly set up.

13.1.2 The ball holder, with a new steel ball, whose nominal diameter has been checked (see 14.1.2), is firmly mounted in the plunger.

13.1.3 The load can be applied and removed witout shock or vibration and in such a manner that the readings are not influenced.

13.2 If the measuring device is integral with the machine, the machine shall be examined to ensure that:

13.2.1 The change from loading to measuring does not influence the readings.

13.2.2 The method of illumination does not affect the readings.

13.2.3 The center of the impression is in the center of the field of view.

14. Verification

14.1 *Separate Verification of Load Application, Indenter, and Measuring Microscope:*

14.1.1 *Load Application*—Brinell hardness testing machines shall be verified at applied loads of 3000, 1500, and 500 kgf. If tests are to be made using other loads, the machine should be verified also at each of these loads. The applied load shall be checked periodically with a proving ring, by the use of dead weights and proving levers, or by an elastic calibration device or springs in the manner described in Practices E 4[4]. A Brinell hardness testing machine for laboratory or referee tests is acceptable for use over a loading range within which the machine error does not exceed ±1 %. A Brinell hardness testing machine used for routine testing is acceptable for use over a loading range within which the machine error does not exceed ±2 %.

14.1.2 *Indenter*—The indenter to be verified shall be a new steel ball selected at random from a lot meeting the hardness requirements specified in Table 3. The diameter of each ball shall be measured to an accuracy of ±0.0005 mm at not less than three positions and the mean of these readings shall not differ from the nominal diameter by more than the tolerance specified in Table 2.

14.1.3 *Measuring Microscope*—The measuring microscope or other device for measuring the diameter of the impression shall be verified at five intervals over the working range by the use of an accurate scale such as a stage micrometer. The adjustment of the micrometer microscope shall be such that, throughout the range covered, the difference between the scale divisions of the microscope and of the calibrating scale does not exceed 0.01 mm.

14.2 *Verification by Standardized Test Block Method:*

14.2.1 A Brinell hardness testing machine used only for routine testing may be checked by making a series of impressions on standardized hardness test blocks (Part C).

14.2.2 If the machine is to be used at conditions other than 10/3000/15, the machine shall also be verified at those other conditions.

14.2.3 The Brinell hardness testing machine shall be considered verified if the mean diameter of any hardness impression differs by no more than 3 % from the mean diameter corresponding to the hardness value of the standardized hardness test block.

C. CALIBRATION OF STANDARDIZED HARDNESS TEST BLOCKS FOR BRINELL HARDNESS TESTING MACHINES

15. Scope

15.1 Part C covers the calibration of standardized hardness test blocks for the verification of Brinell hardness testing machines as described in Part B.

16. Manufacture

16.1 Each metal block to be calibrated shall be not less than ⅝ in. (16 mm) in thickness for 10-mm balls, ½ in. (12 mm) for 5-mm balls, and ¼ in. (6 mm) for smaller balls.

16.2 Each block shall be specially prepared and heat treated to give the necessary homogeneity and stability of structure.

16.3 Each block, if of steel, shall be demagnetized by the manufacturer and maintained demagnetized by the user.

16.4 The supporting surface of the test block shall have a ground finish.

16.5 The test surface shall be free of scratches which would interfere with measurements of the diameters of the impression.

16.5.1 The mean surface roughness height rating shall not exceed 12 μin. (0.0003 mm) center line average for the standard 10-mm ball. For smaller balls a mean surface roughness height rating of 6 μin. (0.00015 mm) is recommended.

16.6 To ensure that no material is subsequently removed from the test surface of the standardized test block, an official mark or the thickness at the time of calibration shall be stamped or engraved on the test surface to an accuracy of ±0.005 in. (±0.1 mm).

17. Standardizing Procedure

17.1 The standardized blocks shall be calibrated on a Brinell hardness testing machine verified in accordance with the requirements of 14.1.

17.2 The mechanism that controls the application of the load shall ensure that the speed of approach immediately before the ball touches the specimen and the speed of penetration does not exceed 1 mm/s.

17.3 The full load shall be applied for 15 s.

18. Indenter

18.1 A new steel ball conforming to the requirements of 14.1.2 shall be used for calibrating standardized hardness test blocks.

19. Number of Impressions

19.1 At least five randomly distributed impressions shall be made on each test surface of

[4] See also Jones, J. L., and Marshall, C. H., "A New Method for Calibrating Brinell Hardness Testing Machines," *Proceedings,* Am. Soc. Testing Mats., Vol. 20, Part II, 1920, p. 392.

the block. If the area of the test block is greater than 100 cm^2 the number of randomly distributed impressions on the test surface should be increased to eight.

20. Measurement of the Diameters of the Impression

20.1 The illuminating system of the measuring microscope shall be adjusted to give uniform intensity over the field of view and maximum contrast between the impressions and the undisturbed surface of the block.

20.2 The measuring microscope shall be graduated to read 0.002 mm for impressions made with balls of 5-mm diameter or larger and 0.001 mm for impressions made with balls of smaller diameter.

20.3 The measuring microscope shall be checked by a stage micrometer, or by other suitable means to ensure that the difference between readings corresponding to any two divisions of the instrument is correct within ±0.001 mm for balls of less than 5-mm diameter and within ±0.002 mm for balls of larger diameter.

20.4 It is recommended that each impression be measured at least by two observers.

21. Repeatability

21.1 If d_1, d_2 ... d_n are the mean values of the measured diameters as determined by one observer and arranged in increasing order of magnitude the repeatability of the hardness readings, on the block under the particular conditions of standardization, is defined as $d_n - d_1$ where n = 5 or 8.

22. Uniformity of Hardness

22.1 Unless the repeatability of the mean diameters of each of the five or eight impressions is within 2 % of the mean value for the five or eight mean readings, the block cannot be regarded as sufficiently uniform for standardization purposes.

23. Marking

23.1 Each block shall be marked with the following:

23.1.1 The arithmetic mean of the hardness values found in the standardizing test (see also 3.2, Note 2),

23.1.2 The name or mark of the supplier,

23.1.3 The serial number of the block,

23.1.4 The thickness of the block or an official mark on the top surface (see 16.6).

NOTE 8—All of the markings except the official mark or thickness shall be placed on the side of the block, the markings being upright when the test surface is the upper face.

TABLE 1 Brinell Hardness Numbers[A]

(Ball 10 mm in Diameter; Applied Loads of 500, 1500, and 3000 kgf)

NOTE—The values given in this table for Brinell hardness numbers are merely solutions of the equation given in the definition in 2.2, and include values for impression diameters outside the ranges recommended in 6.1. These values are indicated by italics.

Diameter of Indentation, mm	500-kgf Load	1500-kgf Load	3000-kgf Load	Diameter of Indentation, mm	500-kgf Load	1500-kgf Load	3000-kgf Load	Diameter of Indentation, mm	500-kgf Load	1500-kgf Load	3000-kgf Load	Diameter of Indentation, mm	500-kgf Load	1500-kgf Load	3000-kgf Load
2.00	*158*	*473*	*945*	2.30	*119*	*356*	*712*	2.60	92.6	278	555	2.90	74.1	222	444
2.01	*156*	*468*	*936*	2.31	*118*	*353*	*706*	2.61	91.8	276	551	2.91	73.6	221	441
2.02	*154*	*463*	*926*	2.32	*117*	*350*	*700*	2.62	91.1	273	547	2.92	73.0	219	438
2.03	*153*	*459*	*917*	2.33	*116*	*347*	*694*	2.63	90.4	271	543	2.93	72.5	218	435
2.04	*151*	*454*	*908*	2.34	*115*	*344*	*688*	2.64	89.7	269	538	2.94	72.0	216	432
2.05	*150*	*450*	*899*	2.35	*114*	*341*	*682*	2.65	89.0	267	534	2.95	71.5	215	429
2.06	*148*	*445*	*890*	2.36	*113*	*338*	*676*	2.66	88.4	265	530	2.96	71.0	213	426
2.07	*147*	*441*	*882*	2.37	*112*	*335*	*670*	2.67	87.7	263	526	2.97	70.5	212	423
2.08	*146*	*437*	*873*	2.38	*111*	*332*	*665*	2.68	87.0	261	522	2.98	70.1	210	420
2.09	*144*	*432*	*865*	2.39	*110*	*330*	*659*	2.69	86.4	259	518	2.99	69.6	209	417
2.10	*143*	*428*	*856*	2.40	*109*	*327*	*653*	2.70	85.7	257	514	3.00	69.1	207	415
2.11	*141*	*424*	*848*	2.41	*108*	*324*	*648*	2.71	85.1	255	510	3.01	68.6	206	412
2.12	*140*	*420*	*840*	2.42	*107*	*322*	*643*	2.72	84.4	253	507	3.02	68.2	205	409
2.13	*139*	*416*	*832*	2.43	*106*	*319*	*637*	2.73	83.8	251	503	3.03	67.7	203	406
2.14	*137*	*412*	*824*	2.44	*105*	*316*	*632*	2.74	83.2	250	499	3.04	67.3	202	404
2.15	*136*	*408*	*817*	2.45	*104*	*313*	*627*	2.75	82.6	248	495	3.05	66.8	200	401
2.16	*135*	*404*	*809*	2.46	*104*	*311*	*621*	2.76	81.9	246	492	3.06	66.4	199	398
2.17	*134*	*401*	*802*	2.47	*103*	*308*	*616*	2.77	81.3	244	488	3.07	65.9	198	395
2.18	*132*	*397*	*794*	2.48	*102*	*306*	*611*	2.78	80.8	242	485	3.08	65.5	196	393
2.19	*131*	*393*	*787*	2.49	*101*	*303*	*606*	2.79	80.2	240	481	3.09	65.0	195	390
2.20	*130*	*390*	*780*	2.50	*100*	*301*	*601*	2.80	79.6	239	477	3.10	64.6	194	388
2.21	*129*	*386*	*772*	2.51	*99.4*	*298*	*597*	2.81	79.0	237	474	3.11	64.2	193	385
2.22	*128*	*383*	*765*	2.52	*98.6*	*296*	*592*	2.82	78.4	235	471	3.12	63.8	191	383
2.23	*126*	*379*	*758*	2.53	*97.8*	*294*	*587*	2.83	77.9	234	467	3.13	63.3	190	380
2.24	*125*	*376*	*752*	2.54	*97.1*	*291*	*582*	2.84	77.3	232	464	3.14	62.9	189	378
2.25	*124*	*372*	*745*	2.55	*96.3*	*289*	*578*	2.85	76.8	230	461	3.15	62.5	188	375
2.26	*123*	*369*	*738*	2.56	*95.5*	*287*	*573*	2.86	76.2	229	457	3.16	62.1	186	373
2.27	*122*	*366*	*732*	2.57	*94.8*	*284*	*569*	2.87	75.7	227	454	3.17	61.7	185	370
2.28	*121*	*363*	*725*	2.58	*94.0*	*282*	*564*	2.88	75.1	225	451	3.18	61.3	184	368
2.29	*120*	*359*	*719*	2.59	*93.3*	*280*	*560*	2.89	74.6	224	448	3.19	60.9	183	366

TABLE 1 Continued

Diameter of Indentation, mm	Brinell Hardness Number 500-kgf Load	1500-kgf Load	3000-kgf Load
3.20	60.5	182	363
3.21	60.1	180	361
3.22	59.8	179	359
3.23	59.4	178	356
3.24	59.0	177	354
3.25	58.6	176	352
3.26	58.3	175	350
3.27	57.9	174	347
3.28	57.5	173	345
3.29	57.2	172	343
3.30	56.8	170	341
3.31	56.5	169	339
3.32	56.1	168	337
3.33	55.8	167	335
3.34	55.4	166	333
3.35	55.1	165	331
3.36	54.8	164	329
3.37	54.4	163	326
3.38	54.1	162	325
3.39	53.8	161	323
3.40	53.4	160	321
3.41	53.1	159	319
3.42	52.8	158	317
3.43	52.5	157	315
3.44	52.2	156	313
3.45	51.8	156	311
3.46	51.5	155	309
3.47	51.2	154	307
3.48	50.9	153	306
3.49	50.6	152	304
3.50	50.3	151	302
3.51	50.0	150	300
3.52	49.7	149	298
3.53	49.4	148	297
3.54	49.2	147	295
3.55	48.9	147	293
3.56	48.6	146	292
3.57	48.3	145	290
3.58	48.0	144	288
3.59	47.7	143	286
3.60	47.5	142	285
3.61	47.2	142	283
3.62	46.9	141	282
3.63	46.7	140	280
3.64	46.4	139	278
3.65	46.1	138	277
3.66	45.9	138	275
3.67	45.6	137	274
3.68	45.4	136	272
3.69	45.1	135	271
3.70	44.9	135	269
3.71	44.6	134	268
3.72	44.4	133	266
3.73	44.1	132	265
3.74	43.9	132	263
3.75	43.6	131	262
3.76	43.4	130	260
3.77	43.1	129	259
3.78	42.9	129	257
3.79	42.7	128	256
3.80	42.4	127	302
3.81	42.2	127	300
3.82	42.0	126	298
3.83	41.7	125	297
3.84	41.5	125	295
3.85	41.3	124	293
3.86	41.1	123	292
3.87	40.9	123	290
3.88	40.6	122	288
3.89	40.4	121	286
3.90	40.2	121	285
3.91	40.0	120	283
3.92	39.8	119	282
3.93	39.6	119	280
3.94	39.4	118	278
3.95	39.1	117	277
3.96	38.9	117	275
3.97	38.7	116	274
3.98	38.5	116	272
3.99	38.3	115	271
4.00	38.1	114	269
4.01	37.9	114	268
4.02	37.7	113	266
4.03	37.5	113	265
4.04	37.3	112	263
4.05	37.1	111	262
4.06	37.0	111	260
4.07	36.8	110	259
4.08	36.6	110	257
4.09	36.4	109	256
4.10	36.2	109	217
4.11	36.0	108	216
4.12	35.8	108	215
4.13	35.7	107	214
4.14	35.5	106	213
4.15	35.3	106	212
4.16	35.1	105	211
4.17	34.9	105	210
4.18	34.8	104	209
4.19	34.6	104	208
4.20	34.4	103	207
4.21	34.2	103	205
4.22	34.1	102	204
4.23	33.9	102	203
4.24	33.7	101	202
4.25	33.6	101	201
4.26	33.4	100	200
4.27	33.2	99.7	199
4.28	33.1	99.2	198
4.29	32.9	98.8	198
4.30	32.8	98.3	197
4.31	32.6	97.8	196
4.32	32.4	97.3	195
4.33	32.3	96.8	194
4.34	32.1	96.4	193
4.35	32.0	95.9	192
4.36	31.8	95.5	191
4.37	31.7	95.0	190
4.38	31.5	94.5	189
4.39	31.4	94.1	188

*Prepared by the Engineering Mechanics Section, National Bureau of Standards.

TABLE 1 *Continued*

Diameter of Indentation, mm	Brinell Hardness Number 500-kgf Load	Brinell Hardness Number 1500-kgf Load	Brinell Hardness Number 3000-kgf Load	Diameter of Indentation, mm	Brinell Hardness Number 500-kgf Load	Brinell Hardness Number 1500-kgf Load	Brinell Hardness Number 3000-kgf Load	Diameter of Indentation, mm	Brinell Hardness Number 500-kgf Load	Brinell Hardness Number 1500-kgf Load	Brinell Hardness Number 3000-kgf Load
4.40	31.2	93.6	187	5.00	23.8	81.4	163	5.30	20.9	62.8	126
4.41	31.1	93.2	186	5.01	23.7	81.0	162	5.31	20.9	62.6	125
4.42	30.9	92.7	185	5.02	23.6	80.7	161	5.32	20.8	62.3	125
4.43	30.8	92.3	185	5.03	23.5	80.3	161	5.33	20.7	62.1	124
4.44	30.6	91.8	184	5.04	23.4	79.9	160	5.34	20.6	61.8	124
4.45	30.5	91.4	183	5.05	23.3	79.6	159	5.35	20.5	61.5	123
4.46	30.3	91.0	182	5.06	23.2	79.2	158	5.36	20.4	61.3	123
4.47	30.2	90.5	181	5.07	23.1	78.9	158	5.37	20.3	61.0	122
4.48	30.0	90.1	180	5.08	23.0	78.5	157	5.38	20.3	60.8	122
4.49	29.9	89.7	179	5.09	22.9	78.2	156	5.39	20.2	60.6	121
4.50	29.8	89.3	179	5.10	22.8	77.8	156	5.40	20.1	60.3	121
4.51	29.6	88.8	178	5.11	22.7	77.5	155	5.41	20.0	60.1	120
4.52	29.5	88.4	177	5.12	22.6	77.1	154	5.42	19.9	59.8	120
4.53	29.3	88.0	176	5.13	22.5	76.8	154	5.43	19.9	59.6	119
4.54	29.2	87.6	175	5.14	22.4	76.4	153	5.44	19.8	59.3	119
4.55	29.1	87.2	174	5.15	22.3	76.1	152	5.45	19.7	59.1	118
4.56	28.9	86.8	174	5.16	22.2	75.8	152	5.46	19.6	58.9	118
4.57	28.8	86.4	173	5.17	22.1	75.4	151	5.47	19.5	58.6	117
4.58	28.7	86.0	172	5.18	22.0	75.1	150	5.48	19.5	58.4	117
4.59	28.5	85.6	171	5.19	21.9	74.8	150	5.49	19.4	58.2	116
4.60	28.4	85.4	170	5.20	21.8	74.4	149	5.50	19.3	57.9	116
4.61	28.3	84.8	170	5.21	21.7	74.1	148	5.51	19.2	57.7	115
4.62	28.1	84.4	169	5.22	21.6	73.8	148	5.52	19.2	57.5	115
4.63	28.0	84.0	168	5.23	21.6	73.5	147	5.53	19.1	57.2	114
4.64	27.9	83.6	167	5.24	21.5	73.2	146	5.54	19.0	57.0	114
4.65	27.8	83.3	167	5.25	21.4	72.8	146	5.55	18.9	56.8	114
4.66	27.6	82.9	166	5.26	21.3	72.5	145	5.56	18.9	56.6	113
4.67	27.5	82.5	165	5.27	21.2	72.2	144	5.57	18.8	56.3	113
4.68	27.4	82.1	164	5.28	21.1	71.9	144	5.58	18.7	56.1	112
4.69	27.3	81.8	164	5.29	21.0	71.6	143	5.59	18.6	55.9	112

TABLE 1 *Continued*

Diameter of Indentation, mm	500-kgf Load	1500-kgf Load	3000-kgf Load	Diameter of Indentation, mm	500-kgf Load	1500-kgf Load	3000-kgf Load	Diameter of Indentation, mm	500-kgf Load	1500-kgf Load	3000-kgf Load	Diameter of Indentation, mm	500-kgf Load	1500-kgf Load	3000-kgf Load
5.60	18.6	55.7	111	5.95	16.2	48.7	97.3	6.30	14.2	42.7	85.5	6.65	12.6	37.7	75.4
5.61	18.5	55.5	111	5.96	16.2	48.5	96.9	6.31	14.2	42.6	85.2	6.66	12.5	37.6	75.2
5.62	18.4	55.2	110	5.97	16.1	48.3	96.6	6.32	14.1	42.4	84.9	6.67	12.5	37.5	74.9
5.63	18.3	55.0	110	5.98	16.0	48.1	96.2	6.33	14.1	42.3	84.6	6.68	12.4	37.3	74.7
5.64	18.3	54.8	110	5.99	16.0	47.9	95.9	6.34	14.0	42.1	84.3	6.69	12.4	37.2	74.4
5.65	18.2	54.6	109	6.00	15.9	47.7	95.5	6.35	14.0	42.0	84.0	6.70	12.4	37.1	74.1
5.66	18.1	54.4	109	6.01	15.9	47.6	95.1	6.36	13.9	41.8	83.7	6.71	12.3	36.9	73.9
5.67	18.1	54.2	108	6.02	15.8	47.4	94.8	6.37	13.9	41.7	83.4	6.72	12.3	36.8	73.6
5.68	18.0	54.0	108	6.03	15.7	47.2	94.4	6.38	13.8	41.5	83.1	6.73	12.2	36.7	73.4
5.69	17.9	53.7	107	6.04	15.7	47.0	94.1	6.39	13.8	41.4	82.8	6.74	12.2	36.6	73.1
5.70	17.8	53.5	107	6.05	15.6	46.8	93.7	6.40	13.7	41.2	82.5	6.75	12.1	36.4	72.8
5.71	17.8	53.3	107	6.06	15.6	46.7	93.4	6.41	13.7	41.1	82.2	6.76	12.1	36.3	72.6
5.72	17.7	53.1	106	6.07	15.5	46.5	93.0	6.42	13.6	40.9	81.9	6.77	12.1	36.2	72.3
5.73	17.6	52.9	106	6.08	15.4	46.3	92.7	6.43	13.6	40.8	81.6	6.78	12.0	36.0	72.1
5.74	17.6	52.7	105	6.09	15.4	46.2	92.3	6.44	13.5	40.6	81.3	6.79	12.0	35.9	71.8
5.75	17.5	52.5	105	6.10	15.3	46.0	92.0	6.45	13.5	40.5	81.0	6.80	11.9	35.8	71.6
5.76	17.4	52.3	105	6.11	15.3	45.8	91.7	6.46	13.4	40.4	80.7	6.81	11.9	35.7	71.3
5.77	17.4	52.1	104	6.12	15.2	45.7	91.3	6.47	13.4	40.2	80.4	6.82	11.8	35.5	71.1
5.78	17.3	51.9	104	6.13	15.2	45.5	91.0	6.48	13.4	40.1	80.1	6.83	11.8	35.4	70.8
5.79	17.2	51.7	103	6.14	15.1	45.3	90.6	6.49	13.3	39.9	79.8	6.84	11.8	35.3	70.6
5.80	17.2	51.5	103	6.15	15.1	45.2	90.3	6.50	13.3	39.8	79.6	6.85	11.7	35.2	70.4
5.81	17.1	51.3	103	6.16	15.0	45.0	90.0	6.51	13.2	39.6	79.3	6.86	11.7	35.1	70.1
5.82	17.0	51.1	102	6.17	14.9	44.8	89.6	6.52	13.2	39.5	79.0	6.87	11.6	34.9	69.9
5.83	17.0	50.9	102	6.18	14.9	44.7	89.3	6.53	13.1	39.4	78.7	6.88	11.6	34.8	69.6
5.84	16.9	50.7	101	6.19	14.8	44.5	89.0	6.54	13.1	39.2	78.4	6.89	11.6	34.7	69.4
5.85	16.8	50.5	101	6.20	14.7	44.3	88.7	6.55	13.0	39.1	78.2	6.90	11.5	34.6	69.2
5.86	16.8	50.3	101	6.21	14.7	44.2	88.3	6.56	13.0	38.9	78.0	6.91	11.5	34.5	68.9
5.87	16.7	50.2	100	6.22	14.7	44.0	88.0	6.57	12.9	38.8	77.6	6.92	11.4	34.3	68.7
5.88	16.7	50.0	99.9	6.23	14.6	43.8	87.7	6.58	12.9	38.7	77.3	6.93	11.4	34.2	68.4
5.89	16.6	49.8	99.5	6.24	14.6	43.7	87.4	6.59	12.8	38.5	77.1	6.94	11.4	34.1	68.2
5.90	16.5	49.6	99.2	6.25	14.5	43.5	87.1	6.60	12.8	38.4	76.8	6.95	11.3	34.0	68.0
5.91	16.5	49.4	98.8	6.26	14.5	43.4	86.7	6.61	12.8	38.3	76.5	6.96	11.3	33.9	67.7
5.92	16.4	49.2	98.4	6.27	14.4	43.2	86.4	6.62	12.7	38.1	76.2	6.97	11.3	33.8	67.5
5.93	16.3	49.0	98.0	6.28	14.4	43.1	86.1	6.63	12.7	38.0	76.0	6.98	11.2	33.6	67.3
5.94	16.3	48.8	97.7	6.29	14.3	42.9	85.8	6.64	12.6	37.9	75.7	6.99	11.2	33.5	67.0

TABLE 2   Brinell Hardness Balls Other Than Standard

Diameter of Ball, mm	Tolerance,[A] mm
from 1 to 3, incl.	±0.0035
over 3 to 6, incl.	±0.004
over 6 to 10, incl.	±0.0045

[A]Steel balls for ball bearings normally satisfy these tolerances.

TABLE 3   Maximum Mean Diagonal of Vickers Hardness Impression on Brinell Hardness Balls

Ball Diameter, mm	Maximum Mean Diagonal of Impression on the Ball Made with Vickers Indenter Under 10-kgf Load, mm
10	0.146
5	0.145
2.5	0.143
2	0.142
1	0.139

TABLE 4   Minimum Thickness Requirements for Brinell Hardness Tests

Minimum Thickness of Specimen		Minimum Hardness for Which the Brinell Test May Safely Be Made		
in.	mm	3000-kgf Load	1500-kgf Load	500-kgf Load
1/16	1.6	602	301	100
1/8	3.2	301	150	50
3/16	4.8	201	100	33
1/4	6.4	150	75	25
5/16	8.0	120	60	20
3/8	9.6	100	50	17

TABLE 5   Standard Test Loads

Ball Diameter, mm	Load, kgf	Recommended Range, HB
10	3000	96 to 600
10	1500	48 to 300
10	500	16 to 100

Designation: E 18 – 84

American Association State Highway and
Transportation Officials Standard
AASHTO No.: T 80

Standard Test Methods for

ROCKWELL HARDNESS AND ROCKWELL SUPERFICIAL HARDNESS OF METALLIC MATERIALS[1]

This standard is issued under the fixed designation E 18; the number immediately following the designation indicates the year of original adoption or, in the case of revision, the year of last revision. A number in parentheses indicates the year of last reapproval. A superscript epsilon (ε) indicates an editorial change since the last revision or reapproval.

These methods have been approved for use by agencies of the Department of Defense to replace method 243.1 of Federal Test Method Standard No. 151b and for listing in DoD Index of Specifications and Standards.

1. Scope

1.1 These test methods cover the determination of the Rockwell hardness (4 to 11) and the Rockwell superficial hardness (12 to 19) of metallic materials, including methods for the verification of machines for Rockwell hardness testing (Part C) and the calibration of standardized hardness test blocks (Part D).

1.2 The information in 3, 4, 5, 12, and 13 is intended to describe and define the type of test that is involved and to outline the limitations of acceptable testing machines. This descriptive material is not mandatory in connection with the test itself.

1.3 The values stated in inch-pound units are to be regarded as the standard.

1.4 *This standard may involve hazardous materials, operations, and equipment. This standard does not purport to address all of the safety problems associated with its use. It is the responsibility of whoever uses this standard to consult and establish safety and health practices and determine the applicability of regulatory limitations prior to use.*

2. Applicable Documents

2.1 *ASTM Standards:*
A 370 Methods and Definitions for Mechanical Testing of Steel Products[2]
B 19 Specification for Cartridge Brass Sheet, Strip, Plate, Bar, and Disks (Blanks)[3]
B 36 Specification for Brass Plate, Sheet, Strip, and Rolled Bar[3]
B 96 Specification for Copper-Silicon Alloy Plate Sheet, Strip, and Rolled Bar for General Purposes and Pressure Vessels[3]
B 97 Specification for Copper-Silicon Alloy Plate, Sheet, Strip, and Rolled Bar for General Purposes[4]
B 103 Specification for Phosphor Bronze Plate, Sheet, Strip, and Rolled Bar[3]
B 121 Specification for Leaded Brass Plate, Sheet, Strip, and Rolled Bar[3]
B 122 Specification for Copper-Nickel-Tin Alloy, Copper-Nickel-Zinc Alloy (Nickel Silver), and Copper-Nickel Alloy Plate, Sheet, Strip, and Rolled Bar[3]
B 130 Specification for Commercial Bronze Strip for Bullet Jackets[3]
B 134 Specification for Brass Wire[3]
B 152 Specification for Copper Sheet, Strip, Plate, and Rolled Bar[3]
B 291 Specification for Copper-Zinc-Manganese Alloy (Manganese Brass) Sheet and Strip[3]
B 370 Specification for Copper Sheet and Strip for Building Construction[3]
E 4 Practice of Load Verification of Testing Machines[5]

[1] These test methods are under the jurisdiction of ASTM Committee E-28 on Mechanical Testing and are the direct responsibility of Subcommittee E28.06 on Indentation Hardness Testing.
Current edition approved April 27, 1984. Published June 1984. Originally published as E 18 – 32 T. Last previous edition E 18 – 79.
[2] *Annual Book of ASTM Standards*, Vol 01.04.
[3] *Annual Book of ASTM Standards*, Vol 02.01.
[4] Discontinued, see *1981 Annual Book of ASTM Standards*, Part 6.
[5] *Annual Book of ASTM Standards*, Vol 03.01.

E 29 Recommended Practice for Indicating
Which Places of Figures are to be Considered Significant in Specified Limiting
Values[6]

E 140 Standard Hardness Conversion Tables
for Metals (Relationship Between Brinell
Hardness, Vickers Hardness, Rockwell
Hardness, Rockwell Superficial Hardness,
and Knoop Hardness)[5]

3. Terminology

3.1 *Rockwell hardness test*—an indentation
hardness test using a calibrated machine to force
a diamond spheroconical penetrator (diamond
penetrator), or hard steel ball under specified
conditions, into the surface of the material under
test in two operations, and to measure the permanent depth of the impression under the specified conditions of minor and major loads.

3.2 *Rockwell superficial hardness test*—same
as the Rockwell hardness test except that smaller
minor and major loads are used.

3.3 *Rockwell hardness number, HR*—a number derived from the net increase in the depth of
impression as the load on a penetrator is increased from a fixed minor load to a major load
and then returned to the minor load.

NOTE 1: *Penetrators*—Penetrators for the Rockwell
hardness test include a diamond sphero-conical penetrator having an included angle of 120° with a spherical
tip having a radius of 0.200 mm and steel balls of
several specified diameters.

NOTE 2—Rockwell hardness numbers are always
quoted with a scale symbol representing the penetrator,
load, and dial used. The hardness number is followed
by the symbol HR and the scale designation.
Examples: 64 HRC = Rockwell hardness number of
64 on Rockwell C scale. 81 HR30N = Rockwell superficial hardness number of 81 on Rockwell 30 N scale.

3.4 *verification*—checking or testing to assure
conformance with the specification.

3.5 *calibration*—determination of the values
of the significant parameters by comparison with
values indicated by a reference instrument or by
a set of reference standards.

A. GENERAL DESCRIPTION AND TEST PROCEDURE FOR ROCKWELL HARDNESS TESTS

4. Apparatus

4.1 *General Principles*—The general principles of the Rockwell hardness test are illustrated
in Fig. 1 (diamond penetrator) and Fig. 2 (ball

penetrator) and the accompanying Tables 1 and
2.

4.2 *Description of Machine and Method of
Test*—The tester for making Rockwell hardness
determinations is essentially a machine that measures hardness by determining the depth of penetration of a penetrator into the specimen under
certain arbitrarily fixed conditions of test. The
penetrator may be either a diamond sphero-conical penetrator or a steel ball. The hardness value,
as read from the dial, is an arbitrary number
which is related to the depth of penetration
caused by two superimposed impressions, and
since the scales are reversed, the number is higher
the harder the material. A minor load of 10 kgf
(98 N) is first applied which causes an initial
penetration which sets the penetrator on the material and holds it in position. The dial is set at
zero on the black-figure scale, and the major load
is applied. This major load is the total applied
and the depth measurement depends solely on
the increase in depth due to the increase from
minor to major load. After the major load is
applied and removed, according to standard procedure, the reading is taken while the minor load
is still in position. The major load is usually 150
kgf (1471 N) when a diamond sphero-conical
penetrator is employed and is customarily 60 kgf
(589 N) or 100 kgf (981 N) when a steel ball is
used as a penetrator, but other loads may be used
when found necessary. The ball penetrator is 1/16
in. (1.588 mm) in diameter normally, but other
penetrators of larger diameter such as 1/8, 1/4, or
1/2 in. (3.175, 6.350, or 12.70 mm) may be employed for soft metals. A variety of loads and
penetrators are thus provided and experience
decides the best combination for use.

4.3 *Rockwell Hardness Scales*—Rockwell
hardness values are usually determined and reported according to one of the standard scales
specified in Table 3. There is no Rockwell hardness value designated by a figure alone because
it is necessary to indicate which penetrator and
load have been employed in making the test. In
all cases the minor load is 10 kgf (98 N) and the
dial is adjusted after applying the minor load so
that the pointer reads at "SET" (C 0 or B 30).
The diamond penetrator is not recommended for
materials giving readings below 20. The use of

---

[6] *Annual Book of ASTM Standards*, Vol 14.02.

ball penetrators for materials that give readings greater than 100 is not recommended primarily because of lack of sensitivity and possible flattening of the ball. For the Rockwell hardness test, one Rockwell number represents 0.00008-in. (0.002-mm) movement of the penetrator. Typical applications of the various scales are shown in Table 3. There may be cases in which more than one scale could be used. It is desirable to employ the smallest ball that can properly be used, because of the loss of sensitivity as the diameter of the ball increases. An exception to this is made when soft nonhomogeneous material is to be tested, in which case it may be preferable to use a larger ball which makes an impression of greater area, thus obtaining more of an average hardness. While the choice of scales is optional, the scale symbol for the combination of penetrator, load, and dial used should be as listed in Table 3.

4.4 *Penetrators:*

4.4.1 The standard penetrators, as have been mentioned in 4.3, are the diamond sphero-conical penetrator and steel balls, $\frac{1}{16}$, $\frac{1}{8}$, $\frac{1}{4}$, and $\frac{1}{2}$ in. (1.588, 3.175, 6.350, and 12.70 mm) in diameter.

4.4.1.1 The shape of the diamond spheroconical penetrator should be a cone forming a 120-° angle with a spherical apex of 0.200-mm radius (Fig. 1).

4.4.1.2 The steel balls used should be free from surface imperfections. The balls should be round and conform to the requirements prescribed in 22.1.3.

4.4.2 An occasional check of the contour of the penetrator should be made by examination with a magnifying glass. This will reveal any chipping of the diamond or flattening of a ball penetrator. If either of these conditions is discovered, the penetrator should be replaced.

4.4.3 Dust, dirt, grease, and scale should not be allowed to accumulate on the penetrator as this will affect the results.

4.5 *Anvils*—An anvil should be used that is suitable for the specimen to be tested. Cylindrical pieces should be tested with a V-notch anvil that will support the specimen with the axis directly under the penetrator, or on hard, parallel, twin cylinders properly positioned and clamped to their base. Flat pieces should be tested on a flat anvil that has a smooth flat bearing surface whose plane is perpendicular to the axis of the penetra-

tor. For thin materials or specimens that are not perfectly flat, a flat anvil having an elevated "spot" about $\frac{1}{4}$ in. (6 mm) in diameter and about $\frac{3}{4}$ in. (19 mm) in height is used. This spot should be polished smooth and flat and should have a Rockwell hardness of at least 60 HRC. The seating and supporting surfaces of all anvils should be free from pits, heavy scratches, dust, dirt, and grease. If the provisions of 5.3 on thickness of specimen are complied with, there will be no danger of indenting the anvil but, if the specimen is so thin that the impression will show through on the under side, it is possible that the anvil may be damaged. Damage may also occur from accidental contacting of the anvil with the penetrator. If the anvil is damaged from any cause, it should be replaced. Anvils showing the least perceptible dent will give inaccurate results on thin material. Very soft material should not be tested on the "spot" anvil because the applied load may cause the penetration of the anvil into the under side of the specimen regardless of its thickness.

4.6 *Test Blocks*—Test blocks meeting the requirements outlined in Part D should be used in periodically checking the accuracy of the hardness tester. The test blocks should be of uniform material, thick enough to be free from "anvil effect," of smooth surface on both top and bottom, and of not more than 4 in.² (26 cm²) surface area on the test surface. They should be prepared to a reasonably fine finish; the surface roughness height rating of the test surface should be 12 or less, that is, 12 µin. (0.0003 mm), center line average, and the blocks should be inspected for freedom from any major surface defects or imperfections that might affect the hardness readings.

4.7 *Verifying Machines for Rockwell Hardness Testing with Standardized Hardness Test Blocks*—It has been found practicable to keep machines for Rockwell hardness tests within the tolerances of the standardized hardness test blocks (see Part D). When machines used in everyday production inspection testing are verified and adjusted once a month, it has been found that they very seldom deviate more than this amount. Daily checking (see Section 8) of such machines assures the operator that the penetrator is in good condition and the machine is operating properly.

4.8 *General Precautions:*

4.8.1 *Protection Against Vibration*—If the

460

bench or table on which the hardness tester is mounted is subject to vibration, such as experienced in the vicinity of other machines, the tester should be mounted on a metal plate on sponge rubber at least 1 in. (25 mm) in thickness or any other type of mounting that will effectually eliminate vibration from the machine, otherwise the penetrator will penetrate farther into the material than when such vibrations are absent.

4.8.2 *Preparation of Specimens*—Specimens employed for hardness determinations shall be prepared with care. Should sheet metal be employed, special care should be taken with material that is curved. The concave side of the curved metal should face toward the penetrator. If such specimens are reversed, an error will be introduced due to the flattening of the metal on the anvil.

4.8.3 *Overhang of Specimens*—Specimens that have sufficient overhang so that they do not balance themselves on the anvil shall be properly supported.

4.8.4 *Penetrator and Anvil*—The penetrator and anvil should not be brought together without a test specimen between them, otherwise the anvil will be indented and the ball flattened.

4.8.5 *Dial Gage Plunger*—The dial gage plunger shall move freely in any position.

5. Test Specimens

5.1 *Form*—Specimens used in Rockwell hardness testing vary greatly in form since it is frequently desirable to make the impression upon a part to be used in the finished product rather than upon a sample test specimen. It is recognized that all of the many conditions of test specimens, size, preparation, etc., cannot be covered specifically and the following paragraphs are intended only as a general guide in the selection of test specimens.

5.2 *Surface Conditions*—Surface conditions have a marked effect on the readings obtained on thin materials and, due to the fact that the thickness of such specimens may influence the results, it cannot be assumed that the indications on the standardized test blocks furnish any reliable measure of the errors to be expected when testing thin material. Standardized test blocks are always thick enough to eliminate the effect of the underlying anvil, but it must be remembered that errors of unknown magnitude may occur when tests are made on material so thin that the impression shows through on the reverse side. The standardized test blocks will indicate the errors of the machine when used to test specimens of similar size, shape, and surface condition, but there is at the present time no satisfactory way of checking the accuracy of the readings taken on thin material nor of evaluating the anvil effect when this is present.

5.3 *Thickness*—Rockwell hardness tests of the highest accuracy are made on specimens of sufficient thickness so that the Rockwell reading is not noticeably affected by the supporting anvil. Absence of a bulge or other marking on the surface of the test specimen opposite the impression is an indication that the specimen is sufficiently thick for precision testing. Commercially acceptable Rockwell hardness readings may be obtained on sheet materials that show some bulging or marking, and in some specifications the sheets to be tested are of a thickness and hardness such that anvil effect will be present. Values of limiting thicknesses at various hardness levels for selected Rockwell scales using the diamond penetrator and the 1/16-in. (1.588-mm) diameter ball are given in Tables 4 and 5 and are presented graphically in Figs. 6 and 7. Rockwell hardness tests on sheet metals are acceptable for hardness specification purposes when the tests are made on thicknesses in accordance with these tables and using the methods defined in these Test Methods E 18. In specifications in which the Rockwell hardness is used as an approximate indication of the tensile strength (Note 2), Tables 4 and 5 do not apply. In these cases the relations between the specification limits for tensile strength and hardness have been established for certain specified thickness limits, hence anvil effects due to testing thin sheets are incorporated in the relationship. For the E scale, material harder than 60 HRE may be tested as thin as 1/8 in. (3.175 mm), but if softer than 60 HRE, the minimum thickness should be 3/16 in. (4.762 mm).

NOTE 3—See the ASTM specifications listed in Appendix X1. To obtain Rockwell hardness values completely independent of anvil effect, minimum thicknesses greater than those given in Tables 4 and 5 may be required, and the specimen should be free of marking on the side opposite the impression.

5.4 *Surface Preparation*—The preparation of the test material should be carefully controlled to avoid any alterations in hardness, such as may be caused by heating during grinding or by work hardening during machining and polishing operations. The test surface of the specimen should be such that the load can be applied normal to it. The surface should be clean, dry, free from scale, pits, and foreign material that might crush or flow under the test pressure and so affect the results. If etching of the test surface is required, it should be no deeper than necessary for metallographic study. The surface in contact with the anvil should be clean, dry and free from any condition which may affect results. In testing coated materials, if a hardness value for the metal is desired, the coatings should be thoroughly removed before determining the hardness, and this should be done in such a manner that the base metal is not affected.

5.5 *Spacing of Indentations*—An error may result if an indentation is spaced closer than 2½ diameters from its center to the edge of the specimen or 3 diameters from another indentation measured center to center.

5.6 *Cylindrical Specimens*—Readings on cylindrical specimens are subject to a correction (see 8.2.9).

6. Verification of Apparatus

6.1 *Verification Methods*—Verify the hardness testing machine as specified in Part C.

6.1.1 Two acceptable methods of verifying machines for Rockwell hardness testing are given in Part C.

7. Adjustment of Apparatus

7.1 *Speed of Load Application*—Adjust the dash pot on the hardness tester so that the operating handle completes its travel in from 4 to 5 s with no specimen on the machine and with the machine set up to apply a major load of 100 kgf (981 N). (Manufacturers of machines without operating handles shall specify an equivalent time cycle and method of adjustment.)

7.2 *Index Lever Adjustment*—Make the following tests (and adjustments, if necessary): Place a piece of material on the anvil and turn the capstan elevating nut to bring the material against the penetrator. Keep turning to elevate the material until the hand feels positive resistance to further turning; this will be felt after the 10-kgf (98-N) minor load has been picked up and when the major load is encountered. When excessive power would have to be used to raise the work higher, take note of the position of the pointer on the dial, after setting the dial so that

C 0 and B 30 are at the top. Then if the pointer stands between B 50 and B 70, no adjustment is necessary; if the pointer stands between B 45 and B 50, adjustment is advisable; and if the pointer stands anywhere else, adjustment is imperative. As the pointer revolves several times when the work is being elevated, the readings mentioned above apply to that revolution of the pointer which occurs either as the reference mark on the gage stem disappears into the sleeve or as the auxiliary hand on the dial passes beyond the zero setting on the dial. The object of this adjustment is to see that the elevation of the specimen to pick up the minor load shall not be carried so far as to cause even a partial application of the major load which, to make a proper test, shall be applied only through the release mechanism.

8. Procedure

8.1 *Determination of Machine Accuracy*—Before using the machine for Rockwell hardness tests, determine its accuracy as described in the following paragraphs:

8.1.1 *Checking Against Standardized Test Block*—Select a standardized test block as near as possible to the hardness of the material being tested (preferably within ±5 hardness numbers for the C scale or scales using the diamond penetrator and ±10 hardness numbers for the B scale using ball penetrators). Make five impressions on the test surface of the block and compare the means thereof with the mean of the five readings made in establishing the hardness value of the block. The difference between these two means is defined as the error of the machine. If the error is more than ±2 hardness numbers, examine and adjust the machine until it comes within this limit before it is used. If the error is less than ±2 hardness numbers, the machine error may be taken into account when comparing the results of two or more machines.

8.1.2 *Reinspection*—In the case of reinspection of material by the manufacturer and the purchaser (Note 4), they shall agree upon standardized test blocks to be used in checking and calibrating the machines used for Rockwell hardness testing.

NOTE 4—When referee tests are to be made, the machine shall be thoroughly examined and all adjustments shall be carefully checked before calibrating it as described above. If readings of high accuracy are necessary, the machine may be calibrated before and after making the tests. If the machine has the same error both times, it can be safely assumed that the correction for this error will give the true Rockwell hardness, except under the conditions described in 5.2.

8.1.3 *Use of Standardized Test Blocks*—Standardized test blocks shall be used only on the test surface because this is the only one that has been checked on a machine of accepted accuracy. Moreover, each impression has a small ridge around it, and if the block is tested on its reverse side, these ridges tend to be flattened under the pressure of the major load and may result in a low reading. Blocks shall never be reground or otherwise resurfaced after being used, because it is the top surface that was originally standardized and this may be of a different hardness from the new surface. The impressions work-harden the block to a considerable depth and this may result in the new test surface being in a work hardened condition and not of the same hardness as the original test surface.

8.1.4 *Use of Single Standardized Test Block*—When a single standardized test block is used, the Rockwell hardness range to be considered verified is that which exceeds the maximum and minimum test block values by ±5 points for the C scale or scales using the diamond penetrator and by ±10 points for the B scale or scales using the ball penetrators.

8.1.5 *Checking Several Ranges*—If tests are made in several ranges of a scale, it is permissible to check that scale at the upper, middle, and lower level. Examples are 20 to 30, 35 to 55, and 59 to 65 when using the C scale; and 40 to 59, 60 to 79, and 80 to 100 when using the B scale.

8.2 *Use of Machine:*

8.2.1 Adjust the machine according to the methods described in 7.1 and 7.2.

8.2.2 Select a suitable scale and use the proper load and penetrator in accordance with 4.3, 4.4, and 5.3.

8.2.3 Select a suitable anvil in accordance with 4.5.

8.2.4 All Rockwell hardness tests shall be made on a single thickness of the material regardless of its thickness. Experience has shown that tests made on more than one thickness of material are unreliable.

8.2.5 The penetrator shall be normal to the surface to be tested.

8.2.6 *Minor Load Application*—Place the specimen to be tested on the anvil and apply the minor load gradually until the proper dial indication is obtained. This shall be understood to be when the pointer has made the proper number of complete revolutions and stands within ±5 divisions of the "SET" position at the top of the dial. The proper number of complete revolutions shall be indicated either by a reference mark on the stem of the gage or by an auxiliary hand on the dial. In bringing the penetrator and the test specimen into contact avoid all impact. The last movement of the elevating or lowering screw shall always be in a direction that will bring penetrator and test specimen together. If the proper setting is overrun, remove the minor load and select a new spot for the test. After the minor load has been applied set the dial pointer at zero on the black-figure scale.

8.2.7 *Major Load Application*—Apply the major load by tripping the operating lever without shock. Remove the major load by bringing the operating lever back to its latched position within 2 s after its motion has stopped, or in accordance with 8.2.7.1 or 8.2.7.2.

8.2.7.1 In the case of materials that exhibit little or no plastic flow after application of the major load, the pointer will come to rest before the motion of the operating lever stops, and in this case the operating lever shall be brought back to its latched position immediately after the pointer stops.

8.2.7.2 In the case of materials that exhibit plastic flow after application of the major load, the pointer will continue to move after the operating lever stops, and in this case the operating lever shall be brought back to its latched position at a specified elapsed time between tripping and removal of load (Note 5). In case the elapsed time after the operating lever stops is other than 2 s, the time shall be recorded, unless the time is specified in the product specification.

NOTE 5—For materials requiring the use of this method the time for application of the major load should be specified in the product specification.

8.2.8 *Reading Scale for Rockwell Hardness*—Take the Rockwell hardness as the reading of the pointer on the proper dial figures after the major load has been removed and while the minor load is still applied. These readings are sometimes estimated to one half of a division or to one tenth of a division, depending on the material being tested.

8.2.9 *Cylindrical Specimens*—Readings on cylindrical specimens are subject to a correction; see Tables 6 and 7.

8.2.10 *Reporting Rockwell Hardness Test Results*—All reports of Rockwell hardness test readings shall indicate the scale used, as described in 4.3. Unless otherwise specified, all readings are to be reported to the nearest whole number, rounding to be in accordance with Recommended Practice E 29.

9. Conversion to Other Hardness Scales or Tensile Strength Values

9.1 There is no general method of converting accurately the Rockwell hardness numbers on one scale to Rockwell hardness numbers on another scale, or to other types of hardness numbers, or to tensile strength values. Such conversions are, at best, approximations and therefore should be avoided, except for special cases where a reliable basis for the approximate conversion has been obtained by comparison tests.

Note 6—Tables E 140 give approximate conversion values for specific materials such as steel, austenitic stainless steel, nickel and high-nickel alloys, and cartridge brass.

10. Report

10.1 The report shall include the following information (see Note 2):

10.1.1 The Rockwell hardness number,

10.1.2 The Rockwell hardness scale, that is, C-scale, B-scale, etc., and

10.1.3 The time of application of the major load (only when the product specification requires that the time be reported).

11. Precision and Bias

11.1 *Precision*—An interlaboratory program is now in progress. When completed, it will be the basis of a statement on precision.

11.2 *Bias*—There is no basis for defining the bias for this method.

B. GENERAL DESCRIPTION AND TEST PROCEDURE FOR ROCKWELL SUPERFICIAL HARDNESS TESTS

12. Apparatus

12.1 *General Principles*—The general principles of the Rockwell superficial hardness test are illustrated in Fig. 3 (diamond penetrator) and Fig. 4 (ball penetrator) and the accompanying Tables 8 and 9.

12.2 *Description of Machine and Method of Test*—The tester for making the Rockwell super-ficial hardness test is a specialized form of the regular tester. It measures hardness by the same principle as the regular test but employs a smaller minor load, smaller major loads, and a more sensitive depth-measuring, system. It is recommended for use where for one reason or another a very shallow impression, or one of small area is desired. The minor load applied in the Rockwell superficial hardness test is 3 kgf (29 N). The major load (total load) is 15, 30, or 45 kgf (147, 294, or 441 N). Minute indentation tests are of value in testing the hardness of thin strip or sheet material, nitrided or lightly carburized pieces, finished pieces on which large test marks would be undersirable, areas near edges, extremely small parts or sections, and shapes that would collapse under a large test load.

12.3 *Rockwell Superficial Hardness Scales*—Rockwell superficial hardness values are usually determined and reported according to one of the standard scales specified in Table 10. In all cases the minor load is 3 kgf (29 N) and the dial is adjusted after applying the minor load so that the pointer reads at "SET." Major loads of 15, 30, and 45 kgf (147, 294, and 441 N) are used. In recording results, the proper scale symbols as shown in Table 10, should be suffixed to the Rockwell superficial hardness numbers obtained by this method. For the Rockwell superficial hardness test, one Rockwell number represents 0.00004-in. (0.001-mm) movement of the penetrator. The "N" scales are used for materials similar to those tested on the Rockwell C, A, and D scales, but of thinner gage or case depth or where a minute indentation is required. The "T" scales are used for materials similar to those tested on the Rockwell B, F, or G scales, but of thinner gage or where a minute indentation is required. The W, X, and Y scales are used for very soft materials. It is desirable to employ the smallest ball that can properly be used, because of the loss of sensitivity as the size of the penetrator increases. An exception to this is when soft nonhomogeneous material is to be tested, in which case it may be preferable to use a larger ball which makes an impression of greater area, thus obtaining more of an average hardness value.

12.4 *Penetrators*—The commonly used penetrators are the superficial diamond sphero-conical penetrator and steel ball 1/16 in. (1.588 mm) in diameter. The shape of the diamond sphero-conical penetrator should be a cone forming a 120-° angle with a spherical apex of 0.200-mm radius. Because of their method of preparation, the normal Rockwell diamond penetrator is not interchangeable with the superficial diamond penetrator. The steel balls used should be free from surface imperfections (see 4.4.1.2). Larger steel ball penetrators, ⅛, ¼, or ½ in. (3.175, 6.350, or 12.70 mm) in diameter, may be used.

12.5 *Anvils*—An anvil should be used that is suitable for the specimen to be tested. Cylindrical pieces should be tested with a V-notch anvil that will support the specimen with the axis directly under the penetrator or on hard, parallel, twin cylinders properly positioned and clamped to their base. Flat pieces should be tested on a flat anvil that has a smooth flat bearing surface whose plane is perpendicular to the axis of the penetrator. For thin materials (see 5.3) or specimens that are not perfectly flat, a flat anvil having an elevated "spot" about ¼ in. (6 mm) in diameter and about ¾ in. (19 mm) in height is used. This spot should be polished smooth and flat and should be free from pits and heavy scratches. This spot should have a Rockwell hardness of at least 60 HRC. When testing thin material with the steel ball penetrator, the use of a diamond spot anvil will present a highly uniform and polished surface to the flow of the material under test load. This diamond spot anvil should never be used with the diamond sphero-conical penetrator because if the material under test should break and both diamonds come in contact with each other, it would probably result in damage to one or both of these parts. When tests are made with the diamond spot anvil this should be so stated in the specifications.

12.6 *Test Blocks*—Test blocks meeting the requirements outlined in Part D should be used in periodically checking the accuracy of the superficial hardness tester. The test blocks should be of uniform material, thick enough to be free from "anvil effect," of smooth surface on both top and bottom and of not more than 4 in.² (26 cm²) surface area on the test surface. They should be prepared to a reasonably fine finish; the surface roughness height rating of the test surface should be 12 or less, that is, 12 μin. (0.0003 mm), center line average, and the blocks should be inspected for freedom from any major surface defects or imperfections that might affect the hardness readings.

12.7 *Verifying Machines for Rockwell Superficial Hardness Testing with Standardized Hardness Test Blocks*—It has been found practicable to keep machines for Rockwell superficial hardness tests within the tolerances of the standardized hardness test blocks (see Part D). When machines used in everyday production inspection testing are verified and adjusted once a month, it has been found that they very seldom deviate more than this amount. Daily checking (see Section 16) of such machines assures the operator that the penetrator is in good condition and the machine is operating properly.

12.8 *Precautions*—The precautions included in 4.8 should be observed in the use of the superficial hardness tester.

13. Test Specimens

13.1 The test specimens should conform to the requirements specified in Section 5 except that smoother surfaces are desirable for the Rockwell superficial hardness impressions and greater care should be observed to avoid overhang of the specimens on the anvil.

13.2 Commercially acceptable Rockwell superficial hardness readings may be obtained on sheet materials that show some bulging or marking, and in some specifications the sheets to be tested are of a thickness and hardness such that anvil effect will be present. Tables of limiting thicknesses at various hardness levels for selected Rockwell superficial hardness scales using the diamond penetrator and the ¹⁄₁₆-in. (1.588-mm) diameter ball are given in Tables 11 and 12 and are presented graphically in Figs. 6 and 7. In the case of tin mill products, the Rockwell superficial test is used for specimens thinner than those recommended in Table 12. Thickness limitations for these products appear in the product specifications. Rockwell superficial hardness tests on sheet metals are acceptable for hardness specification purpose when the tests are made on thicknesses in accordance with these tables and using the methods defined in Test Methods E 18. In specifications in which the Rockwell superficial hardness is used as an approximate indication of the tensile strength (Note 3 and Appendix X1) Tables 11 and 12 do not apply. In these cases the relations between the specification limits for tensile strength and hardness have been established for certain specified thickness limits, hence anvil effects due to testing thin sheets are incorporated in the relationship. To obtain Rockwell superficial hardness values completely independent of anvil effect, minimum thicknesses greater than those given in Tables 11 and 12 may be required, and the specimen should be free of marking on the side opposite the impression.

13.3 Readings on cylindrical specimens are subject to correction (see 16.10).

14. Verification of Apparatus

14.1 *Verification Methods*—Verify the hardness testing machine as specified in Part C.

14.1.1 Two acceptable methods of verifying the machine for Rockwell superficial hardness testing are given in Part C.

15. Adjustment of Apparatus

15.1 *Speed of Load Application*—Adjust the dash pot on the superficial hardness tester so that the operating handle completes its travel in 5 to 7 s with no specimen on the machine and with the machine set up to apply a major load of 30 kgf (294 N). (Manufacturers of superficial hardness testing machines without operating handles shall specify an equivalent time cycle and method of adjustment.)

15.2 *Index Lever Adjustment*—Make the following tests (and adjustments, if necessary): Place a piece of material on the anvil and turn the capstan elevating nut to bring the material against the penetrator. Keep turning to elevate the material until the hand feels positive resistance to further turning; this will be felt after the 3-kgf (29-N) minor load has been picked up and when the major load is encountered. When excessive power would have to be used to raise the work higher, take note of the position of the pointer on the dial, after setting the dial so that "SET" is at the top. Then if the pointer stands between 45 and 55, adjustment is not necessary; if the pointer stands anywhere else, adjustment is imperative. As the pointer revolves several times when the work is being elevated, the readings mentioned above apply to that revolution of the pointer which occurs as the auxiliary hand on the dial passes beyond the zero setting on the dial. The object of this adjustment is to see that the elevation of the specimen to pick up the minor load shall not be carried so far as to cause even a partial application of the major load which, to make a proper test, shall be applied only through the release mechanism.

16. Procedure

16.1 Before using the machine, determine its accuracy as described in 8.1.

16.2 Adjust the machine in accordance with the methods described in 15.1 and 15.2.

16.3 Select a suitable scale and use the proper load and penetrator in accordance with 12.3, 12.4, and 5.3.

16.4 Select a suitable anvil in accordance with 12.5.

16.5 All Rockwell superficial hardness tests shall be made on a single thickness of the material regardless of thickness.

16.6 *Minor Load Application*—Place the specimen to be tested on the anvil and apply the minor load gradually until the proper dial indication is obtained. This shall be understood to be when the pointer has made the proper number of complete revolutions and stands within ±5 divisions of the "SET" positions at the top of the dial. The proper number of complete revolutions shall be indicated either by a reference mark on the stem of the gage or by an auxiliary hand on the dial. In bringing the penetrator and the specimen into contact avoid all impact. The last movement of the elevating screw shall always be in a direction which will bring penetrator and specimen together. If the proper setting is overrun, remove the minor load and select a new spot for the test. After the minor load has been applied, set the dial pointer at zero on the black-figure scale.

16.7 *Major Load Application*—Apply the major load by tripping the operating lever without shock. Remove the major load by bringing the operating lever back to its latched position within 2 s after its motion has stopped, or in accordance with 16.7.1 or 16.7.2.

16.7.1 In the case of materials that exhibit little or no plastic flow after application of the major load, the pointer will come to rest before the motion of the operating level stops, and in this case the operating level shall be brought back to its latched position immediately after the pointer stops.

16.7.2 In the case of materials that exhibit plastic flow after application of the major load, the pointer will continue to move after the operating lever stops, and in this case the operating lever shall be brought back to its latched position at a specified elapsed time between tripping and removal of load (Note 7). In case the elapsed time is other than 2 s, the time shall be recorded, unless the time is specified in the product specification.

NOTE 7—For materials requiring the use of this method the time for application of the major load should be specified in the product specification.

16.8 *Reading Scale for Rockwell Superficial Hardness*—Take the Rockwell superficial hardness number as the reading of the pointer on the dial after the major load has been removed and while the minor load is still applied. These readings are sometimes estimated to one half of a division or to one tenth of a division, depending on the material being tested.

16.9 Readings on cylindrical specimens are subject to correction; see Tables 13 and 14.

16.10 *Reporting Rockwell Superficial Hardness Test Results*—All reports of Rockwell superficial hardness test readings shall indicate the scale used, as described in 12.3. Unless otherwise specified, all readings are to be reported to the nearest whole number, rounding to be in accordance with Recommended Practice E 29.

17. Conversion to Other Hardness Scales or Tensile Strength Values

17.1 There is no general method for accurately converting the Rockwell hardness numbers on one scale to hardness numbers on another Rockwell scale, or to other types of hardness numbers, or to tensile strength values. Such conversions are, at best, approximations and therefore should be avoided, except for special cases where a reliable basis for the approximate conversion has been obtained by comparison tests.

NOTE 8—Tables E 140 give approximate conversion values for specific materials such as steel, nickel and high-nickel alloys, and cartridge brass.

18. Report

18.1 The report shall include the following information (see Note 2):

18.1.1 The Rockwell superficial hardness number,

18.1.2 The Rockwell superficial hardness scale, that is, 30N-scale, 30T-scale, etc., and

18.1.3 The time of application of the major load (only when the product specification requires that the time be reported).

19. Precision and Bias

19.1 *Precision*—An interlaboratory comparison program is now in progress. When completed, it will be the basis of a statement on precision.

19.2 *Bias*—There is no basis for defining the bias for this method.

C. VERIFICATION OF MACHINES FOR ROCKWELL HARDNESS AND ROCKWELL SUPERFICIAL HARDNESS TESTING

20. Scope

20.1 Part C covers two procedures for the verification of machines for Rockwell hardness and Rockwell superficial hardness testing and a procedure which is recommended for use to confirm that the machine has not become maladjusted in the intervals between the periodical routine checks made by the user. The two methods of verification are:

20.1.1 Separate verification of load application, penetrator, and the depth measuring device followed by a performance test (22.2). This method shall be used for new and rebuilt machines.

20.1.2 Verification by standardized test block method. This method shall be used in referee, laboratory, or routine testing to assure the operator that the machine for Rockwell hardness testing is operating properly (see 22.2).

21. General Requirements

21.1 Before a hardness testing machine is verified, the machine shall be examined to ensure that:

21.1.1 The machine is properly set up.

21.1.2 The dial gage plunger on the depth measuring device moves freely in any position.

21.1.3 The penetrator holder is properly seated in the plunger.

21.1.4 When the penetrator is a steel ball, the holder shall be fitted with a new ball whose diameter has been checked (see 22.1.3).

21.1.5 The diamond penetrator is free from cracks or flaws which would lead to incorrect readings.

21.1.6 The load can be applied and removed without shock or vibration in such a manner that the readings are not influenced.

22. Verification

22.1 *Separate Verification of Load Application, Penetrator, and Depth Measuring Device:*

22.1.1 *Load Application*—Machines for

Rockwell hardness testing shall be verified at loads of 10, 60, 100, and 150 kgf (98, 589, 981, and 1472 N). Machines for Rockwell superficial hardness testing shall be verified at loads of 3, 15, 30, and 45 kgf (29, 147, 294, and 441 N). The applied load shall be checked by the use of standardized dead weights (masses), by the use of standardized dead weights and proving levers, or by the use of an elastic proving device or springs in the manner described in Methods E 4. Each testing machine for the determination of Rockwell hardness shall be verified at the minor load (10 ± 0.2 kgf) (98 ± 2.0 N) before application and after removal of the additional load. Each testing machine for the determination of Rockwell superficial hardness shall be verified at the minor load (3 ± 0.060 kgf) (29 ± 0.59 N) before application and after removal of the additional load. The load application shall be considered verified if the mean for three readings, at each of three positions of the power level, for each load falls within the tolerance listed in Table 15.

22.1.2 *Diamond Sphero-Conical Penetrator (Diamond Penetrator)*— The verification of the form of the diamond penetrator shall be made by direct measurement of its shape or by measurement of its projection on a screen. Verification shall be made at not less than four sections. The diamond penetrator shall have an included angle of 120° ±30 min and shall have its axis in line with the axis of the penetrator within a tolerance of ±30 min. The tip of the cone shall have a nominal radius of 0.200 mm. The contour of the tip of the cone shall lie within a band defined by two concentric arcs parallel to the nominal tip contour radius (0.200 mm) but displaced from it by 0.002 mm as shown in Fig. 5. In Fig. 5(a) the tolerance band is shown at a magnification of 500×. In Fig. 5(b) possible minimum and maximum variations are shown schematically (not to scale). The surface of the cone shall blend in a tangential manner with the surface of the spherical tip. The penetrator shall be polished to such an extent that no unpolished part of its surface makes contact with the test specimen when it penetrates to a depth of 0.012 in. (0.3 mm). Since the hardness values given by a testing machine do not depend upon these dimensions alone, but also on the surface roughness, the position of the crystallographic axes of the diamond, and the seating of the diamond in the holder, a performance test is required. In this test the sphero-conical penetrator shall be used in a machine in which the load application and the depth-measuring device have been verified. Five impressions shall be made on a standardized hardness test block calibrated with a verified penetrator as prescribed in Part D. The mean of these five hardness readings shall not differ from the average of the standard test block by more than the amount shown in Table 16.

22.1.3 *Ball Penetrator*—The mean of three diameters measured on a new steel ball selected at random from a lot shall not differ from the nominal by more than the amounts shown in Table 17. The permissible difference between the largest diameter and the smallest diameter measurable on one ball shall be not more than 0.00004 in. (0.0010 mm). The steel ball penetrator shall have a Vickers hardness of at least 850 HV using a 10-kgf (98-N) load. Therefore, the maximum mean diagonal of the Vickers impression made on a steel ball penetrator shall not exceed the value shown in Table 18.

22.1.4 Depth-Measuring   Device—The depth-measuring device shall be verified over not less than three ranges, including the ranges corresponding to the lowest and highest hardness for which the scale is normally used, by making known incremental movements of the penetrator or dial gage plunger. The depth-measuring device shall correctly indicate the Rockwell hardness within ±0.5 of a scale unit over each range, that is, within ±0.001 mm depth reading. The depth-measuring device shall correctly indicate the Rockwell superficial hardness within ±0.5 of a scale unit over each range, that is, within ±0.0005 mm depth reading.

22.2 *Verification by Standardized Test Block Method:*

22.2.1 A machine for Rockwell hardness testing or Rockwell superficial hardness testing used for referee, laboratory, or routine testing or machines which are verified in service may be checked by making a series of impressions on standardized hardness test blocks (Part D).

22.2.2 A minimum of five hardness readings shall be taken on the test surface of at least three blocks having different levels of hardness, as shown in Table 19, using the following test loads:

Rockwell C scale	150 kgf (1472 N) load
Rockwell B scale	100 kgf (981 N) load
Rockwell 30N scale	30 kgf (294 N) load
Rockwell 30T scale	30 kgf (294 N) load

22.2.3 When tests are made in several ranges

of a scale, it is permissible to check that scale at the lower, middle, and upper levels. Examples are 20 to 30, 35 to 55, and 59 to 65 when using the C scale; and 40 to 59, 60 to 79, and 80 to 100 when using the B scale.

22.2.4 Machines for Rockwell hardness and Rockwell superficial hardness tests shall be considered verified if the results meet the requirements of Sections 24 and 25.

23. Procedure for Periodic Checks by the User

23.1 Verification by the standardized test block method (22.2) is too lengthy for daily use. Instead, the following is recommended:

23.1.1 Make at least one routine check each day that the testing machine is used.

23.1.2 Before making the check, make at least two preliminary indentations to ensure that the hardness testing machine is working freely and that the test block, penetrator, and anvil are reading correctly. The results of these preliminary indentations should be ignored.

23.1.3 Make at least five hardness readings on a standardized hardness test block on the scale and at the hardness level at which the machine is being used. If the values fall within the range of the standardized hardness test block the machine may be regarded as satisfactory. If not the machine should be verified as described in 22.2.

24. Repeatability and Error

24.1 *Repeatability:*

24.1.1 For each standardized block let R_1, $R_2 \ldots R_5$ be the hardness readings of the five indentations arranged in an increasing order of magnitude.

24.1.2 The repeatability of the hardness testing machine under the particular verification conditions is expressed by the quantity $R_5 - R_1$.

24.2 *Error:*

24.2.1 The error of the hardness testing machine under the particular verification conditions is expressed by the quantity $\bar{R} - R$, where

$$\bar{R} = (R_1 + R_2 \ldots R_5)/5$$

$R_1, R_2 \ldots R_5$ are the individual hardness values, and R is the stated hardness of the standardized hardness test block.

25. Assessment of Verification

25.1 *Repeatability*—The repeatability of the hardness testing machine is considered satisfac-

tory if it satisfies the conditions given in Table 20.

25.2 *Error*—The mean hardness value for five impressions should not differ from the mean corresponding to the hardness of the standardized test block by more than the tolerance of the latter.

D. CALIBRATION OF STANDARDIZED HARDNESS TEST BLOCKS FOR MACHINES USED FOR ROCKWELL AND ROCKWELL SUPERFICIAL HARDNESS TESTING

26. Scope

26.1 Part D covers the calibration of standardized hardness test blocks for the verification of machines used for Rockwell and Rockwell superficial hardness testing as described in Part C.

27. Manufacture

27.1 Each metal block to be standardized shall be not less than 1/4 in. (6 mm) in thickness.

27.2 Each block shall be specially prepared and heat treated to give the necessary homogeneity and stability of structure.

27.3 Each block, if of steel, shall be demagnetized by the manufacturer and maintained demagnetized by the user.

27.4 The lower surface of the test block shall have a fine ground finish.

27.5 The test (upper) surface shall be polished or fine ground and free from scratches which would influence the depth of the impression.

27.5.1 The mean surface roughness height rating of the test surfaces shall not exceed 12 μin. (0.0003 mm) center line average.

27.6 To ensure that no material is subsequently removed from the test surface of the standardized test block, an official mark or the thickness to an accuracy of ±0.005 in. (±0.1 mm) or at the time of calibration shall be marked on the test surface.

28. Standardizing Procedure

28.1 The standardizing hardness test blocks shall be calibrated on a hardness testing machine verified in accordance with the requirements of 20.1.1.

28.2 The major load shall be removed by bringing the operating lever back to its latched position according to one of the alternative methods given in 8.2.8.1 or 8.2.8.2 and 16.8.1 or 16.8.2.

29. Number of Indentations

29.1 At least five randomly distributed indentations shall be made on each test block.

29.2 The dial indicator of the depth measuring device shall be read to ±0.1 unit, that is, to the nearest 0.1 Rockwell hardness or Rockwell superficial hardness number.

30. Repeatability

30.1 Let $R_1, R_2 \ldots R_n$ be the observed Rockwell hardness or Rockwell superficial hardness number as determined by one observer, arranged in increasing order of magnitude.

30.2 The repeatability of the hardness readings on the block is defined as $R_n - R_1$.

31. Uniformity of Hardness

31.1 Unless the repeatability of the hardness readings is within the limits given in Table 21, the block cannot be regarded as sufficiently uniform for standardization purposes.

32. Marking

32.1 Each block shall be marked with the following:

32.1.1 Arithmetic mean of the hardness values found in the standardization test prefixed by the scale designation and followed by the tolerance range.

32.1.2 The name or mark of the supplier,

32.1.3 The serial number of the block, and

32.1.4 The thickness of the test block or an official mark on the test surface (see 27.6).

NOTE 9—All of the markings except the official mark or thickness shall be placed on the side of the block, the markings being upright when the test surface is the upper face.

TABLE 1   Symbols and Designations Associated with Fig. 1

Number	Symbol	Designation
1	. . .	Angle at the top of the diamond penetrator (120°)
2	. . .	Radius curvature at the tip of the cone (0.200 mm)
3	P_0	Minor load = 10 kgf (98 N)
4	P_1	Additional load = 90 or 140 kgf (883 or 1373 N)
5	P	Major load = $P_0 + P_1$ = 100 or 150 kgf (981 or 1472 N)
6	. . .	Depth of impression under minor load before application of additional load
7	. . .	Increase in depth of impression under additional load
8	e	Permanent increase in depth of impression under minor load after removal of additional load, the increase being expressed in units of 0.002 mm
9	xx HRA	Rockwell A hardness = $100 - e$
10	xx HRC	Rockwell C hardness = $100 - e$

TABLE 2   Symbols and Designations Associated with Fig. 2

Number	Symbol	Designation
1	D	Diameter of ball = $\frac{1}{16}$ in. (1.588 mm)
3	P_0	Minor load = 10 kgf (98 N)
4	P_1	Additional load = 90 kgf (883 N)
5	P	Major load = $P_0 + P_1$ = 10 + 50, 90, or 140 = 60, 100, or 150 kgf (589, 981, or 1472 N)
6	. . .	Depth of impression under minor load before application of additional load
7	. . .	Increase in depth of impression under additional load
8	e	Permanent increase in depth of impression under minor load after removal of the additional load, the increase being expressed in units of 0.002 mm
9	xx HRF	Rockwell F hardness = $130 - e$
10	xx HRB	Rockwell B hardness = $130 - e$
11	xx HRG	Rockwell G hardness = $130 - e$

TABLE 3 Rockwell Hardness Scales

Scale Symbol	Penetrator	Major Load, kgf	Dial Figures	Typical Applications of Scales
B	¹⁄₁₆-in. (1.588-mm) ball	100	red	Copper alloys, soft steels, aluminum alloys, malleable iron, etc.
C	diamond	150	black	Steel, hard cast irons, pearlitic malleable iron, titanium, deep case hardened steel, and other materials harder than B 100.
A	diamond	60	black	Cemented carbides, thin steel, and shallow case-hardened steel.
D	diamond	100	black	Thin steel and medium case hardened steel, and pearlitic malleable iron.
E	⅛-in. (3.175-mm) ball	100	red	Cast iron, aluminum and magnesium alloys, bearing metals.
F	¹⁄₁₆-in. (1.588-mm) ball	60	red	Annealed copper alloys, thin soft sheet metals.
G	¹⁄₁₆-in. (1.588-mm) ball	150	red	Malleable irons, copper-nickel-zinc and cupro-nickel alloys. Upper limit G 92 to avoid possible flattening of ball.
H	⅛-in. (3.175-mm) ball	60	red	Aluminum, zinc, lead.
K	⅛-in. (3.175-mm) ball	150	red	
L	¼-in. (6.350-mm) ball	60	red	
M	¼-in. (6.350-mm) ball	100	red	Bearing metals and other very soft or thin materials. Use
P	¼-in. (6.350-mm) ball	150	red	smallest ball and heaviest load that does not give anvil
R	½-in. (12.70-mm) ball	60	red	effect.
S	½-in. (12.70-mm) ball	100	red	
V	½-in. (12.70-mm) ball	150	red	

TABLE 4 A Guide for Selection of Scales Using the Diamond Penetrator (see Fig. 6)

Note—For a given thickness, any hardness greater than that corresponding to that thickness can be tested. For a given hardness, material of any greater thickness than that corresponding to that hardness can be tested on the indicated scale.

Thickness		Rockwell Scale		
		A		C
in.	mm	Dial Reading	Approximate Hardness C-Scale[A]	Dial Reading
0.014	0.36	...	...	...
0.016	0.41	86	69	...
0.018	0.46	84	65	...
0.020	0.51	82	61.5	...
0.022	0.56	79	56	69
0.024	0.61	76	50	67
0.026	0.66	71	41	65
0.028	0.71	67	32	62
0.030	0.76	60	19	57
0.032	0.81	...	...	52
0.034	0.86	...	...	45
0.036	0.91	...	...	37
0.038	0.96	...	...	28
0.040	1.02	...	...	20

[A] These approximate hardness numbers are for use in selecting a suitable scale and should not be used as hardness conversions. If necessary to convert test readings to another scale, refer to Hardness Conversion Tables E 140 (Relationship Between Brinell Hardness, Vickers Hardness, Rockwell Hardness, Rockwell Superficial Hardness, and Knoop Hardness).

TABLE 5 A Guide for Selection of Scales Using the ¹⁄₁₆-in. (1.588-mm) Diameter Ball Penetrator (see Fig. 7)

Note—For a given thickness, any hardness greater than that corresponding to that thickness can be tested. For a given hardness, material of any greater thickness than that corresponding to that hardness can be tested on the indicated scale.

Thickness		Rockwell Scale		
		F		B
in.	mm	Dial Reading	Approximate Hardness B-Scale[A]	Dial Reading
0.022	0.56	...	...	...
0.024	0.61	98	72	94
0.026	0.66	91	60	87
0.028	0.71	85	49	80
0.030	0.76	77	35	71
0.032	0.81	69	21	62
0.034	0.86	...	...	52
0.036	0.91	...	...	40
0.038	0.96	...	...	28
0.040	1.02	...	...	...

[A] These approximate hardness numbers are for use in selecting a suitable scale and should not be used as hardness conversions. If necessary to convert test readings to another scale refer to Hardness Conversion Tables E 140 (Relationship Between Brinell Hardness, Vickers Hardness, Rockwell Hardness, Rockwell Superficial Hardness and Knoop Hardness).

TABLE 6 Corrections to Be Added to Rockwell C, A, and D Values Obtained on Cylindrical Specimens[A] of Various Diameters

Dial Reading	Diameters of Cylindrical Specimens								
	¼ in. (6.4 mm)	⅜ in. (10 mm)	½ in. (13 mm)	⅝ in. (16 mm)	¾ in. (19 mm)	⅞ in. (22 mm)	1 in. (25 mm)	1¼ in. (32 mm)	1½ in. (38 mm)
	Corrections to be Added to Rockwell C, A, and D Values[B]								
20	6.0	4.5	3.5	2.5	2.0	1.5	1.5	1.0	1.0
25	5.5	4.0	3.0	2.5	2.0	1.5	1.0	1.0	1.0
30	5.0	3.5	2.5	2.0	1.5	1.5	1.0	1.0	0.5
35	4.0	3.0	2.0	1.5	1.5	1.0	1.0	0.5	0.5
40	3.5	2.5	2.0	1.5	1.0	1.0	1.0	0.5	0.5
45	3.0	2.0	1.5	1.0	1.0	1.0	0.5	0.5	0.5
50	2.5	2.0	1.5	1.0	1.0	0.5	0.5	0.5	0.5
55	2.0	1.5	1.0	1.0	0.5	0.5	0.5	0.5	0
60	1.5	1.0	1.0	0.5	0.5	0.5	0.5	0	0
65	1.5	1.0	1.0	0.5	0.5	0.5	0.5	0	0
70	1.0	1.0	0.5	0.5	0.5	0.5	0.5	0	0
75	1.0	0.5	0.5	0.5	0.5	0.5	0	0	0
80	0.5	0.5	0.5	0.5	0.5	0	0	0	0
85	0.5	0.5	0.5	0	0	0	0	0	0
90	0.5	0	0	0	0	0	0	0	0

[A] When testing cylindrical specimens, the accuracy of the test will be seriously affected by alignment of elevating screw, V-anvil, penetrators, surface finish, and the straightness of the cylinder.

[B] These corrections are approximate only and represent the averages to the nearest 0.5 Rockwell number, of numerous actual observations.

TABLE 7 Corrections to Be Added to Rockwell B, F, and G Values Obtained on Cylindrical Specimens[A] of Various Diameters

Dial Reading	Diameters of Cylindrical Specimens						
	¼ in. (6.4 mm)	⅜ in. (10 mm)	½ in. (13 mm)	⅝ in. (16 mm)	¾ in. (19 mm)	⅞ in. (22 mm)	1 in. (25 mm)
	Corrections to be Added to Rockwell B, F, and G Values[B]						
0	12.5	8.5	6.5	5.5	4.5	3.5	3.0
10	12.0	8.0	6.0	5.0	4.0	3.5	3.0
20	11.0	7.5	5.5	4.5	4.0	3.5	3.0
30	10.0	6.5	5.0	4.5	3.5	3.0	2.5
40	9.0	6.0	4.5	4.0	3.0	2.5	2.5
50	8.0	5.5	4.0	3.5	3.0	2.5	2.0
60	7.0	5.0	3.5	3.0	2.5	2.0	2.0
70	6.0	4.0	3.0	2.5	2.0	2.0	1.5
80	5.0	3.5	2.5	2.0	1.5	1.5	1.5
90	4.0	3.0	2.0	1.5	1.5	1.5	1.0
100	3.5	2.5	1.5	1.5	1.0	1.0	0.5

[A] When testing cylindrical specimens, the accuracy of the test will be seriously affected by alignment of elevating screw, V-anvil, penetrators, surface finish, and the straightness of the cylinder.

[B] These corrections are approximate only and represent the averages to the nearest 0.5 Rockwell number, of numerous actual observations.

472

TABLE 8 Symbols and Designations Associated with Fig. 3

Number	Symbol	Designation
1	...	Angle at the tip of the diamond penetrator (120°)
2	...	Radius of curvature at the tip of the cone (0.200 mm)
3	P_0	Minor load = 3 kgf (29 N)
4	P_1	Additional load = 12, 27, or 42 kgf (118,265, or 412 N)
5	P	Major load = $P_0 + P_1$ = 3 + 12, 27, or 42 = 15, 30, or 45 kgf (147, 294, or 441 N)
6	...	Depth of impression under minor load before application of additional load
7	...	Increase in depth of impression under additional load
8	e	Permanent increase in depth of impression under minor load after removal of additional load, the increase being expressed in units of 0.001 mm
9	xx HR15N	Rockwell 15N hardness = 100 − e
	xx HR30N	Rockwell 30N hardness = 100 − e
	xx HR45N	Rockwell 45N hardness = 100 − e

TABLE 9 Symbols and Designations Associated with Fig. 4

Number	Symbol	Designation
1	D	Diameter of ball = $\frac{1}{16}$ in. (1.588 mm)
3	P_0	Minor load = 3 kgf (29 N)
4	P_1	Additional load = 12, 27, or 42 kgf (118, 265, or 412 N)
5	P	Major load = $P_0 + P_1$ = 3 + 12, 27, or 42 = 15, 30, or 45 kgf (147, 294, or 441 N)
6	...	Depth of impression under minor load before application of additional load
7	...	Increase in depth of impression under additional load
8	e	Permanent increase in depth of impression under minor load after removal of the additional load, the increase being expressed in units of 0.001 mm
9	xx HR15T	Rockwell 15T hardness = 100 − e
	xx HR30T	Rockwell 30T hardness = 100 − e
	xx HR45T	Rockwell 45T hardness = 100 − e

TABLE 10 Rockwell Superficial Hardness Scales

Major Load, kgf (N)	Scale Symbols				
	N Scale, Diamond Penetrator	T Scale, $\frac{1}{16}$-in. (1.588-mm) Ball	W Scale, $\frac{1}{8}$-in. (3.175-mm) Ball	X Scale, $\frac{1}{4}$-in. (6.350-mm) Ball	Y Scale, $\frac{1}{2}$-in. (12.70-mm) Ball
15 (147)	15N	15T	15W	15X	15Y
30 (294)	30N	30T	30W	30X	30Y
45 (441)	45N	45T	45W	45X	45Y

TABLE 11 A Guide for Selection of Scales Using the Diamond Penetrator (see Fig. 6)

NOTE—For a given thickness, any hardness greater than that corresponding to that thickness can be tested. For a given hardness, material of any greater thickness than that corresponding to that hardness can be tested on the indicated scale.

Thickness		Rockwell Superficial Scale					
		15N		30N		45N	
in.	mm	Dial Reading	Approximate Hardness C-Scale[A]	Dial Reading	Approximate Hardness C-Scale[A]	Dial Reading	Approximate Hardness C-Scale[A]
0.006	0.15	92	65	...	...	...	...
0.008	0.20	90	60	...	...	...	...
0.010	0.25	88	55	...	...	...	...
0.012	0.30	83	45	82	65	77	69.5
0.014	0.36	76	32	78.5	61	74	67
0.016	0.41	68	18	74	56	72	65
0.018	0.46	...	...	66	47	68	61
0.020	0.51	...	...	57	37	63	57
0.022	0.56	...	...	47	26	58	52.5
0.024	0.61	...	...	...	...	51	47
0.026	0.66	...	...	...	...	37	35
0.028	0.71	...	...	...	...	20	20.5
0.030	0.76	...	...	...	...	...	...

[A] These approximate hardness numbers are for use in selecting a suitable scale, and should not be used as hardness conversions. If necessary to convert test readings to another scale, refer to Hardness Conversion Tables E 140 (Relationship Between Brinell Hardness, Vickers Hardness, Rockwell Hardness, Rockwell Superficial Hardness and Knoop Hardness).

TABLE 12 A Guide for Selection of Scales Using the ¹⁄₁₆ in. (1.588 mm) Diameter Ball Penetrator (see Fig. 7)

NOTE—For a given thickness, any hardness greater than that corresponding to that thickness can be tested. For a given hardness, material of any greater thickness than that corresponding to that hardness can be tested on the indicated scale.

Thickness		Rockwell Superficial Scale					
		15T		30T		45T	
in.	mm	Dial Reading	Approximate Hardness B-Scale[A]	Dial Reading	Approximate Hardness B-Scale[A]	Dial Reading	Approximate Hardness B-Scale[A]
0.010	0.25	91	93	...	...	...	...
0.012	0.30	86	78	...	...	...	...
0.014	0.36	81	62	80	96	...	...
0.016	0.41	75	44	72	84	71	99
0.018	0.46	68	24	64	71	62	90
0.020	0.51	...	...	55	58	53	80
0.022	0.56	...	...	45	43	43	70
0.024	0.61	...	...	34	28	31	58
0.026	0.66	...	...	...	...	18	45
0.028	0.71	...	...	...	...	4	32
0.030	0.76	...	...	...	...	...	...

[A] These approximate hardness numbers are for use in selecting a suitable scale, and should not be used as hardness conversions. If necessary to convert test readings to another scale refer to Hardness Conversion Tables E 140 (Relationship Between Brinell Hardness, Vickers Hardness, Rockwell Hardness, Rockwell Superficial Hardness and Knoop Hardness).

TABLE 13 Corrections to Be Added to Rockwell Superficial 15N, 30N, and 45N Values Obtained on Cylindrical Specimens of Various Diameters[A]

Dial Reading	Diameters of Cylindrical Specimens					
	⅛ in. (3.2 mm)	¼ in. (6.4 mm)	⅜ in. (10 mm)	½ in. (13 mm)	¾ in. (19 mm)	1 in. (25 mm)
	Corrections to be Added to Rockwell Superficial 15N, 30N, and 45N Values[B]					
20	6.0	3.0	2.0	1.5	1.5	1.5
25	5.5	3.0	2.0	1.5	1.5	1.0
30	5.5	3.0	2.0	1.5	1.0	1.0
35	5.0	2.5	2.0	1.5	1.0	1.0
40	4.5	2.5	1.5	1.5	1.0	1.0
45	4.0	2.0	1.5	1.0	1.0	1.0
50	3.5	2.0	1.5	1.0	1.0	0.5
55	3.5	2.0	1.5	1.0	0.5	0.5
60	3.0	1.5	1.0	1.0	0.5	0.5
65	2.5	1.5	1.0	0.5	0.5	0.5
70	2.0	1.0	1.0	0.5	0.5	0.5
75	1.5	1.0	0.5	0.5	0.5	0
80	1.0	0.5	0.5	0.5	0	0
85	0.5	0.5	0.5	0.5	0	0
90	0	0	0	0	0	0

[A] When testing cylindrical specimens the accuracy of the test will be seriously affected by alignment of elevating screw, V-anvil, penetrators, surface finish, and the straightness of the cylinder.

[B] These corrections are approximate only and represent the averages, to the nearest 0.5 Rockwell superficial number, of numerous actual observations.

TABLE 14 **Corrections to Be Added to Rockwell Superficial 15T, 30T, and 45T Values Obtained on Cylindrical Specimens[A] of Various Diameters**

	Diameters of Cylindrical Specimens						
Dial Reading	⅛ in. (3.2 mm)	¼ in. (6.4 mm)	⅜ in. (10 mm)	½ in. (13 mm)	⅝ in. (16 mm)	¾ in. (19 mm)	1 in. (25 mm)
	Corrections to be Added to Rockwell Superficial 15T, 30T, and 45T Values[B]						
20	13.0	9.0	6.0	4.5	4.5	3.0	2.0
30	11.5	7.5	5.0	3.5	3.5	2.5	2.0
40	10.0	6.5	4.5	3.5	3.0	2.5	2.0
50	8.5	5.5	4.0	3.0	2.5	2.0	1.5
60	6.5	4.5	3.0	2.5	2.0	1.5	1.5
70	5.0	3.5	2.5	2.0	1.5	1.0	1.0
80	3.0	2.0	1.5	1.5	1.0	1.0	0.5
90	1.5	1.0	1.0	0.5	0.5	0.5	0.5

[A] When testing cylindrical specimens, the accuracy of the test will be seriously affected by alignment of elevating screw, V-anvil, penetrators, surface finish, and the straightness of the cylinder.

[B] These corrections are approximate only and represent the averages, to the nearest 0.5 Rockwell superficial number, of numerous actual observations.

TABLE 15   Tolerances on Applied Loads

Load, kgf (N)	Tolerance, kgf (N)
10 (98)	±0.20 (±1.96)
60 (589)	±0.45 (±4.41)
100 (981)	±0.65 (±4.57)
150 (1472)	±0.90 (±8.83)
3 (29)	±0.060 (±0.589)
15 (147)	±0.100 (±0.981)
30 (294)	±0.200 (±1.961)
45 (441)	±0.300 (±2.943)

TABLE 16   Allowable Deviation in Hardness Readings for Verified Diamond Penetrators

For Hardness Readings in Range of:	Allowable Deviation, Rockwell Units
C 63	±0.5
C 25	±1.0
30N 80	±0.5
30N 45	±1.0

TABLE 17   Tolerances for Rockwell Hardness Ball Penetrators

Diameter of Ball		Tolerance[A]	
in.	mm	in.	mm
¹⁄₁₆	1.588	±0.0001	±0.0025
⅛	3.175	±0.0001	±0.0025
¼	6.350	±0.0001	±0.0025
½	12.700	±0.0001	±0.0025

[A] For balls in the range of diameters specified, these tolerances and the permissible variation in the diameter of any one ball, as specified in 19.1.3, are met be Grade 25 steel balls of the Anti-Friction Bearing Manufacturers' Association (AFBMA).

TABLE 18   Maximum Mean Diagonal of Vickers Hardness Impression on Rockwell Hardness Balls

Ball Diameters		Maximum Mean Diagonal of Impression on the Ball Made with Vickers Indenter Under 10-kgf (98-N) Load, mm
in.	mm	
¹⁄₁₆	1.588	0.141
⅛	3.175	0.144
¼	6.350	0.145
½	12.700	0.147

TABLE 19   Hardness Ranges Used in Verification by Standardized Test Block Method

Rockwell Scale	Hardness Ranges
C	20 to 30 35 to 55 59 to 65
B	40 to 59 60 to 79 80 to 100
30N	40 to 50 55 to 73 75 to 80
30T	43 to 56 57 to 70 incl over 70 to 82

TABLE 21   Repeatability of Hardness Readings

Nominal Hardness of Standardized Test Block	The Repeatability of the Test Block Readings Shall Be Not Greater Than:
C Scale:	
60 and greater	0.5
Below 60	1.0
B Scale:	
60 to 100, incl	1.0
Below 60 to 40, incl	1.5
30N Scale:	
41.5 and greater	1.0
30T Scale:	
43 to 82	1.0

TABLE 20   Repeatability of Machines

Range of Standardized Hardness Test Blocks	The Repeatability[A] of the Machine Shall Be Not Greater Than:
Rockwell C Scale:	
20 to 30	2.0
35 to 55	1.5
59 to 65	1.0
Rockwell B Scale:	
40 to 59	2.5
60 to 79	2.0
80 to 100	2.0
Rockwell 30N Scale:	
40 to 50	2.0
55 to 73	1.5
75 to 80	1.0
Rockwell 30T Scale:	
43 to 56	2.5
57 to 70, incl	2.0
Over 70 to 82	2.0

[A] The repeatability of machines on Rockwell or Rockwell superficial hardness scales other than those given in Table 20 shall be the equivalent converted difference in hardness for those scales, except for the 15N and 15T scales. In the case of the 15N and 15T scales, the repeatability shall be no greater than 1.0 for all ranges.

Example—At C 60, typical readings of a series of impressions might range from 59 to 60, 59.5 to 60.5, 60 to 61, etc. Thus, converted A-scale values corresponding to C 59 to 60 (see Table II of Standard Tables E 140) would be A 80.7 to 81.2 and the repeatability for the A-scale would be 0.5.

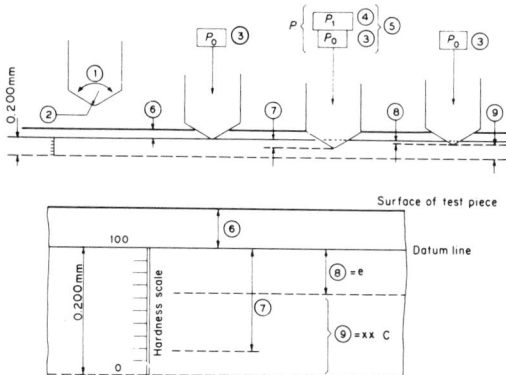

FIG. 1   Rockwell Hardness Test with Diamond Penetrator (Rockwell C) (Table 1)

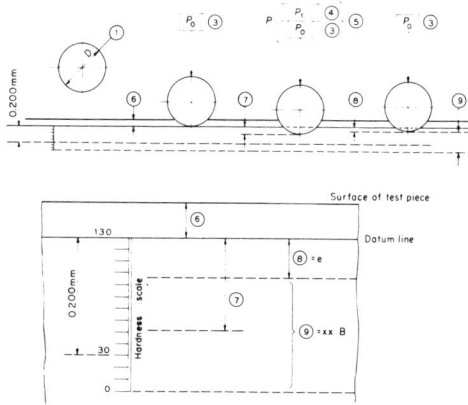

FIG. 2   Rockwell Hardness Test with Steel Ball Penetrator (Rockwell B) (Table 2)

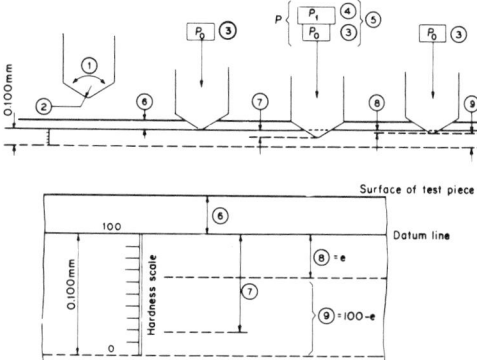

FIG. 3   Rockwell Superficial Hardness Test with Diamond Penetrator (Rockwell 30N) (Table 8)

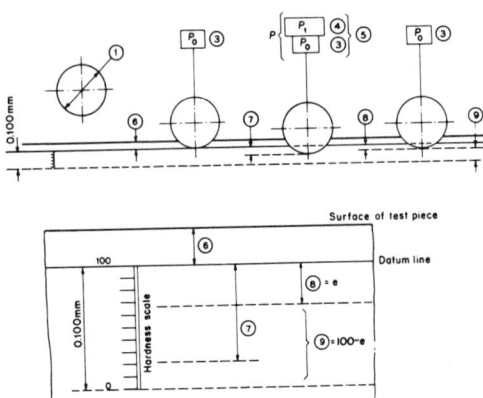

FIG. 4   Rockwell Superfical Hardness Test with Steel Ball Penetrator (Rockwell 30T) (Table 9)

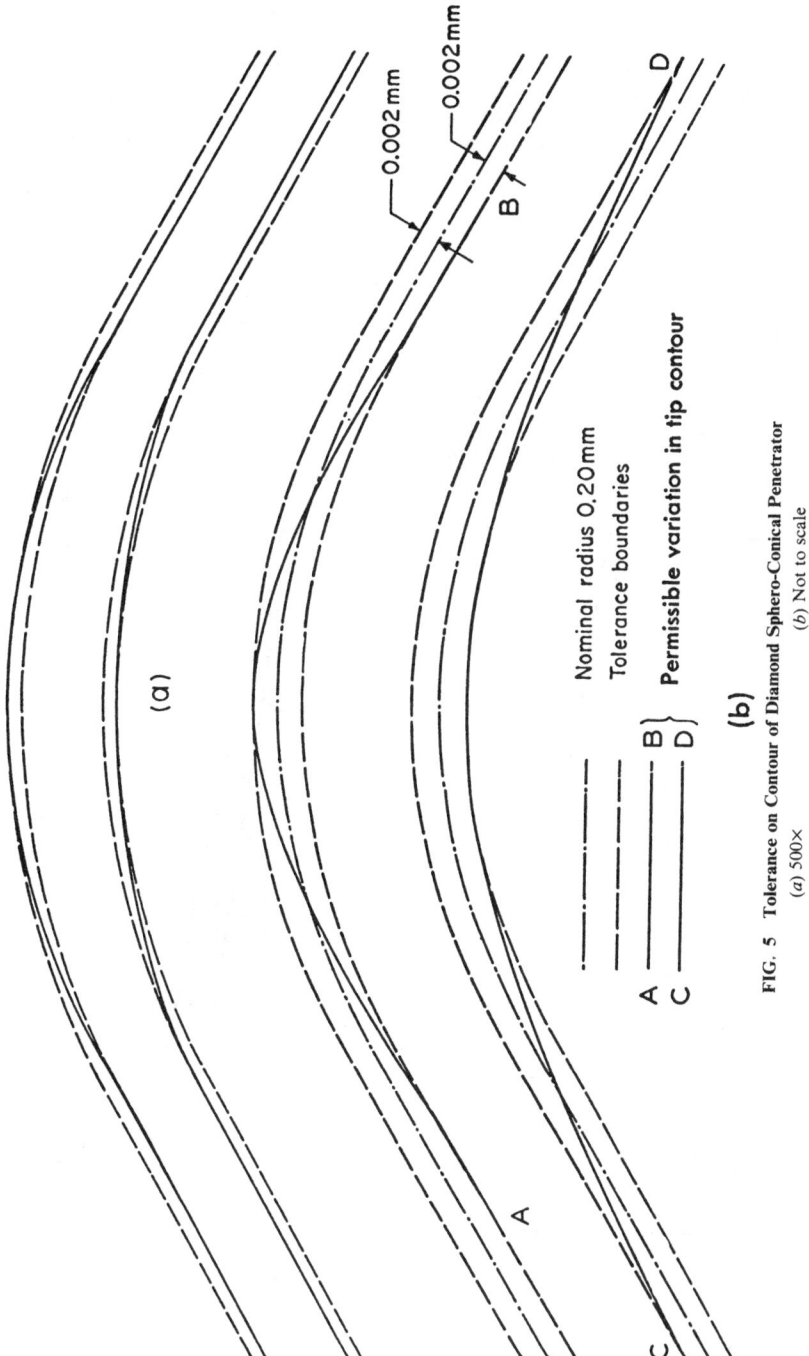

0.002 mm

0.002 mm

(a)

B

D

A

B

C

D

Nominal radius 0.20mm

Tolerance boundaries

A
B } Permissible variation in tip contour
C
D }

(b)

FIG. 5   Tolerance on Contour of Diamond Sphero-Conical Penetrator
(a) 500×   (b) Not to scale

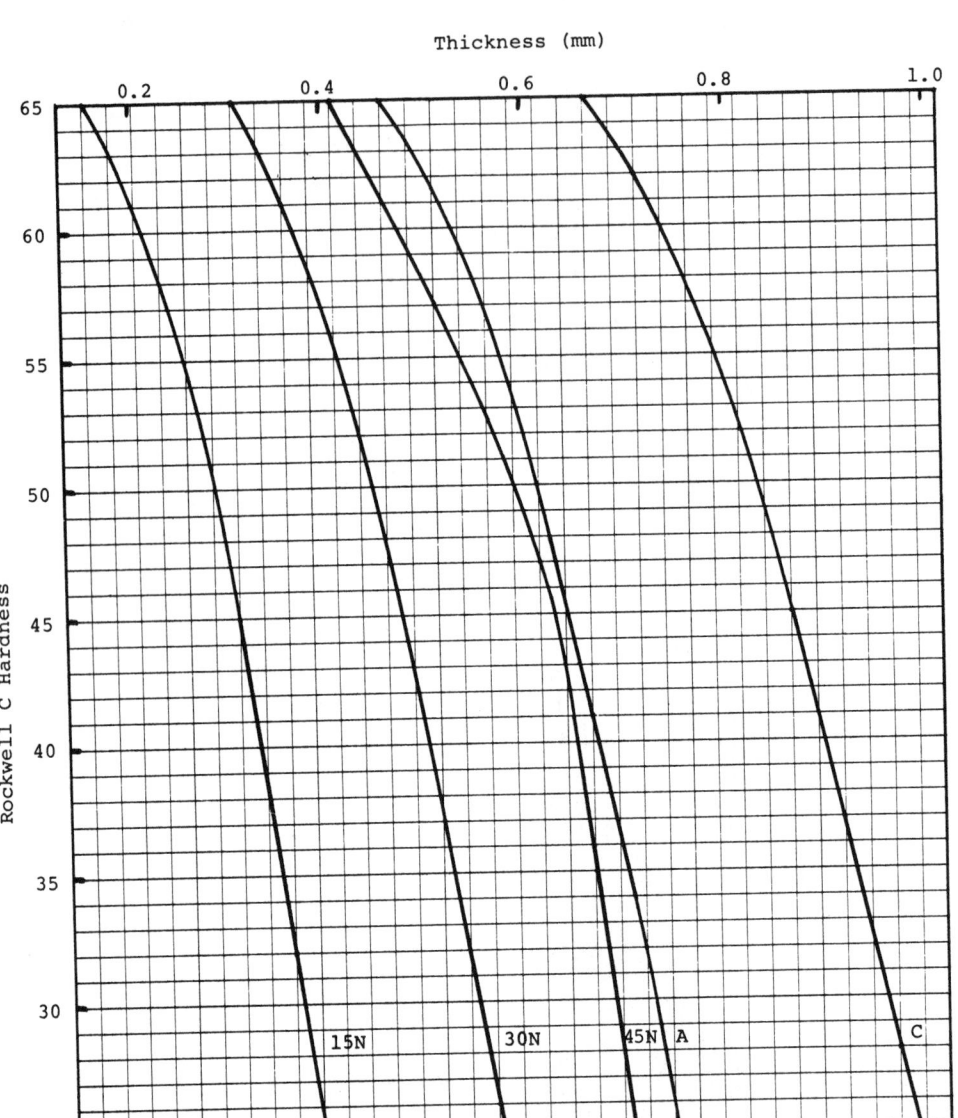

NOTE—Locate a point corresponding to the thickness-hardness combination to be tested. Only scales falling to the left of this point may be used to test this combination.

FIG. 6 Thickness Limits for Rockwell Hardness Testing Using the Diamond Penetrator

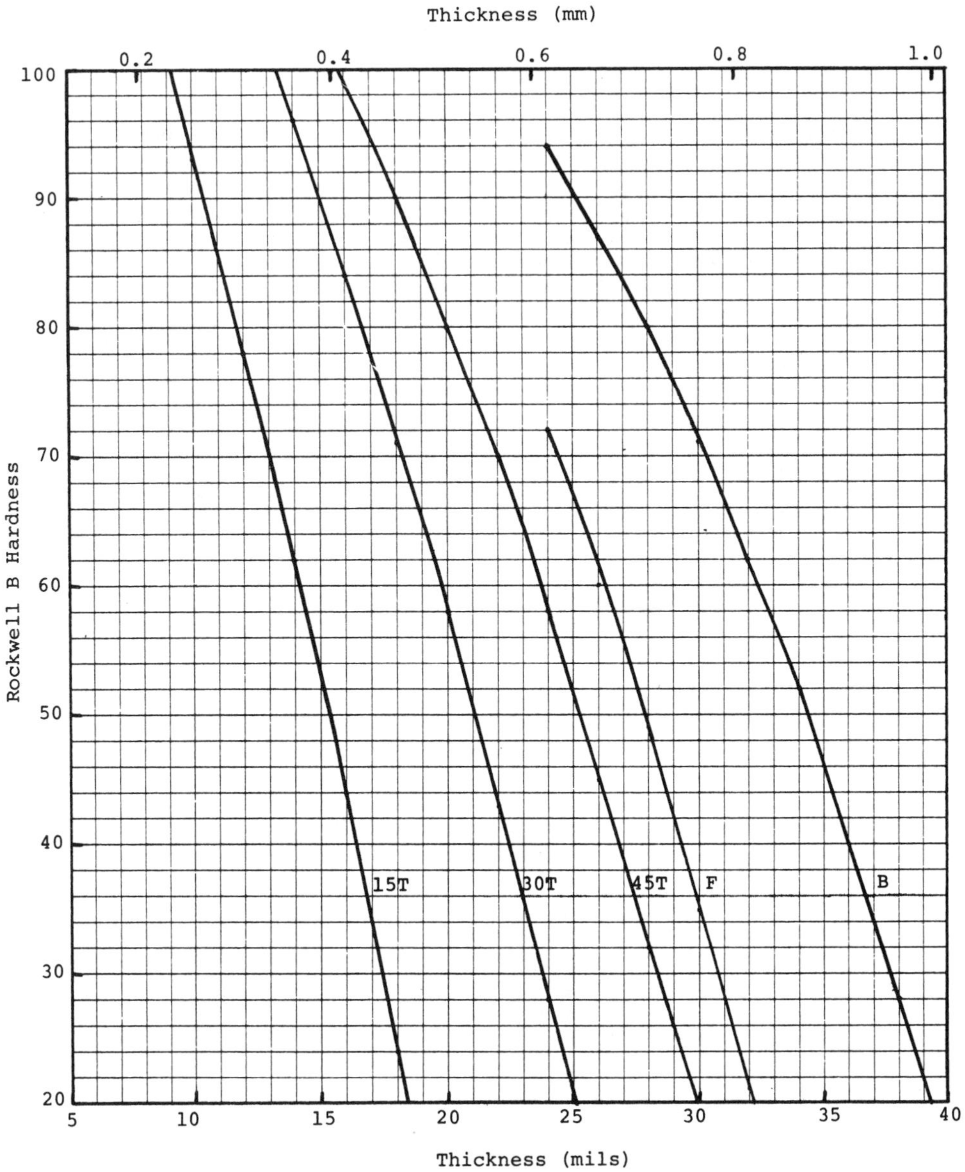

Thickness (mm)

Rockwell B Hardness

15T 30T 45T F B

Thickness (mils)

NOTE—Locate a point corresponding to the thickness-hardness combination to be tested. Only scales falling to the left of this point may be used to test this combination.

FIG. 7   Thickness Limits for Rockwell Hardness Testing Using the ¹⁄₁₆-in. (1.588-mm) Diameter Ball Penetrator

APPENDIX

(Nonmandatory Information)

X1. LIST OF ASTM SPECIFICATIONS GIVING HARDNESS VALUES CORRESPONDING TO TENSILE STRENGTH

X1.1 The following ASTM standards give approximate Rockwell hardness or Rockwell superficial hardness values corresponding to the tensile strength values specified for the materials covered: Methods and Definitions A 370 and Specifications B 19, B 36, B 96, B 97, B 103, B 12ᵢ, B 122, B 130, B 134, B 152, B 291, and B 370.

Standard Methods for
NOTCHED BAR IMPACT TESTING OF METALLIC MATERIALS[1]

This standard is issued under the fixed designation E 23; the number immediately following the designation indicates the year of original adoption or, in the case of revision, the year of last revision. A number in parentheses indicates the year of last reapproval. A superscript epsilon (ϵ) indicates an editorial change since the last revision or reapproval.

These methods have been approved for use by agencies of the Department of Defense to replace method 221.1 of Federal Test Method Standard No. 151b and for listing in the DoD Index of Specifications and Standards.

NOTE—Figures 2, 3, 4, 5, 6, 7, 11, 12, 13, 14, 15, and 16 were editorially corrected, and the designation date was changed March 5, 1982.

1. Scope

1.1 These methods describe notched-bar impact testing of metallic materials by the Charpy (simple-beam) apparatus and the Izod (cantilever-beam) apparatus. They give: (*a*) a description of apparatus, (*b*) requirements for inspection and calibration, (*c*) safety precautions, (*d*) sampling, (*e*) dimensions and preparation of specimens, (*f*) testing procedures, (*g*) precision and accuracy, and (*h*) appended notes on the significance of notched-bar impact testing. These methods will in most cases also apply to tests on unnotched specimens.

1.2 The values stated in SI units are to be regarded as the standard.

2. Summary of Methods

2.1 The essential features of an impact test are: (*a*) a suitable specimen (specimens of several different types are recognized), (*b*) an anvil or support on which the test specimen is placed to receive the blow of the moving mass, (*c*) a moving mass of known kinetic energy which must be great enough to break the test specimen placed in its path, and (*d*) a device for measuring the energy absorbed by the broken specimen.

3. Significance

3.1 These methods of impact testing relate specifically to the behavior of metal when subjected to a single application of a load resulting in multiaxial stresses associated with a notch, coupled with high rates of loading and in some cases with high or low temperatures. For some materials and temperatures, impact tests on notched specimens have been found to predict the likelihood of brittle fracture better than tension tests or other tests used in material specifications. Further information on significance appears in the Appendix.

4. Apparatus

4.1 *General Requirements:*

4.1.1 The testing machine shall be a pendulum type of rigid construction and of capacity more than sufficient to break the specimen in one blow.

4.1.2 The machine frame shall be equipped with a bubble level or a machined surface suitable for establishing levelness. The machine shall be level to within 3:1000 and securely bolted to a concrete floor not less than 150 mm (6 in.) thick or, when this is not practical, the machine shall be bolted to a foundation having a mass not less than 40 times that of the pendulum. The bolts shall be tightened as specified by the machine manufacturer.

4.1.3 The machine shall be furnished with scales graduated either in degrees or directly in energy on which readings can be estimated in increments of 0.25 % of the energy range or less. The scales may be compensated for wind-

---

[1] These methods are under the jurisdiction of ASTM Committee E-28 on Mechanical Testing and are the direct responsibility of Subcommittee E28.07 on Impact Testing.

Current edition approved March 5, 1982. Published July 1982. Originally published as E 23 – 33 T. Last previous edition E 23 – 81.

age and pendulum friction. The error in the scale reading at any point shall not exceed 0.2 % of the range or 0.4 % of the reading, whichever is larger. (See 5.2.6.2 and 5.2.7.)

4.1.4 The total friction and windage losses of the machine during the swing in the striking direction shall not exceed 0.75 % of the scale range capacity, and pendulum energy loss from friction in the indicating mechanism shall not exceed 0.25 % of scale range capacity.

4.1.5 The dimensions of the pendulum shall be such that the center of percussion of the pendulum is at the center of strike within 1 % of the distance from the axis of rotation to the center of strike. When hanging free, the pendulum shall hang so that the striking edge is within 2.5 mm (0.10 in.) of the position where it would just touch the test specimen. When the indicator has been positioned to read zero energy in a free swing, it shall read within 0.2 % of scale range when the striking edge of the pendulum is held against the test specimen. The plane of swing of the pendulum shall be perpendicular to the transverse axis of the Charpy specimen anvils or Izod vise within 3:1000.

4.1.6 Transverse play of the pendulum at the striker shall not exceed 0.75 mm (0.030 in.) under a transverse force of 4 % of the effective weight of the pendulum applied at the center of strike. Radial play of the pendulum bearings shall not exceed 0.075 mm (0.003 in.). The tangential velocity (the impact velocity) of the pendulum at the center of the strike shall not be less than 3 nor more than 6 m/s (not less than 10 nor more than 20 ft/s).

4.1.7 Before release, the height of the center of strike above its free hanging position shall be within 0.4 % of the range capacity divided by the pendulum weight, measured as described in 5.2.3.3. If windage and friction are compensated for by increasing the height of drop, the height of drop may be increased by not more than 1 %.

4.1.8 The mechanism for releasing the pendulum from its initial position shall operate freely and permit release of the pendulum without initial impulse, retardation, or side vibration. If the same lever that is used to release the pendulum is also used to engage the brake, means shall be provided for preventing the brake from being accidentally engaged.

4.2 *Specimen Clearance*—To ensure satisfactory results when testing materials of different strengths and compositions, the test specimen shall be free to leave the machine with a minimum of interference and shall not rebound into the pendulum before the pendulum completes its swing. Pendulums used on Charpy machines are of two basic designs, as shown in Fig. 1. When using a C-type pendulum, the broken specimen will not rebound into the pendulum and slow it down if the clearance at the end of the specimen is at least 13 mm (0.5 in.) or if the specimen is deflected out of the machine by some arrangement as is shown in Fig. 1. When using the U-type pendulum, means shall be provided to prevent the broken specimen from rebounding against the pendulum (Fig. 1). In most U-type pendulum machines, the shrouds should be designed and installed to the following requirements: (*a*) have a thickness of approximately 1.5 mm (0.06 in.), (*b*) have a minimum hardness of 45 HRC, (*c*) have a radius of less than 1.5 mm (0.06 in.) at the underside corners, and (*d*) be so positioned that the clearance between them and the pendulum overhang (both top and sides) does not exceed 1.5 mm (0.06 in.).

NOTE 1—In machines where the opening within the pendulum permits clearance between the ends of a specimen (resting on the anvil supports) and the shrouds, and this clearance is at least 13 mm (0.5 in.) requirements (*a*) and (*d*) need not apply.

4.3 *Charpy Apparatus:*

4.3.1 Means shall be provided (Fig. 2) to locate and support the test specimen against two anvil blocks in such a position that the center of the notch can be located within 0.25 mm (0.010 in.) of the midpoint between the anvils (see 11.2.1.2).

4.3.2 The supports and striking edge shall be of the forms and dimensions shown in Fig. 2. Other dimensions of the pendulum and supports should be such as to minimize interference between the pendulum and broken specimens.

4.3.3 The center line of the striking edge shall advance in the plane that is within 0.40 mm (0.016 in.) of the midpoint between the supporting edges of the specimen anvils. The striking edge shall be perpendicular to the longitudinal axis of the specimen within 5:1000. The striking edge shall be parallel within 1:1000 to the face of a perfectly square test specimen held against the anvil.

4.3.4 Specimen supports shall be square with anvil faces within 2.5:1000. Specimen supports shall be coplanar within 0.125 mm (0.005 in.) and parallel within 2:1000.

4.4 *Izod Apparatus:*

4.4.1 Means shall be provided (Fig. 3) for clamping the specimen in such a position that the face of the specimen is parallel to the striking edge within 1:1000. The edges of the clamping surfaces shall be sharp angles of 90 ± 1° with radii less than 0.40 mm (0.016 in.). The clamping surfaces shall be smooth with a 2-μm (63-μin.) finish or better, and shall clamp the specimen firmly at the notch with the clamping force applied in the direction of impact. For rectangular specimens, the clamping surfaces shall be flat and parallel within 0.025 mm (0.001 in.). For cylindral specimens, the clamping surfaces shall be contoured to match the specimen and each surface shall contact a minimum of $\pi/2$ rad (90°) of the specimen circumference.

4.4.2 The dimensions of the striking edge and its position relative to the specimen clamps shall be as shown in Fig. 3.

4.5 *Energy Range*—Energy values above 80 % of the scale range are inaccurate and shall be reported as approximate. Ideally an impact test would be conducted at a constant impact velocity. In a pendulum-type test, the velocity decreases as the fracture progresses. For specimens that have impact energies approaching the capacity of the pendulum, the velocity of the pendulum decreases during fracture to the point that accurate impact energies are no longer obtained.

5. Inspection

5.1 *Critical Parts:*

5.1.1 *Specimen Anvils and Supports or Vise*— These shall conform to the dimensions shown in Fig. 2 or 3. To ensure a minimum of energy loss through absorption, bolts shall be tightened as specified by the machine manufacturer.

NOTE 2—The impact machine will be inaccurate to the extent that some energy is used in deformation or movement of its component parts or of the machine as a whole; this energy will be registered as used in fracturing the specimen.

5.1.2 *Pendulum Striking Edge*—The striking edge (tup) of the pendulum shall conform to the dimensions shown in Figs. 2 or 3. To ensure a minimum of energy loss through absorption, the striking edge bolts shall be tightened as specified by the machine manufacturer. The

pendulum striking edge (tup) shall comply with 4.3.3 (for Charpy tests) or 4.4.1 (for Izod tests) by bringing it into contact with a standard Charpy or Izod specimen.

5.2 *Pendulum Operation:*

5.2.1 *Pendulum Release Mechanism*—The mechanism for releasing the pendulum from its initial position shall comply with 4.1.8.

5.2.2 *Pendulum Alignment*—The pendulum shall comply with 4.1.5 and 4.1.6. If the side play in the pendulum or the radial plays in the bearings exceeds the specified limits, adjust or replace the bearings.

5.2.3 *Potential Energy*—Determine the initial potential energy using the following procedure when the center of strike of the pendulum is coincident with the line from the center of rotation through the center of percussion. If the center of strike is more than 2.5 mm (0.1 in.) from this line, suitable corrections in elevation of the center of strike must be made in 5.2.3.2, 5.2.3.3, 5.2.6.1, and 5.2.7, so that elevations set or measured correspond to what they would be if the center of strike were on this line.

5.2.3.1 For Charpy machines place a half-width specimen (see Fig. 4) 10 by 5 mm (0.394 by 0.197 in.) in test position. With the striking edge in contact with the specimen, a line scribed from the top edge of the specimen to the striking edge will indicate the center of strike on the striking edge.

5.2.3.2 For Izod machines, the center of strike may be considered to be the contact line when the pendulum is brought into contact with a specimen in the normal testing position.

NOTE 3—A method of accurately determining the centers of strike of Izod machines is to place a specimen, so machined that the distance from the center of the notch to the top of the specimen is 22.66 mm (0.892 in.), in test position. With the striking edge in contact with the specimen, a line scribed from the top edge of the specimen to the striking edge will indicate the center of strike on the striking edge.

5.2.3.3 Support the pendulum horizontally to within 15:1000 with two supports, one at the bearings (or center of rotation) and the other at the center of strike on the striking edge (see Fig. 5). Arrange the support at the striking edge to react upon some suitable weighing device such as a platform scale or balance, and determine the weight to within 0.4 %. Take care to minimize friction at either point of support.

Make contact with the striking edge through a round rod crossing the edge at a 90° angle. The weight of the pendulum is the scale reading minus the weights of the supporting rod and any shims that may be used to maintain the pendulum in a horizontal position.

5.2.3.4 Measure the height of pendulum drop for compliance with the requirement of 4.1.7. On Charpy machines measure the height from the top edge of a half-width (or center of a full-width) specimen to the elevated position of the center of strike to 0.1 %. On Izod machines measure the height from a distance 22.66 mm (0.892 in.) above the vise to the release position of the center of strike to 0.1 %.

5.2.3.5 The potential energy of the system is equal to the height from which the pendulum falls, as determined in 5.2.3.4, times the weight of the pendulum, as determined in 5.2.3.3.

5.2.4 *Impact Velocity*—Determine the impact velocity, v, of the machine, neglecting friction, by means of the following equation:

$$v = \sqrt{2\,gh}$$

where:

v = velocity, m/s (or ft/s),
g = acceleration of gravity, m/s² (or ft/s²), and
h = initial elevation of the striking edge, m (or ft).

5.2.5 *Center of Percussion*—To ensure that minimum force is transmitted to the point of rotation, the center of percussion shall be at a point within 1 % of the distance from the axis of rotation to the center of strike in the specimen. Determine the location of the center of percussion as follows:

5.2.5.1 Using a stop watch or some other suitable time-measuring device, capable of measuring time to within 0.2 s, swing the pendulum through a total angle not greater than 15° and record the time for 100 complete cycles (to and fro).

5.2.5.2 Determine the center of percussion by means of the following equation:

$$l = 0.2484p^2, \text{ to determine } l \text{ in metres}$$
$$l = 0.815p^2, \text{ to determine } l \text{ in feet}$$

where:

l = distance from the axis to the center of percussion, m (or ft), and
p = time of a complete cycle (to and fro) of the pendulum, s.

5.2.6 *Friction*—The energy loss from friction and windage of the pendulum and friction in the recording mechanism, if not corrected, will be included in the energy loss attributed to breaking the specimen and can result in erroneously high impact values. In machines recording in degrees, normal frictional losses are usually not compensated for by the machine manufacturer, whereas they are usually compensated for in machines recording directly in energy by increasing the starting height of the pendulum. Determine energy losses from friction as follows:

5.2.6.1 Without a specimen in the machine, and with the indicator at the maximum energy reading, release the pendulum from its starting position and record the energy value indicated. This value should indicate zero energy if frictional losses have been corrected by the manufacturer. Raise the pendulum so it just contacts the pointer at the value obtained in the free swing. Secure the pendulum at this height and determine the vertical distance from the center of strike to the top of a half-width specimen positioned on the specimen rests (see 5.2.3.1). Determine the weight of the pendulum as in 5.2.3.2 and multiply by this distance. The difference in this value and the initial potential energy is the total energy loss in the pendulum and indicator combined. Without resetting the pointer, repeatedly release the pendulum from its initial position until the pointer shows no further movement. The energy loss determined by the final position of the pointer is that due to the pendulum alone. The frictional loss in the indicator alone is then the difference between the combined indicator and pendulum losses and those due to the pendulum alone.

5.2.6.2 To ensure that friction and windage losses are within tolerances allowed (see 4.1.4), a simple weekly procedure may be adopted for direct-reading machines. The following steps are recommended: (*a*) release the pendulum from its upright position without a specimen in the machine, and the energy reading should be 0 J (0 ft·lbf); (*b*) without resetting the pointer, again release the pendulum and permit it to swing 11 half cycles; and after the pendulum starts its 11th cycle, move the pointer to between 5 and 10 % of scale range capacity and record the value obtained. This value, divided by 11, shall not exceed 0.4 % of scale range capacity. If this value does exceed 0.4 %, the bearings should be cleaned or replaced.

486

5.2.7 *Indicating Mechanism*—To ensure that the scale is recording accurately over the entire range, check it at graduation marks corresponding to approximately 0, 10, 20, 30, 50, and 70 % of each range. With the striking edge of the pendulum scribed to indicate the center of strike, lift the pendulum and set it in a position where the indicator reads, for example, 13 J (10 ft·lbf). Determine the height of the pendulum to within 0.1 %. The height of the pendulum multiplied by its weight, as determined in 5.2.3.3, is the residual energy. Increase this value by friction and windage losses in accordance with 5.2.6 and subtract from the potential energy determined in 5.2.3. Make similar calculations at other points of the scale. The scale pointer shall not overshoot or drop back with the pendulum. Make test swings from various heights to check visually the operation of the pointer over several portions of the scale.

5.2.8 The impact value shall be taken as the energy absorbed in breaking the specimen and is equal to the difference between the energy in the striking member at the instant of impact with the specimen and the energy remaining after breaking the specimen.

6. Precaution in Operation of Machine

6.1 *Safety Precautions*—Precautions should be taken to protect personnel from the swinging pendulum, flying broken specimens, and hazards associated with specimen warming and cooling media.

7. Sampling

7.1 Specimens shall be taken from the material as specified by the applicable specification.

8. Test Specimens

8.1 *Material Dependence*—The choice of specimen depends to some extent upon the characteristics of the material to be tested. A given specimen may not be equally satisfactory for soft nonferrous metals and hardened steels; therefore, a number of types of specimens are recognized. In general, sharper and deeper notches are required to distinguish differences in the more ductile materials or with lower testing velocities.

8.1.1 The specimens shown in Figs. 6 and 7 are those most widely used and most generally satisfactory. They are particularly suitable for ferrous metals, excepting cast iron.[2]

8.1.2 The specimen commonly found suitable for die cast alloys is shown in Fig. 8.

8.1.3 The specimens commonly found suitable for powdered metals (P/M) are shown in Figs. 9 and 10. The specimen surface may be in the as-produced condition or smoothly machined, but polishing has proven generally unnecessary. Unnotched specimens are used with P/M materials. In P/M materials, the impact test results will be affected by specimen orientation. Therefore, unless otherwise specified, the position of the specimen in the machine shall be such that the pendulum will strike a surface that is parallel to the compacting direction.

8.2 *Sub-Size Specimen*—When the amount of material available does not permit making the standard impact test specimens shown in Figs. 6 and 7, smaller specimens may be used, but the results obtained on different sizes of specimens cannot be compared directly (X1.3). When Charpy specimens other than the standard are necessary or specified, it is recommended that they be selected from Fig. 4.

8.3 *Supplementary Specimens*—For economy in preparation of test specimens, special specimens of round or rectangular cross section are sometimes used for cantilever beam test. These are shown as Specimens X, Y, and Z in Figs. 11 and 12. Specimen Z is sometimes called the Philpot specimen after the name of the original designer. In the case of hard materials, the machining of the flat surface struck by the pendulum is sometimes omitted. Types Y and Z require a different vise from that shown in Fig. 3, each half of the vise having a semicylindrical recess that closely fits the clamped portion of the specimen. As previously stated, the results cannot be reliably compared to those obtained using specimens of other sizes or shapes.

8.4 *Specimen Machining:*

8.4.1 When heat-treated materials are being evaluated, the specimen shall be finish machined, including notching, after the final heat treatment, unless it can be demonstrated that

---

[2] For testing cast iron, see 1933 Report of Subcommittee XV on Impact Testing of Committee A-3 on Cast Iron, *Proceedings*, Am. Soc. Testing Mats., Vol 33, Part 1, 1933.

there is no difference when machined prior to heat treatment.

8.4.2 Notches shall be smoothly machined but polishing has proven generally unnecessary. However, since variations in notch dimensions will seriously affect the results of the tests, it is necessary to adhere to the tolerances given in Fig. 6 (X1.2 illustrates the effects from varying notch dimensions on Type A specimens). In keyhole specimens, the round hole shall be carefully drilled with a slow feed. The slot may be cut by any feasible method. Care must be exercised in cutting the slot to see that the surface of the drilled hole opposite the slot is not marked.

8.4.3 Identification marks shall not be placed on any surface of the specimen that contacts the striking edge or specimen supports. All stamping shall be done in a way that avoids cold deforming of the specimen at the notch root or at any other portion of the specimen that is visibly deformed during fracture.

9. Preparation of Apparatus

9.1 *Daily Checking Procedure*—After the testing machine has been ascertained to comply with Sections 4 and 5, the routine daily checking procedures shall be as follows:

9.1.1 Prior to testing a group of specimens and before a specimen is placed in position to be tested, check the machine by a free swing of the pendulum. With the indicator at the maximum energy position, a free swing of the pendulum shall indicate zero energy on machines reading directly in energy, which are compensated for frictional losses. On machines recording in degrees, the indicated values when converted to energy shall be compensated for frictional losses that are assumed to be proportional to the arc of swing.

10. Verification of Charpy Machines

10.1 Verification consists of inspecting those parts subjected to wear to ensure that the requirements of Sections 4 and 5 are met and the testing of standardized specimens (Notes 4 to 6). It is not intended that parts not subjected to wear (such as pendulum and scale linearity) need to be remeasured during verification unless a problem is evident. The average value at each energy level determined for the standardized specimens shall correspond to the nominal

values of the standardized specimens within 1.4 J (1.0 ft·lbf) or 5.0 %, whichever is greater.

NOTE 4—Standardized specimens are available for Charpy machines only.

NOTE 5—Information pertaining to the availability of standardized specimens may be obtained by addressing: Director, Army Materials and Mechanics Research Center, ATTN: DRXMR-MQ, Watertown, Mass. 02172.

NOTE 6—The Army Materials and Mechanics Research Center has for many years conducted a Charpy machine qualification program whereby standardized specimens are used to certify the machines of laboratories using the test as an inspection requirement on government contracts.[3] If the user desires, the results of tests with the standardized specimens will be evaluated. Participants desirous of the evaluation should complete the questionnaire provided with the standardized specimens. The questionnaire provides for information such as testing temperature, the dimensions of certain critical parts, the cooling and testing techniques, and the results of the test. The broken standardized specimens are to be returned along with the completed questionnaire for evaluation (see Note 5 for address). Upon completion of the evaluation, the Army Materials and Mechanics Research Center will return a report. If a machine is producing values outside the standardized specimen tolerances, the report may suggest changes in machine design, repair or replacement of certain machine parts, a change in testing techniques, etc.

10.2 *Frequency of Verification*—Charpy machines shall be verified within one year prior to the time of testing. Charpy machines shall, however, be verified immediately after replacing parts, making repairs or adjustments, after they have been moved, or whenever there is reason to doubt the accuracy of the results, without regard to the time interval.

11. Procedure

11.1 The Daily Checking Procedure (Section 9) shall be performed at the beginning of each day or each shift.

11.2 *Charpy Test Procedure*—The Charpy test procedure may be summarized as follows: the test specimen is removed from its cooling (or heating) medium, if used, and positioned on the specimen supports; the pendulum is released without vibration, and the specimen is broken within 5 s after removal from the medium. Information is obtained from the machine and from the broken specimen. The details are described as follows:

[3] Driscoll, D. E., "Reproducibility of Charpy Impact Test," *Symposium on Impact Testing, ASTM STP 176*, Am. Soc. Testing Mats., 1955, p. 170.

11.2.1 *Temperature of Testing*—In most materials, impact values vary with temperature. Unless otherwise specified, tests shall be made at 15 to 32°C (60 to 90°F). Accuracy of results when testing at other temperatures requires the following procedure: For liquid cooling or heating fill a suitable container, which has a grid raised at least 25 mm (1 in.) from the bottom, with liquid so that the specimen when immersed will be covered with at least 25 mm (1 in.) of the liquid. Bring the liquid to the desired temperature by any convenient method. The device used to measure the temperature of the bath should be placed in the center of a group of the specimens. Verify all temperature-measuring equipment at least twice annually. When using a liquid medium, hold the specimens in an agitated bath at the desired temperature within ±1°C (±2°F) for at least 5 min. When using a gas medium, position the specimens so that the gas circulates around them and hold the gas at the desired temperature within ±1°C (±2°F) for at least 30 min. Leave the mechanism used to remove the specimen from the medium in the medium except when handling the specimens.

NOTE 7—Temperatures up to +260°C (+500°F) may be obtained with certain oils, but "flash-point" temperatures must be carefully observed.

11.2.2 *Placement of Test Specimen in Machine*—It is recommended that self-centering tongs similar to those shown in Fig. 13 be used in placing the specimen in the machine (see 4.3.1). The tongs illustrated in Fig. 13 are for centering V-notch specimens. If keyhole specimens are used, modification of the tong design may be necessary. If an end-centering device is used, caution must be taken to ensure that low-energy high-strength specimens will not rebound off this device into the pendulum and cause erroneously high recorded values. Many such devices are permanent fixtures of machines, and if the clearance between the end of a specimen in test position and the centering device is not approximately 13 mm (0.5 in.), the broken specimens may rebound into the pendulum.

11.2.3 *Operation of the Machine:*

11.2.3.1 Set the energy indicator at the maximum scale reading; take the test specimen from its cooling (or heating) medium, if used; place it in proper position on the specimen anvils; and release the pendulum smoothly. This entire sequence shall take less than 5 s if a cooling or heating medium is used.

11.2.3.2 If any specimen fails to break, do not repeat the blow but record the fact, indicating whether the failure to break occurred through extreme ductility or lack of sufficient energy in the blow. Such results of such tests shall not be included in the average.

11.2.3.3 If any specimen jams in the machine, disregard the results and check the machine thoroughly for damage or maladjustment, which would affect its calibration.

11.2.3.4 To prevent recording an erroneous value caused by jarring the indicator when locking the pendulum in its upright position, read the value from the indicator prior to locking the pendulum for the next test.

11.2.4 *Information Obtainable from the Test*:

11.2.4.1 *Impact Energy*—The amount of energy required to fracture the specimen is determined from the machine reading.

11.2.4.2 *Lateral Expansion*—The method for measuring lateral expansion must take into account the fact that the fracture path seldom bisects the point of maximum expansion on both sides of a specimen. One half of a broken specimen may include the maximum expansion for both sides, one side only, on neither. The technique used must therefore provide an expansion value equal to the sum of the higher of the two values obtained for each side by measuring the two halves separately. The amount of expansion on each side of each half must be measured relative to the plane defined by the undeformed portion of the side of the specimen, Fig. 16. Expansion may be measured by using a gage similar to that shown in Figs. 17 and 18. Measure the two broken halves individually. First, though, check the sides perpendicular to the notch to ensure that no burrs were formed on these sides during impact testing; if such burrs exist, they must be removed, for example, by rubbing on emery cloth, making sure that the protrusions to be measured are not rubbed during the removal of the burr. Next, place the halves together so that the compression sides are facing one another. Take one half and press it firmly against the reference supports, with the protrusion against the gage anvil. Note the reading, then repeat this step with the other broken half, ensuring that the same side of the specimen is measured. The larger of the two values is the expansion of that side of the specimen. Next, repeat this procedure to measure the protrusions on the opposite side, then add the larger values obtained for each side. Measure each specimen.

NOTE 8—Examine each fracture surface to ascertain that the protrusions have not been damaged by contacting the anvil, machine mounting surface, etc. Such specimens should be discarded since this may cause erroneous readings.

11.2.4.3 *Fracture Appearance*—The percentage of shear fracture may be determined by any of the following methods: (*1*) measure the length and width of the cleavage portion of the fracture surface, as shown in Fig. 14, and determine the percent shear from either Table 1 or Table 2 depending on the units of measurement; (*2*) compare the appearance of the fracture of the specimen with a fracture appearance chart such as that shown in Fig. 15; (*3*) magnify the fracture surface and compare it to a precalibrated overlay chart or measure the percent shear fracture by means of a planimeter; or (*4*) photograph the fracture surface at a suitable magnification and measure the percent shear fracture by means of a planimeter.

NOTE 9—Because of the subjective nature of the evaluation of fracture appearance, it is not recommended that it be used in specifications.

11.3 *Izod Test Procedure*—The Izod test procedure may be summarized as follows: the test specimen is positioned in the specimen-holding fixture and the pendulum is released without vibration. Information is obtained from the machine and from the broken specimen. The details are described as follows:

11.3.1 *Temperature of Testing*—The specimen-holding fixture for Izod specimens is in most cases part of the base of the machine and cannot be readily cooled (or heated). For this reason, Izod testing is not recommended at other than room temperature.

11.3.2 Clamp the specimen firmly in the support vise so that the centerline of the notch is in the plane of the top of the vise within 0.125 mm (0.005 in.). Set the energy indicator at the maximum scale reading, and release the pendulum smoothly. Sections 11.2.3.2 to 11.2.3.4 inclusively, also apply when testing Izod specimens.

11.3.3 *Information Obtainable from the Test*—The impact energy, lateral expansion, and fracture appearance, may be determined as described in 11.2.4.

12. Report

12.1 For commercial acceptance testing, the following is considered sufficient:

12.1.1 Type of specimen used (and size if not the standard size).

12.1.2 Temperature of the specimen.

12.1.3 When required any or all of the following shall be reported:

12.1.3.1 Energy absorbed,

12.1.3.2 Lateral expansion, and

12.1.3.3 Fracture appearance (see Note 9).

13. Precision and Accuracy

13.1 The precision and accuracy of these methods are being established.

TABLE 1 Percent Shear for Measurements Made in Millimetres

NOTE—100 % shear is to be reported when either A or B is zero.

Dimension A, mm

Dimension B, mm	1.0	1.5	2.0	2.5	3.0	3.5	4.0	4.5	5.0	5.5	6.0	6.5	7.0	7.5	8.0	8.5	9.0	9.5	10
1.0	99	98	98	97	96	96	95	94	94	93	92	92	91	91	90	89	89	88	88
1.5	98	97	96	95	94	93	92	92	91	90	89	88	87	86	85	84	83	82	81
2.0	98	96	95	94	92	91	90	89	88	86	85	84	82	81	80	79	77	76	75
2.5	97	95	94	92	91	89	88	86	84	83	81	80	78	77	75	73	72	70	69
3.0	96	94	92	91	89	87	85	83	81	79	77	76	74	72	70	68	66	64	62
3.5	96	93	91	89	87	85	82	80	78	76	74	72	69	67	65	63	61	58	56
4.0	95	92	90	88	85	82	80	77	75	72	70	67	65	62	60	57	55	52	50
4.5	94	92	89	86	83	80	77	75	72	69	66	63	61	58	55	52	49	46	44
5.0	94	91	88	85	81	78	75	72	69	66	62	59	56	53	50	47	44	41	37
5.5	93	90	86	83	79	76	72	69	66	62	59	55	51	47	44	40	36	33	29
6.0	92	89	85	81	77	74	70	66	62	59	55	51	47	43	39	35	31	27	23
6.5	92	88	84	80	76	72	67	63	59	55	51	47	43	39	34	30	26	21	17
7.0	91	87	82	78	74	69	65	61	56	52	47	43	39	34	30	25	20	16	11
7.5	91	86	81	77	72	67	62	58	53	48	44	39	34	30	25	20	15	10	5
8.0	90	85	80	75	70	65	60	55	50	45	40	35	30	25	20	15	10	5	0

TABLE 2 Percent Shear for Measurements Made in Inches

NOTE—100 % shear is to be reported when either A or B is zero.

Dimension B, in.	Dimension A, in.																
	0.05	0.10	0.12	0.14	0.16	0.18	0.20	0.22	0.24	0.26	0.28	0.30	0.32	0.34	0.36	0.38	0.40
0.05	98	96	95	94	94	93	92	91	90	90	89	88	87	86	85	85	84
0.10	96	92	90	89	87	85	84	82	81	79	77	76	74	73	71	69	68
0.12	95	90	88	86	85	83	81	79	77	75	73	71	69	67	65	63	61
0.14	94	89	86	84	82	80	77	75	73	71	68	66	64	62	59	57	55
0.16	94	87	85	82	79	77	74	72	69	67	64	61	59	56	53	51	48
0.18	93	85	83	80	77	74	72	68	65	62	59	56	54	51	48	45	42
0.20	92	84	81	77	74	72	68	65	61	58	55	52	48	45	42	39	36
0.22	91	82	79	75	72	68	65	61	57	54	50	47	43	40	36	33	29
0.24	90	81	77	73	69	65	61	57	54	50	46	42	38	34	30	27	23
0.26	90	79	75	71	67	62	58	54	50	46	41	37	33	29	25	20	16
0.28	89	77	73	68	64	59	55	50	46	41	37	32	28	23	18	14	10
0.30	88	76	71	66	61	56	52	47	42	37	32	27	23	18	13	9	3
0.31	88	75	70	65	60	55	50	45	40	35	30	25	20	18	10	5	0

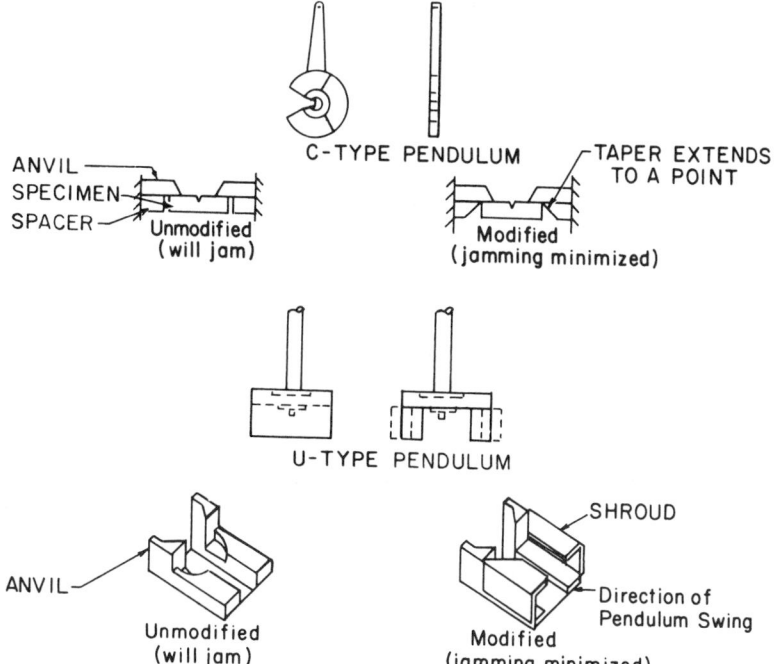

ANVIL
SPECIMEN
SPACER
Unmodified (will jam)

C-TYPE PENDULUM

TAPER EXTENDS TO A POINT

Modified (jamming minimized)

U-TYPE PENDULUM

ANVIL

Unmodified (will jam)

SHROUD

Direction of Pendulum Swing

Modified (jamming minimized)

FIG. 1 Typical Pendulums and Anvils for Charpy Machines, Shown with Modifications to Minimize Jamming

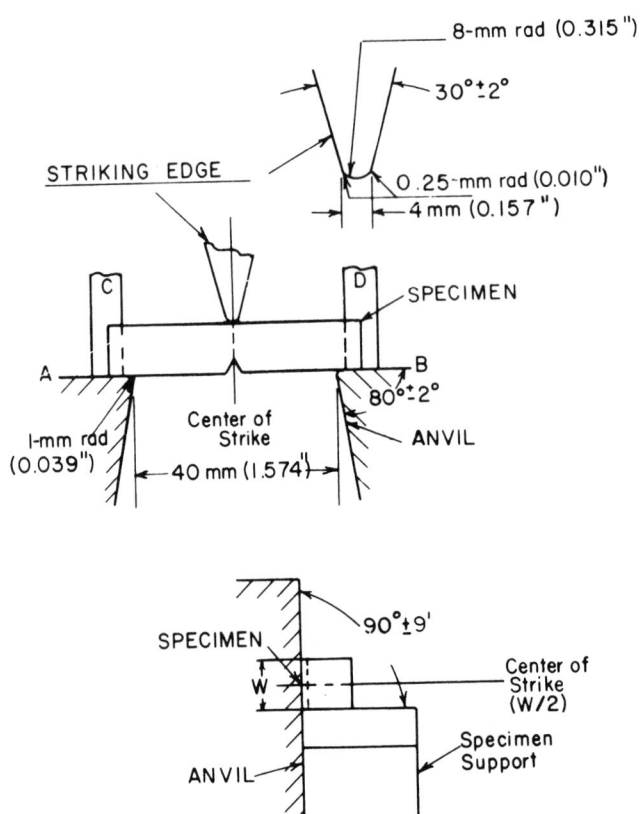

All dimensional tolerances shall be ±0.05 mm (0.002 in.) unless otherwise specified.

NOTE 1—A shall be parallel to B within 2:1000 and coplanar with B within 0.05 mm (0.002 in.).
NOTE 2—C shall be parallel to D within 2.0:1000 and coplanar with D within 0.125 mm (0.005 in.).
NOTE 3—Finish on unmarked parts shall be 4 μm (125 μin.).

FIG. 2 Charpy (Simple-Beam) Impact Test

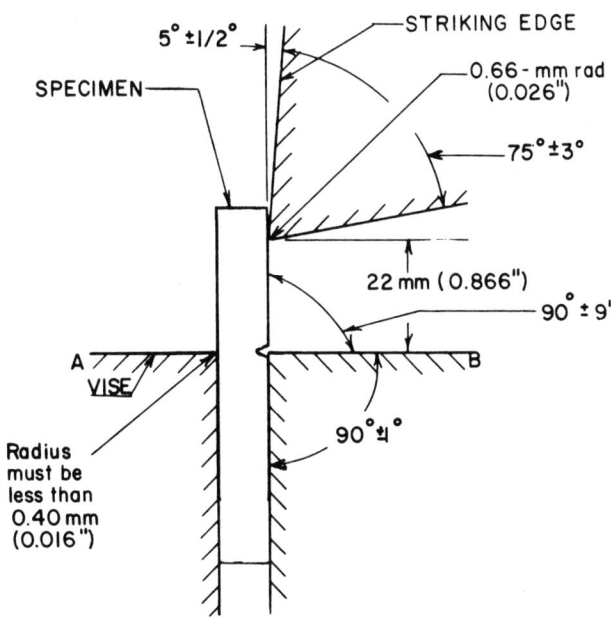

All dimensional tolerances shall be ±0.05 mm (0.002 in.) unless otherwise specified.

NOTE 1—The clamping surfaces of A and B shall be flat and parallel within 0.025 mm (0.001 in.).
NOTE 2—Finish on unmarked parts shall be 2 μm (63 μin.).
NOTE 3—Striker width must be greater than that of the specimen being tested.

FIG. 3   Izod (Cantilever-Beam) Impact Test

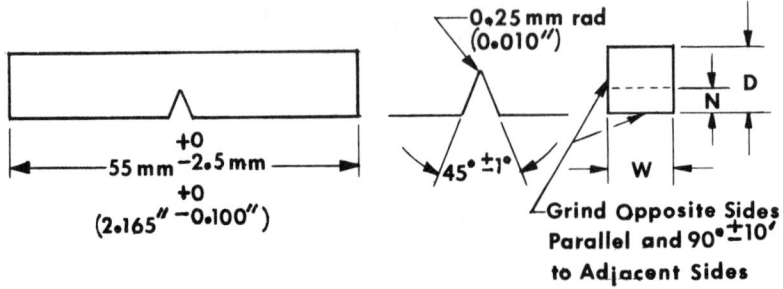

On subsize specimens the length, notch angle, and notch radius are constant (see Fig. 6); depth (D), notch depth (N), and width (W) vary as indicated below.

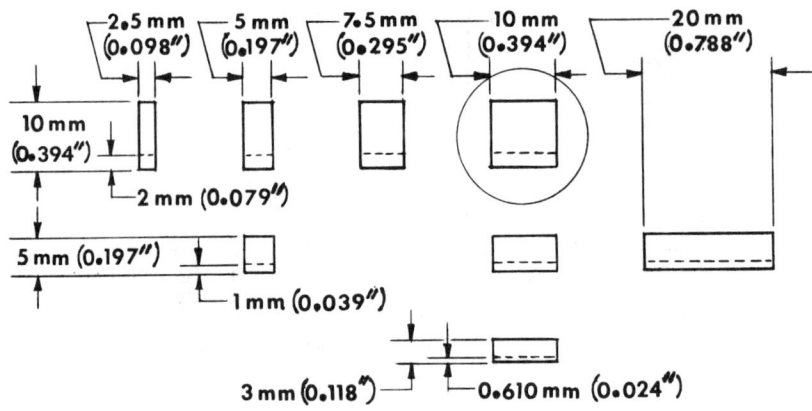

NOTE 1—Circled specimen is the standard specimen (see Fig. 6).
NOTE 2—Permissible variations shall be as follows:

Cross-section dimensions	±1% or ±0.075 mm (0.003 in.), whichever is smaller
Radius of notch	±0.025 mm (0.001 in.)
Depth of notch	±0.025 mm (0.001 in.)
Finish requirements	2 μm (63 μin.) on notched surface and opposite face; 4 μm (125 μin.) on other two surfaces

FIG. 4   Charpy (Simple-Beam) Subsize (Type A) Impact Test Specimens

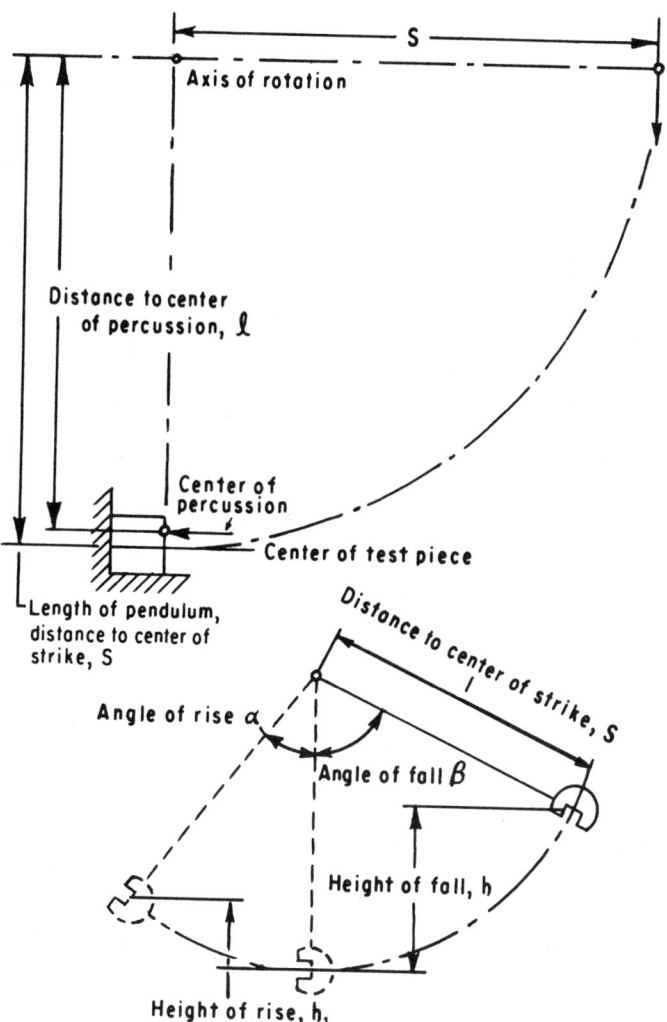

FIG. 5 Dimensions for Calculations

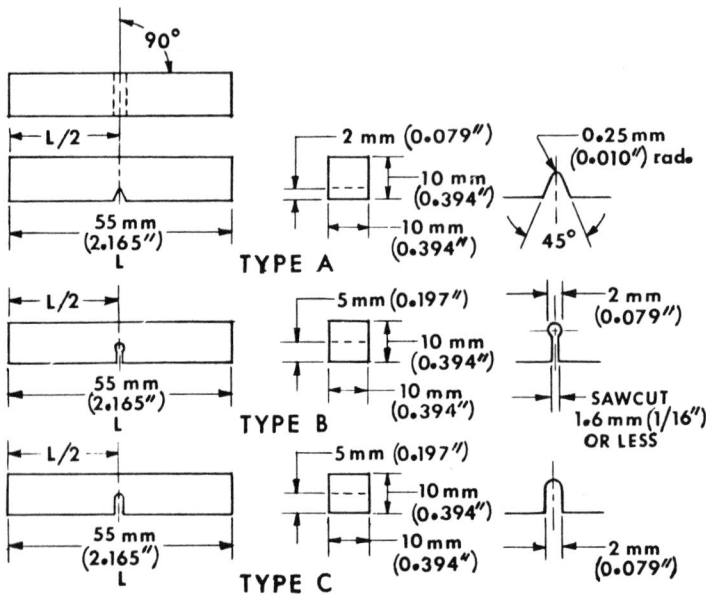

NOTE—Permissible variations shall be as follows:

Notch length to edge	±2°
Adjacent sides shall be at	90° ±10 min
Cross-section dimensions	±0.075 mm (±0.003 in.)
Length of specimen (L)	+0, −2.5 mm (+0, −0.100 in.)
Centering of notch ($L/2$)	±1 mm (±0.039 in.)
Angle of notch	±1°
Radius of notch	±0.025 mm (±0.001 in.)
Notch depth:	
Type A specimen	±0.025 mm (±0.001 in.)
Types B and C specimen	±0.075 mm (±0.003 in.)
Finish requirements	2 μm (63 μin.) on notched surface and opposite face; 4 μm (125 μin.) on other two surfaces

FIG. 6 Charpy (Simple-Beam) Impact Test Specimens, Types A, B, and C

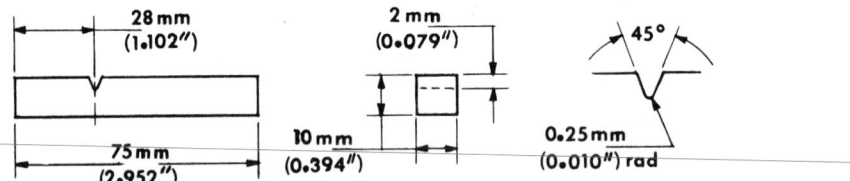

NOTE—Permissible variations shall be as follows:

Notch length to edge	90 ±2°
Cross-section dimensions	±0.025 mm (±0.001 in.)
Length of specimen	+0, −2.5 mm (±0, −0.100 in.)
Angle of notch	±1°
Radius of notch	±0.025 mm (±0.001 in.)
Notch depth	±0.025 mm (±0.001 in.)
Adjacent sides shall be at	90° ± 10 min
Finish requirements	2 μm (63 μin.) on notched surface and opposite face; 4 μm (125 μin.) on other two surfaces

FIG. 7 Izod (Cantilever-Beam) Impact Test Specimen, Type D

496

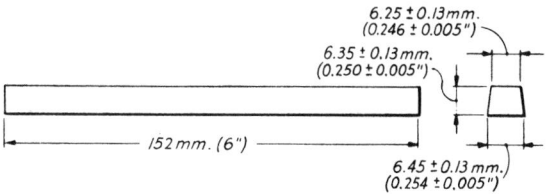

NOTE 1—Two test specimens may be cut from this bar.
NOTE 2—Blow shall be struck on narrowest face.

FIG. 8 Simple Beam Impact Test Bar for Die Castings Alloys

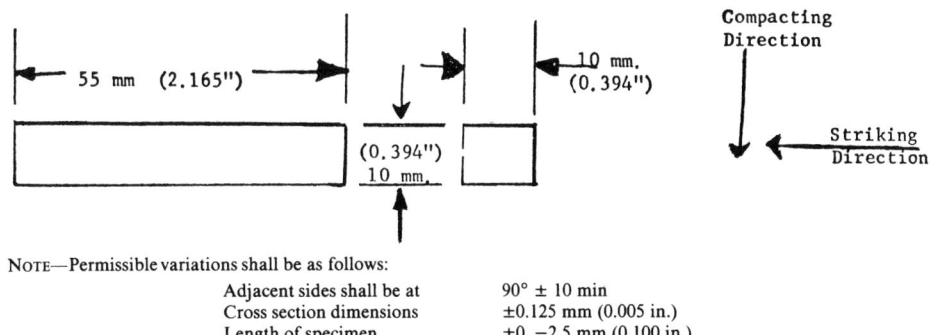

NOTE—Permissible variations shall be as follows:

    Adjacent sides shall be at   90° ± 10 min
    Cross section dimensions   ±0.125 mm (0.005 in.)
    Length of specimen   ±0, −2.5 mm (0.100 in.)

FIG. 9 Charpy (Simple Beam) Impact Test Specimens for Metal Powder Structural Parts

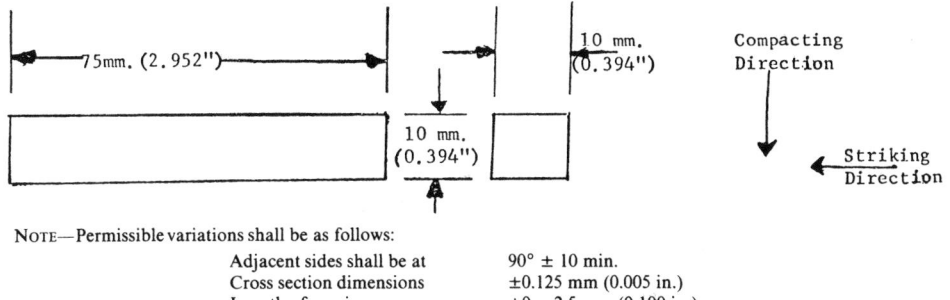

NOTE—Permissible variations shall be as follows:

    Adjacent sides shall be at   90° ± 10 min.
    Cross section dimensions   ±0.125 mm (0.005 in.)
    Length of specimens   +0, −2.5 mm (0.100 in.)

FIG. 10 Izod (Cantilever-Beam) Impact Test Specimen for Metal Powder Structural Parts

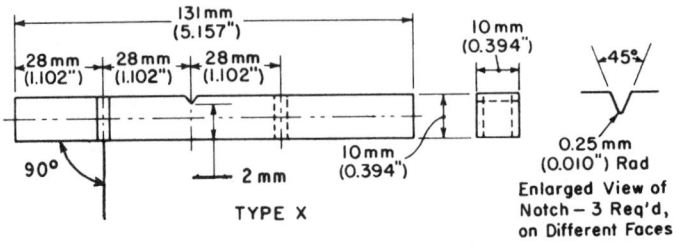

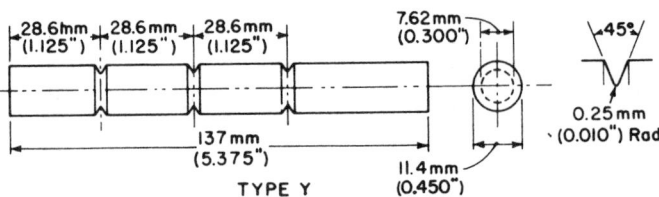

NOTE—Permissible variations shall be as follows:

Notch length to edge	±2 mm
Adjacent sides shall be at	90° ±10 min
Cross-section dimensions	±0.025 mm (±0.001 in.)
Lengthwise dimensions	+0, −2.5 mm (±0.100 in.)
Angle of notch	±1°
Radius of notch	±0.025 mm (±0.001 in.)
Notch depth of Type X specimen	±0.025 mm (±0.001 in.)
Notch diameter of Type Y specimen	±0.025 mm (±0.001 in.)

FIG. 11 Izod (Cantilever-Beam) Impact Test Specimens, Types X and Y

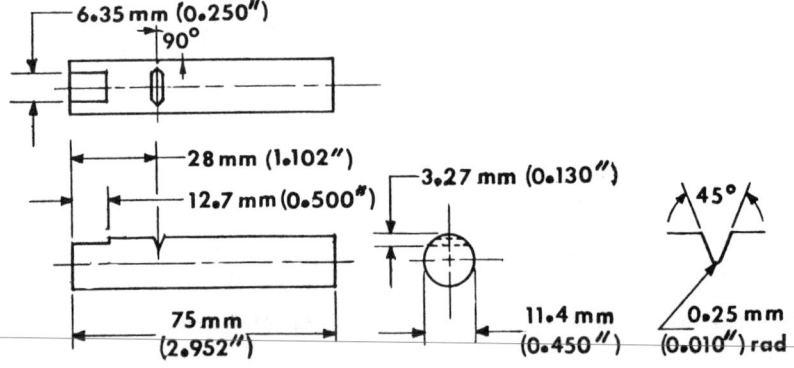

The flat shall be parallel to the longitudinal centerline of the specimen and shall be parallel to the bottom of the notch within 2:1000.

NOTE—Permissible variations shall be as follows:

Notch length to longitudinal centerline	±2°
Cross-section dimensions	±0.025 mm (−0.001 in.)
Length of specimen	+0, −2.5 mm (+0 −0.100 in.)
Angle of notch	±1°
Radius of notch	±0.025 mm (±0.001 in.)
Notch depth	±0.025 mm (.130 ±0.001 in.)

FIG. 12 Izod (Cantilever-Beam) Impact Test Specimen (Philpot), Type Z

498

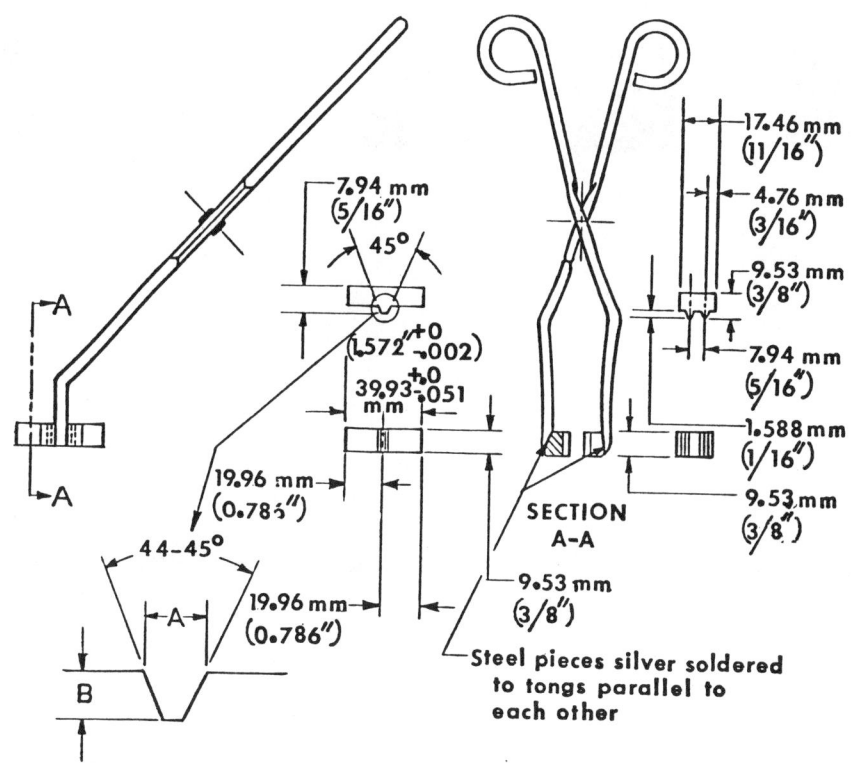

Specimen Depth, mm (in.)	Base Width (A), mm (in.)	Height (B), mm (in.)
10 (0.394)	1.60 to 1.70 (0.063 to 0.067)	1.52 to 1.65 (0.060 to 0.065)
5 (0.197)	0.74 to 0.80 (0.029 to 0.033)	0.69 to 0.81 (0.027 to 0.032)
3 (0.118)	0.45 to 0.51 (0.016 to 0.020)	0.36 to 0.48 (0.014 to 0.019)

FIG. 13   Centering Tongs for V-Notch Charpy Specimens

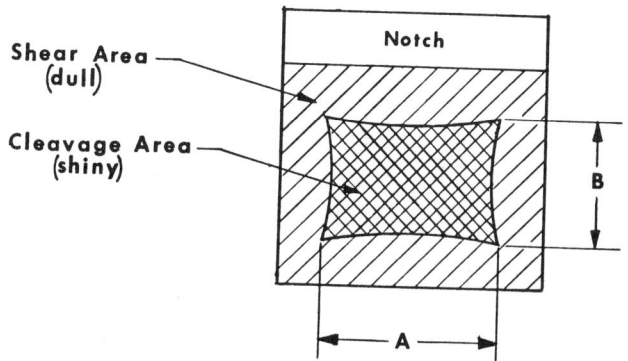

NOTE 1—Measure average dimensions *A* and *B* to the nearest 0.5 mm or 0.02 in.

NOTE 2—Determine the percent shear fracture using Table 1 or Table 2.

FIG. 14   Determination of Percent Shear Fracture

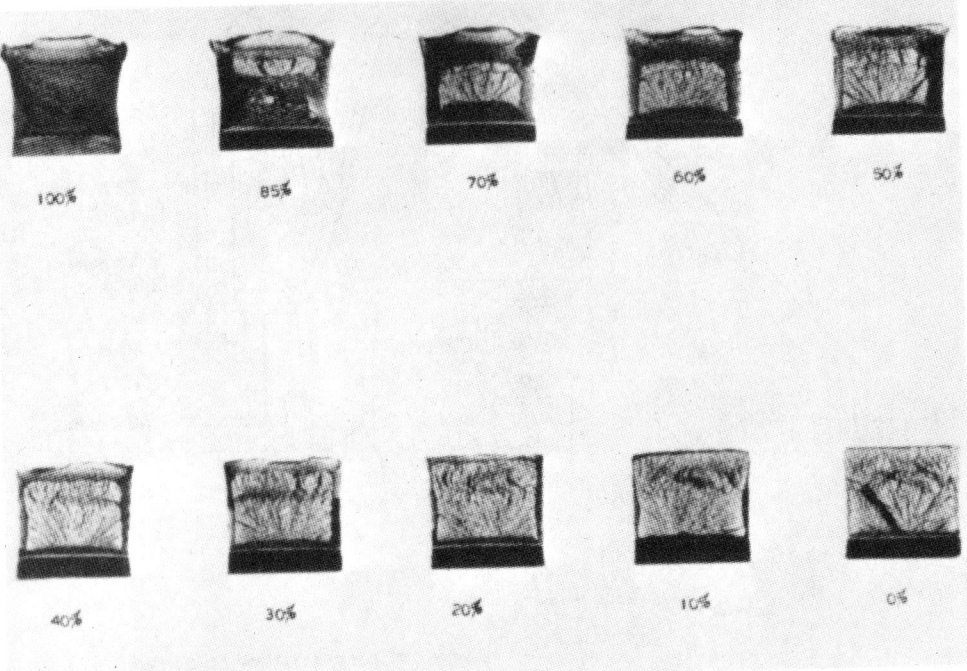

(a) Fracture Appearance Charts and Percent Shear Fracture Comparator

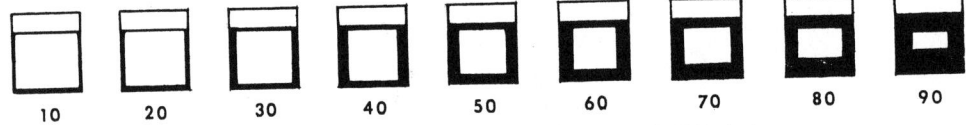

(b) Guide for Estimating Fracture Appearance Using SulAG Method

FIG. 15   Fracture Appearance

FIG. 16 Halves of Broken Charpy V-Notch Impact Specimen Positioned to Illustrate the Measurement of Lateral Expansion, Dimension *A* and Original Width, Dimension *W*

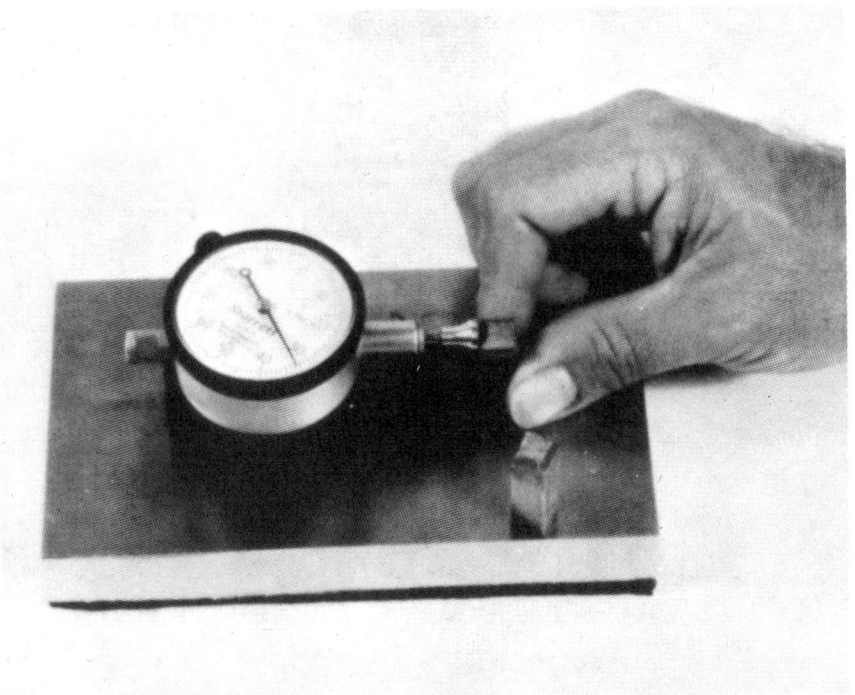

FIG. 17 Lateral Expansion Gage for Charpy Impact Specimens

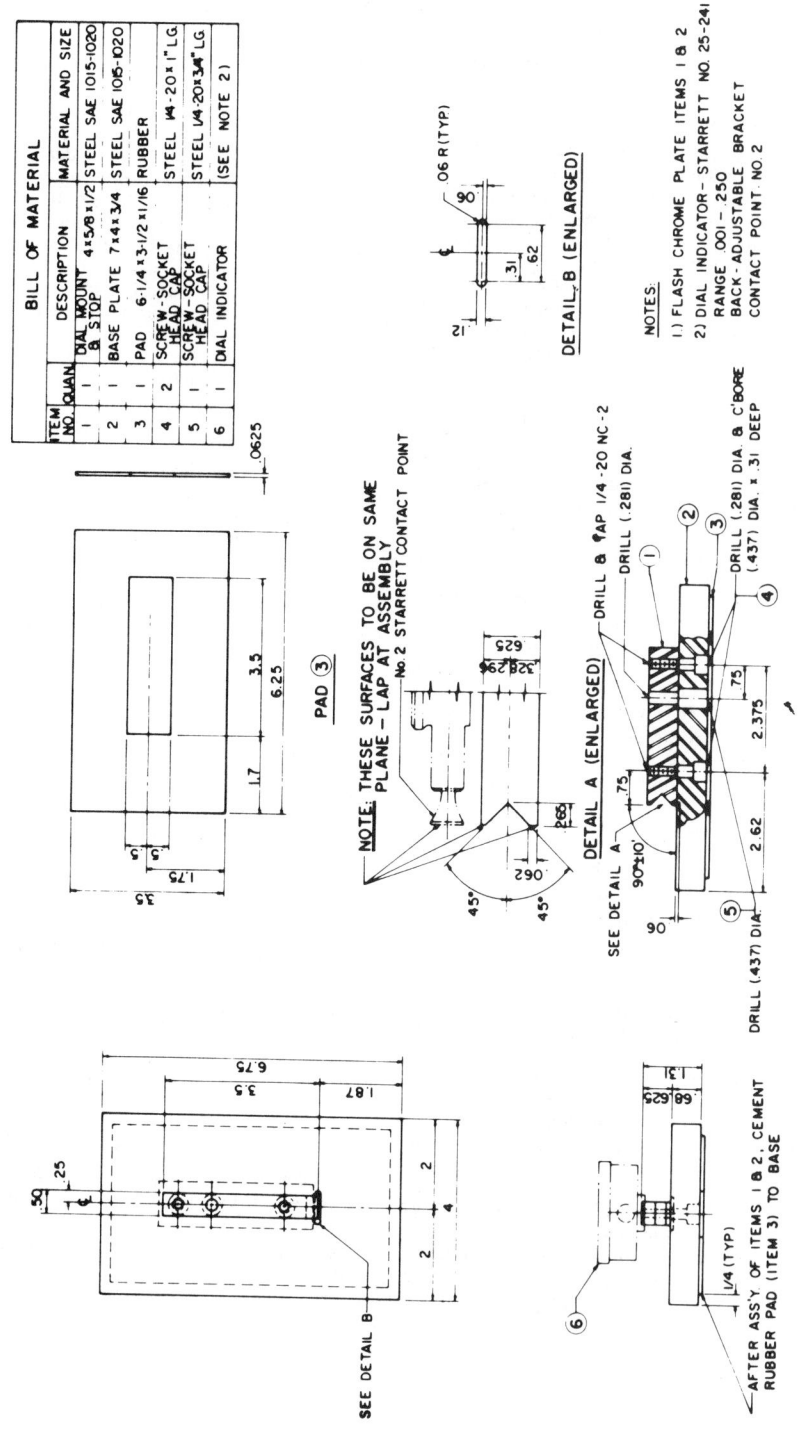

FIG. 18 Assembly and Details for Lateral Expansion Gage

BILL OF MATERIAL

ITEM NO.	QUAN.	DESCRIPTION	MATERIAL AND SIZE
1	1	DIAL MOUNT & STOP	STEEL SAE 1015-1020 4×5/8×1 1/2
2	1	BASE PLATE	STEEL SAE 1015-1020 7×4×3/4
3	1	PAD	RUBBER 6-1/4×3-1/2×1/16
4	2	SCREW-SOCKET HEAD CAP	STEEL 1/4-20×1" LG
5	1	SCREW-SOCKET HEAD CAP	STEEL 1/4-20×3/4" LG
6	1	DIAL INDICATOR	(SEE NOTE 2)

NOTES:
1) FLASH CHROME PLATE ITEMS 1 & 2
2) DIAL INDICATOR– STARRETT NO. 25-241
   RANGE .001–.250
   BACK-ADJUSTABLE BRACKET
   CONTACT POINT NO. 2

DETAIL B (ENLARGED)

PAD ③

NOTE: THESE SURFACES TO BE ON SAME PLANE – LAP AT ASSEMBLY
No 2 STARRETT CONTACT POINT

DETAIL A (ENLARGED)

SEE DETAIL A

DRILL & TAP 1/4-20 NC-2
DRILL (.281) DIA.
DRILL (.281) DIA. & C'BORE (.437) DIA. × .31 DEEP
DRILL (.437) DIA.

SEE DETAIL B

AFTER ASSY OF ITEMS 1 & 2, CEMENT RUBBER PAD (ITEM 3) TO BASE

503

APPENDIX

X1. NOTES ON SIGNIFICANCE OF NOTCHED-BAR IMPACT TESTING

X1.1 Notch Behavior

X1.1.1 The Charpy and Izod type tests bring out notch behavior (brittleness versus ductility) by applying a single overload of stress. The energy values determined are quantitative comparisons on a selected specimen but cannot be converted into energy values that would serve for engineering desgin calculations. The notch behavior indicated in an individual test applies only to the specimen size, notch geometry, and testing conditions involved and cannot be generalized to other sizes of specimens and conditions.

X1.1.2 The notch behavior of the face-centered cubic metals and alloys, a large group of nonferrous materials and the austenitic steels can be judged from their common tensile properties. If they are brittle in tension they will be brittle when notched, while if they are ductile in tension they will be ductile when notched, except for unusually sharp or deep notches (much more severe than the standard Charpy or Izod specimens). Even low temperatures do not alter this characteristic of these materials. In contrast, the behavior of the ferritic steels under notch conditions cannot be predicted from their properties as revealed by the tension test. For the study of these materials the Charpy and Izod type tests are accordingly very useful. Some metals that display normal ductility in the tension test may nevertheless break in brittle fashion when tested or when used in the notched condition. Notched conditions include restraints to deformation in directions perpendicular to the major stress, or multiaxial stresses, and stress concentrations. It is in this field that the Charpy and Izod tests prove useful for determining the susceptibility of a steel to notch-brittle behavior though they cannot be directly used to appraise the serviceability of a structure.

X1.2 Notch Effect

X1.2.1 The notch results in a combination of multiaxial stresses associated with restraints to deformation in directions perpendicular to the major stress, and a stress concentration at the base of the notch. A severely notched condition is generally not desirable, and it becomes of real concern in those cases in which it initiates a sudden and complete failure of the brittle type. Some metals can be deformed in a ductile manner even down to the low temperatures of liquid air, while others may crack. This difference in behavior can be best understood by considering the cohesive strength of a material (or the property that holds it together) and its relation to the yield point. In cases of brittle fracture, the cohesive strength is exceeded before significant plastic deformation occurs and the fracture appears crystalline. In cases of the ductile or shear type of failure, considerable deformation precedes the final fracture and the broken surface appears fibrous instead of crytalline. In intermediate cases the fracture comes after a moderate amount of deformation and is part crystalline and part fibrous in appearance.

X1.2.2 When a notched bar is loaded, there is a normal stress across the base of the notch which tends to initiate fracture. The property that keeps it from cleaving, or holds it together, is the "cohesive strength." The bar fractures when the normal stress exceeds the cohesive strength. When this occurs without the bar deforming it is the condition for brittle fracture.

X1.2.3 In testing, though not in service because of side effects, it happens more commonly that plastic deformation precedes fracture. In addition to the normal stress, the applied load also sets up shear stresses which are about 45° to the normal stress. The elastic behavior terminates as soon as the shear stress exceeds the shear strength of the material and deformation or plastic yielding sets in. This is the condition for ductile failure.

X.1.2.4 This behavior, whether brittle or ductile, depends on whether the normal stress exceeds the cohesive strength before the shear stress exceeds the shear strength. Several important facts of notch behavior follow from this. If the notch is made sharper or more drastic, the normal stress at the root of the notch will be increased in relation to the shear stress and the bar will be more prone to brittle fracture (see Table X1.1). Also, as the speed of deformation increases, the shear strength increases and the likelihood of brittle fracture increases. On the other hand, by raising the temperature, leaving the notch and the speed of deformation the same, the shear strength is lowered and ductile behavior is promoted, leading to shear failure.

X1.2.5 Variations in notch dimensions will seriously affect the results of the tests. Tests on E 4340 steel specimens[4] have shown the effect of dimensional variations on Charpy results (see Table X1.1).

X1.3 Size Effect

X1.3.1 Increasing either the width or the depth of the specimen tends to increase the volume of metal subject to distortion, and by this factor tends to

---

[4] N. H. Fahey, "Effects of Variables in Charpy Impact Testing," *Materials Research & Standards*, Vol 1, No. 11, November 1961, p. 872.

504

increase the energy absorption when breaking the specimen. However, any increase in size, particularly in width, also tends to increase the degree of restraint and by tending to induce brittle fracture, may decrease the amount of energy absorbed. Where a standard-size specimen is on the verge of brittle fracture, this is particularly true, and a doublewidth specimen may actually require less energy for rupture than one of standard width.

X1.3.2 In studies of such effects where the size of the material precludes the use of the standard specimen, as for example when the material is 6.35 mm (0.25-in.) plate, subsize specimens are necessarily used. Such specimens (Fig. 4) are based on the Type A specimen of Fig. 6.

X1.3.3 General correlation between the energy values obtained with specimens of different size or shape is not feasible, but limited correlations may be established for specification purposes on the basis of special studies of particular materials and particular specimens. On the other hand, in a study of the relative effect of process variations, evaluation by use of some arbitrarily selected specimen with some chosen notch will in most instances place the methods in their proper order.

X1.4 Temperature Effect

X1.4.1 The testing conditions also affect the notch behavior. So pronounced is the effect of temperature on the behavior of steel when notched that comparisons are frequently made by examining specimen fractures and by plotting energy value and fracture appearance versus temperature from tests of notched bars at a series of temperatures. When the test temperature has been carried low enough to start cleavage fracture, there may be an extremely sharp drop in impact value or there may be a relatively gradual falling off toward the lower temperatures. This drop in energy value starts when a specimen begins to exhibit some crystalline appearance in the fracture. The transition temperature at which this embrittling effect takes place varies considerably with the size of the part or test specimen and with the notch geometry.

X1.5 Testing Machine

X1.5.1 The testing machine itself must be sufficiently rigid or tests on high-strength low-energy materials will result in excessive elastic energy losses either upward through the pendulum shaft or downward through the base of the machine. If the anvil supports, the pendulum striking edge, or the machine foundation bolts are not securely fastened, tests on ductile materials in the range from 108 J (80 ft·lbf) may actually indicate values in excess of 122 to 136 J (90 to 100 ft·lbf)

X1.5.2 A problem peculiar to Charpy-type tests occurs when high-strength, low-energy specimens are tested at low temperatures. These specimens may not leave the machine in the direction of the pendulum swing but rather in a sidewise direction. To ensure that the broken halves of the specimens do not rebound off some component of the machine and contact the pendulum before it completes its swing, modifications may be necessary in older model machines. These modifications differ with machine design. Nevertheless the basic problem is the same in that provisions must be made to prevent rebounding of the fractured specimens into any part of the swinging pendulum. Where design permits, the broken specimens may be deflected out of the sides of the machine and yet in other designs it may be necessary to contain the broken specimens within a certain area until the pendulum passes through the anvils. Some low-energy high-strength steel specimens leave impact machines at speeds in excess of 15.2 m/s (50 ft/s) although they were struck by a pendulum traveling at speeds approximately 5.2 m/s (17 ft/s). If the force exerted on the pendulum by the broken specimens is sufficient, the pendulum will slow down and erroneously high energy values will be recorded. This problem accounts for many of the inconsistencies in Charpy results reported by various investigators within the 14 to 34-J (10 to 25-ft·lb) range. Figure 1 illustrates a modification found to be satisfactory in minimizing jamming.

X1.6 Velocity of Straining

X1.6.1 Velocity of straining is likewise a variable that affects the notch behavior of steel. The impact test shows somewhat higher energy absorption values than the static tests above the transition temperature and yet, in some instances, the reverse is true below the transition temperature.

X1.7 Correlation with Service

X1.7.1 While Charpy or Izod tests may not directly predict the ductile or brittle behavior of steel as commonly used in large masses or as components of large structures, these tests can be used as acceptance tests or tests of identity for different lots of the same steel or in choosing between different steels, when correlation with reliable service behavior has been established. It may be necessary to make the tests at properly chosen temperatures other than room temperature. In this, the service temperature or the transition temperature of full-scale specimens does not give the desired transition temperatures for Charpy or Izod tests since the size and notch geometry may be so different. Chemical analysis, tension, and hardness tests may not indicate the influence of some of the important processing factors that affect susceptibility to brittle fracture nor do they comprehend the effect of low temperatures in inducing brittle behavior.

	High-Energy Specimens, J (ft·lbf)	High-Energy Specimens, J (ft·lbf)	Low-Energy Specimens, J (ft·lbf)
Specimen with standard dimensions	103.0 ± 5.2 (76.0 ± 3.8)	60.3 ± 3.0 (44.5 ± 2.2)	16.9 ± 1.4 (12.5 ± 1.0)
Depth of notch, 2.13 mm (0.084 in.)[A]	97.9 (72.2)	56.0 (41.3)	15.5 (11.4)
Depth of notch, 2.04 mm (0.0805 in.)[A]	101.8 (75.1)	57.2 (42.2)	16.8 (12.4)
Depth of notch, 1.97 mm (0.0775 in.)[A]	104.1 (76.8)	61.4 (45.3)	17.2 (12.7)
Depth of notch, 1.88 mm (0.074 in.)[A]	107.9 (79.6)	62.4 (46.0)	17.4 (12.8)
Radius at base of notch 0.13 mm (0.005 in.)[B]	98.0 (72.3)	56.5 (41.7)	14.6 (10.8)
Radius at base of notch 0.38 mm (0.015 in.)[B]	108.5 (80.0)	64.3 (47.4)	21.4 (15.8)

[A] Standard 2.0 ± 0.025 mm (0.079 ± 0.001 in.).
[B] Standard 0.25 ± 0.025 mm (0.010 ± 0.001 in.).

Index

A

AASHTO classification, 17, 20
Abram's water cement law, 120
abrasion, 66, 161
absorption, 75, 76, 125, 127, 161, 237
accelerating admixtures, 168
ACI method of mix design, 122
activity, 14
admixture, 49, 137, 167, 169
adsorption, 12, 13
aggregate blending, 54
aggregate gradation, 54, 55, 63
aggregate microcrack, 153
aggregates, 52, 57, 114, 116, 128, 136
air content, 137
air drying, 176
air entraining admixtures, 169
air entraining agents, 169
air entrainment, 138
algae, 114
alkali carbonate reaction, 158, 160
alkalies, 84
alkali reactivity, 74

alkali silica reaction, 158, 159, 160
allotropic behavior, 216
alloy, 218
alloying, 218
alloy steels, 260
alumina, 83
alumina cement, 79, 95
anionic emulsions, 202
anisotropic, 25
annealing, 253
antibacterial cement, 79, 96
ants, 185, 187
apparent cohesion, 38
applied load, 105
argillaceous, 85
argillocalcareous, 85
asphalt, 52, 194
asphalt cement, 194, 195
asphalt emulsions, 202
asphalt testing, 196
asphaltenes, 195
asphaltic mixes, 205
ASTM, 133
Attenberg Limits, 13
austenite, 250